Lecture Notes in Networks and Systems

Volume 74

Series Editor

Janusz Kacprzyk, Systems Research Institute, Polish Academy of Sciences,
Warsaw, Poland

Advisory Editors

Fernando Gomide, Department of Computer Engineering and Automation—DCA,
School of Electrical and Computer Engineering—FEEC, University of Campinas—
UNICAMP, São Paulo, Brazil
Okyay Kaynak, Department of Electrical and Electronic Engineering, Bogazici
University, Istanbul, Turkey
Derong Liu, Department of Electrical and Computer Engineering, University of
Illinois at Chicago, Chicago, USA;
Institute of Automation, Chinese Academy of Sciences, Beijing, China
Witold Pedrycz, Department of Electrical and Computer Engineering, University of
Alberta, Alberta, Canada;
Systems Research Institute, Polish Academy of Sciences, Warsaw, Poland
Marios M. Polycarpou, Department of Electrical and Computer Engineering, KIOS
Research Center for Intelligent Systems and Networks, University of Cyprus,
Nicosia, Cyprus
Imre J. Rudas, Óbuda University, Budapest, Hungary
Jun Wang, Department of Computer Science, City University of Hong Kong,
Kowloon, Hong Kong

The series "Lecture Notes in Networks and Systems" publishes the latest developments in Networks and Systems—quickly, informally and with high quality. Original research reported in proceedings and post-proceedings represents the core of LNNS.

Volumes published in LNNS embrace all aspects and subfields of, as well as new challenges in, Networks and Systems.

The series contains proceedings and edited volumes in systems and networks, spanning the areas of Cyber-Physical Systems, Autonomous Systems, Sensor Networks, Control Systems, Energy Systems, Automotive Systems, Biological Systems, Vehicular Networking and Connected Vehicles, Aerospace Systems, Automation, Manufacturing, Smart Grids, Nonlinear Systems, Power Systems, Robotics, Social Systems, Economic Systems and other. Of particular value to both the contributors and the readership are the short publication timeframe and the world-wide distribution and exposure which enable both a wide and rapid dissemination of research output.

The series covers the theory, applications, and perspectives on the state of the art and future developments relevant to systems and networks, decision making, control, complex processes and related areas, as embedded in the fields of interdisciplinary and applied sciences, engineering, computer science, physics, economics, social, and life sciences, as well as the paradigms and methodologies behind them.

** Indexing: The books of this series are submitted to ISI Proceedings, SCOPUS, Google Scholar and Springerlink **

More information about this series at http://www.springer.com/series/15179

H. S. Saini · Rishi Sayal ·
Aliseri Govardhan · Rajkumar Buyya
Editors

Innovations in Computer Science and Engineering

Proceedings of the Sixth ICICSE 2018

 Springer

Editors
H. S. Saini
Guru Nanak Institutions
Ibrahimpatnam, Telangana, India

Rishi Sayal
Guru Nanak Institutions
Ibrahimpatnam, Telangana, India

Aliseri Govardhan
JNTUH College of Engineering Hyderabad
Hyderabad, Telangana, India

Rajkumar Buyya
CLOUDS Laboratory
The University of Melbourne
Melbourne, VIC, Australia

ISSN 2367-3370   ISSN 2367-3389   (electronic)
Lecture Notes in Networks and Systems
ISBN 978-981-13-7081-6   ISBN 978-981-13-7082-3   (eBook)
https://doi.org/10.1007/978-981-13-7082-3

Library of Congress Control Number: 2019933701

This Springer imprint is published by the registered company Springer Nature Singapore Pte Ltd.
The registered company address is: 152 Beach Road, #21-01/04 Gateway East, Singapore 189721, Singapore

Organizing Committee

Patrons

Sardar Tavinder Singh Kohli
Sardar Gagandeep Singh Kohli

Conference Chair

Dr. H. S. Saini

Conference Co-chairs

Dr. M. Ramalinga Reddy
Dr. S. Sreenatha Reddy
Dr. Rishi Sayal

Conveners

Dr. J. Rajeshwar
Dr. S. Senthil Kumar
Prof. V. Deva Sekhar
Dr. M. I. Thariq Hussan
Dr. J. Mercy Geraldine
Dr. Ch. Subbalakshmi

Co-conveners

Dr. M. Venkata Narayana
Dr. Stalin Alex
Mr. S. Madhu
Mr. Lalu Nayak
Ms. D. Shireesha
Mr. D. Saidulu
Mr. A. Ravi
Mr. A. Ugendhar

Committees

Conference Committee: Dr. Rishi Sayal

Mr. S. Siva Sankar Rao
Mr. S. Sreekanth
Mrs. C. Sangeetha
Mr. Manik Rao Patil

Publicity Chair International: Dr. M. I. Thariq Hussan

Mr. J. Naresh Kumar
Mrs. Sumitra Mallick
Mr. M. Yadaiah
Ms. B. Mamatha
Mr. Mohammed Imran Sheikh

Publicity Chair National: Dr. S. Senthil Kumar/Prof. V. Deva Sekhar

Mr. B. Nandan
Mrs. K. Annapoorna
Mrs. Padma Rajani

Program and Publication Chair: Dr. J. Rajeshwar/Dr. CH. Subbalakshmi

Mr. D. S. S. Veeresh
Mr. I. Phani Raja
Mr. L. Srinivas
Mrs. P. Srilakshmi
Mr. Devi Prasad Mishra
Ms. Rajashree Sutware
Mr. K. Chandra Shekar
Mr. P. Dathatreya

Accommodation Chair: Dr. M. Venkata Narayana

Mr. Arun Singh
Mr. Nusrath Khan

Advisory Board International

Dr. San Murugesan, Australia
Prof. Rajkumar Buyya, Australia
Dr. Hemant Pendharkar, USA
Dr. Anuj Sharma, USA
Dr. Chandrashekar Commuri, USA
Dr. S. R. Subramanya, USA

Mr. Maliyanath Sundaramurthy, USA
Dr. Sitalakshmi Venkataraman, Australia
Mr. Kannan Govindaraj, Texas, USA
Dr. Hamid R. Arbania, USA
Dr. Anitha Thangasamy, Ethiopia
Dr. K. M. Sharavana Raju, Saudi Arabia
Dr. Lipo Wang, Professor, Singapore

Advisory Board National

Dr. Raj Kamal, India
Mr. Sanjay Mohapatra, India
Dr. Aliseri Govardhan, India
Dr. D. V. L. N. Somayajulu, India
Dr. Aruna Malapadi, India
Dr. Uday Bhaskar Vemulapati, India
Dr. R. B. V. Subramanyam, India
Mr. K. Mohan Raidu, India
Dr. Vasumathi, India
Dr. P. Premchand, India
Dr. D. D. Sarma, India
Dr. Nicholas Savarimuthu, India
Mr. Mohamed Kasim Khan, India
Dr. G. Narsimha, India
Dr. T. Venugopal, India
Dr. G. Vadivu, India
Dr. K. V. Ramana, India
Dr. P. Natarajan, India
Dr. E. Logashanmugam, India
Dr. S. V. Ranganayakulu, India
Dr. L. Sumalatha, India
Dr. V. Parthasarathy, India
Dr. Sasikala, India
Dr. P. Nirmal Kumar, India
Dr. M. P. Vani, India

Preface

This volume contains 68 papers that were presented in the Sixth International Conference on Innovations in Computer Science and Engineering (ICICSE 2018) held during August 17 and 18, 2018, at Guru Nanak Institutions (GNI) in association with Computer Society of India (CSI) and funding from Defense Research and Development Organization (DRDO).

The aim of ICICSE 2018 was to provide a platform for researchers, scientists, technocrats, academicians, and engineers to exchange their innovative ideas and new research findings in the fields of computer science and engineering till the end of 2018. The conference will boost excellent innovations in terms of day-to-day life and academics.

This conference received a vast number of research publications from different fields of computer science and engineering. All papers were peer-reviewed with the help of our core committee members and external reviewers. The final acceptance of 71 papers came through with an acceptance ratio of 0.28.

ICICSE 2018 was inaugurated and attended by top dignitaries such as Mr. K. Mohan Raidu, Chairman, CSI Hyderabad; Lt. Col. Khairol Amali bin Ahmad, Dean of the Engineering Faculty, Malaysia; K. Suchendar, Scientist 'G', Director, DSTARS, DRDL; and Dr. C. Krishna Mohan, Department of Computer Science and Engineering, IIT Hyderabad. The conference was chaired by international and national personalities: Dr. K. M. Sharavana Raju, Professor, Department of Computer Science and Engineering, College of Computer Science and Information Systems, Jazan University, Jazan, Saudi Arabia; Dr. Anitha Thangasamy, Associate Professor, Department of Computer Science and Engineering, Wollega University, Nekemte, Ethiopia; Dr. P. Natrajan, Associate Professor, SCOPE, VITU, Vellore, Tamil Nadu; and Dr. S. Krishna Mohan Rao, Professor, Principal, GIFT, Bhubaneswar, Odisha. Further conference keynote speakers were Dr. P. Krishna Reddy, Data Sciences and Analytics Center (DSAC), International Institute of Information Technology Hyderabad (IIITH), and Mr. Bala Prasad, Chairman, Hyderabad Section of IEEE, Member Managing Committee of CSI Hyderabad, and Faculty TCS.

The organizing committee of ICICSE 2018 takes an opportunity to thank the keynote speakers, session chairs, and reviewers for their excellent support in making ICICSE 2018 a grand success.

The quality of all these research papers is a courtesy from respective authors and reviewers to come up to the desired level of excellence. We are indebted to the program committee members and external reviewers in producing the best-quality research papers in a short span of time. We also thank CSI delegates, DRDO, toward their valuable suggestions and funding in making this event a grand success.

Hyderabad, India	H. S. Saini
Hyderabad, India	Rishi Sayal
Hyderabad, India	Aliseri Govardhan
Melbourne, Australia	Rajkumar Buyya

Contents

Editors and Contributors

About the Editors

Dr. H. S. Saini Managing Director for Guru Nanak Institutions, obtained his Ph.D. in the field of Computer Science. He has over 28 years of experience at University/College level in teaching UG/PG students and has guided several B.Tech., M.Tech. projects. He has published/presented high-quality research papers in International, National Journals and proceedings of International Conferences. He has two books to his credit. Dr. Saini is a lover of innovation and is an advisor for NBA/NAAC accreditation process to many institutions in India and abroad. He is chief editor of many innovative journals and chairing various international conferences.

Dr. Rishi Sayal Associate Director, Guru Nanak Institutions Technical Campus has completed his B.E. (CSE), M. Tech. (IT), Ph.D. (CSE). He has obtained his Ph.D. in Computer Science and Engineering in the field of data mining from prestigious Mysore University of Karnataka state. He has over 26 years of experience in training, consultancy, teaching and placements. His current areas of research interest include Data Mining, Network Security and Databases. He has published wide number of research papers in International Conferences & Journals. He has guided many UG and PG research projects and he is recipient of many research grants from Government funding agencies. He is co-editor of various innovative journals and convened international conferences.

Dr. Aliseri Govardhan is presently a Professor of Computer Science & Engineering, Rector, JNTUH and Executive Council Member, Jawaharlal Nehru Technological University Hyderabad (JNTUH), India. He did his Ph.D. from Jawaharlal Nehru Technological University, Hyderabad. He is a member on the Editorial Boards for twelve International Journals. He is a Member on Advisory

Boards & Academic Boards and Technical Program Committee Member for more than 65 International and National Conferences. He has two monographs and ten book chapters published.

Dr. Rajkumar Buyya is a Redmond Barry Distinguished Professor and Director of the Cloud Computing and Distributed Systems (CLOUDS) Laboratory at the University of Melbourne, Australia. He is also serving as the founding CEO of Manjrasoft Pvt. Ltd., a spin-off company of the University, commercializing its innovations in Cloud Computing. He served as a Future Fellow of the Australian Research Council during 2012–2016. He received a Doctor of Philosophy (Ph.D.) in Computer Science and Software Engineering from Monash University, Melbourne, Australia in 2002. Dr. Buyya has authored/co-authored over 625 publications. He has co-authored five text books, and edited proceedings of over 25 international conferences.

Contributors

Ugendhar Addagatla Department of Computer Science and Engineering, Guru Nanak Institutions Technical Campus, Ranga Reddy, Telangana, India

T. Aditya Sai Srinivas Department of Computer Science and Engineering, VIT University, Vellore, Tamil Nadu, India

A. Aishwarya School of Information Technology and Engineering, Vellore Institute of Technology, Vellore, India

M. Aishwarya School of Information Technology and Engineering, Vellore Institute of Technology, Vellore, India

A. K. Alzo'ubi Department of Civil Engineering, Abu Dhabi University, Al Ain, UAE

M. Anand Kumar Amrita School of Engineering, Center for Computational Engineering and Networking (CEN), Amrita Vishwa Vidyapeetham, Coimbatore, India

C. M. Ananda Council of Scientific and Industrial Research-National Aerospace Laboratories, Bengaluru, India

B. P. Aniruddha Prabhu Department of Computer Science and Engineering, Siddaganga Institute of Technology, Tumkur, Karnataka, India

Bairi Anjaneyulu Department of Computer Science and Engineering, RGUKT, Basar, Hyderabad, India

J. Anuradha SCOPE Vellore Institute of Technology, Vellore, Tamil Nadu, India

Tarique Anwar Khan Department of Computer Science and Engineering, Siddaganga Institute of Technology, Tumkur, Karnataka, India

S. Aravindharamanan Department of Computer Science and Engineering, VIT University, Chennai, TamilNadu, India

B. P. Ashwini Department of Computer Science and Engineering, Siddaganga Institute of Technology, Tumkur, Karnataka, India

Kale Suhash Babasaheb Balbhim College Beed, Beed, India

G. N. Balaji CVR College of Engineering, Hyderabad, India

G. Balakrishna Department of Computer Science and Engineering, Koneru Lakshmaiah Education Foundation, Guntur, India

Gautam Bandyopadhyay Department of Management Studies, NIT Durgapur, Durgapur, India

Anish Batra Department of Computer Science, Bharati Vidyapeeth's College of Engineering, New Delhi, Delhi, India

A. Bepari Nawazish Bangalore Institute of Technology, Bengaluru, India

Parag Bhalchandra School of Computational Sciences, SRTM University, Nanded, MS, India

Kumbham Bhargavi Department of Computer Science, Keshav Memorial Institute of Technology, Hyderabad, India

Namrata Bhattacharya Department of Computer Science and Engineering, University of Calcutta, Kolkata, India

Dharmendra G. Bhatti Uka Tarsadia University, Bardoli, Gujarat, India

Harlieen Bindra Department of Computer Science, Bharati Vidyapeeth's College of Engineering, New Delhi, Delhi, India

Sanjib Biswas Calcutta Business School, Calcutta, India

Sai Jyothi Bolla Department of Computer Science, Sri Padmavathi Mahila Viswa Vidyalayam, Tirupati, India

C. R. Byra Reddy Bangalore Institute of Technology, Bengaluru, India

Vineetha Rebecca Chacko Center for Computational Engineering and Networking (CEN), Amrita School of Engineering, Amrita Vishwa Vidyapeetham, Coimbatore, India

Priti Chandra ASL, DRDO, Hyderabad, India

K. Chandra Shekar Department of Computer Science and Engineering, JNTUH, Hyderabad, India;
Department of Computer Science and Engineering, GNITC, Hyderabad, India

L. Chandra Sekhar Reddy Department of Computer Science and Engineering, Shri JJT University, Jhunjhunu, Rajasthan, India

Omprakash Chandrakar Uka Tarsadia University, Bardoli, Gujarat, India

Arup Das Department of Computer Science and Engineering, Siddaganga Institute of Technology, Tumkur, Karnataka, India

Indrani Das Department of Computer Science, Assam University, Silchar, Assam, India

B. Deena Narayan Department of Computer Science and Engineering, NIT Andhra Pradesh, Tadepalligudam, Andhra Pradesh, India

C. K. Deepa Department of Computer Science and Engineering, Ramaiah Institute of Technology, Bengaluru, India

V. Devasekhar Department of Computer Science and Engineering, Guru Nanak Institutions Technical Campus, Hyderabad, Telangana, India

Robinson Devasia Indraprastha Institute of Information Technology Delhi, New Delhi, Delhi, India

Hemant Kumar Dua Department of Computer Science, Bharati Vidyapeeth's College of Engineering, New Delhi, Delhi, India

Ranjith Gandhasiri Department of Computer Science and Engineering, RGUKT, Basar, Hyderabad, India

M. L. Garg Department of Computer Science Engineering, DIT University, Dehradun, India

J. Geetha Reddy Department of Computer Science and Engineering, Ramaiah Institute of Technology, Bangalore, India

Kailash Gogineni School of Computer Science and Engineering, Vellore Institute of Technology, Vellore, Tamil Nadu, India

C. Y. Gopinath Bangalore Institute of Technology, Bengaluru, India

K. Govinda Department of Computer Science and Engineering, VIT University, Vellore, Tamil Nadu, India;
Department of Computer Science and Engineering, VIT University, Chennai, Tamil Nadu, India

Aman Gupta Continental Pvt. Ltd., Bengaluru, Karnataka, India

Sayan Gupta Department of Management Studies, NIT Durgapur, Durgapur, India

Sudarshan Gurram Department of Computer Science and Engineering, RGUKT, Basar, Hyderabad, India

Mohammed Abdul Habeeb AlMusanna College of Technology, Musannah, Sultanate of Oman

V. Hariharan Amrita School of Engineering, Center for Computational Engineering and Networking (CEN), Amrita Vishwa Vidyapeetham, Coimbatore, India

Nagaratna P. Hedge Vasavi College of Engineering, Hyderabad, India

Farid Ibrahim Department of Information Technology, Abu Dhabi University, Al Ain, UAE

Mohammed Ghouse Ul Islam Department of Electrical and Electronics, Hyderabad, India

Vandana Jagtap Department of Computer Engineering, MAEER'S Maharashtra Institute of Technology, Pune, Maharashtra, India

Alind Jain Department of Computer Science, Bharati Vidyapeeth's College of Engineering, New Delhi, Delhi, India

Rachna Jain Department of Computer Science, Bharati Vidyapeeth's College of Engineering, New Delhi, Delhi, India

Shubham Jain Department of Information Technology, DTU Delhi, Rohini, India

V. Janaki Department of Computer Science and Engineering, Vaagdevi Engineering College, Warangal, Telangana, India

Ch. Jayanth Babu Department of Computer Science and Engineering, Kakatiya Institute of Technology and Science, Warangal, Telangana, India

S. Jayanthi Samskruti College of Engineering and Technology, Hyderabad, Telangana, India

K. Jeevan Suma SE, School of Information Technology, JNTUH, Hyderabad, Telangana, India

Bela Joglekar Department of Information Technology, Maharashtra Institute of Technology, Pune, India

S. Jyothi Department of Computer Science, Sri Padmavathi Mahila Viswa Vidyalayam, Tirupati, India

Muskaan Kalra Department of Computer Science, Bharati Vidyapeeth's College of Engineering, New Delhi, Delhi, India

Shruti Kant Department of Computer Engineering, MAEER'S Maharashtra Institute of Technology, Pune, Maharashtra, India

Shridevi Karande Department of Computer Engineering, Maharashtra Institute of Technology, Pune, India

Shradha Katyal Department of Computer Science, Bharati Vidyapeeth's College of Engineering, New Delhi, Delhi, India

Amarjeet Kaur Department of Computer Science and Technology, SNDT Women's University, UMIT, Mumbai, India

Mohd Niyaz Ali Khan Department of Electrical and Electronics, Hyderabad, India

Munsifa Firdaus Khan Department of Computer Science, Assam University, Silchar, Assam, India

Fariha Khatoon Department of Electrical and Electronics, Hyderabad, India

Sunirmal Khatua Department of Computer Science and Engineering, University of Calcutta, Kolkata, India

Govind Kulkarni School of Computational Sciences, SRTM University, Nanded, MS, India

Pradnya Kulkarni Department of Computer Engineering, Maharashtra Institute of Technology, Pune, India

U. P. Kulkarni SDM College of Engineering and Technology, Dharwad, Karnataka, India

Morampudi Mahesh Kumar Department of Computer Science and Engineering, NIT Warangal and IDRBT, Warangal, Telangana, India

Shrawan Kumar Department of Computer Science and Engineering, Vardhaman College of Engineering, Shamshabad, India

B. K. S. P. Kumar Raju Alluri Department of Computer Science and Engineering, NIT Andhra Pradesh, Tadepalligudam, Andhra Pradesh, India

S. K. Lokesh Naik Department of Computer Science and Engineering, Vardhaman College of Engineering, Shamshabad, Telangana, India

S. Lomte Santosh Matoshri Pratishthan Group of Institutions, Nanded, India

K. Madhavi Department of Computer Science and Engineering, JNTUA College of Engineering Ananthapuramu, Anantapur, Andhra Pradesh, India

Aruna Mailavaram TKR College of Engineering and Technology(K9), Hyderabad, India

Sahaj Singh Maini Department of Computer Science and Engineering, VIT University, Vellore, Tamilnadu, India

Deepika Mallampati Department of Computer Science and Engineering, Sreyas IET, Hyderabad, India

Dudimetla Mallesh Department of Computer Science and Engineering, RGUKT, Basar, Hyderabad, India

Jangam J. S. Mani Department of Computer Applications, K.T.S. Government Degree College, Rayadurg, Anantapuramu, Andhra Pradesh, India

K. Manikandan Vellore Institute of Technology, Vellore, Tamil Nadu, India

Anisha R. Maske Department of Information Technology, Maharashtra Institute of Technology, Pune, India

Mallikarjun M. Math Gogte Institute of Technology, Belagavi, Karnataka, India

Jasraj Meena Department of Information Technology, DTU Delhi, Rohini, India

J. Mercy Geraldine Guru Nanak Institute of Technology, Hyderabad, Telangana, India

Pooja Mishra Department of Electronics and Communication Engineering, Inderprastha Engineering College, Ghaziabad, India

Burhanuddin Mohammad AlMusanna College of Technology, Musannah, Sultanate of Oman

Sudip Mondal Department of Computer Science and Engineering, University of Calcutta, Kolkata, India

S. Md. Mujeeb Jawaharlal Nehru Technological University Anantapur, Anantapur, Andhra Pradesh, India

Aniket Muley School of Mathematical Sciences, SRTM University, Nanded, MS, India

D. Murali Department of Computer Science and Engineering, Vemu Institute of Technology, Tirupathi, India

T. Murali Mohan Department of Computer Science and Engineering, Swarnandhra Institute of Engineering and Technology, Narsapur, West Godavari, Andhra Pradesh, India

P. Naga Sravanthi Department of Information Technology, Gudlavalleru Engineering College, Gudlavalleru, India

Moparthi Nageshwara Rao Department of Computer Science and Engineering, Koneru Lakshmaiah Education Foundation, Guntur, India;
Velagapudi Ramakrishna Siddhartha Engineering College, Vijayawada, India

Siva Prasad Nandyala Model Based Design, ABU, Tate Elxsi Limited, Bangalore, India

Shaik Naseera School of Computer Science and Engineering (SCOPE), VIT University, Vellore, Tamil Nadu, India

P. Natarajan School of Computer Science and Engineering, Vellore Institute of Technology, Vellore, Tamil Nadu, India

Nithya Chidambaram Department of Electronics and Communication Engineering, SASTRA Deemed University, Thanjavur, Tamil Nadu, India

B. Padmaja Rani JNTUH College of Engineering, Hyderabad, India

Janwale Asaram Pandurang Balbhim College Beed, Beed, India

Twinkle Pardeshi Department of Computer Engineering, Maharashtra Institute of Technology, Pune, India

P. Praveen Department of Computer Science and Engineering, SR Engineering College, Warangal, India;
Department of Computer Science and Engineering, Kakatiya University, Warangal, Telangana, India

R. Praveen Sam Department of Computer Science and Engineering, G. Pulla Reddy Engineering College, Kurnool, Andhra Pradesh, India

Sudhakar Putheti Department of Computer Science and Engineering, Vasireddy Venkatadri Institute of Technology, Guntur, Andhra Pradesh, India

Imran Qureshi AlMusanna College of Technology, Musannah, Sultanate of Oman

T. Raghunadha Reddy Department of Information Technology, Vardhaman College of Engineering, Shamshabad, Telangana, India

J. Rajeshwar Department of Computer Science and Engineering, Guru Nanak Institutions Technical Campus, Ibrahimpatnam, India

S. Rajeswari Department of Computer Science and Engineering, VR Siddhartha Engineering College, Vijayawada, India

Udayabhanu N. P. G. Raju Sri Satya Sai University of Technology and Medical Sciences, Sehore, Madhya Pradesh, India

S. Ramani Department of Computer Science and Engineering, Ramaiah Institute of Technology, Bengaluru, India

Somula Ramasubbareddy Department of Computer Science and Engineering, VIT University, Vellore, Tamil Nadu, India;
Department of Computer Science and Engineering, VIT University, Chennai, TamilNadu, India

P. Ramya Department of Computer Science and Engineering, Gudlavalleru Engineering College, Gudlavalleru, India

Shilpa Rani Department of Computer Science, Sreyas Institute of Engineering and Technology, Hyderabad, Telangana, India

Neeru Rathee Maharaja Surajmal Institute of Technology, New Delhi, India

K. Ravikanth Department of Computer Science and Engineering, RGUKT, Basar, India

Saba Department of Computer Engineering, Maharashtra Institute of Technology, Pune, India

P. Sachidhanandam Department of Computer Science and Engineering, Knowledge Institute of Technology, Salem, Tamil Nadu, India

Meka Poorna Sai Department of Electronics and Communication Engineering, SASTRA Deemed University, Thanjavur, Tamil Nadu, India

S. Sai Sri Charan Department of Electronics and Communication Engineering, SASTRA Deemed University, Thanjavur, Tamil Nadu, India

D. Saidulu Department of Computer Science and Engineering, Guru Nanak Institutions Technical Campus, Hyderabad, Telangana, India

R. Saidulu Department of Electronics and Communication Engineering, RGUKT, Basar, India

Jatinderkumar R. Saini Narmada College of Computer Application, Bharuch, Gujarat, India

M. Sakthivel Department of Computer Science and Engineering, United Institute of Technology, Coimbatore, Tamil Nadu, India

Rashmi R. Salavi Gogte Institute of Technology, Belagavi, Karnataka, India

Sangeeta Gupta Vardhaman College of Engineering, Hyderabad, India

M. Sasi Kumar Research and Development, Centre for Development of Advanced Computing, Mumbai, India

Krishna Sehgal Department of Computer Science, Bharati Vidyapeeth's College of Engineering, New Delhi, Delhi, India

Suri Shanmukh Department of Electronics and Communication Engineering, SASTRA Deemed University, Thanjavur, Tamil Nadu, India

Sapna Sharma Indraprastha Institute of Information Technology Delhi, New Delhi, Delhi, India

Vishal Sharma Department of Computer Science, Bharati Vidyapeeth's College of Engineering, New Delhi, Delhi, India

Akash Shrivastava Department of Computer Science Engineering, DIT University, Dehradun, India

Reena Singh Department of Computer Science, Bharati Vidyapeeth's College of Engineering, New Delhi, Delhi, India

Saurav Singh Rochester Institute of Technology, New York, USA

J. Sirisha Devi Institute of Aeronautical Engineering, Hyderabad, India

G. Siva Shankar Vellore Institute of Technology, Vellore, Tamil Nadu, India

K. P. Soman Amrita School of Engineering, Center for Computational Engineering and Networking (CEN), Amrita Vishwa Vidyapeetham, Coimbatore, India

M. Sreenivasa Rao Department of Computer Science and Engineering, School of Information Technology, JNTUH, Hyderabad, Telangana, India

H. S. Sreeyuktha Department of Computer Science and Engineering, Ramaiah Institute of Technology, Bangalore, India

M. N. Sri Harsha Department of Computer Science and Engineering, Vasireddy Venkatadri Institute of Technology, Guntur, Andhra Pradesh, India

M. Srikanth Department of Computer Science and Engineering, Swarnandhra Institute of Engineering and Technology, Narsapuram, Andhra Pradesh, India

Vaidya Srivani School of Information Technology and Engineering, Vellore Institute of Technology, Vellore, India

P. Subramanian Guru Nanak Institute of Technology, Chennai, India

V. Sunnydayal Vellore Institute of Technology AP, Amaravathi, Andhra Pradesh, India

M. Surender Department of Computer Science and Engineering, RGUKT, Basar, India

Y. Sushma CMR College of Engineering & Technology, Hyderabad, India

A. Swathi Department of Computer Science, Sreyas Institute of Engineering and Technology, Hyderabad, Telangana, India

V. Swathi Department of Computer Science and Engineering, Guru Nanak Institutions Technical Campus, Hyderabad, Telangana, India

Ishrath Tabassum Bangalore Institute of Technology, Bengaluru, India

S. Tejasree VEMU Institute of Technology, Chittoor, AP, India;
School of Computer Science and Engineering (SCOPE), VIT University, Vellore, Tamil Nadu, India

Narina Thakur Department of Computer Science, Bharati Vidyapeeth's College of Engineering, New Delhi, Delhi, India

Sudhir Tirumalasetty Department of Computer Science and Engineering, Vasireddy Venkatadri Institute of Engineering, Guntur, India

K. C. Tripathi Department of Computer Science and Engineering, Inderprastha Engineering College, Ghaziabad, India

B. K. Uday School of Computer Science and Engineering, Vellore Institute of Technology, Vellore, Tamil Nadu, India

V. Uma Rani Department of Computer Science and Engineering, School of Information Technology, JNTUH, Hyderabad, Telangana, India

Arun Upadhyay NSHM Business School, Durgapur, India

Para Upendar Department of Computer Science and Engineering, Vardhaman College of Engineering, Shamshabad, Telangana, India

Anirudh Vattikuti School of Computer Science and Engineering, Vellore Institute of Technology, Vellore, Tamil Nadu, India

B. Venkat Raman Osmania University, Hyderabad, India;
Department of Computer Science and Engineering, RGUKT, Basar, Hyderabad, India

B. Venkatesh SCOPE Vellore Institute of Technology, Vellore, Tamil Nadu, India

K. S. Venkatesh Council of Scientific and Industrial Research-National Aerospace Laboratories, Bengaluru, India

R. Venkateswara Reddy Department of Computer Science and Engineering, Shri JJT University, Jhunjhunu, Rajasthan, India

K. Venugopala Rao GNITS, Hyderabad, India

P. Vineetha Department of Computer Science and Engineering, NIT Andhra Pradesh, Tadepalligudam, Andhra Pradesh, India

A. Vishnuvardhan Department of Computer Science and Engineering, Vasireddy Venkatadri Institute of Technology, Guntur, Andhra Pradesh, India

R. Vivekanandam Sri Satya Sai University of Technology and Medical Sciences, Sehore, Madhya Pradesh, India

Prediction of Employee Attrition Using GWO and PSO Optimised Models of C5.0 Used with Association Rules and Analysis of Optimisers

Krishna Sehgal, Harlieen Bindra, Anish Batra and Rachna Jain

Abstract Prediction of employee attrition based on five selected attributes which are Gender, Distance from Home, Environment Satisfaction, Work–Life Balance and Education Field out of 36 variables present in the data-set. Application of Grey Wolf Optimisation (GWO) Algorithm and Particle Swarm Optimisation (PSO) on the model of Decision Tree Algorithm "C5.0" which is fed in the inputs of Associated Rules, using this optimised algorithm for the prediction of employee attrition using IBM Watson Human Resource Employee Attrition Data. After comparing the efficiency of GWO and PSO, we have come to a conclusion that time to predict an employee attrition and consumption of RAM have been optimised with GWO. Employee Attrition is one of the major problems faced by companies nowadays. Sometimes, when the long-term working employees leave the company, it affects the relationship of the company with the client and in turn affects the revenue of the company if the person replacing the old employee is not able manage a good rapport with the client. The paper can be used to frame better work policies which will help both the employer and employee. It can be seen as a mirror to the working conditions of the employees.

Keywords Apriori algorithm · Association technique · C5.0 · Data mining · Decision tree · Employee attrition · Entropy · IBM Watson HR · Information gain · Grey Wolf Optimization · Particle swarm optimisation

K. Sehgal (✉) · H. Bindra · A. Batra · R. Jain
Department of Computer Science, Bharati Vidyapeeth's College of Engineering,
New Delhi, Delhi, India
e-mail: krishnasehgal2108@gmail.com

H. Bindra
e-mail: harlieenbindra@gmail.com

A. Batra
e-mail: anish.batra.yo@gmail.com

R. Jain
e-mail: rachna.jain@bharatividyapeeth.edu

© Springer Nature Singapore Pte Ltd. 2019
H. S. Saini et al. (eds.), *Innovations in Computer Science
and Engineering*, Lecture Notes in Networks and Systems 74,
https://doi.org/10.1007/978-981-13-7082-3_1

1

1   Introduction

Employee Attrition is one of the major problems faced by companies nowadays. Loss of employees from a company is actually the loss of all the training and efforts put in by the company in the employee. Sometimes, when the long term working employees leave the company, it affects the relationship of the company with the client and, in turn, affects the revenue of the company if the person replacing the old employee is not able to manage a good rapport with the client. Also, finding an immediate replacement for the leaving employee is difficult and the company has to put in the time and efforts in hiring new people leading to loss of valuable time and resources. Determining the attrition rate helps a company compare it with the industry average and work towards reducing the attrition rate. It helps the company in knowing the reasons for the attrition of employees so that they can improve as a company and keep its employees satisfied and content. Here comes the role of Data Analytics that helps us use various factors like Work–Life Balance, Environment Satisfaction, and other factors to predict the attrition of an employee. In our Research Work, we have used Data Mining techniques to predict the attrition of an employee using a model which is optimised with the help of Grey Wolf Algorithm.

2   Literature Review

In 2015, Seyedali Mirjalili [1] with other research scholars proposed a new technique inspired by grey wolves called Grey Wolf Optimizer hereby shortened to GWO. This algorithm is inspired by the hunting methodology of grey wolves and their way of leadership in nature. For the purpose of simulation, alpha, beta, omega, and delta type of grey wolves are taken. Also the three steps, that is, searching, encircling, and attacking the prey, utilised for the purpose of hunting by wolves are implemented in the algorithm. After that, benchmarking was done against 29 test functions. In 2015, Dr. Sudhir Sharma [2] with other research scholars presented a new way to provide a solution for economic load–dispatch problem (convex). This was done by using a Grey Wolf inspired metaheuristic known as Grey Wolf optimization. In 2011 [3], Dian Palupi Rini, Siti Mariyam Shamsuddin, and Siti Sophiyati Yuhaniz studied about Particle swarm optimisation or PSO in short is an optimisation and computational search method which draws its inspiration from biology. Its development took place in 1995. It was developed by Eberhart and Kennedy. The social behaviours of fish schooling and birds flocking formed its basis. The algorithm draws inspiration from the behaviour of those animals who do not have any leader in their swarm or group, such as fish schooling or bird flocking. The reason behind that lies in the fact that such flocks with no leaders find food at random and then will follow an animal of the group that is nearest to the source of food, i.e. a solution (potential).

3   Materials and Methods

We have analysed IBM Watson Human Resource Employee Attrition Data (source—Kaggle) [4] set to predict the employee attrition based on 5 selected bases classes, which are Gender, Distance from Home, Environment Satisfaction, Work–Life Balance, and Education Field variables out of the set of 36 variables.

Tools used are Microsoft Visual Studio [5] and Microsoft SQL Server [6] on core i7, and 7th Generation processor with 16 GB RAM.

4   Proposed Research Scheme

The dataset used has been acquired from Kaggle. After acquiring the dataset, we selected base classes for the employee attrition which are: Gender, Distance from Home, Environment Satisfaction, Work–Life Balance, and Education Field. Then, we have cleaned and transformed the dataset according to the selected attributes. During cleaning, we have removed entries that contain redundant values and incomplete tuples. Then to use our new approach that is C5.0 with association [7, 8], we have applied association rule mining using Apriori [9] algorithm to form association rules using selected attributes. Then, using these association rules we have trained the C5.0 decision tree [10]. Using this model, we have then predicted the attrition of employee and then matched the predicted results with the actual attrition to evaluate the efficiency of the proposed algorithm on this dataset. Now, to further optimise our model, we have used Grey Wolf Optimiser. Then using this optimised algorithm, we have again predicted the attrition of employee and then matched the predicted results with the actual attrition to evaluate the efficiency of the optimised model on this dataset. To compare Grey Wolf Optimiser with another optimisation algorithm, we have also used Particle Swarm Optimisation and again predicted the attrition of employee.

5   Performance Comparison

After acquiring the dataset from Kaggle, we are using GWO and PSO to optimise further C5.0 with association algorithm. We have observed that the time and memory consumption is least with GWO, then with PSO and then finally with Association Rules in comparison to traditional C5.0. Therefore, Grey Wolf Optimised C5.0 with association algorithm is more efficient in time and memory consumption as compared to others techniques with C5.0 [11] as shown in Table 1.

Table 1 It shows improved efficiency when Grey Wolf Optimised C5.0 Association Algorithm is used as compared to the PSO Optimised and C5.0 with Association Rule Algorithm

Basis	GWO optimised	PSO optimised	C5.0 with association rules	Traditional C5.0
Processing time consumption (ms)	0.046	0.14	0.2	2
RAM consumption (MB)	30.9	32.5	39.6	48

6 Results and Discussion

After comparing the efficiency of different algorithms, we have come to a conclusion that time to predict employee attrition and consumption of RAM have been optimised* with GWO. Figure 1 shows the transformed data set after redundancies are removed. Figures 2 and 3 show Entropy and Attrition values while training and testing, respectively. Figure 5 shows actual and predicted values of Attrition (Figs. 4, 6, and 7).

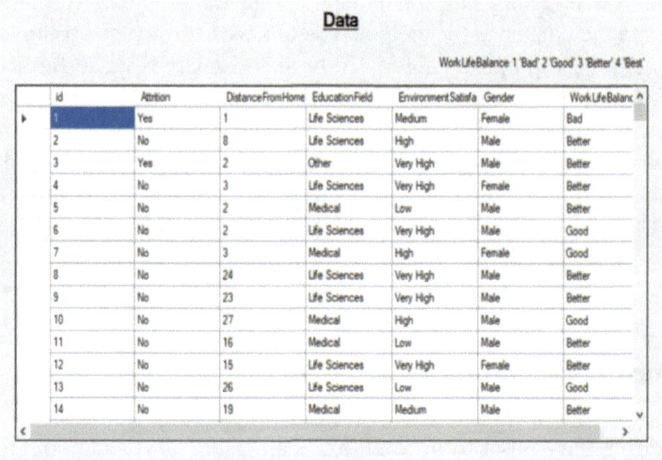

id	Attrition	DistanceFromHome	EducationField	EnvironmentSatisfa	Gender	WorkLifeBalanc
1	Yes	1	Life Sciences	Medium	Female	Bad
2	No	8	Life Sciences	High	Male	Better
3	Yes	2	Other	Very High	Male	Better
4	No	3	Life Sciences	Very High	Female	Better
5	No	2	Medical	Low	Male	Better
6	No	2	Life Sciences	Very High	Male	Good
7	No	3	Medical	High	Female	Good
8	No	24	Life Sciences	Very High	Male	Better
9	No	23	Life Sciences	Very High	Male	Better
10	No	27	Medical	High	Male	Good
11	No	16	Medical	Low	Male	Better
12	No	15	Life Sciences	Very High	Female	Better
13	No	26	Life Sciences	Low	Male	Good
14	No	19	Medical	Medium	Male	Better

Fig. 1 Transformed dataset

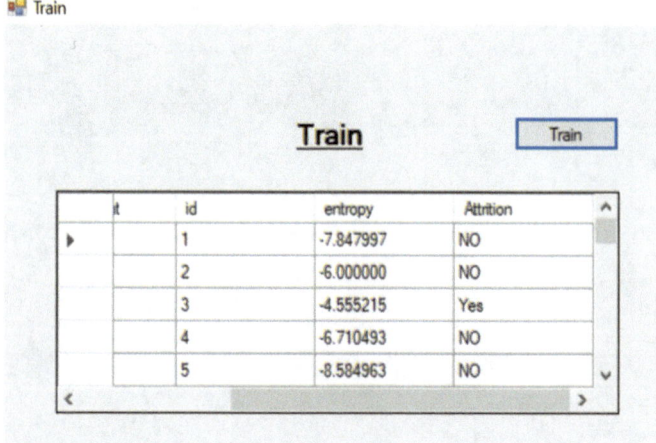

Fig. 2 Entropy and attrition values while training

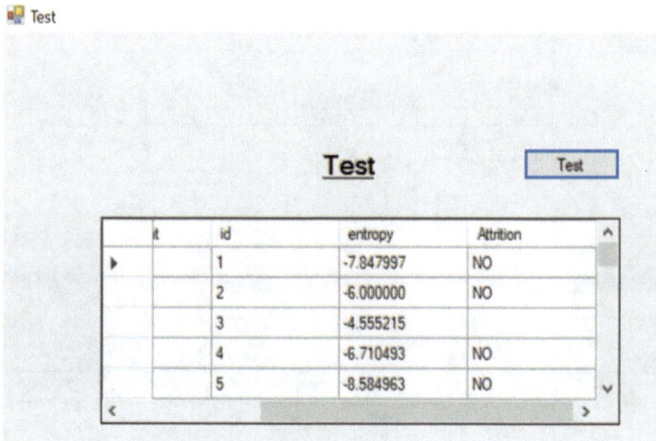

Fig. 3 Computation of entropy and attrition values while testing

7 Future Work

The work done in this paper holds immense scope for extension in future endeavours. This is because a number of nature-inspired algorithms are currently active areas of research. Furthermore, the Grey Wolf optimizer algorithm utilised in this paper itself has the potential to be superseded by another nature based algorithm.

Nature-based algorithms is a very prolific research area. This is because the problems with which we are normally familiar are getting further complicated and complex due to size and other aspects. In addition, new problems are cropping up in which

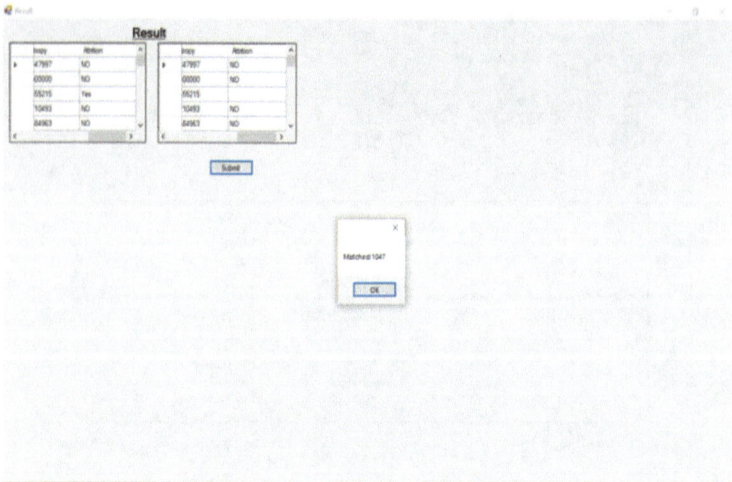

Fig. 4 Correctly predicted instances by the Grey Wolf optimised algorithm

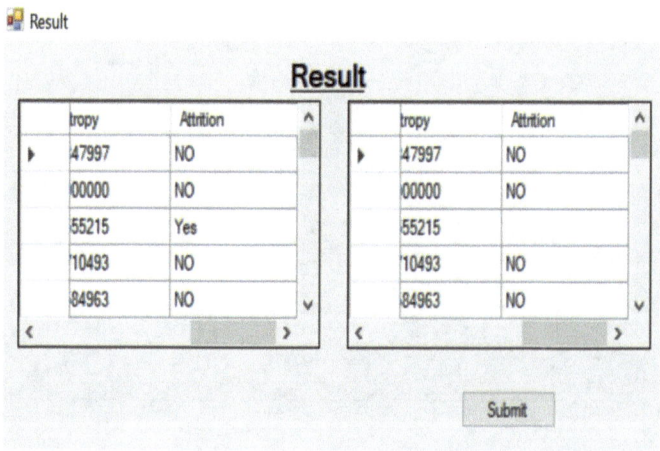

Fig. 5 Result of prediction made by Grey Wolf optimised algorithm. The Grid View on the left of the Image shows the actual values of attrition while the Grid View on the right shows the predicted values of attrition. Blank cell in the column shows that the prediction was incorrect

existing methods are not effective. Nature seems to have faced similar problems and solved them in due course of time. That is the reason we get a lot of inspiration from it. Some of the recent nature-inspired algorithms are artificial bee colony algorithm, the firefly algorithm, the social spider algorithm, the bat algorithm, the strawberry algorithm, the plant propagation algorithm and so on. These are very effective com-

Fig. 6 Processing time consumption

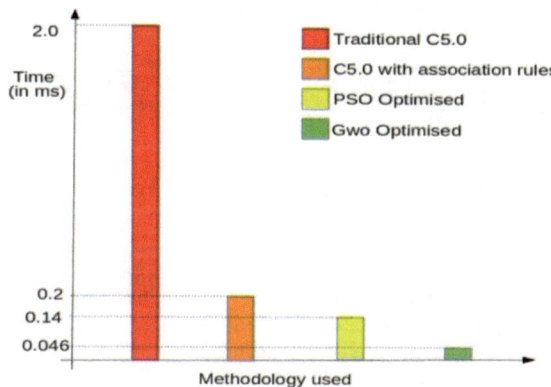

Fig. 7 RAM consumption

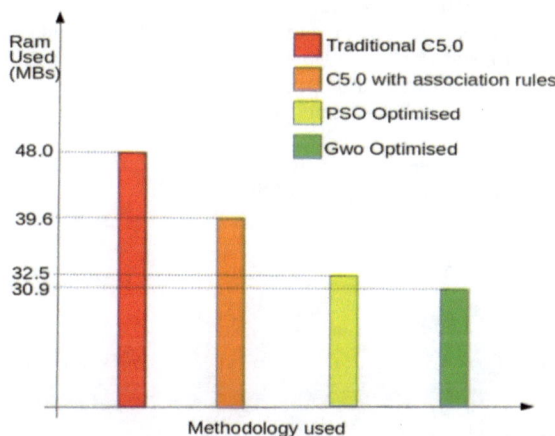

pared to early nature-inspired algorithms such as the genetic algorithm, ant colony and swarm optimization and so on. Such algorithms have very few parameters that need arbitrary setting.

8 Conclusion

After comparing the efficiency of Grey Wolf Optimised C5.0 with association algorithm to C5.0 with association rule and Particle Swarm optimisation, it is observed that the time to predict an employee attrition and consumption of RAM is optimised hence it can be concluded that efficiency of algorithm is improved when C5.0 Association rule algorithm is used with Grey Wolf.

References

1. Mirjalili S, Mirjalili SM, Lewis A (2014) Grey Wolf optimizer. Adv Eng Softw 69:46–61
2. Sharma S, Mehta S, Chopra N et al (2015) Int J Eng Res Appl 5(4, Part-6):128–132. ISSN 2248-9622
3. Mehta S, Shukla D (2015) Optimization of C5.0 classifier using Bayesian theory. In: IEEE International conference on computer, communication and control
4. https://www.ibm.com/communities/analytics/watson-analytics-blog/hr-employee-attrition/. Accessed 30 Oct 2017
5. https://en.wikipedia.org/wiki/Microsoft_Visual_Studio. Accessed 30 Oct 17
6. https://en.wikipedia.org/wiki/Microsoft_SQL_Server. Accessed 30 Oct 17
7. Agrawal R, Imieliński T, Swami A (1993) Mining association rules between sets of items in large databases. In: Proceedings of the 1993 ACM SIGMOD international conference on management of data, SIGMOD '93, pp 207–216
8. Sethi M, Jindal R (2016) Distributed data association rule mining: tools and techniques. In: 3rd international conference on computing for sustainable global development (INDIACom)
9. Yuan X (2017) An improved Apriori algorithm for mining association rules. In: AIP conference proceedings, vol 1820, Issue 1, Mar 2017
10. Du W, Zhan Z (2002) Building decision tree classifier on private data. In: Proceedings of the IEEE international conference on privacy, security and data mining, CRPIT '14, vol 14, pp 1–8
11. Bujlow T, Riaz T, Pedersen JM (2012) A method for classification of network traffic based on C5.0 machine learning algorithm. In: International conference on computing, networking and communications (ICNC). IEEE

An Ensemble Classifier Characterized by Genetic Algorithm with Decision Tree for the Prophecy of Heart Disease

K. Chandra Shekar, Priti Chandra and K. Venugopala Rao

Abstract The prediction of heart disease is critically significant for diagnosis of diseases and treatment. The data mining techniques that can be applied in medicine, and in particular some machine learning techniques including the mechanisms that make them better suited for the analysis of medical databases. Extensive amounts of data gathered in medical databases require specialized tools for storing and accessing data, for data analysis, and for effective use of data. In particular, the increase in data volume causes great difficulties in extracting useful information and also consumes time for decision support. Intuitively, this large amount of stored data contains valuable hidden information, which could be used to improve the decision making process of an organization. For prediction of heart disease, many researchers have used some machine learning algorithms like Bayesian Classification, Neural Networks, Support Vector Machines, and K-nearest neighbor algorithms. We propose a hybrid technique of ensemble classifier to provide a better solution for the classification problem. The output from this hybrid scheme gives the optimized feature. This output is then given as the input to the decision tree classifier for predicting the occurrence and possibly obtaining the type of heart disease. Here, the features are initialized through the decision tree and fitness is evaluated via genetic algorithm.

Keywords Ensemble classifier · Genetic algorithm · Decision tree · Fitness function · Selection · Crossover · Mutation · Reproduction

K. Chandra Shekar (✉)
JNTUH, Hyderabad, Telangana, India
e-mail: chandhra2k7@gmail.com

P. Chandra
ASL, DRDO, Hyderabad, India
e-mail: priti_murali@yahoo.com

K. Venugopala Rao
GNITS, Hyderabad, India
e-mail: kvgrao1234@gmail.com

© Springer Nature Singapore Pte Ltd. 2019   9
H. S. Saini et al. (eds.), *Innovations in Computer Science
and Engineering*, Lecture Notes in Networks and Systems 74,
https://doi.org/10.1007/978-981-13-7082-3_2

1   Introduction

In recent years, data mining has been extensively used in the areas of bioinformatics, science and engineering, genetics, and medicine [1]. Data mining is an interdisciplinary field of study in databases, machine learning, and visualization. Data mining is the research domain which deals with discovering the relationships and global patterns that exist hidden among large amounts of data [2, 3]. Most of the healthcare organizations are facing a major challenge of providing quality services to their patients like accurate automated diagnosis and administering treatment at affordable costs [4]. Data mining helps in identifying the patterns from successful medical case sheets for different illnesses and it also aims to find knowledge which is useful for the diagnostics [5]. It is a collection of various techniques and algorithms, through which we can extract informative patterns from raw data [6]. It plays a vital role in tackling the data overload in medical diagnostics. Data mining technology provides a deep insight providing a user oriented approach to discover novel and hidden patterns in the data. This helps in evaluating the effectiveness of medical treatments [7]. The data generated by healthcare transactions is enormous. This medical data containing patients' symptoms is analyzed to perform medical research [8].

With the development of information technology, extensive medical data is available. Medical data classification plays a significant role in various medical applications [9–11]. Medical classification can be widely used in hospitals for the statistical analysis of diseases and therapies [12, 13]. It addresses the problems of diagnosis, analysis and teaching purposes in medicine [14–16]. Medical data has made a great progress over the past decades in the development and use of classification algorithms [17–19]. In healthcare, these medical data can be transformed into aggregations to calculate average values per patient and compare with ranges/other values, to group data into clusters of similar data, etc. [20–22].

Ensemble Methods are the methods that use a combination of models to improve classifier and predictor accuracy. Bagging and Boosting are the two such general strategies. According to the Wolpert's no free lunch theorem, a classifier may perform well in few specific domains, but never in all application domains. Therefore, by combining the outputs of multiple classifiers, the ensemble of classifiers strategically extends the power of aggregated method to achieve better prediction accuracy.

2   Related Work

Akhil Jabbar [1] had proposed an algorithm which combines K-Nearest Neighbor with genetic algorithm for effective classification. Muthukaruppan et al. [23] had proposed particle swarm optimization (PSO), which is based on fuzzy expert system involving four stages. Lahsasna et al. [24] proposed a fuzzy rule-based system (FRBS) to serve as a decision support system for Coronary heart disease (CHD) diagnosis that not only considers the decision accuracy of the rules but also their transparency at the same time. Yilmaz et al. [25] had presented a new data preparation method based on clustering algorithms for the diagnosis of heart and diabetes diseases. Kim et al. [26] had proposed a Fuzzy Rule-based Adaptive Coronary Heart Disease Prediction Support Model (FbACHD_PSM), which gives content recommendation to coronary heart disease patients.

3 Proposed Method

The proposed methodology integrates with supervised machine learning technique which is based on a hybrid approach for providing a better decision system using dual decision tree and genetic algorithm. Genetic algorithms are one of the best methods for search and optimization problems.

A decision tree is a tree structure classifier that includes a root node, branches, and leaf nodes. Each internal node denotes a test on an attribute, each branch denotes the outcome of a test, and each leaf node holds a class label.

Pros of Decision Trees (DTs):

- DTs do not require any domain knowledge.
- DTs are easy to comprehend.
- The learning and classification steps of a DT are simple and fast.

Tree pruning is performed in order to remove anomalies in the training data due to noise or outliers. The pruned trees are smaller and less complex. Tree Pruning can be done through two approaches:

- Pre-pruning—The tree is pruned by halting its construction early.
- Post-pruning—This approach removes a sub-tree from a fully grown tree.

The cost complexity of a decision tree is measured by two parameters, the number of leaves in the tree and the error rate of the tree.

Genetic algorithms (GA) were invented by John Holland in 1975. Genetic algorithms can be applied for search and optimization problems. GA uses genetics approach as its model for problem solving. Each solution in genetic algorithm is represented through chromosomes. Chromosomes are made up of genes, which are individual elements that represent the problem. The collection of all chromosomes is called the population [1, 27].

In general, there are three operators that can be applied in GA.

(1) **Selection**:
   This operator is used in selecting individuals for reproduction with the help of fitness function. Fitness function in GA is the value of an objective function for its phenotype. The chromosome has to be first decoded, for calculating the fitness function.

(2) **Crossover**:
   This is the process of taking two parent chromosomes and producing a child from them. This operator is applied to create better string.

Fig. 1 Proposed system architecture for an ensemble classifier characterized by genetic algorithm with dual decision tree

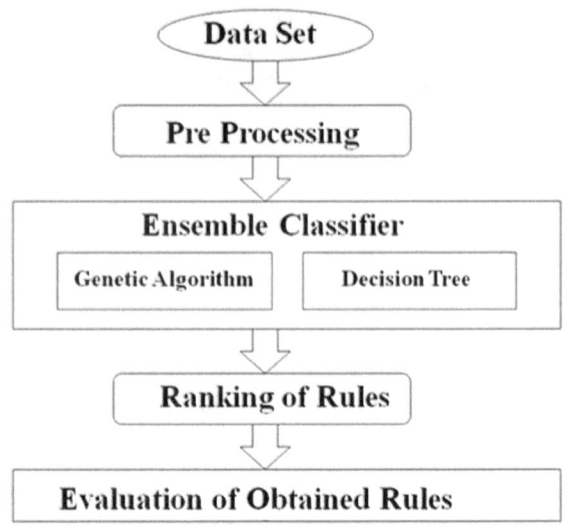

(3) **Mutation**:

This operator is used to alter the new solutions in the search for better solution. Mutation prevents the GA to be trapped in a local minimum.

The proposed system architecture (Fig. 1) consists of an ensemble classifier characterized by genetic algorithm with dual decision tree facilitates as follows, in the first stage multiple risk factors such as age, hypercholesterolemia, hypertension, diabetes, obesity, stress level, alcohol taken, etc., are taken as input. This input is preprocessed to fill up the missing values, remove noise and inconsistencies if any in the data and then is given to the hybrid scheme which consists of genetic algorithm and decision tree. Here, the features are initialized through decision tree and fitness is evaluated via genetic algorithm. The output from this hybrid scheme gives the optimized feature. This output is then given as the input to the decision tree classifier for obtaining the type of heart disease.

In decision tree, both training and testing phases are carried out. In the training phase, a classifier known as iterative dichotomizer or random forest classifier can be utilized. This classifier makes use of number of decision trees at training stage in order to enhance classification rate. This random classifier contains two steps namely oob (out-of-bag) and permutation to avoid classification error and to measure the importance of variable. This classifier has a combined group of techniques to process such as randomized node optimization, bagging and CART model. In random optimization algorithm, the best tree model is given as output and in bagging it repeatedly selects the random sample with the replacement of the training set and fit trees. After that CART (classification and regression) is done for recognizing the type of attack and finally the output is displayed using a tree structure. The output displays the type of heart attack for the patient to occur. This can be determined in the classification step by comparing the information stored in the database. If

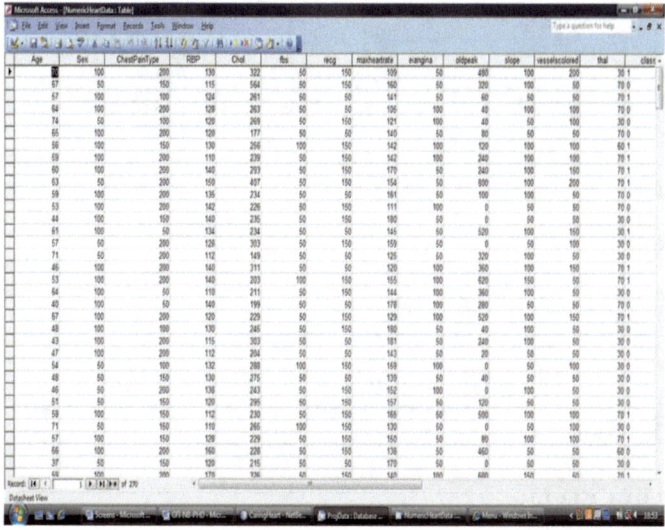

Fig. 2 Heart disease dataset

there is any type of attack possibility is predicted then it will show the prediction by percentage value by the utilization of the regression method. Decision tree has four major advantages for predictive analytics namely it implicitly performs feature selection, it needs relatively very less effort from users for data preparation, the nonlinear relationships between parameters do not affect tree performance, and it is very simple to explain.

In our proposed hybrid technique, we can predict the accurate type of heart attack and optimum feature selection for reducing dimensionality, training time, and over-fitting. The proposed methodology can be implemented using MATLAB platform and the experimental results can be analyzed and compared with the conventional methods.

4   Results and Discussions

The experimental results attained from the proposed method are compared with the existing methods in terms of classification accuracy and time complexity with respect to the heart disease dataset (Fig. 2).

The proposed approach generates the optimized features through genetic algo-rithm. The classification accuracy is higher when compared with the existing methods (Table 1).

The reduction of time complexity is expected due to the optimization performed on the features (Table 2).

Table 1 Accuracy analysis

Classification method	Accuracy (%)
Random subspace	78.91
Decision tree	66.67
Multilayer perceptron	79.25
Proposed approach	85.37

Table 2 Accuracy analysis

Classification method	Time complexity (s)
Random subspace	0.18
Decision tree	0.17
Multilayer perceptron	27.88
Proposed approach	0.12

Thus, it is natural to realize the efficient of the proposed approach as the accuracy has increased and the time complexity has reduced significantly.

5 Conclusion and Future Directions

Majority of the health care organizations are facing a severe challenge in the provision of quality services like diagnosing patients correctly and administering treatment at reasonable costs. Data mining helps to identify the patterns of successful medical therapies for different illnesses and also it aims to find useful information from large collections of data. With the development of information technology, extensive medical data is available. Medical data classification plays an essential role in most of the medical applications. Ensemble Methods are the methods that use a combination of models, to improve classifier and predictor accuracy. The purpose of this research is to enhance performance of the heart disease prediction system by avoiding mis-prediction rate. We further plan to compare the performance of our ensemble classifier with the existing and traditional classifiers. We would also like to move towards hybrid generic intelligent systems to further improve the predictive accuracy.

References

1. Akhil Jabbar M, Deekshatulu BL, Chandra P (2013) Classification of heart disease using K-nearest neighbor and genetic algorithm. Procedia Technol 10:85–94
2. Karaolis MA et al (2010) Assessment of the risk factors of coronary heart events based on data mining with decision trees. IEEE Trans Inf Technol Biomed 559–566
3. Oztekin A, Delen D, Kong ZJ (2009) Predicting the graft survival for heart–lung transplantation patients: an integrated data mining methodology. Int J Med Inform e84–e96

4. Kurt I, Ture M, Turhan Kurum A (2008) Comparing performances of logistic regression, classification and regression tree, and neural networks for predicting coronary artery disease. Expert Syst Appl 366–374
5. Tsipouras MG et al (2008) Automated diagnosis of coronary artery disease based on data mining and fuzzy modeling. IEEE Trans Inf Technol Biomed 447–458
6. Das R, Turkoglu I, Sengur A (2009) Effective diagnosis of heart disease through neural networks ensembles. Expert Syst Appl 7675–7680
7. Kahramanli H, Allahverdi N (2008) Design of a hybrid system for the diabetes and heart diseases. Expert Syst Appl 82–89
8. Huang Y et al (2007) Feature selection and classification model construction on type 2 diabetic patients' data. Artif Intell Med 251–262
9. Ramon J et al (2007) Mining data from intensive care patients. Adv Eng Inform 243–256
10. Cho BH et al (2008) Application of irregular and unbalanced data to predict diabetic nephropathy using visualization and feature selection methods. Artif Intell Med 37–53
11. Sarkar BK, Sana SS, Chaudhuri K (2012) A genetic algorithm-based rule extraction system. Appl Soft Comput 238–254
12. Karaboga D, Ozturk C (2011) A novel clustering approach: Artificial Bee Colony (ABC) algorithm. Appl Soft Comput 652–657
13. Wright A, Chen ES, Maloney FL (2010) An automated technique for identifying associations between medications, laboratory results and problems. J Biomed Inform 891–901
14. Chandra Shekar K, Sree Kanth K, Ravi Kanth K (2012) Improved algorithm for prediction of heart disease using case based reasoning technique on non-binary datasets. Int J Res Comput Commun Technol 1(7)
15. Deepika N, Chandra Shekar K, Sujatha D (2011) Association rule for classification of heart-attack patients. Int J Adv Eng Sci Technol 11(2):253–257
16. Polat K, Güneş S (2007) An expert system approach based on principal component analysis and adaptive neuro fuzzy inference system to diagnosis of diabetes disease. Digit Signal Process 702–710
17. Yap BW, Ong SH, Mohamed Husain NH (2011) Using data mining to improve assessment of credit worthiness via credit scoring models. Expert Syst Appl 13274–13283
18. Özçift A (2011) Random forests ensemble classifier trained with data resampling strategy to improve cardiac arrhythmia diagnosis. Comput Biol Med 265–271
19. Pal D et al (2012) Fuzzy expert system approach for coronary artery disease screening using clinical parameters. Knowl-Based Syst 162–174
20. Seera M, Lim CP (2014) A hybrid intelligent system for medical data classification. Expert Syst Appl 2239–2249
21. Acharya UR et al (2013) Automated classification of patients with coronary artery disease using grayscale features from left ventricle echocardiographic images. Comput Methods Programs Biomed 624–632
22. Exarchos TP et al (2007) A methodology for the automated creation of fuzzy expert systems for ischaemic and arrhythmic beat classification based on a set of rules obtained by a decision tree. Artif Intell Med 187–200
23. Muthukaruppan S, Er MJ (2012) A hybrid particle swarm optimization based fuzzy expert system for the diagnosis of coronary artery disease. Expert Syst Appl 11657–11665
24. Lahsasna A et al (2012) Design of a fuzzy-based decision support system for coronary heart disease diagnosis. J Med Syst 3293–3306
25. Yilmaz N, Inan O, Uzer MS (2014) A new data preparation method based on clustering algorithms for diagnosis systems of heart and diabetes diseases. J Med Syst
26. Kim J-K et al (2014) Adaptive mining prediction model for content recommendation to coronary heart disease patients. Clust Comput 881–891
27. Vijay Bhasker G, Chandra Shekar K, Lakshmi Chaitanya V (2011) Mining frequent itemsets for non binary data set using genetic algorithm. Int J Adv Eng Sci Technol 11(1):143–152

A Novel Approach for Predicting Nativity Language of the Authors by Analyzing Their Written Texts

Para Upendar, T. Murali Mohan, S. K. Lokesh Naik
and T. Raghunadha Reddy

Abstract The technique of predicting the profiling characteristics like gender, age, nativity language, and location of an anonymous author's text by examining the author's style of writing is called Author Profiling. The differences in authors writing styles play a crucial role in Author Profiling. Several researchers extracted different types of stylistic features to discriminate the authors writing styles. Most of the researchers used a standard Bag Of Words model for document representation in Author Profiling approaches. This model has some problems like high dimensionality of features, sparsity in representation of a document and the relationship between the features were not captured in document representation. In this work, a new approach is proposed for document representation. In this approach, the weights of documents were used to generate the vectors for documents. This approach also solves the problems faced in Bag Of Words model. In this work, we concentrated on prediction of nativity language of the authors from the corpus collected from hotel reviews. Different classification algorithms were used from WEKA tool to predict the accuracy of nativity language of an author's text. The obtained results were good when compared with the state-of-the-art approaches for nativity language prediction in Author Profiling.

Keywords Author profiling · Bag Of Words model · Nativity language prediction · Document weight measure · Term weight measure

P. Upendar (✉) · S. K. Lokesh Naik
Department of CSE, Vardhaman College of Engineering, Shamshabad, Telangana, India
e-mail: upendar.para@gmail.com

S. K. Lokesh Naik
e-mail: sklokeshnaik@gmail.com

T. Murali Mohan
Department of CSE, Swarnandhra Institute of Engineering and Technology, Narsapur,
West Godavari, Andhra Pradesh, India
e-mail: drtmm512@gmail.com

T. Raghunadha Reddy
Department of IT, Vardhaman College of Engineering, Shamshabad, Telangana, India
e-mail: raghu.sas@gmail.com

© Springer Nature Singapore Pte Ltd. 2019
H. S. Saini et al. (eds.), *Innovations in Computer Science
and Engineering*, Lecture Notes in Networks and Systems 74,
https://doi.org/10.1007/978-981-13-7082-3_3

1   Introduction

The data in the World Wide Web is increasing exponentially day by day mainly through blogs, reviews, social media, and Twitter tweets. Extracting useful information and finding correct information from this huge amount of data is a challenging problem for the researchers. Authorship Analysis is a popular text processing technique attracted by the researchers to extract correct details from the data [1]. Author Profiling is a type of Authorship Analysis technique to extract the genuine details of the text like the author's age, gender, location, nativity language, educational background, and personality traits [2].

In Author Profiling, linguistic features are used to find the profile of an author and the most common techniques that are used are different kinds of machine learning techniques [3]. Many applications like marketing, forensic analysis and security used Author Profiling techniques [4]. In Author Profiling approaches, the standard document representation technique Bag Of Words model is used by several researchers for document vector representation [5]. In this work, a new approach is proposed for document vector representation and also addresses the drawbacks of BOW model.

This paper is organized into seven sections. The existing work in Author Profiling for nativity language prediction was explained in Sect. 2. The corpus characteristics and performance evaluation measures were explained in Sect. 3. Section 4 describes the standard BOW model for representing the document. Section 5 describes the proposed approach. The experimental results of BOW and PDW models were discussed in Sect. 6. Section 7 concludes this paper with conclusions.

2   Related Work

In general, the researchers extracted different types of stylistic features for predicting the demographic characteristics of the authors in Author Profiling approaches [6]. Adamearcia et al., proposed [7] a simple classification method based on similarity among objects. They considered various features for document representation. They predicted gender and language variety from English language tweets. It was observed that they got good results for gender prediction and low accuracy for language variety prediction.

Argamon used [8] the corpus of ICLE and corpus was tested to predict the age, gender, and native language. He also used 251 essays for neurotism prediction. While predicting the age, gender, nativity language, and neurotism, they observed that style based features gave an accuracy of 65.1%, content-based features gave an accuracy of 0.823, and both style-based and content-based features together gave an accuracy of 0.793.

Table 1 The characteristics of Hotel Reviews English corpus for Nativity Language prediction

Number of authors	Name	Labels	Label distribution	
3600	Hotel Reviews	Native Language	India (IN)	600
			China (CH)	600
			Pakistan (PK)	600
			Australia (AU)	600
			United States (US)	600
			Great Britain (GB)	600

Ciobanu et al., used [9] character and word n-grams as features and multiple linear SVM as classifier for predicting gender and language variety of authors in different languages. They obtained 75% accuracy for language variety prediction in English language.

3 Performance Evaluation Measures and Dataset Characteristics

In general, when evaluating the results in the experiments in Author Profiling approaches, the researchers used accuracy, precision, recall, and F1-score was taken into consideration [10]. Accuracy is the most commonly used performance measure which measures the proportion of all predictions that are correct [11]. In this work, the accuracy measure was used to evaluate the performance of the classifiers. Table 1 shows the characteristics of the corpus.

In this work, Hotel reviews of six different nativity language countries such as India (IN), China (CH), Pakistan (PK), Australia (AU), United States (US,) and Great Britain (GB) were collected from TripAdvisor.com for experimentation. The distribution of the classes is balanced. In total, the corpus consisted of 3600 authors reviews with a distribution of 600 reviews for each author.

4 Bag Of Words (BOW) Model

In BOW model, first, preprocessing methods like stop word removal and stemming were applied on the training dataset. Stop word removal eliminates the words which are ineffective for differentiating the text. Stemming converts different forms of the same word into its root word. Extract the most frequent terms from the updated corpus. In this work, 8000 terms were identified as frequent terms. The document vectors were represented with these 8000 terms. The vector value is computed using the term frequency in the document. The Naïve Bayes Multinomial and Random

Forest classifiers from WEKA tool used for generating classification model by using these document vectors. The nativity language of anonymous document is predicted using this classification model.

5 Nativity Language Specific Document Weighted Approach

In this work, a new approach namely Nativity Language specific Document Weighted (NLDW) approach is proposed to represent the document vectors. In this model, first, the preprocessing methods like stop word removal and stemming were applied on the corpus to prepare the content for further analysis. The most frequent terms were extracted from the corpus. These term weights were calculated specific to the every nativity language country using term weight measure. The document weight measure is used to calculate the document weights specific to every nativity language country of documents. Finally, the document vectors were represented with these document weights and different classification algorithms were used these document vectors to produce the classification model and this model is used to predict the nativity language of unknown document. In this approach, finding appropriate weight measures is very important to improve the accuracy prediction of nativity language of a document.

5.1 Term Weight Measure (TWM)

The TWM assign suitable weights to the terms. In this work, we used a TWM proposed in [12]. The TWM is represented in Eq. (1). In this measure, tf(t_i, d_k) is the term frequency t_i in document d_k, DF_k is the total number of terms in a document d_k. $DC(t_i, p_{IN})$ is the number of documents in India profile, which contains term t_i. $DC(t_i \varepsilon p_{IN})$ and $DC(t_i, \neq p_{IN})$ are the number of documents in India profile and other than India profile, which contains the term t_i.

$$W(t_i, d_k \in p_{IN}) = \frac{tf(t_i, d_k)}{DF_k} * \frac{\sum_{x=1,d_x \in p_{IN}}^{m} tf(t_i, d_x)}{1 + \sum_{y=1,d_y \notin p_{IN}}^{m} tf(t_i, d_y)} * \frac{DC(t_i, \in p_{IN})}{1 + DC(t_i, \notin p_{IN})}$$
(1)

5.2 Document Weight Measure

The document weight measure computes the document weight by aggregating the weights of the terms in that document. In this work, a document weight measure [13]

Table 2 The accuracy of BOW and NLDW model for nativity language of authors prediction using different classifiers

Classifiers/Number of terms	Naïve Bayes multinomial		Random forest	
	BOW	NLDW	BOW	NLDW
1000	58.18	76.27	61.19	80.27
2000	59.23	78.33	63.38	81.33
3000	61.41	79.39	64.42	83.39
4000	62.89	79.86	64.85	84.86
5000	62.94	80.77	65.51	85.77
6000	63.06	81.69	66.07	87.69
7000	63.59	82.71	66.59	87.71
8000	64.77	84.70	68.26	88.53

is used to compute the weights of the documents. Equation (2) shows the document weight d_k specific to a nativity language profile p_j.

$$W_{dkj} = \sum_{t_i \in d_k, d_k \in p_j} TF(t_i, d_k) \cdot W_{tij} \tag{2}$$

where, $TF(t_i, d_k)$ is the term frequency term t_i in document d_k, W_{tij} is the weight of the term t_i in Profile p_j.

In this work, the NLDW model used only 6 (the number of nativity language countries in the corpus) values to represent the document vector. This avoids the problem of high dimensionality of BOW model. The sparsity representation (most of the zeros in document vector representation) was also avoided in NLDW model. The relationships between terms were captured when document weight is computed.

6   Experimental Results

The accuracies of nativity language prediction in BOW and PDW model using different classifiers are shown in Table 2.

In this work, 8000 most frequent terms were extracted from the corpus. By analyzing the corpus of language variety profile, it was observed that the content-based features like the terms they used in their writings are different for different nativity language countries authors. It was understood that the selection of words to write a review is almost the same for the users of one nativity language country. With this assumption, most frequent 8000 terms were extracted from the corpus as features. Different classification algorithms such as Random Forest and Naïve Bayes Multinomial were used to produce the classification model. The proposed model obtained good results for nativity language prediction when compared with BOW model and most of the existing solutions of Author Profiling for nativity language prediction.

The Random Forest classifier obtained the highest accuracy of 88.53% for nativity language prediction when NLDW model is used. It was observed that the NLDW obtained good accuracies for nativity language prediction when compared with BOW model.

7 Conclusions

This work was delimited to author profiling of native language on text written in English language. In addition, it was observed that how the NLDW approach performed on different classifiers by comparing with standard BOW model for author profiling. Out of two evaluated classifiers, the results showed that the overall best performing classifier was the Random Forest classifier which outperformed the other classifier. The Random Forest classifier obtained 88.53% accuracy for nativity language prediction when 8000 terms were used as features.

References

1. Raghunadha Reddy T, Vishnu Vardhan B, Vijayapal Reddy P (2016) Author profile prediction using pivoted unique term normalization. Indian J Sci Technol 9(46)
2. Raghunadha Reddy T, Vishnu Vardhan B, Vijayapal Reddy P (2017) N-gram approach for gender prediction. In: 7th IEEE International advanced computing conference, Hyderabad, Telangana, 5–7 Jan 2017, pp 860–865
3. Raghunadha Reddy T, Vishnu Vardhan B, GopiChand M, Karunakar K (2018) Gender prediction in author profiling using ReliefF feature selection algorithm. Proc Adv Intell Syst Comput 695:169–176
4. Raghunadha Reddy T, Gopichand M, Hemanath K (2017) Location prediction of anonymous text using author profiling technique. Int J Civil Eng Technol (IJCIET) 8(12):339–345
5. Buddha Reddy P, Raghunadha Reddy T, Gopi Chand M, Venkannababu A (2018) A new approach for authorship attribution. Adv Intell Syst Comput 701:1–9
6. Raghunadha Reddy T, Vishnu Vardhan B, Vijayapal Reddy P (2016) A survey on authorship profiling techniques. Int J Appl Eng Res 11(5):3092–3102
7. Adame-Arcia Y, Castro-Castro D, Bueno RO, Muñoz R (2017) Author profiling, instance-based similarity classification. In: Proceedings of CLEF 2017 evaluation labs
8. Argamon S, Koppel M, Pennebaker JW, Schler J (2009) Automatically profiling the author of an anonymous text. Commun ACM 52(2):119
9. Ciobanu AM, Zampieri M, Malmasi S, Dinu LP (2017) Including dialects and language varieties in author profiling. In: Proceedings of CLEF 2017 evaluation labs
10. Swathi Ch, Karunakar K, Archana G, Raghunadha Reddy T (2018) A new term weight measure for gender prediction in author profiling. Proc Adv Intell Syst Comput 695:11–18
11. Bhanu Prasad A, Rajeswari S, Venkannababu A, Raghunadha Reddy T (2018) Author verification using rich set of linguistic features. Proc Adv Intell Syst Comput 701:197–203
12. Raghunadha Reddy T, Vishnu Vardhan B, Vijayapal Reddy P (2016) Profile specific document weighted approach using a new term weighting measure for author profiling. Int J Intell Eng Syst 9(4):136–146. https://doi.org/10.22266/ijies2016.1231.15
13. Raghunadha Reddy T, Vishnu Vardhan B, Vijayapal Reddy P (2017) A document weighted approach for gender and age prediction. Int J Eng 30(5):647–653

A New Approach for Authorship Verification Using Information Retrieval Features

Shrawan Kumar, S. Rajeswari, M. Srikanth and T. Raghunadha Reddy

Abstract Authorship Verification is the process of verifying an author by checking whether the document is written by the suspected author or not. The performance of the Authorship Verification mainly depends on the features used for differentiating the writing style in the documents. The researchers extracted various types of stylistic features for author identification based on the writing style of a particular author. In this work, a new approach is proposed by using the information retrieval features for author identification. In this approach, the test document was treated as a query and the training data as a document database. Cosine similarity measure is used to compute the similarity between the test document and training data. The accuracy of author verification depends on the scores of the similarity measures. The proposed approach performance is good when compared with most of the state-of-the-art approaches in Authorship Verification.

Keywords Author Verification · Stylistic features · Information retrieval features · Cosine similarity measure · Term weight measure

S. Kumar (✉)
Department of CSE, Vardhaman College of Engineering, Shamshabad, India
e-mail: shrawan_60@yahoo.com

S. Rajeswari
Department of CSE, VR Siddhartha Engineering College, Vijayawada, India
e-mail: rajeswari.setti@gmail.com

M. Srikanth
Department of CSE, Swarnandhra Institute of Engineering and Technology, Narsapuram, Andhra Pradesh, India
e-mail: srikanth.mandela@gmail.com

T. R. Reddy
Department of IT, Vardhaman College of Engineering, Shamshabad, Telangana, India
e-mail: raghu.sas@gmail.com

© Springer Nature Singapore Pte Ltd. 2019
H. S. Saini et al. (eds.), *Innovations in Computer Science and Engineering*, Lecture Notes in Networks and Systems 74,
https://doi.org/10.1007/978-981-13-7082-3_4

23

1   Introduction

The Internet is growing exponentially with a vast amount of data every day. Such a high growth rate brings some problems like fraudulent, stolen, or unidentified data. These problems can be dangerous and serious problems in places like the government sector, public websites, forensics, and schools. Because of these threats, and in detection of truth, it is important to analyze the authorship of a text. Authorship Analysis is one such technique which is used to find the authorship of a document.

Authorship Analysis is categorized as three classes such as Authorship Attribution, Authorship Profiling, and Authorship Verification [1]. Figure 1 shows the procedure of Authorship Verification. The task of authorship verification is to assess whether the text in dispute was written by the same author or not [2].

Author Verification approaches were important in various applications [3]. In cyberspace application, digital documents were used as evidence to prove whether the suspect is a criminal or not by analyzing their documents. If the suspect authors are unknown, i.e., there is no suspect, thus this is commonly known as an authorship identification problem. However, there are also some cases when the identification of the author is not necessary, i.e., it is enough just to know if the dispute document was written by the suspected author or not. This is a problem faced by many forensic linguistic experts which is called as authorship verification problem.

This paper is structured in 5 sections. Section 2 reviews the existing approaches proposed to authorship verification. The approach using information retrieval features is explained in Sect. 3. The experimental results of Authorship Verification were analyzed in Sect. 4. Section 5 concludes this work along with future directions.

2   Literature Review

Author Verification is a type of author identification task, which deals with the identification of whether two documents were written by the same author or not [4]. The researchers proposed different types of stylistic features to discriminate the authors'

Fig. 1 Authorship
verification

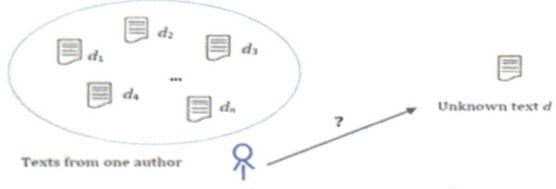

style of writing [5]. Raghunadha Reddy et al., experimented [6] with three types of linguistic features to improve the accuracy of author verification. They used Naïve Bayes Multinomial classifier to build the classification model and good accuracy is achieved for Author Verification.

Feng et al., experimented [7] with stylometric features and coherence features of 568 features for Greek, 538 for English and 399 for the Spanish language. They adopted an unmasking approach to enhance the features quality which is used in constructing weak classifiers. They understand that their work achieved less accuracy for Greek texts and good accuracy for Spanish and English texts. Vilariño et al., used [8] a support vector machine to generate the classification model. They extracted lexical, syntactic, and graph-based features for document vector representation.

Veenman et al., proposed [9] three approaches like two class classifications in compression prototype space, nearest neighbor with compression distances and boot-strapped document samples for author verification task. They computed the compression distance between the documents using compression dissimilarity measure. It was identified that they achieved the best accuracy between the PAN 2013 competition submissions. van Dam used [10] the common N-gram (CNG) method to normalize the distance between unbalance and short texts. In CNG method, the character n-grams were exploited for document vector representation. They observed that CNG approach fails for Greek language but achieved good accuracy for Spanish and English languages. Seidman proposed [11] a method which compares the similarity between given textual documents. This approach obtained overall first rank in the competition.

Bobicev used [12] the prediction by partial matching (PPM) technique depends on statistical n-gram model. They proposed an approach to detect the author of a given text automatically when the dataset contains small training sets of known authors. The PPM method collects total information from the original corpus without feature engineering. It was identified that when the document length was increased their system performance measure F-measure is not increased.

3 Proposed Approach

In this work, the experimentation was carried on PAN 2014 competition author verification dataset. Table 1 shows the characteristics of the PAN 2014 Competition dataset for Authorship Verification.

In Authorship Verification, most of the researchers used different types of performance evaluation Measures such as precision, recall, and accuracy to evaluate the accuracy of their author verification approach. In this work, the accuracy measure is used to evaluate the performance of the proposed approach. Accuracy is the ratio of number of test documents correctly verified their author to total number of test documents considered [13].

Table 1 Dataset characteristics of PAN 2014 competition authorship verification

Features	Training data	Testing data
Number of authors	100	100
Number of documents	500	100
Vocabulary size	41,583	12,764
Number of documents per author	5	1
Average words per sentence	25	21
Average words per document	1135	1121

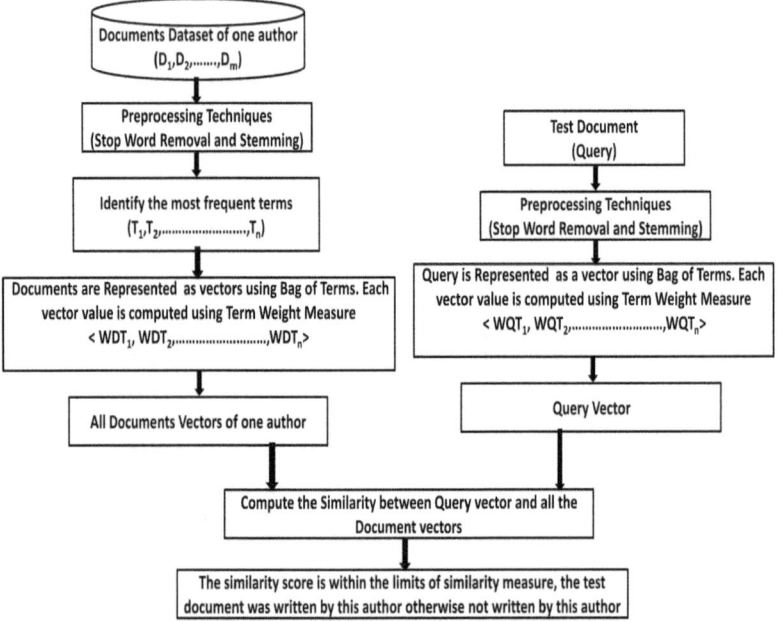

Fig. 2 The proposed approach

The proposed approach for author verification is presented in Fig. 2. In this approach, first, the preprocessing methods such as stemming and stop words removal were performed on the documents of one author. Extract the most frequent terms. The document vectors are generated by using extracted terms from the corpus. The vector value is computed using a term weight measure. The test document represented as a vector using extracted most frequent terms. The similarity is computed among test document vectors and the document vectors of one author using cosine similarity measure. If the similarity scores of similarity measure are within the limits, then the test document was written by this particular author otherwise the test document was not written by this author. The next sections explain the term weight measure used to compute the vector value in document vector and cosine similarity measure.

3.1 NDTW (Nonuniform Distributed Term Weight) Measure

In general, Authorship Verification approaches easily analyze and verify the author of a document when the document contains large amount of text. For small documents, it is difficult to verify the author. The terms in documents were not distributed uniformly. In the NDTW measure, more weight was assigned to the terms which are distributed nonuniformly across the documents [14]. Equation (1) shows the NDTW measure.

$$W_{t_{ij}} = W\left(t_i, A_j\right) = \log(TOTF_{ti}) - \sum_{k=1}^{m} \left(\frac{tf(t_i, d_k)}{TOTF_{ti}} \log\left[\frac{1 + tf(t_i, d_k)}{1 + TOTF_{ti}}\right]\right) \quad (1)$$

where $TOTF_{ti}$ is total number of occurrences of the term t_i in author group A_j and $tf(t_i, d_k)$ is the term frequency of term t_i in d_k document.

3.2 Cosine Similarity Measure (CSM)

In vector space model, the sets of documents and queries are viewed as vectors. Cosine similarity measure is a popular method for calculating the similarity value between the vectors [15]. With document and queries being represented as vectors, similarity signifies the proximity between the two vectors. Cosine similarity measure computes similarity as a function of the angle made by the vectors. If two vectors are close, the angle formed between them would be small and if the two vectors are distant, the angle formed between them would be large. The cosine value varies from $+1$ to -1 for angles ranging from $0°$ to $180°$, respectively, making it the ideal choice for these requirements. A score of 1 evaluates to the angle being $0°$, which means the documents are similar. While a score of 0 evaluates to the angle being $90°$, which means the documents are entirely dissimilar.

The cosine weighting measure is implemented on length normalized vectors for making their weights comparable. Equation (2) gives the formula for Cosine Similarity.

$$CSM(q, d_j) = \frac{\sum_{i=1}^{m} w(t_i, q) \times w(t_i, d_j)}{\sqrt{\sum_{i=1}^{m} w(t_i, q)^2} \times \sqrt{\sum_{i=1}^{m} w(t_i, d_j)^2}} \quad (2)$$

where $w(t_i, q)$ and $w(t_i, d_j)$ are the weights of the term t_i in query q and document d_j, respectively.

4 Empirical Evaluations

In this work, 500 documents of 100 authors were considered as a training dataset
and 100 documents of 100 authors were considered as a test data. The experiment
performed with 100 test documents as queries. The 100 test documents were taken
from 100 authors, 1 document from each author. The test documents were considered
as queries. For every test document, compute the similarity score between the test
document and training data of one particular author using cosine similarity measure.
The similarity scores were acceptable range of cosine similarity measure, it was
conformed that the document was written by that particular author otherwise the
document was not written by this author. In this work, we obtained 92.50% accuracy
for verifying the author. The proposed approach obtained better results than most of
the existing approaches in Authorship Verification.

5 Conclusion and Future Scope

In general, the stylistic features play a vital role in the author identification process.
In this work, the experimentation was performed with information retrieval features
of cosine similarity measure. Our approach obtained an accuracy of 92.50% for
verifying the author of a document.

In future, it was planned to find an efficient similarity measure to compute the similarity between query document and test document. It was also planned to implement
deep learning approaches to improve the accuracy of author verification.

References

1. Sreenivas M, Raghunadha Reddy T, Vishnu Vardhan B (2018) A novel document representation approach for authorship attribution. Int J Intell Eng Syst 11(3):261–270
2. Raghunadha Reddy T, Vishnu Vardhan B, Vijayapal Reddy P (2017) N-gram approach for gender prediction. In: 7th IEEE International advanced computing conference, Hyderabad, Telangana, 5–7 Jan 2017, pp 860–865
3. Raghunadha Reddy T, Vishnu Vardhan B, Vijayapal Reddy P (2016) A survey on author profiling techniques. Int J Appl Eng Res 11(5):3092–3102
4. Raghunadha Reddy T, Vishnu Vardhan B, Vijayapal Reddy P (2017) A document weighted approach for gender and age prediction. Int J Eng-Trans B: Appl 30(5):647–653
5. Raghunadha Reddy T, Vishnu Vardhan B, Vijayapal Reddy P (2016) Author profile prediction using pivoted unique term normalization. Indian J Sci Technol 9(46)
6. Bhanu Prasad A, Rajeswari S, Venkanna Babu A, Raghunadha Reddy T (2018) Author verifi- cation using rich set of linguistic features. Proc Adv Intell Syst Comput 701:197–203
7. Feng VW, Hirst G (2013) Authorship verification with entity coherence and other rich linguistic features. In: Proceedings of CLEF 2013 evaluation labs,
8. Vilariño D, Pinto D, Gómez H, León S, Castillo E (2013) Lexical-syntactic and graph-based features for authorship verification. In: Proceedings of CLEF 2013 evaluation labs

9. Veenman CJ, Li Z (2013) Authorship verification with compression features. In: Proceedings of CLEF 2013 evaluation labs
10. van Dam M (2013) A basic character N-gram approach to authorship verification. In: Proceedings of CLEF 2013 evaluation labs
11. Seidman S (2013) Authorship verification using the impostors method. In: Proceedings of CLEF 2013 evaluation labs
12. Bobicev V (2013) Authorship detection with PPM. In: Proceedings of CLEF 2013 evaluation labs
13. Raghunadha Reddy T, Vishnu Vardhan B, GopiChand M, Karunakar K (2018) Gender prediction in author profiling using ReliefF feature selection algorithm. Proc Adv Intell Syst Comput 695:169–176
14. Raghunadha Reddy T, Vishnu Vardhan B, Vijayapal Reddy P (2016) Profile specific document weighted approach using a new term weighting measure for author profiling. Int J Intell Eng Syst 9(4):136–146. https://doi.org/10.22266/ijies2016.1231.15S
15. Pradeep Reddy K, Raghunadha Reddy T, Apparao Naidu G, Vishnu Vardhan B (2018) Term weight measures influence in information retrieval. Int J Eng Technol 7(2):832–836

An Algorithmic Approach for Mining Customer Behavior Prediction in Market Basket Analysis

Anisha R. Maske and Bela Joglekar

Abstract Market basket analysis is the search for data that contain customer purchasing items. Market basket analysis is a process of showing the correlation between the data with respect to support and confidence. Support indicates that how frequently items appear in the database and confidence indicate that rules must be generated based on the frequent items. Data analysis in a supermarket database means to understand each transaction available in the dataset that contains customer purchasing pattern to determine how the product should be located on shelves. Product arrangement is the most important aspect to get supermarket profit. The dataset of the retailer contains transaction of the items which is purchased by the customer and also comment regarding that product whatever they fill regarding that product. Apriori algorithm is used to find frequent items and association rule based on customer transactions. Frequent items calculate with respect to support and Association rule determines with respect to confidence. This paper tells about how customer behavior predicted based on the customer purchase items. This technique generally used in the agricultural field, marketing, and education field.

Keywords Data mining · Market basket analysis · Customer behavior · Apriori algorithm · Association rule · Layout · Support · Confidence

1 Introduction

Market basket analysis is one of the techniques that analyze customer purchasing habits by finding the different relationship between the different items that can be stored in customer shopping baskets. Association rule can help retailers to produce effective marketing strategies by gaining items frequently, purchased together by

A. R. Maske (✉) · B. Joglekar
Department of Information Technology, Maharashtra Institute of Technology, Pune, India
e-mail: anishamaske@gmail.com

B. Joglekar
e-mail: bela.joglekar@mitpune.edu.in

© Springer Nature Singapore Pte Ltd. 2019
H. S. Saini et al. (eds.), *Innovations in Computer Science and Engineering*, Lecture Notes in Networks and Systems 74,
https://doi.org/10.1007/978-981-13-7082-3_5

customers [1]. Data mining is the understanding of large datasets to find the irrelevant association and summarize the data in proper ways both are understandable and useful to the retailer [2]. Knowledge discovery database is discovering informative knowledge from a large amount of complex data. The knowledge discovery database is a process of interactive and iterative data form from the large database [3]. It contains different steps such as selection, preprocessing, transformation, data mining, and interpretation or evaluation. Each step performs their own role to discover informative knowledge from the database [4].

Market basket analysis is an example of elaborating association rule mining. It is one of the technique that all the retailer in any kind of shop or departmental stores would like to gain knowledge about the purchasing behavior of every customer. These results help to guide retailer to make a plan for marketing or advertising approach [5]. Market basket analysis will also help managers to propose a new way of arrangement in store. Based on this analysis, items are regularly purchased together that can be placed in close proximity with the purpose of further promoting the sale of such items together [6]. If consumers who purchase computers also likely to purchase anti-virus software at the same time then placing the hardware display close to the software display will help to enhance the sales of both of these items.

Market basket analysis is an example of extracting association rule mining. It is a fact that all the managers in any kind of shop or departmental stores would like to gain knowledge about the buying behavior of every customer [7]. Association rules are if-then statements that help to uncover the relationship between seemingly unrelated data in a relational database or other information [8].

2 Related Work

The work in [1] describes the support and confidence that have been calculated by the generic formulae and it does not give the complete information of the association rule. The database contains all the transaction of items.

The researchers in [2] describe the product which is the relationship with one another finding with the help of market basket analysis is located in the store layout.

In another survey, authors [3], Information system containing the relation between each customer purchasing item that is helpful to get the future decision.

The work in [4] describes it used in sports company regarding purchasing the sports items through the customer. It identifies the purchasing pattern of sports items which is present in the database.

Researchers discovered that [5], Market basket analysis is used to discover purchasing patterns of customers by extracting associations from different store transactional data.

3 Proposed Approach

3.1 Dataset

The dataset is a relational set of files describing customer's orders. The input data for a Market Basket Analysis is normally a list of sales transactions where each has two dimensions one represents a product and the other represents a customer.

3.2 Data Preprocessing

All the items in the transaction are sorted in descending order in reference to their frequencies. The algorithm does not depend upon the specific order of the frequencies of items sorting in descending order may lead to much less execution time than ordered randomly (Fig. 1).

3.3 Apriori Algorithm

Apriori algorithm generates sets of large items-sets that find each support size of items. The complexity of an apriori algorithm is always high. Frequent item-sets are extended one item at a time and group of candidates is tested against data. It operates all the transaction which is present in the database.

Fig. 1 Market basket analysis system

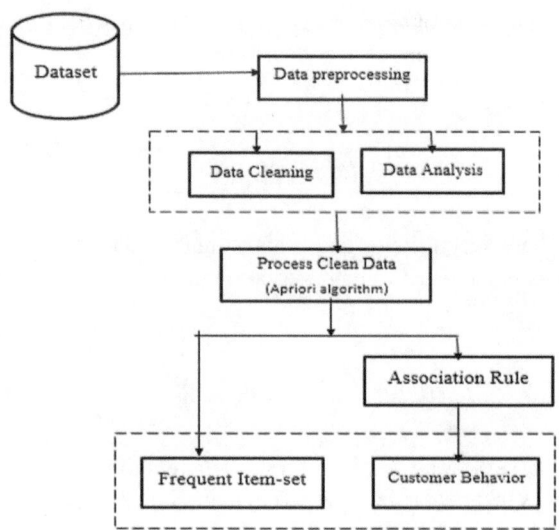

Findings

Input: Database containing items
Output: Frequent Item-sets

Algorithm

1: S is a dataset containing the item. Minimum support is less than 1 and greater than 0. Minimum support is real.
2: Take a transaction of the customer.
3: Calculate support for each item.
4: Take the first transaction and so on.
5: Calculate support for the first item which is the ratio of the number of transaction containing the item and a total number of the transaction.
6: **Compare item support with minimum support. Item support is greater than or equal to the minimum support**.
7: It generates frequent item-set
8: **Again go to Step 4 and calculate all frequent item-set**.

3.4   Association Rule

It contains if-then rules which support the data. Market basket analysis is an association rule which deals with the content of point-of-sale transaction of large retailers. It identifies the relationship among the attribute which is present in the database. It assigns the relationship of one item with another item. It is a fact that all the managers in any kind of shop or departmental stores would like to gain knowledge about the buying behavior of every customer.

3.5   Frequent Item-Set

There are 'n' items and it gives multiple combinations of 'n' items and at last, the customer selects the proper combination of items according to their own choice.

3.6   Customer Behavior

Market Basket Analysis allows retailers to identify relationships between the products that people buy. Targeting market must send promotional coupons to customers for products related to items they recently purchased. Most of the customer buy the same product according to their requirement.

3.7 Classification Rule Mining

Classification rule mining is an effective strategy to generate customer behavior. Obtaining comprehensible classifiers may be as important as achieving high accuracy that helps to make an effective decision in business. The complexity of classification rule mining is $O(pCp \ n)$ where p is number of items in classification rule and n is number customer transactions. It contains more than three items to generate the rules.

Classification method is better than the association rules. Dataset contains different fields of items purchasing by the customer. Association rule contains two items so it cannot give complete count based on whole dataset instead of that classification rule mining give count based on whole dataset using different fields of items.

4 Result

Figure 2 shows that association rule mining generated from the apriori algorithm. From that figure, the rule must be generated according to that rule we get customer behavior. Customer behavior is items based on rules generated from the dataset. Dataset contains items purchased by the customer with their quantity, unit price, and so on. This rule contains two items. Based on the rules, graph must be generated that shows smaller and larger circle contain large or smaller amount of customer purchasing such items. Together how many customers purchasing those items based on rules must be generated hence it gives limited count as compare to classification rule.

```
> topRules <- rules[1:10]
> inspect(rules[1:10])
     lhs                          rhs                  support     confidence lift         cou
[1]  {PINK   SPOTS}            => {SWISS ROLL TOWEL} 0.001013896 1          356.74468 17
[2]  {WOBBLY CHICKEN}          => {DECORATION}       0.001669947 1          335.34000 28
[3]  {WOBBLY CHICKEN}          => {METAL}            0.001669947 1          335.34000 28
[4]  {DECOUPAGE}               => {GREETING CARD}    0.001371742 1          304.85455 23
[5]  {BILLBOARD FONTS DESIGN}  => {WRAP}             0.001729588 1          540.87097 29
[6]  {WOBBLY RABBIT}           => {DECORATION}       0.002027793 1          335.34000 34
[7]  {WOBBLY RABBIT}           => {METAL}            0.002027793 1          335.34000 34
[8]  {BLACK TEA}               => {SUGAR JARS}       0.002385638 1          220.61842 40
[9]  {BLACK TEA}               => {COFFEE}           0.002385638 1          57.42123  40
[10] {ART LIGHTS}              => {FUNK MONKEY}      0.002266357 1          441.23684 38
```

Fig. 2 Association rules

Fig. 3 Customer behavior based on association rule

Figure 3 shows the customer behavior generated from the rules.

Figure 4 shows the classification based association which is generated by the association rule. It gives better performance than the association rul e. It reduces the complexity of apriori algorithm and improves the performance. Classification rule mining is better as compared to association rule mining.

Figure 5 shows that the customer behavior generated from the classification rule. This rule generated from the country so it gives actual count based on complete dataset.

Table 1 shows the accuracy of each algorithm. From that, we can say that classification rule mining gives better accuracy than the other algorithms.

```
> inspect(rule[1:10])
    lhs                                                rhs                                  support    confidence    lift      count
[1] {Quantity=[2,6),
    UnitPrice=[-1.11e+04,1.45)}                   => {Country=United Kingdom} 0.06867020  0.9878159  1.080384  37213
[2] {Quantity=[-8.1e+04,2),
    InvoiceDate=2010-12-01 08:26:00}              => {Country=United Kingdom} 0.10341404  0.9828823  1.074988  56041
[3] {Quantity=[2,6),
    UnitPrice=[1.45,3.29)}                        => {Country=United Kingdom} 0.09760864  0.9828313  1.074932  52895
[4] {Quantity=[-8.1e+04,2),
    InvoiceDate=2010-12-01 08:26:00,
    UnitPrice=[3.29,3.9e+04]}                     => {Country=United Kingdom} 0.05327463  0.9816723  1.073664  28870
[5] {Quantity=[-8.1e+04,2),
    UnitPrice=[-1.11e+04,1.45)}                   => {Country=United Kingdom} 0.05803188  0.9798411  1.071662  31448
[6] {Quantity=[-8.1e+04,2),
    UnitPrice=[1.45,3.29)}                        => {Country=United Kingdom} 0.08857576  0.9798318  1.071651  48000
[7] {Quantity=[-8.1e+04,2)}                       => {Country=United Kingdom} 0.28667913  0.9779857  1.069632  155354
[8] {Quantity=[-8.1e+04,2),
    UnitPrice=[3.29,3.9e+04]}                     => {Country=United Kingdom} 0.14007149  0.9760570  1.067523  75906
[9] {Quantity=[-8.1e+04,2),
    InvoiceDate=2011-09-26 15:28:00}              => {Country=United Kingdom} 0.09532781  0.9757659  1.067204  51659
[10] {Quantity=[-8.1e+04,2),
    InvoiceDate=2011-05-09 10:24:00}              => {Country=United Kingdom} 0.08793727  0.9746789  1.066016  47654
```

Fig. 4 Classification based association

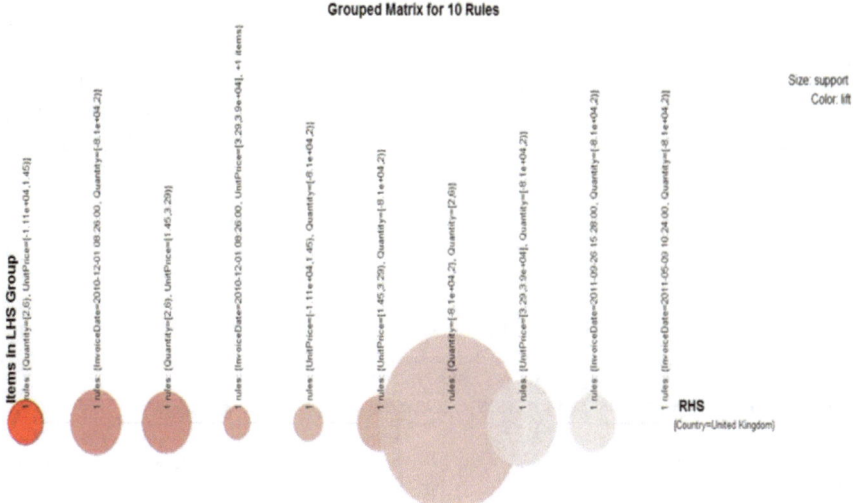

Fig. 5 Customer behavior based on classification rule

Table 1 Accuracy of algorithm

Algorithms	Accuracy (%)
Association rule mining	74.58
Classification rule mining	81.33

5 Conclusion

This paper shows that market basket analysis is an important tool to get frequent item and relationship between the items. It generates the frequencies of the item. It generates the frequencies of the items, based on their frequency items placed into the store layout. The item that is frequently purchased by the customer that is first placed on the layout. Items placed one after another that are helpful to the customer to search the items easily. Apriori algorithm helps to generate association rule and frequent item-set. The apriori algorithm helps to get association rule mining algorithms for market basket analysis will help in better classification of the huge amount of data. The apriori algorithm can be modified effectively with respect to classification rule mining that reduces the time complexity and enhances the accuracy. Association rule helps to get customer behavior based on the product. Both association rule and classification rule able to get the customer behavior based on products different field.

References

1. Pukach P, Shakhovska K (2017) The mathematical method development of decisions supporting concerning products placement based on analysis of market basket content. In: 2017 IEEE 14th International conference the experience of designing and application of CAD systems in microelectronics (CADSM), Lviv, pp 347–350
2. Mirajkar AM, Pradnyavant AR (2016) Data mining based store layout architecture for supermarket. Int Res J Eng Technol 03
3. Nengsih W (2015) A comparative study on market basket analysis and apriori association technique. In: 2015 IEEE 3rd international conference on information and communication technology (ICoICT), Nusa Dua, pp 461–464
4. Abbas WF, Ahmad ND, Zaini NB (2013) Discovering purchasing pattern of sports items using market basket analysis. In: 2013 IEEE international conference on advanced computer science applications and technologies, Kuching, pp 120–125
5. Gajalakshmi V, Murali Dhar MS (2013) A survey on algorithms for market basket analysis. Int J Adv Res Comput Sci I
6. Aguinis H, Forcum LE, Joo H (2013) Using market basket analysis in management research. J Manag 39(7)
7. Setiabudi DH, Budhi GS, Purnama IWJ, Noertjahyana A (2011) Data mining market basket analysis' using hybrid-dimension association rules, case study in Minimarket X. In: 2011 IEEE international conference on uncertainty reasoning and knowledge engineering, Bali, pp 196–199
8. Dhanabhakyam M, Punithavalli M (2011) A survey on data mining algorithm for market basket analysis. Global J Comput Sci Technol XI(Version I)

Optimizing Association Rule Mining Using Walk Back Artificial Bee Colony (WalkBackABC) Algorithm

Imran Qureshi, Burhanuddin Mohammad and Mohammed Abdul Habeeb

Abstract Association rule mining is considered to be the major task in data mining by most of the researchers where the process will find interesting relationships among various items in itemsets of huge database or a dataset. In this process, Aproiri algorithm is considered to be the familiar algorithm for performing association rule mining for implementing frequent itemset generation by providing minimum threshold value and we have explored advantages and disadvantages of association rule mining. In this paper, we have proposed WalkBackABC framework that optimizes ARM by increasing the exploration area and optimizes association rules which is further compared individually with apriori, FP growth and ABC algorithms in our proposed results where the rules generated are simple and comprehensible.

Keywords Data mining · Association rule mining · Artificial bee colony · Random · Repository

1 Introduction

Many data mining tools and techniques have been proposed and projected by many researchers in this field that proposes generation of association rules or generation of decision trees or implementation of neural networks which has become an attractive point in their researches since the past decade. Data mining is a technique that is implemented to discover hidden patterns or prefigure from any source such as flat files or databases or data repositories or any of such combinations.

I. Qureshi (✉) · B. Mohammad · M. A. Habeeb
AlMusanna College of Technology, Musannah, Sultanate of Oman
e-mail: imran@act.edu.om

B. Mohammad
e-mail: burhanuddin@act.edu.om

M. A. Habeeb
e-mail: habeeb@act.edu.om

© Springer Nature Singapore Pte Ltd. 2019
H. S. Saini et al. (eds.), *Innovations in Computer Science
and Engineering*, Lecture Notes in Networks and Systems 74,
https://doi.org/10.1007/978-981-13-7082-3_6

Association rule mining (ARM) is the process for finding meaningful frequent patterns or correlations or associations or causal structures from data sets found in distinct types of databases such as relational databases or transactional databases or data repositories or a combination of any of these, as the association rule mining aims to identify or generate rules that lead to predicting the occurrence of a specific item based on the occurrences of the other items in any specific transaction [1].

They provide rules or results in the form of if-then statements or rules and mainly based on two aspects called as support and confidence that are provided as input by the user for generation of association rules, for example: A → B, where A is called as an antecedent and B is called as consequent which is the count of occurrence of B. Many algorithms have been proposed by researchers such as Apriori algorithm and FP tree based algorithm.

Apriori algorithm is one of the most renowned algorithms for generating association rule based mining which is based on breadth-first search technique using downward closure property in a bottom-up strategy for counting itemset support and generation of rules and candidate sets. And FP-growth uses ad FP-prefix tree to store database data in a compressed manner for frequent pattern growth generation [2].

The Artificial bee colony (ABC) algorithm is population-based swarm intelligence algorithm which is proposed by Karaboga in 2005 where a set of honey bees called swarms communicate with one another to resolve a problem with social cooperation [3].

2   Literature Survey

The process of association mining by utilizing the Apriori algorithm for acquiring good results and we already know that due to increase in database size there will be a decrease in performance attained, it happens because for processing the data entire database is to be scanned for every transaction [3].

For attaining the large data set apriori algorithms scans the entire database using breadth-first search approach as the major issue is it cannot be implemented directly on a huge dataset whereas the other better approach to do the same is the FP growth algorithm which is implemented using the divide-and-conquer technique [4] which is a better approach than apriori.

In the process of identification of frequent itemset mining that includes generation of candidate sets for acquiring all the maximal frequent itemsets which are upward closed and requires to identify various maximum frequent itemsets (MFI) that comprises of maximal frequent itemset that reduces total search time [5].

The process of maximal FP max algorithm that tends to reduce total number of subsets that stores and check the patterns that are based on possible conditional maximal frequent itemsets [6]. FP tree can be further reduced for attaining the efficiency QFP algorithm is used for scanning the database only once and generates the QFP tree by processing the transaction database for generating tree based association rule mining using FP array technology [7].

Another approach of dynamic tree restructuring is implemented using CP-tree restructuring technique which generates a prefix tree that reduces the mining time but the problem is we can implement only on relatively small tree due to which the performance attained is remarkably small [8].

3 Association Rule Mining

The concept of association rule mining was first introduced in the reference provided by Agarwal R [4]. where an association rule can be defined as: let I = {i1, i2, i3, i4, i5, …} be an items set in a D database where a set of transactions are available and each transaction T is considered to be a subset of I and an association rule is an inference of the form X → Y where X and Y are subsets of I and X ∩ Y = φ and the set of items X is called antecedent and Y is called as the consequent as the two properties: confidence and support are basically measured in association rule mining where the freq(X) considered to be number of rows containing X itemset in the given transactional database and the support of the itemset X is identified to a fraction of all rows containing the itemset frequency(X)/D.

Calculation of support of an association rule is the union of the antecedent X and the consequent Y where the support (X → Y) = (XUY)/D.

Calculation of confidence of an association rule is the percentage of rows in D containing itemset X along with itemset Y is: Confidence (X → Y) = P(X/Y) = support(XUY)/support(X).

Each itemset is considered to be frequent if its support is equal to or more than support specified by user input and the ARM is used to check whether all rules through user-specified constraints imposed such as minimum support and minimum confidence is attained or not.

The first algorithm to generate large itemsets on a transaction database is AIS algorithm [7] which is used to determine qualitative rules by performing multiple passes over database and the only issue is it generates many candidates and the data structure utilized is also not specified. The Apriori algorithm [1] is implemented based on the rule called as "from what comes before and after" using breadth-first search technique utilizing lesser memory and it only explains absence or presence of an item and the number of candidates generated is larger. Many of the issues in apriori are overcome in FP growth algorithm which is based on tree generation structure and only two passes are required to generate frequent itemsets on a compressed database without candidate generation process which tends to implement at a faster pace but the disadvantage is it cannot be implemented in incremental or interactive mining system.

The below table specifies the advantages and disadvantages of ARM-based on performance and various properties (Table 1).

Hence, we propose below Algorithm 1 that overcomes disadvantages of ARM and implements advantages of ARM algorithms.

Table 1 Advantages and disadvantages of ARM

Association rule	Advantages	Disadvantages
AIS	Large candidate itemsets cannot be used as the pruning step of the algorithm can handle small transactions only Small cardinality sparse transaction database can only be implemented	Major limitation is algorithm can accommodate only one item in the consequent Single pass cannot be implemented to attain the data needs multiple passes over the database where time complexity increases Larger data structures are required to generate candidate itemsets
Apriori	Memory consumption is very less Implementation step is very easy Pruning step utilizes apriori property for due to which itemsets are very less to check minimum support	Requires huge database scans to be implemented Only one minimum threshold value can be provided Can be implemented hassle free on small database Only availability or nonavailability of a data element can be explained
FP-growth	Quickest association rule mining algorithm Can also be implemented on compressed database Database scan is implemented only once	More Memory is consumed Cannot be implemented for incremental mining or interaction based mining Generated FP growth tree is not unique even for similar logical database

Algorithm 1: proposed Association Rule Mining-Frequent itemset mining

Step 1: Start

Step 2: generate frequent itemset with possible combinations by considering all intermediate nodes

Step 3: generate frequency =q.F+C

Step 4: if q.F < q. S then goto Step 2

Step 5: if q.F != q.S then goto Step 8

Step 6: generate frequent itemset with possible combinations by considering nodes having higher frequency

Step 7: generate frequency = q.F

Step 8: generate frequent itemset with possible combinations by considering all possibilities and parent node

Step 9: generate frequency = q.F

Step 10: Stop

In the above algorithm, F represents frequent itemsets, C represents a candidate set and q represents the itemset available in query set. We need to execute the above algorithm for every element in itemset based on which association rules will be generated and a FP tree is constructed.

4 Artificial Bee Colony (ABC) Algorithm

The Artificial bee colony (ABC) algorithm is proposed by Karaboga in 2005 which is an optimization algorithm centered on the simulation process of foraging activities of artificial honey bees (a set of honey bees are used to implement the system) which is successfully implemented and researched by many researchers and has been successfully applied to various practical problems.

The term Swarm is used to represent set of honey bees that are used for implementing a predefined task by using intercommunication process where bees communicate with each other. As per ABC algorithm honey bees are classified into three types called as:

- **Employed bees**: The employed bees are also called as On-duty bees that search for food in various food generating sources as per their knowledge and upon finding will share the food availability information to onlooker bees, these bees are equal to number of food sources found and these bees are called as scouts when food sources are exhausted.
- **Onlooker bees**: these bees will perform quality check, in other words, will be selecting quality food sources from all the list of information provided by employed bees. The main task is to select only those sources which have higher quality (fitness) and ignores the food sources that comprises of lower quality.
- **Scout bees**: are the special categories of employed bees that have abandon their food sources and start new searches for attaining food.

As per the count is concerned total bees are divided into two types where the first half of the swarm consists of employed bees and the rest are onlooker bees where these bees are equal to the number of swarms available or swarm size denoted by SN.

The initial distribution process of sources is randomly generated using below equation:

$$X_{ij} = X_{min.j} + rand(0, 1)\left(X_{max.j} - X_{min.j}\right) \text{ for all } i = \{1, 2, \ldots, CS/2\} \text{ and } j = \{1, 2, \ldots, D\}$$
(1)

where CS represents colony size and D represents number of design variable then rand(0, 1) represents generation of random number in between 0 to 1 and each employee is allocated to X_{ij} sources.

Candidate solutions are generated in the below equation that is V_{ij} for initiating exploitation process:

$$V_{ij} = X_{ij} + \Phi_{ij}(X_{ij} - X_{kj}) \text{ for all } i = \{1, 2, \ldots, CS/2\} \text{ and } j = \{1, 2, \ldots, D\} \text{ and } K \neq i \quad (2)$$

where Φ is a random number between -1 and $+1$, CS represents colony size and D represents number of design variable.

The main components of ABC algorithm are:

- **Rule Format**: in a dataset where each attribute comprises of lower bound or the lowest value and the upper bound or highest value along with three other values associated for generation of fitness function are predictive class, the fitness value, and cover percentage of rule generated or impose.
- **Fitness Function**: will be used for performing classification instead of measuring the total nectar amount collected that comprises of {false positive (FP), true positive (TP)} or {false negative (FN), true negative (TN)} associated with every rule. FP denotes total number of records covered by rule without predicting the rule class, TP denotes total number of records covered by rule by predicting the rule class, FN denotes total number of records not covered by rule by predicting the rule class, TN denotes total number of records not covered by the rule and does not comprises of any class.
- **Exchanged Local Search Strategy**: when an employed bee cannot attain the target assigned to meet maximum cycle number need to move to a new food source by implementing local search strategy which leads to high consumption of time as the data set consists of inappropriate data which is to be removed and loaded again called as exchanged with matching data set.
- **Rule Discovery**: a set of rules are identified while performing rule mining for identification of a specific class for implementing rule discovery phase which iteratively discovers all the rules or its instances.
- **Rule Pruning**: this phase all the rules will be prepared for implementing pruning step after all the rule set is generated by eliminating all the duplicate rules or rule sets for improving the accuracy of algorithm.
- **Prediction Policy**: using the pruned results set as input through the classes are unknown prediction is performed by calculating the prediction values based on test data record or based on the different possible class or based on the highest prediction value in the final class attained.

The below are the advantages and disadvantages of ABC algorithm are (Table 2).

5 Optimization of Association Rule Mining with ABC Algorithm

For performing optimization of association rule mining, we propose a framework which implements random walk back on ABC algorithm so we name it as Walk-BackABC which consists of four phases: Primary phase is initialization phase then next is employed bee phase then next is onlooker bee phase and finally, the last one is scout bee phase. In the whole process, the onlooker bees will be performing the

Table 2 Advantages and disadvantages of ABC algorithm

Advantages	Disadvantages
ABC is very easy to implement	Secondary information will not be used about the problem gradients
Utilization of discrete or mixed variables is possible due to its broader applicability on complex functions that are implemented continuously	–
Due to its high flexibility modifications can be easily done for creating a new knowledge of a specific problem by analyzing the nature of problem	Requires new parameters to implement algorithm for improving the performance to attain requirements of new fitness test
ABC does not require the main objective function to be represented mathematically or continuous or differentiable	Optimization of all the functions leads to loosing of behavior or relevant information
Initialization process is very robust with respect to feasibility or distribution of basic defaults population	Maximum number of objective count of evolutions
Saves maximum processing time due to parallel processing of algorithm structure	Process slows down very drastically when implemented on sequential processing
Very easy to create regions of parity by implementing parallelism in implementation of populations of solution	Requires more memory and more number of iterations due to the slowdown of computational cost parameter and increase in the population of solutions
Local solutions can be easily explored	Accurate solutions can be attained at a very slower pace
Effective search can be implemented using global optimizer under high complexity with lower risk to attain premature convergence	Higher accuracy can be attained by implementing only deterministic methods for identifying solutions without sticking in local minimum values attained

random walk back process for performing the local optimization process while performing the search of food process in neighborhood and tends to select the highest probable food source with quality food.

One of the disadvantages of the conventional ABC algorithm leads to slow convergence due to implementation of local optima rate so our framework will be implemented in onlooker bee phase by performing all the above mentioned four phases by performing random walk back phase based on the random colony size that is generated with available dimensions and their limits with maximum cycle being implemented are considered to be important factors of the proposed framework that comprises of loudness and pulse rate impedance.

Our proposed framework is applied on the various rules that are generated using apriori algorithm for optimizing all the association rules where the termination condition is set at first hand in the proposed algorithm due to which the rules that are generated using apriori or FP growth algorithm can easily evolve fitness value for each rule that is generated for supporting the output set that comprises of execution of fitness function till the desired criteria is met or reached and below are the steps involved in our proposed methodology for optimizing association rules are:

1. Fetch and load all tuples in dataset from the repository.
2. Using apriori algorithm or FP growth algorithm generates frequent itemsets.
3. Using frequent itemsets generated list out the association rules.
4. We need to assign the termination condition based on the dataset being optimized.
5. Apply our proposed algorithm on selected members to generate association rules on optimal basis.
6. Execute and generate the fitness value for each and every rule generated.
7. Add the rules to output set when fitness value meets the required condition.
8. Repeat the above rules till the termination condition satisfies.


```
Algorithm 2: proposed WalkBackABC

Step 1: Start

Step 2: randomly generate the initial population from
[1,2,3,... i]

Step 3: assign gen.max= random()

Step 4: assign gen=1, i=1

Step 5: when gen < gen.max then goto Step 3

Step 6:  Vij=Xij +φij (Xij-Xkj)

Step 7: i = i+1

Step 8: j=j+1

Step 9: if i < max then goto Step

Step 10: find best solution among all generated Vij

Step 11: i=j=0

Step 12: Xij=Xmin.j + rand(0,1) (Xmax.j - Xmin.j )

Step 13: i=i +1

Step 14: j=j+1

Step 15: evaluate_rank(X, i, j)

Step 16: gen=gen+1;
```

Repeat the above algorithm till all the cycles are met.

6 Implementation Results

We have implemented the proposed WalkBackABC algorithm in Matlab software and generated the performance analysis by considering student data, retail data, and mortgage data through a repository and then compared with apriori, FP growth, and ABC algorithms (Fig. 1).

We have taken data set related to three areas such as student, retail, and bank and implemented in ARM algorithms and generated the number of rules attained. As per the comprehensibility measures attained through proposed work yields that is WalkBackABC which comprises of simplicity and easy understand ability of the rules and result attained is shown in Fig. 2.

As per the above two figures, the WalkBackABC framework has optimized when they are compared with individual algorithms such as apriori, FP growth, and ABC algorithms. Hence, we propose that the proposed framework is better than the three algorithms.

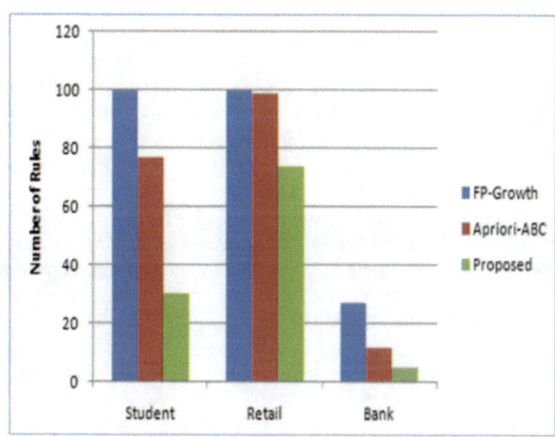

Fig. 1 Performance of the proposed WalkBackABC algorithm

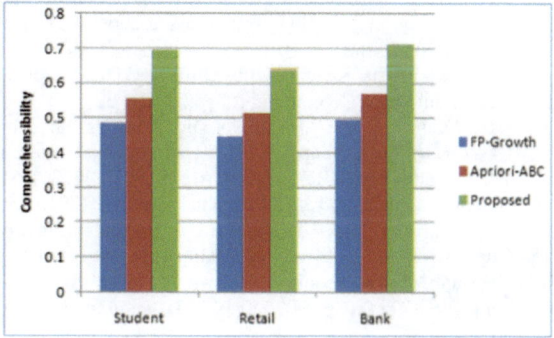

Fig. 2 Comprehensibility measures attained through the proposed work

7 Conclusion

In this paper, we have proposed Association rule mining and Artificial Bee Colony algorithms working, advantages and disadvantages and we worked on WalkBack-ABC algorithm which randomized the implementation and concentrated on the performance of proposed work which is compared with ABC then FP-growth and apriori algorithms and generated results that illustrates that the proposed framework is an optimized one and all the rules and methodologies generated based on the proposed framework are comprehend and simple when implemented and generated. In our future work, we would like to apply further modifications on the Apriori algorithm to implement on larger datasets.

References

1. https://www.techopedia.com/definition/30306/association-rule-mining
2. Sharma P, Tiwari S (2015) Optimize association rules using artificial bee colony algorithm with mutation. In: International conference on computing communication control and automation. IEEE, pp 370–373. 978-1-4799-6892-3/15. https://doi.org/10.1109/iccubea.2015.77
3. Neelima S, Satyanarayana N, Krishna Murthy P (2017) Optimization of association rule mining using hybridized artificial bee colony (ABC) with BAT algorithm. In: 2017 IEEE 7th international advance computing conference (IACC)
4. Narvekar M, Fatma Syed S (2015) An optimized algorithm for association rule mining using FP tree. Procedia Comput Sci
5. Ilayaraja M, Meyyappan T (2015) Efficient data mining method to predict the risk of heart diseases through frequent itemsets. Procedia Comput Sci
6. Narvekar M, Fatma Syed S (2015) An optimized algorithm for association rule mining using FP tree. In: International conference on advanced computing technologies and applications (ICACTA-2015), pp 101–110
7. Agrawal R, Imielinski T, Swami A (1993) Mining association rules between sets of items in large database. In: Proceeding of the 1993 ACM SIGMOD international conference on management of data, Dec 1993. ACM Press, pp 207–216
8. Dunham MH, Xiao Y, Gruenwald L, Hossain Z (1997) A survey of association rules
9. Bhadoriya VS, Dutta U (2015) Improved association rules optimization using modified ABC algorithm. Int J Comput Appl (0975-8887) 122(13):23–26
10. Sahota S, Verma P (2016) Improved association rule mining based on ABC. Int J Comput Appl (0975-8887) 135(10):6–10
11. Kotsiantis Sotiris, Kanellopoulos Dimitris (2006) Association rules mining: a recent overview. GESTS Int Trans Comput Sci Eng 32(1):71–82
12. Han J, Kamber M (2000) Data mining: concepts and techniques, 2nd edn
13. Raval R, Rajput IJ, Gupta V (2013) Survey on several improved Apriori algorithms. IOSR J Comput Eng (IOSR-JCE) 9(4):57–61. e-ISSN: 2278-0661. pISSN: 2278-8727
14. Neelima S, Satyanarayana N, Krishna Murthy P (2015) Enhanced Apriori algorithm using top down approach. Int J Adv Res Comput Sci Softw Eng 5(6):1016–1019
15. Gupta MK, Sikka G (2013) Association rules extraction using multi-objective feature of genetic algorithm. In: Proceedings of the world congress on engineering and computer science 2013, vol II, WCECS 2013, 23–25 Oct 2013, San Francisco, USA

Predicting Election Result with Sentimental Analysis Using Twitter Data for Candidate Selection

B. P. Aniruddha Prabhu, B. P. Ashwini, Tarique Anwar Khan and Arup Das

Abstract The aim of this paper is to provide a well and effective solution to distribute political party's tickets during the election. The sole parameter of the current distribution of political party's tickets is based on the money power and person's references from their superior leaders. Our proposal is to highlight the discrepancy between the real candidate who is predicted to win the election based on their popularity with other parameters and those who have only references with money power. We will choose the deserving candidate by analyzing parameters such as social work, criminal records, educational qualification, and his/her popularity on social media (twitter).

Keywords Sentimental analysis · Big data · Data mining · Text mining · Data dictionary · Election prediction

1 Introduction

In the recent trend of technical field, there are plenty of data that are generated every moment. We can use such data for any kind of analysis. In this paper, we are using such kind of data that are generated from social media (twitter) to make the analysis of candidates that will contest election. The current distribution of political party's

B. P. Aniruddha Prabhu (✉) · B. P. Ashwini · T. Anwar Khan · A. Das
Department of Computer Science and Engineering, Siddaganga Institute of Technology, Tumkur, Karnataka, India
e-mail: aniprabhubp@gmail.com

B. P. Ashwini
e-mail: ashvinibp@gmail.com

T. Anwar Khan
e-mail: tarique.1si14cs125@gmail.com

A. Das
e-mail: das12997@gmail.com

© Springer Nature Singapore Pte Ltd. 2019
H. S. Saini et al. (eds.), *Innovations in Computer Science and Engineering*, Lecture Notes in Networks and Systems 74, https://doi.org/10.1007/978-981-13-7082-3_7

tickets is simply based on the candidate's asset and the references he is using but they are ignoring all the necessary aspects. As a result, wrong candidates get the tickets and they may lose the election. If somehow, they won, then they hold the important position in government offices and take the advantage of their power, they will never worry about the development of their constituency instead their main focus is to make money to contest next election, as a result, many of the places are suffering and in the name of development they will do nothing. So, we came up with an alternative idea that will make the use of JSP, CSS, Naive Bayes Algorithm, and Coding techniques to overcome the flaws from the current system [1, 2].

1.1 Definitions

In this section, we provide definitions for some of the terminologies used in this paper,

Naïve Bayes Algorithm: Naive Bayes classifiers are a family of simple probabilistic classifiers based on applying Bayes' theorem. Naive Bayes is a simple technique for constructing classifier's models that assign class labels to problem instances, represented as vectors of feature values, where the class labels are drawn from some finite set [3, 4].

Text Mining: Text mining also referred to as text data mining is the process of deriving high-quality information from text. The overarching goal is essentially, to turn text into data for analysis, via application of natural language processing (NLP) and analytical methods.

Sentimental Analysis: Sentiment analysis is a type of natural language processing for tracking the mood of the public about a particular product or topic. Sentiment analysis, which is also called opinion mining, involves in building a system to collect and examine opinions about the product made in blog posts, comments, reviews, or tweets [5].

Big Data: Big data is a set of data that are voluminous and complex. It is the set of data that are generated every moment from various sources that can be structured and unstructured.

Text Mining: Text mining is the process of analysis of data from natural language text. Through this, we can convert the words into quantitative and actionable insights [6, 7].

Data Dictionary: A data dictionary is the collection of the description of words or data object in a data model for the benefit of program.

2 Related Work

Sentimental analysis is very effective analysis that will help any domain to grow and develop their industry. If we come to know about the negative response of public early, it will help how to target specific aspects of their product in order to increase its audience satisfaction and it will also protect the business from loss [8]. Opinion polls are the most valuable thing of politics that are used for the analysis of the government's work and political party as well to know about their popularity among the citizen of our country. The 'Wisdom of Crowds' concept turns conventional predictions on their head. It assumes that any crowd that conforms to a core set of principles is capable of delivering a more accurate prediction than the smartest people within it. UK, showing that the Wisdom of Crowds approach used by ICM Research in 2010 [9]. Electoral Analysis from twitter data is challenging research topic in the current world. It is relatively straightforward and optimistic. It focuses on overall aspect of data whether it is used for analysis or optimization and it will provide a fair decision based on the opinion of public. The analysis must have certain algorithm and rules. Every parameter must be validated, the tweet's dates are also considered and the place from which it is obtained [10].

3 Methodology

The overall system architecture is as shown in Fig. 1.

The three phases of methodology are as explained below.

3.1 Analysis Phase

The detailed description of analysis phase is as follows,

Our goal of analysis phase is to determine the candidate based on their popularity on social media like Twitter. Our focus is to gather information about the candidate. We design an interface where we get the tweets related to the candidate. We collect the tweets into a file and then we have our own dictionary where we store some words that will match against those tweets if those words describe good attitude of that candidate we will provide positive points to them, if those words have negative meaning then we will give negative points to them and we use some mathematical formula to calculate the percentage of their popularity. It is as shown in Fig. 2.

Formula Used for analysis

$$\left(\text{Positive tweets} \,/\, \left(\text{Positive Tweets} + \text{Negative Tweets}\right)\right) * 100$$

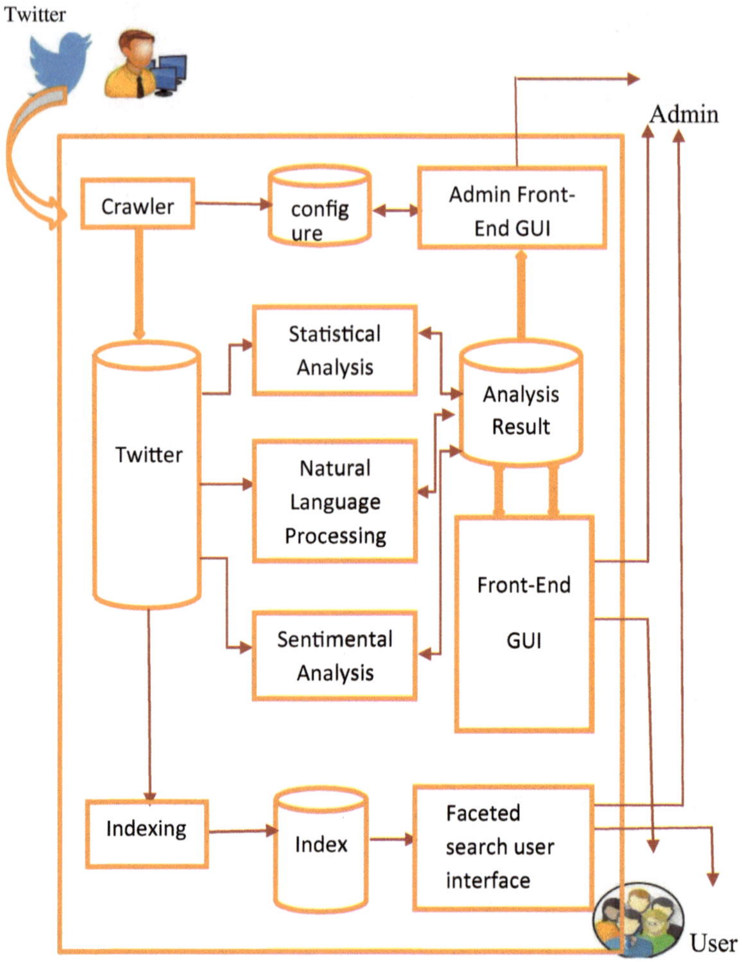

Twitter

Admin

Crawler → config ure ↔ Admin Front-End GUI

Twitter

Statistical Analysis

Analysis Result

Natural Language Processing

Sentimental Analysis

Front-End GUI

Indexing → Index → Faceted search user interface

User

Fig. 1 System architecture

Fig. 2 Flow diagram of analysis Phase

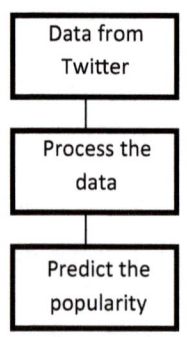

Data from Twitter

Process the data

Predict the popularity

3.2 Combining Analysis Phase with Parameters

The main focus of this phase is to club all the parameters and analysis phase to get the result. The candidate judges according to the following parameters:

- Educational Qualification
- Criminal Records
- Social work
- Social Status and Popularity
- Previous Election Records

We will give appropriate weightage to each and every parameter to get the result. In this phase, we collect all the related information about the candidate and do the analysis.

3.3 Prediction

This is the final phase of this paper in this phase, we will combine all parameters and sentimental analysis. Once we get all the information related to the candidate, we will use Naive Bayes Algorithm to predict the most deserving candidate among all the candidates of that particular constituency. Various parameter value used in prediction is shown in Table 1.

Formula Used for Prediction

$$(\textbf{Educational Qualification}) + (\textbf{Criminal Records}) + (\textbf{Social work})$$
$$+ (\textbf{Social Status and Popularity}) + (\textbf{Previous Election Records})$$
$$+ ((\textbf{Points get from analysis phase}) / 100)$$

Table 1 Parameter table

Parameters	Points
Educational qualification	+8 for PHD, +6 for UG, +4 for PUC, +2 for Metric and +1 for the rest
Criminal record	−10
Social work	+8
Social status and popularity	+15
Previous election record	If won +8 else 0

Algorithm/Pseudocode

Step 1:   Gather people's opinion about the candidates that cover all the parameter.
Step 2:   Use Java API to get the twitter data into a file.
Step 3:   Use database to store information about the candidates.
Step 4:   Make appropriate set of positive, negative, and neutral keywords that will be used for performing text mining on the dataset.
Step 5:   Use formula to get the popularity by getting negative and positive tweets.
Step 6:   Store the result into a file.
Step 7:   Collect the data about the politician.
Step 8:   Assign proper points to each parameter.
Step 9:   Store all data in the form of tables in Sql.
Step 10: Perform integration activity and add the required field useful for analysis.
Step 11: Use Naive Bayes Algorithm.
Step 12: Using the weightage, judge the candidate on the basis of the parameters and Generate score for each candidate.
Step 13: The candidate with the highest score is the deserving candidate.
Step 14: Direct this output to the .txt file.
Step 15: The .txt file will show the difference between the deserved candidate and the candidate predicted to win the election.

As shown in the algorithm complete analysis is done with the use of Twitter analysis and combining the related parameters of the candidate's behavior with their overall contribution toward their constituency. If we use to implement such things then one who wins the election will definitely make developments and connect with people and try their best to solve the problems of their constituency. They will never misuse their power and always try to develop a good relation with their people.

4   Conclusion

The loopholes in current system are clearly highlighted in this paper. After implementing these logics, we can clearly say that current distribution of political party's ticket is not suitable. This is why eligible candidate lose their election and corrupt opponent wins. Hence by observing current Scenario, we have proposed this solution that can be implemented by the Government as early as possible by making required changes. This type of analytic based approach can be used for any type of election in India and elsewhere. This paper also highlights the power of data analytics in the Computer Science field and demonstrates the use of social data to get more clarity for the common man as well.

References

1. Laney D (2001) 3D data management: controlling data volume velocity and variety
2. Beulke D (2011) http://davebeulke.com/big-data-impacts-data-management-the-five-vs-of-big-data/
3. Preety SD (2017) Sentiment analysis using SVM and Naive Bayes algorithm, Dept of CSE International Institute of Technology, Sonipat, India
4. Vadivukarassi M, Puviarasan N, Aruna P (2017) Sentimental analysis of tweets using Naive Bayes algorithm. Annamalai University, Tamil Nadu, India
5. Moujahid A (2014) http://adilmoujahid.com/posts/2014/07/twitter-analytics/. Accessed 20 Aug 2017
6. Pratiba D, Samrudh J, Dadapeer, Srikanth J (2017) Elicitation of student learning experiences from twitter using data mining techniques
7. Herzog A (2018) https://business.twitter.com/en/blog/7-useful-insights-Twitter-analytics.html/. Accessed 22 Feb 2018
8. Wagh B (2016) Sentimental analysis on twitter data using Naive Bayes. Department of Computer Science, 12 Dec 2016
9. Boon M (2012) Predicting election a 'Wisdom of Crowds' approach. ICM Res
10. Gayo-Avello D (2012) A meta-analysis of state-of-the-art electoral prediction from Twitter data. University of Oviedo Spain

Performance Analysis of LSA for Descriptive Answer Assessment

Amarjeet Kaur and M. Sasi Kumar

Abstract Latent Semantic Analysis (LSA), in general, can be considered as an excellent information retrieval technique, but for this specific task of Descriptive Answer Assessment (DAA) some more explorations are required and it is still considered as an open problem. This paper discusses and evaluates the performance of LSA for DAA, through experimentation and deep investigations. Several state-of-the-art claim that LSA correlates with the human assessor's way of assessment. With this as background, we investigated assessment of students' descriptive answer using Latent Semantic Analysis (LSA). In the course of research, it was discovered that LSA has limitations like: LSA research usually involves heterogeneous text (text from various domains) which may include irrelevant terms that are highly susceptible to noisy, missing and inconsistent data. The experiments revealed that the general behavior of LSA has an adverse impact on DAA. It has also been observed that factors which influence the performance of LSA are corpus preprocessing, the creation of term-document matrix with and without weight function and choice of dimensionality.

Keywords Latent semantic analysis · Descriptive answer · Assessment · Dimension reduction · Feature extraction · Evaluation

1 Introduction

The purpose of education is to make student learn specific topic or domain, so that the student can apply that knowledge and information in practice. This is possible only if the student is able to grasp the knowledge properly, which is checked by

A. Kaur (✉)
Computer Science & Technology, SNDT Women's University, UMIT, Mumbai, India
e-mail: dhariwal.amarjeet@gmail.com

M. Sasi Kumar
Research & Development, Centre for Development of Advanced Computing, Mumbai, India
e-mail: thelitttlesasi@gmail.com

© Springer Nature Singapore Pte Ltd. 2019
H. S. Saini et al. (eds.), *Innovations in Computer Science and Engineering*, Lecture Notes in Networks and Systems 74,
https://doi.org/10.1007/978-981-13-7082-3_8

evaluating the student through the periodic examination process set up by institutes. Evaluation of students' knowledge can be done by analyzing their response for a given question in the examination. Response can be objective or descriptive. Descriptive response is believed to give clearer idea about the students' knowledge. Descriptive answer given by the student is evaluated by human assessor to identify the extent of learning. While evaluating descriptive answers, the assessor is looking for important aspects and an overall understanding of the student for a specific topic. For making judgments about the student's understanding, the human assessor uses cognitive skills, natural language understanding and her knowledge about a specific domain. For automating this process, the machine should be intelligent enough to behave like a human assessor. One of the biggest advantages of automatic assessment is its ability to provide helpful and immediate feedback to the learner without putting much burden on human assessors. Automated assessments will be helpful in distance education too. Considerable work has been done in the area of using LSA to evaluate essays and to provide content-based feedback, but evaluating descriptive answers is still an open problem.

Various student essays evaluation systems have been under development since 1960s. A National network of US universities supported the development of system to grade essays for thousands of high school students' essays. It scores essays by processing a number of essays on the same topic, each scored by two or more human assessors. In 1960s, computer technology was not stable enough or accessible enough to expand into large scale. Currently, researchers are emphasizing more on descriptive answers in exams and its evaluation.

The paper is organized as follows: Sect. 2 explains research works related to the field of automation of descriptive answer assessment using LSA. Section 3 explains methodology used to determine the semantic similarity between two texts. This section also explains the implementation of LSA and its performance evaluation. Section 4 lists several issues, conclusion, and areas of improvement that future studies will address.

2 Literature Review

In this section, research work related to the field of Descriptive Answer Assessment (DAA) has been discussed. Methods and techniques implemented so far for automation of DAA process are discussed. Several state-of-the-art short answer graders require manually designed patterns which have to be matched with the student's response; if matched, implies correct response. One of the information extraction-based system [9] is developed by the Oxford University to fulfill the need of the University of Cambridge Local Examinations Syndicate (UCLES) as many of the UCLES exam questions are short answers questions. In this system, handcrafted patterns are filtered from the training datasets by human experts and the student responses were matched with these patterns.

The Intelligent Essay Assessor (IEA) [4, 7] used Latent Semantic Analysis (LSA) model to extract semantic similarity of words and passages from the textual sentence. The comparison-based approach is applied in which the student's essay is compared with the reference essay. Cosine similarity measure and proximity measure (Euclidean distance between the two) have been used to find out the similarity between two units. LSA is first trained on domain-specific corpus and derives a semantic representation of the information. In this research work, the vectors of student essays are compared with the vectors of pre-graded essays. The students' essay which appear to be highly unusual and irrelevant to topic will be directed to human assessor for evaluation. When IEA is applied to content scoring, human–machine correlation is 0.73 [8].

Autotutor [5], considering Pearson correlation as a performance measure LSA works well in evaluating the student answers. The correlation between LSA's answer quality scores and the mean quality scores of four human experts was 0.49. This correlation of 0.49 is closer to the correlation between the ratings of the two intermediate experts which is 0.51. Select-a Kibitzer [6] like Apex, generates outlines of essays without referring the reference text. The program uses clustering methods on the LSA semantic space to identify disconnected chunks in the corpus. Its content correlation with teacher grade, $r = 0.62$. Some of the systems, such as, Intelligent Essay Assessor, State of essence, Summary Street, Apex, Autotutor, and Select-a Kibitzer; though differing in subject domain and the similarities, all are LSA-based. All such systems claim that LSA correlates with the human assessors. This was one of the motivations of looking at LSA for our research.

3 Methodology

To evaluate the performance of LSA for descriptive answer assessment, LSA has been applied to set of students' descriptive answers. The similarity between student and standard/expert descriptive answer are determined by using LSA and its performances are evaluated using three performance measures such as average score difference between computer assessor (marks calculated using LSA technique) and Human Assessor (HA), standard deviation and a Pearson correlation. Broadly, the system has three major modules: the standard answer representation, the student answer representation, and a comparison unit (using LSA) (Fig. 1).

The approach is to represent student's descriptive answer and the standard answer in a textual form and analyze it with the cosine similarity measure. There are various categories of questions asked by the teachers in the exam question paper of universities/institutions like explain, describe, what, why, how, justify, define, elaborate, short notes, etc. The categories of questions are also analyzed to identify the syntactic structure of descriptive answer. Some categories of question like draw and calculate are excluded from the list because diagrams and mathematical expressions are out of the scope of this research work.

Fig. 1 The broad approach

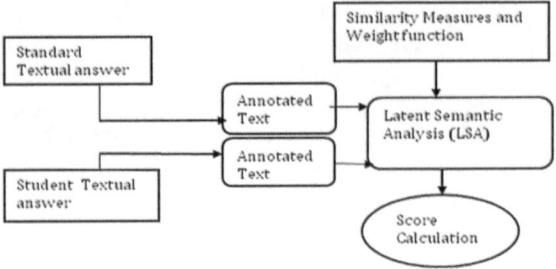

Latent Semantic Analysis (LSA) is a natural language processing technique for analyzing relationships between a set of documents [1]. LSA assumes that the unigrams occur in the same set of text are semantically similar. A term-document matrix is created by considering each unigram as a row and each document as column of the matrix. The mathematical technique called Singular Value Decomposition (SVD) is used for dimension reduction to eliminate unwanted noise from the information. Words are represented as semantic vectors in the semantic space and then compared by using the cosine of the angle between the two vectors. Values ranges between 1 and 0 represents, its similarity level from similar to dissimilar words [2, 3]. The data for this experiment consisted of student's answers (1440 samples) in electronic form. The samples of student descriptive answer used for this experiment are free-form text and are in a range of 5–6 grammatically correct English sentences (approx. 80–100 words). The samples are collected by conducting an online examination using Moodle software. Multiple categories of questions and volume of answers are considered from multiple domains. From Computer Engineering Domain: the question is "why is java known as a platform independent language?" and "what is a distributed computing system?" Electronics Domain: the question "What is the need of timing diagram?" and "write disadvantages of EDF method". Student's answers have been assessed by four experienced human assessors. Average of these marks is used in this analysis considering the variation in marks between different human assessors. For performance analysis, LSA score would be compared with the average human assessor score (HA (MEAN)).

3.1 Implementation

The general steps of LSA technique are implemented using python 2.7:

Step 1: Consider all the students' descriptive answers as documents in term-document matrix and standard answer as query matrix.

Step 2: Parse all the documents and create a dictionary of index words as unigrams. Index words are the words which occurs in more than one document or whose frequency >1 across the documents.

Table 1 Results obtained for why, what, and write type of questions of sample size: 1440 descriptive answers

S. no.	Marks allocated	Sample size	Domain	Question category	ASD	SD	PC between HA and CA
1	5	320	Computer	Why	1.73	1.07	0.26
2	5	320	Computer	What	2.36	1.45	−0.01
3	1	400	Electronic	What	0.23	0.17	0.66
4	2	400	Electronic	Write	0.76	0.64	0.20

Step 3: Create the frequency count matrix.

Step 4: Modify the count matrix, by applying Term Frequency-Inverse Document Frequency (TF-IDF) weight function [1] to each cell of the term-document matrix.

Step 5: Apply Singular value decomposition (SVD), to decompose the term-document matrix into three parts.

Step 6: Dimension reduction (to reduce the noisy data). The SVD operation, along with this reduction, has the effect of preserving the most important semantic information in the text while reducing.

Step 7: For statistical similarity measure, calculate the cosine similarity between the standard answer and student answer.

Step 8: Score calculation, Student score = cosine similarity * total marks allocated.

3.2 Results and Discussion

Three statistical performance measures are used to analyze the performance of LSA for DAA such as Average Score Difference (ASD), Standard Deviation (SD), and a Pearson Correlation (PC) between Human Assessor and marks calculated using LSA The data for this experiment (shown in Table 1) consisted of student's descriptive answers (1440 samples) in electronic form. Student's descriptive answers have been assessed by four experienced human assessors such as HA1, HA2, HA3, and HA4.

In Table 1, it has been observed that there is a significant gap between the results obtained according to LSA (computer assessor) and human assessor (HA). Considering various categories of questions and different sample sizes, the experiments revealed that factors which influence the performance of LSA are:

a. Creation of term-document matrix with and without weight function:

Importance or relevance factor of a term in a corpus of matching algorithm actually affects the overall performance of information retrieval system. Choice of effective term-weighting scheme is also a matter of consideration (Fig. 2).

Fig. 2 Implemented LSA with and without weight function for 90 samples

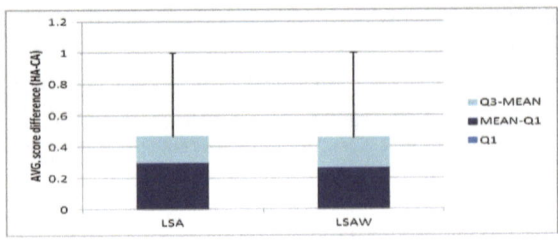

Fig. 3 Choice of dimensionality affects the performance of LSA

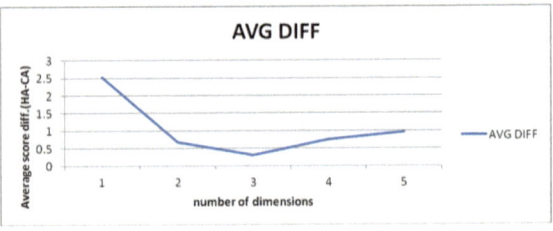

b. Choice of dimensionality:

Choice of dimension affects the recall and precision level of LSA. Dimension greater than the optimum one can add irrelevant features and the smaller may eradicate important features, consequently, the algorithm will give faulty results. Different choices of dimensionality have been tried (dimension 1, 2, 3, 4, 5) for set of 5 descriptive answers. The result shown in Fig. 3. (sample size 5), that the optimum dimension is 3 for these set of documents, as the average score difference between human assessor (HA) and LSA (CA, Computer Assessor) is minimum.

4 Conclusion and Future Scope

After deep investigation and experimentation, it has been observed that there is a significant gap between the human assessor judgments and LSA for descriptive answers assessment. The general behavior of LSA has been explored and the experiments revealed that behavior of LSA has an adverse impact on descriptive answer assessment (DAA). It has been observed that factors which influence the performance of LSA are corpus preprocessing, creation of term-document matrix with and without weight function and choice of dimentionality. The impact of the general behavior of LSA on DAA is similarity through co-occurrences of words across the documents which may include irrelevant terms as a part of index words pool hence giving false result, use of Bag of Words (BOW) technique which leads to loss of contextual meaning, ordering of words are ignored, and furthermore LSA has high recall but less precision. In Future studies, LSA should be modified and a novel technique can be introduced to solve all such issues.

Acknowledgements This research was supported by The Department of Science & Technology, Ministry of Science & Technology, under Women Scientist Scheme (WOS-A)—SR/WOS-A/ET-1064/2014(G).

References

1. Burstein J, Kukich K, Wolff S, Lu C, Chodorow M (1998) Enriching automated scoring using discourse marking. In: Proceedings of the workshop on discourse elations and discourse marking, 36 annual meeting of the association for computational linguistics and the 17 international conference on computational linguistic
2. Burstein J, Marcu D (2000) Benefits of modularity in an automated scoring system, Proceedings of the workshop on using toolsets and architectures to build NLP systems. In: 18th International conference on computational linguistic
3. Da Silva AA, Padilha N, Siqueira S, Baiao F, Revoredo K (2012) Using concept maps and ontology alignment for learning assessment. IEEE Technol Eng Educ 7(3)
4. Gliozzo Alfio M, New York, NY (US) (2015) Latent semantic analysis for application in a question answer system. Patent no: US 9,020,810 B2
5. Hennig L, Labor TD (2009) Topic-based multi-document summarization with probabilistic latent semantic analysis. In: Proceedings of the recent advances in natural language processing conference, RANLP-2009, pp 144–149
6. Kashi A, Shastri S, Deshpande AR, Doreswamy J, Srinivasa G (2016) A score recommendation system towards automating assessment in professional courses. In: 2016 IEEE 8th international conference on technology for education, pp 140–143
7. Leacock C, Chodorow M (2003) C-rater: automated scoring of short-answer questions. Comput Humanit 37(4):389–405
8. Lemaire B, Dessus P (2001) A system to assess the semantic content of student essays. J Educ Comput Res 24(3):305–320
9. Sukkarieh JZ, Pulman SG, Raikes N (2005) Auto-marking 2: an update on the UCLES-Oxford University research into using computational linguistics to score short free text responses. Paper presented at the 30th annual conference of the International Association for Educational Assessment (IAEA)

Efficient Retrieval
from Multi-dimensional Dataset Based
on Nearest Keyword

Twinkle Pardeshi and Pradnya Kulkarni

Abstract Search engine deals with the user provided query to deliver the informative results to users. The volume of data associated with these search engines is very vast and it becomes very difficult to handle just data. These data are dynamic, increase day by day, and hence many techniques have been proposed to handle such dynamic data. Existing Tree-based techniques are mostly applicable to queries such as spatial queries, range queries, and many more. These techniques are useful only for the queries that have coordinates. But these techniques are not applicable to the queries which do not have coordinates. Keyword-based search has been considered more helpful in many applications and tools. The paper considers objects, i.e., images, tagged with number of keywords and those will be embedded into vector space for evaluation. The main aim here is to achieve higher scalability with the increasing data and speedup of retrieval of results for the users. A query is been implemented in the paper called as nearest keyword set query that deals with multi-dimensional datasets. Hash-based index structure is implemented along with a new algorithm called as ProMiSH, i.e., Projection and Multi-scale Hashing.

Keywords Data mining · Nearest keyword set query · Inverted index · Hashing

1 Introduction

A directory can be thought of as indexing results of documents and the documents classified in respective categories. A huge number of search tool are available to search and retrieve particular information. Study of keyword search is increasing day by day. Keywords are used when you are unaware of the contents to get searched by search engines. Keyword search has been proved very useful to provide relevant

T. Pardeshi (✉) · P. Kulkarni
Computer Engineering, Maharashtra Institute of Technology, Pune, India
e-mail: twinkle.pardeshi@gmail.com

P. Kulkarni
e-mail: pradnya.kulkarni@mitpune.edu.in

© Springer Nature Singapore Pte Ltd. 2019
H. S. Saini et al. (eds.), *Innovations in Computer Science
and Engineering*, Lecture Notes in Networks and Systems 74,
https://doi.org/10.1007/978-981-13-7082-3_9

search results to the users. Keywords can be thought of as a bridge between what to search for and information retrieved to fulfill the need. The right keyword you choose the right the contents you will get. Ongoing research on queries deals with other queries, other than spatial queries such as nearest neighbor queries, range queries, and spatial joins [1]. Queries on spatial objects are associated with textual information even have received significant attention. The main focus is on the new type of query that is the Nearest Keyword Set (NKS) [2]. In NKS query, user provides a set of keywords and query retrieves set of points that contain those keywords.

2   Motivation

1. Information can be searched at search engine to get relevant results of the search. Search engines allow the user to navigate through different page or site. But there can be a case where user wants to navigate between documents for particular keywords.
2. Category searches provide search among sites while keyword searches deal with searching among textual documents, which makes them more beneficial.
3. For various queries those have coordinates such as spatial queries, range queries, and many more, tree-based techniques are adopted. But these techniques cannot be adopted for queries without coordinates [3].
4. Thus, there comes a need to develop a technique that will improve the performance of the queries without coordinates and provide efficient result set. There are many techniques that use tree-based indexes for NKS queries, but the performance of these techniques decreases when the size of dataset increases [2].

3   Review of Tree-Based Techniques

See Table 1.

Table 1 Review of tree-based techniques

Techniques	Description
IR Tree	For approximation algorithm, greedy algorithm is used and for exact algorithm, dynamic algorithm is used [4]
IUR Tree	Text-based clustering is used with outlier detection
BR* Tree	It is the combination of R*tree and bitmap indexing [5]
Tree	It is the combination of R tree and signature files

4 Existing System

1. Web and GIS systems have adopted location-specific keyword queries, which are executed with the help of R-Tree and inverted index [6, 7].
2. Felipe et al. have developed an efficient tree-based technique for ranking set of objects of datasets that is spatial [8]. Those were based on the combinations of the location distance and relevant keyword i.e., textual description of the object.
3. Survey reviews that Cong et al. studied Felipe et al. discovery and then combined R-tree and inverted index to provide an answer for the similar query to Felipe et al. using a different function for ranking.

5 Mathematical Model

Set theory:

$$s = \{I, F_1, F_2, F_3, o, \phi\}$$

where,
S: Implemented System
I: Set of Inputs
F: Set of Functions
O: Set of Outputs/Final Output
Φ = Failures and Success Conditions

- Input the Query Keyword

$$I = \left\{ q = \left(V_1, \ldots, V_q \right) \right\}$$

where,
q = query keyword
v = keyword

- Find the points having Query Keyword

$$F_1 = bs[o] \leftarrow true$$

$$f' \leftarrow f' \cup o$$

where,
bs = bit set to track the points having query keywords

- Check for the Candidate Duplication

$$F_2 = \left\{ \text{checkduplicationcand}\left(f', hc\right) = \text{ture} \right\}$$

where,

f' = output of F_1
hc = hashtable that checks duplication of candidates

then

- Perform element wish matching
  $F_{2.1}$ = {elementwishmatch (f', hc[h])}
  where,
  h = hash key generated by concatenating two hash values of points.

- Subset search
  F_3 = searchinsubset (f', pq)
  Where,
  pq = priority queue containing top-k results

- Output is the subset that contains those points which contains minimum one query keyword
  O = {f'}
  where,
  f' = subset of points that contain the query keyword.

6   Proposed System Architecture

An algorithm called Projection and Multi-scale Hashing mostly known as ProMiSH is implemented that make use of hash table's sets and inverted indexes for performing a search that is localized. It enables faster processing for keyword queries and retrieves the relevant top-k results. A spatial database deals with multi-dimensional objects and facilitates quick access to the objects (Fig. 1).

A user of the system enters the keywords in the search engine. Use of keyword search has been proven very useful in many applications. Keywords are used when a user is not aware of any specific search to perform. Database of the system consists of multi-dimensional database. In our case, we have textual document and images. Images are the real world databases that are crawled online from real-world image website called Flickr. Flickr contains various images with keywords tagged with all images. There are various tags available at Flickr websites for images. Each image is treated as an object and keywords associated with objects are treated as data points. Inverted index has been implemented to provide efficient retrieval of data from database. User can also take advantage of getting a histogram of every image to use it for comparison with other images. The main aim of the system is to retrieve

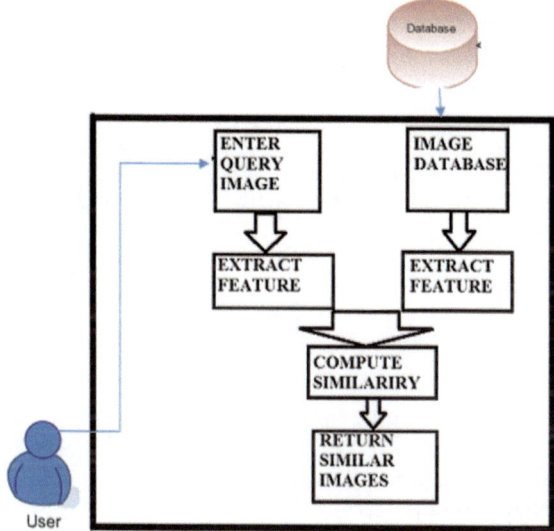

Fig. 1 System architecture for user

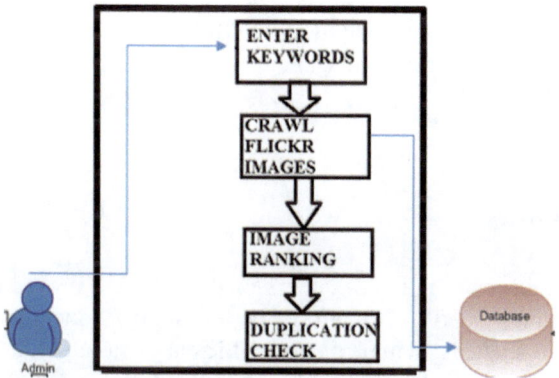

Fig. 2 System architecture for admin

result set that consists of nearest points. Objects containing all the keywords that are provided by the user, then that objects are treated as the nearest of the available objects. Euclidean distance is calculated between objects to retrieve nearest among all as shown in Fig. 2.

Graph 1 Response time of algorithm

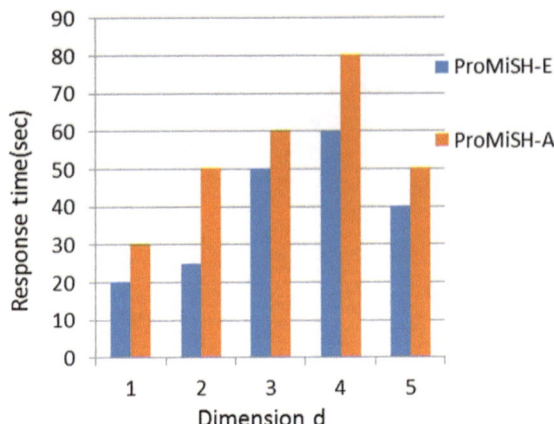

7 Experimental Setup

The implementation of the system is done using Java with eclipse jee-indigo-SR2-win64 and Apache-tomcat-7.0.42 server. The testing of the system is conducted on Intel(R) Core(TM)i3 processor@2 GB and 4 GB RAM. The system makes use My-SQL query browser, a database management system that uses SQL for management of the database (Graph 1).

8 Conclusion

Search based on keyword in multi-Dimensional database is proved useful in many applications in today's world. Therefore, the technique for the Top-k nearest keyword set search is studied and implemented. Efficient search algorithms have studied that work with indexes for fast processing. A novel model called as Projection and multi-scale Hashing (ProMiSH) which makes the use of random projection with index structures which are hash based and achieves higher scalability and speedup is implemented along with inverted indexing and hashing.

References

1. Cao X, Cong G, Jensen CS, Ooi BC (2011) Collective spatial keyword querying. In: Proceedings of ACM SIGMOD international conference on management of data, pp 373–384
2. Singh V, Zong B, Singh AK (2016) Nearest keyword set search in multi-dimensional dataset. IEEE Trans Knowl Data Eng 28(3)

3. Li W, Chen CX (2008) Efficient data modeling and querying system for multi-dimensional spatial data. In: Proceedings of 16th ACM SIGSPATIAL international conference on advanced geographic information systems, pp 58:1–58:4
4. Zhang D, Ooi BC, Tung AKH (2010) Locating mapped resources in web 2.0. In: Proceedings of IEEE 26th international conference on data engineering, pp 521–525
5. Singh V, Venkatesha S, Singh AK (2010) Geo-clustering of images with missing geotags. In: Proceedings of IEEE international conference on granular computing, pp 420–425
6. Zhang D, Chee YM, Mondal A, Tung AKH, Kitsuregawa M (2009) Keyword search in spatial databases: towards searching by document. In: Proceedings of IEEE 25th international conference on data engineering, pp 688–699
7. Hariharan R, Hore B, Li C, Mehrotra S (2007) Processing spatial keyword (SK) queries in geographic information retrieval (GIR) systems. In: Proceedings of 19th international conference on scientific and statistical database management, p 16
8. I. De Felipe, V. Hristidis, and N. Rishe (2008) Keyword search on spatial databases. In: Proceedings of IEEE 24th international conference on data engineering, pp 656–665

A Hybrid Bio-inspired Clustering Algorithm to Social Media Analysis

Akash Shrivastava and M. L. Garg

Abstract Particle swarm optimization algorithm is known as a population-based algorithm which actually maintains a population of particles. The particle plays a significant role which represents an effective solution to an optimization problem. The study proposed in the paper intends to integrate PSO with the artificial bee colony (ABC) algorithm. The research inspired from the intelligent biological behavior of swarms where it involves the merits of both the algorithm to perform experimental analysis on the social media data. The hybrid bio-inspired clustering approach is being proposed to apply on social media data which is known to be highly categorical in nature. The result shows that clustering analysis is helpful to classify high dimensional categorical data. Social media analysis effectively can be achieved through clustering which is being demonstrated in the proposed hybrid approach.

Keywords Particle swarm optimization · Artificial bee colony · Clustering analysis · Social media analysis

1 Introduction

In the mine of large amount of datasets, it is highly tangible to discriminate among the data. The amount of data not only large in number but also it is highly unstructured and uncertain when encounters through various resources. Social media data like tweets, YouTube comments, e-commerce companies data which are being generated through the activities/reactions expressed over the social forum. Clustering analysis is observed to apply across several problem domains like data mining [1], machine learning [2], and pattern recognition [3] over the years. The common objective of several approaches applied under clustering is to group those of the data

A. Shrivastava · M. L. Garg (✉)
Computer Science Engineering Department, DIT University, Dehradun, India
e-mail: dr.ml.garg@dituniversity.edu.in

A. Shrivastava
e-mail: akash.10may@gmail.com

© Springer Nature Singapore Pte Ltd. 2019
H. S. Saini et al. (eds.), *Innovations in Computer Science and Engineering*, Lecture Notes in Networks and Systems 74,
https://doi.org/10.1007/978-981-13-7082-3_10

objects associated with the similar properties in term of features, statistics, zone, nature from different clusters. Clustering methods are mostly based on two types of popular methods of clustering. These known clustering methods are hierarchical and partitional [4]. Partitional clustering then categorized into supervised and unsupervised learning. The two of the categories are differentiated on the fact that supervised learning required to mention the number of clusters. Although, the algorithms lies under the clustering essentially can be categorized as hard, fuzzy, possibilistic, and probabilistic [5].

Particle swarm optimization is known as a population-based approach which is the optimize way to resolve the different optimization problems. This can be applied to those problems also which can easily be transformed to function optimization problems [6]. In particle swarm optimization, the approach is being implemented which is responsible to maintain the population of particles. Although, earlier PSO had been integrated with K-means by [7]. In recent times, categorical datasets are frequently used to encounter through various mediums. On the other hand, karaboga and bastruk proposed an artificial bee colony (ABC) algorithm [8]. The performance of the proposed approach then compared with other algorithms like genetic algorithm (GA), particle swarm optimization algorithm (PSOA), and particle swarm inspired evolutionary algorithm (PS-EA) [9]. The experiment carried over the various datasets and the obtained results outperformed the other algorithms. The limitation of clustering algorithms lies in the fact that it may fall into local optima. In artificial bee colony, this limitation is being avoided through the process wherein each iteration, it conducts local search. This process significantly increases the probability of finding optimal solutions [10].

Related Work

The process of clustering analysis is divided into two classes including hierarchical and partitional clustering. The objective functions involved in the fuzzy c-means has been transformed and reformulated by particle swarm optimization: PSO-V and PSOU [11]. The approach is also compared with the alternating optimization and ant colony optimization. Later, in [12], an ant colony clustering algorithm has been proposed for optimally clustering N objects into K clusters. The proposed algorithm works in a way where global pheromone updating takes place while the heuristic information has been used to construct required clustering solutions. The uniform crossover operator has also been constructed to improve the efficiency of solutions discovered by ants. In clustering analysis, fuzzy approach plays an effective role by proposing fuzzy c-means algorithm but it also has a limitation which had been overcome by presenting PSO-based fuzzy clustering algorithm [13]. The potential of PSO lies in its capability of global search which has been utilized in PSO algorithm to resolve the shortcoming in FCM. Later, the fuzzy k-modes algorithm is integrated with genetic approach which is being implemented over categorical datasets as genetic fuzzy k-modes algorithm [14].

2   Particle Swarm Optimization

The intelligent foraging behavior of social animal attracted the research and scholars in recent years. Genetic algorithms, ant bee colony, swarm intelligence, and particle swarm optimization are the classical evident approaches developed in this domain. PSO originally proposed in [15], which elaborated the approach as a population-based stochastic optimization technique whose design and evolution have been motivated by bird flocking and fish schooling. The algorithm also includes the concept of fitness function; fitness function is required to justify the potential of particle's positions. Here using a fitness function, the fitness value associated with each particle's position is determined. The PSO algorithm derives two parameters as personal best position refers as *pbest* and global best position refers as *gbest*. These parameters are being utilized to update the velocity of each particle [16] (fuzzy c-means and fuzzy swarm). A particle's velocity and positions are updated as follows:

$$V(t+1) = w.V(t) + c_1 r_1 (pbest(t) - X(t)) + c_2 r_2 (gbest(t) - X(t)); \qquad (1)$$

$$X(t+1) = X(t) + V(t+1) \qquad (2)$$

where X and V are position and velocity, respectively. w is the initial weight, c_1 and c_2 are positive constants, these are called acceleration coefficients which monitor the impact of *pbest* and *gbest* on the search process, and r_1 and r_2 are the randomly chosen vectors from the population.

Pseudocode for FUZZY PSO for fuzzy clustering:

Step 1:   Here, P considers as a population size, c_1, c_2, w parameters has been initialized along with the maximum iterative count.

Step 2:   Swarm with particles P (X, pbest, gbest, and V are n x c matrices) has been created in this step.

Step 3:   gbest for swarm and X, V, pbest for each particle has been initialized.

Step 4:   cluster center of each particle is calculated as

$$z_j = \frac{\sum_{i=1}^{n} \mu_{ij}^m o_i}{\sum_{i=1}^{n} \mu_{ij}^m}$$

Step 5:   Fitness function of each particle is being calculated using Eq. (5).

Step 6:   pbest for each particle is calculated.

Step 7:   gbest for the swarm is calculated.

Step 8:   Velocity matrix for each particle is being updated using Eq. (3).

Step 9:   position matrix for each particle is being updated using Eq. (4).

Step 10:   If terminating condition is not met, then go to step 4 and start again.

3   Artificial Bee Colony (ABC)

According to [8, 17], the intelligent foraging behavior of ant has been analyzed and research with the intention to put it into clustering perspective. In ABC, the bees are categorized into three groups as employed bees, onlookers, and scouts. The approach works in two parts in which employed bees play a role in the first and latter part includes onlookers. One employed bee is there for each food source which ensures the fact that number of food sources is equal to the number of employed bees surrounded the hive. As derived in the approach, there are following parameters involved in the main approach which actually consider as control parameters used in the ABC algorithm.

(i)   The number of employed or onlooker bees (SN)
(ii) The value of limit
(iii) The maximum cycle number (MCN).

3.1   Pseudo Code of the ABC Algorithm

Step 1:   Population of solutions has initialized and evaluated.
Step 2:   Employed bees sent to food sources and nectar amount has been evaluated.
Step 3:   Probability of all food sources is being evaluated chosen by the onlooker bees by the following method:

$$p_i = \frac{f_i t_i}{\sum_{n=1}^{SN} f_i t_n}$$

where

$f_i t_i$ The fitness value of the solution i
i Position of the food source
SN   Number of food source

Step 4:   Onlookers sent to food sources where food sources are being chosen by onlookers on the basis of the probabilities of food source calculated in Step 3.
Step 5:   Nectar amount has been calculated and greedy selection process applied at this stage.
Step 6:   If the food source has been exhausted then terminate the exploitation process of an employed bee, then this employed bee becomes scout.
Step 7:   Perform random search for new food source by sending scout into search space.
Step 8:   The best food source recorded which has been found so far and get memorized.
Step 9:   If the desired output is achieved, the present output else go to step 2.

4 Proposed Hybrid Bio-inspired Clustering Approach

The merits of both PSO and ABC have taken into the account and implemented to design and develop the hybrid approach. The popular research perspective that is being developed from recent years regarding the intelligent behavior of social animals has been focused. The activities adopted and exhibit by the swarms, ants, and bees are studied and approaches have been proposed by eminent researchers in the same domain. The approaches were undertaken on the basis of the performances of these two algorithms reflected in their past experiments. They majorly implemented over the numerical or categorical data which had basically involved only biological datasets.

4.1 Pseudo Code of Hybrid Bio-inspired Clustering Algorithm

Step 1: Initialize the parameters which include population size of particles P, c_1, c_2, w, m, and population of solutions.

Step 2: Swarm with particles P (X, pbest, gbest, and V are n x c matrices) has been created in this step.

Step 3: The parameters which include X, V, and pbest for each particle and gbest for each swarm have been initialized.

Step 4: Employed bees sent to food sources and nectar amount has been evaluated.

Step 5: FPSO algorithm:

Step 5.1: Cluster center of each particle is calculated as

$$z_j = \frac{\sum_{i=1}^{n} \mu_{ij}^m o_i}{\sum_{i=1}^{n} \mu_{ij}^m}$$

Step 5.2: Fitness function of each particle is being calculated using Eq. (5)

Step 5.3: pbest for each particle is calculated.

Step 5.4: gbest for each swarm is calculated.

Step 5.5: Velocity matrix for each particle is being updated using Eq. (3)

Step 5.6: Position matrix for each particle is being updated using Eq. (4)

Step 5.7: If terminating condition is not met, then go to Step 4 and start again.

Step 6: ABC algorithm

Step 6.1: Probability of all food sources is being evaluated chosen by the onlooker bees by the following method:

$$p_i = \frac{f_i t_i}{\sum_{n=1}^{SN} f_i t_n}$$

where $f_i t_i$ = The fitness value of the solution I, i = position of the food source, SN = Number of food source

Step 6.2: Onlookers sent to food sources where food sources are being chosen by onlookers on the basis of the probabilities of food source calculated in Step 3.

Step 6.3: Nectar amount has been calculated and greedy selection process applied at this stage.

Step 6.4: If the food source has been exhausted then terminate the exploitation process of an employed bee, then this employed bee become scout.

Step 6.5: Perform random search for new food source by sending scout into search space;

Step 6.6: The best food source recorded which has been found so far and get memorized;

Step 6.7: If the desired output is achieved, present the output else go to step 2.

Step 7: If FPSO-ABC terminating condition is not met, go to step 5.

5 Experiments Results

For evaluating the FPSO, ABC, and FPSO-ABC methods, well-known real-time streaming datasets have been considered. The implementation of algorithms has been applied on the most dynamic dataset which is highly categorical in nature. The real-time streaming twitter datasets has been utilized to perform the execution of proposed hybrid bio-inspired clustering algorithm, which is the integration of fuzzy PSO and ABC algorithm. The pseudocode of proposed hybrid algorithm has been implemented in python language and executed on Intel core i7, 3.7 GHz, 64 GB RAM computer. In all of the experiments that are being carried out, the parameters of clustering algorithm set as: S = 40, MCN = 200, X = 5 and V = 5. The experiments have been carried out where the three parameters where Yang's accuracy measure [18] has been adopted for the research carried out. In Yang's method, accuracy (AC), precision (PR), and recall (RE) are given as follows:

$$AC = \frac{\sum_{i=1}^{k} a_i}{n}, \tag{3}$$

$$PR = \frac{\sum_{i=1}^{k} \frac{a_i}{a_i+b_i}}{k}, \tag{4}$$

$$RE = \frac{\sum_{i=1}^{k} \frac{a_i}{a_i+c_i}}{k} \qquad (5)$$

where,

a_i The number of data objects that are correctly allocated to class C_i,
b_i The number of data objects that are incorrectly allocated to class C_i,
c_i The number of data objects that are incorrectly denied from class C_i,
k The total number of class contained in a dataset, and
n The total number of data objects in a dataset.

In the above measures, the AC has the same meaning as the clustering accuracy r defined in [19]. Given a dataset $X = \{x_1, x_2, \ldots, x_n\}$ as well as two partitions of this dataset: $Y = \{y_1, y_2, y_{t1}\}$ and $Y' = \{y_1', y_2', \ldots, y_{t2}'\}$, the Rand Index (RI) [20] is given by

$$RI = \frac{\sum_{i=1, j=2; i<j}^{n} a_{ij}}{\binom{n}{2}} \qquad (6)$$

where

$$\alpha_{ij} = \begin{cases} 1, & if\ there\ exist\ t\ and\ t'\ such\ that\ both\ x_i\ and\ x_j\ are\ in\ both\ y_t\ and\ y_t', \\ 1, & if\ there\ exist\ t\ and\ t'\ such\ that\ x_i\ is\ in\ both\ y_t\ and\ y_t', \\ & while\ x_j\ is\ in\ neither\ y_t\ or\ y_t', \\ & 0,\ otherwise \end{cases}$$

The value of RI is evaluated through the true clustering and the clustering process which is obtained from a clustering algorithm. The number of cluster k is being set as per the number of classes provided by the class information of the dataset. There is a fact that needs to be mentioned that the class information involved in the clustering process except those number of classes that will not be used in the clustering phenomenon (Figs. 1, 2 and 3, Tables 1, 2 and 3).

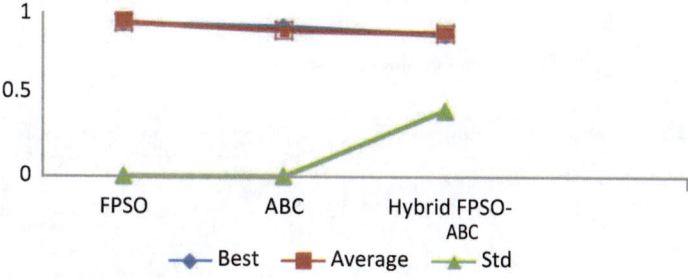

Fig. 1 Graph for the value of AC of the three algorithms on the Twitter Dataset

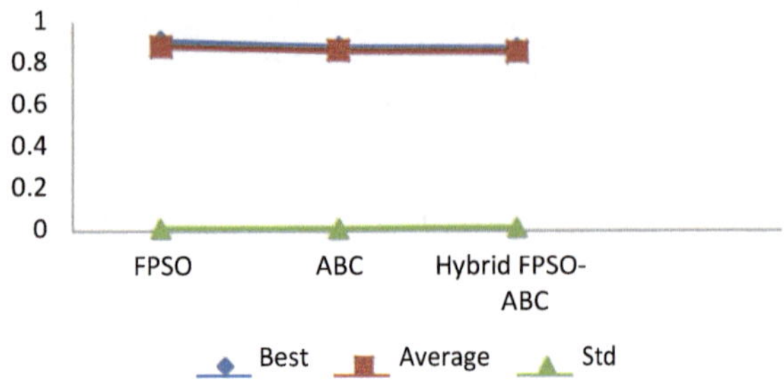

Fig. 2 Graph for the value
of PR of the three algorithms
on the Twitter Dataset

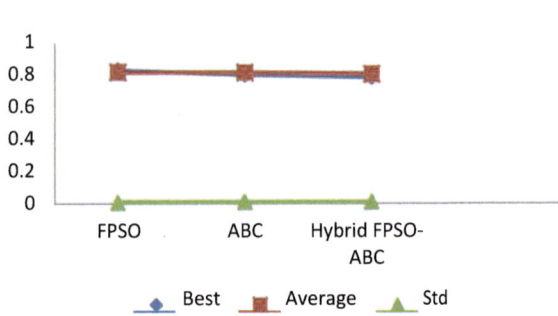

Fig. 3 Graph for the value
of RE of the three algorithms
on the Twitter dataset

Table 1 The AC of the three
algorithms on the Twitter
dataset

Algorithms	AC		
	Best	Avg	Std
FPSO	0.9205	0.9253	0.0106
ABC	0.9003	0.8786	0.0108
Hybrid FPSO-ABC	0.8534	0.8634	0.0402

Table 2 The PR of the three
algorithms on the Twitter
dataset

Algorithms	PR		
	Best	Avg	Std
FPSO	0.9076	0.8786	0.0087
ABC	0.8812	0.8634	0.0092
Hybrid FPSO-ABC	0.8734	0.8562	0.0123

Table 3 The RE of the three
algorithms on the Twitter
dataset

Algorithms	RE		
	Best	Avg	Std
FPSO	0.8273	0.8132	0.0074
ABC	0.7983	0.8119	0.0087
Hybrid FPSO-ABC	0.7821	0.8017	0.0079

The results obtained after executing the three algorithms reflects an effective and better improvement in the best, average, and lower standard values in AC, PR, and RE. This justifies the fact that the proposed hybrid approach is capable to provide better clustering performance while applied over the social media dataset which is highly categorical in nature. There are a series of newly designed and developed approaches for clustering categorical data which have been tested and experimented over various datasets. Social media data is highly categorical in nature and required to be classified for decision-making process in many of the domains.

6 Conclusion

Social media analysis is the current research trend which aims to classify the massive amount of unstructured data accumulated over the web. Twitter forum is highly popular and credible in terms of the analytics of genuine data. Clustering analysis plays a vital role to classify the datasets for numerical optimization problems and now becoming very useful for categorical datasets. In the research carried out in this paper is significant evidence in the same series. The performance in terms of computation, execution, and classification is far better than the existing approaches for clustering. There is a need for the development of more bio-inspired approaches in this series of social media data analysis.

References

1. Tan PN, Steinbach M, Kumar V (2005) Introduction to data mining. Addison-Wesley, Boston
2. Alpaydin E (2004) Introduction to machine learning. MIT Press, Cambridge
3. Webb A (2002) Statistical pattern recognition. Wiley, New Jersey
4. Han J (2006) Kamber M data mining: concepts and techniques, 2nd edn. Morgan Kaufmann, Menlo Park
5. Hathway RJ, Bezdek J (1995) Optimization of clustering criteria by reformulation. IEEE Trans Fuzzy Syst 241–245
6. Kennedy J, Eberhart R (2001) Swarm intelligence. Morgan Kaufmann
7. Van der Merwe DW, Engelbrecht AP (2003) Data clustering using particle swarm optimization. In: The 2003 congress on evolutionary computation, pp 215–220
8. Basturk B, Karaboga D (2006) An artificial bee colony (ABC) algorithm for numeric function optimization. In: IEEE swarm intelligence symposium 2006, 12–14 May 2006, Indianapolis, Indiana, USA
9. Karaboga D, Basturk B (2007) A powerful and efficient algorithm for numerical function optimization: artificial bee colony (ABC) algorithm. J Global Optimiz 39(3):459–471
10. Xu Chunfan, Duan Haibin (2010) Artificial bee colony (ABC) optimized edge potential function (EPF) approach to target recognition for low-altitude aircraft. Pattern Recogn Lett 31(13):1759–1772
11. Runkler TA, Katz C (2006) Fuzzy clustering by particle swarm optimization. In: 2006 IEEE international conference on fuzzy systems, Canada, pp 601–608
12. Zhao B (2007) An ant colony clustering algorithm. In: Proceedings of the sixth international conference on machine learning and cybernetics, Hong Kong, pp 3933–3938

13. Li L, Liu X, Xu M (2007) A novel fuzzy clustering based on particle swarm optimization. In: First IEEE international symposium on information technologies and applications in education, pp 88–90
14. Gan G, Wu J, Yang Z (2009) A genetic fuzzy k-modes algorithm for clustering categorical data. Expert Syst Appl 36:1615–1620
15. Kennedy J, Eberhart R (1995) Particle swarm optimization. In: Proceedings of the IEEE international conference on neural networks, pp 1942–1948
16. Pang W, Wang K, Zhou C, Dong L (2004) Fuzzy discrete particle swarm optimization for solving traveling salesman problem. In: Proceedings of the fourth international conference on computer and information technology, IEEE CS Press, pp 796–800
17. Karaboga D (2005) An idea based on honey bee swarm for numerical optimization. Technical Report-TR06, Erciyes University, Engineering Faculty, Computer Engineering Department
18. Yang Y (1999) An evaluation of statistical approaches to text categorization. J Inform Retr 1:67–88
19. Huang Z (1998) Extensions to the k-means algorithm for clustering large data sets with categorical values. Data Min Knowl Disc 2:283–304
20. Rand WM (1971) Objective criteria for the evaluation of clustering methods. J Am Stat Assoc 66:846–850

Privacy Preservation in Health care by Process Mining

Y. Sushma, J. Rajeshwar and S. Tejasree

Abstract Procedure mining may be a special variety of statistics mining to extract understanding from event logs recorded by approach of Associate in nursing records gizmo. In e-healthcare system, mining technique is employed for the analysis of medical statistics based on event logs. Process mining targets the potency and ability of ways by providing techniques and tools for locating complete technique, facts and social structures from event logs. Key capability of process mining approach is to produce us with valuable knowledge hidden in event logs. The facts are limitless in event logs. Facts mining in scientific space for distinctive upset analysis are trending. Procedure or process mining methods can build e-fitness care machine bigger effective. This survey has specific attention on procedure mining algorithms, techniques, gear, methodologies, manner, and mining views that are utilized in aid and conjointly explains exploitation technique mining to boost the aid technique.

Keywords Process mining · Health care · Hospital information system (HIS) · Event logs

Y. Sushma (✉)
CMR College of Engineering &Technology, Hyderabad, India
e-mail: y.sushmareddy@cmrcet.org

J. Rajeshwar
CSE, Guru Nanak Institutions Technical Campus, Hyderabad, Telangana, India
e-mail: prof.rajeshwar@gmail.com

S. Tejasree
VEMU Institute of Technology, Chittoor, AP, India
e-mail: tejareddysamakoti@gmail.com

© Springer Nature Singapore Pte Ltd. 2019
H. S. Saini et al. (eds.), *Innovations in Computer Science and Engineering*, Lecture Notes in Networks and Systems 74,
https://doi.org/10.1007/978-981-13-7082-3_11

83

1 Introduction

Process mining could be very younger research vicinity. Process mining is a technique to retrieve the useful data from event logs [1]. The facts are limitless in event logs. Event logs can be produced by using Process Aware Information System (PAIS). Purchaser relationship management gadgets, employer resource planning, and medical institution system are the examples for Process Aware Information System. In Hospital Information System (HIS) big quantities of facts are stored about the care system. Process mining is a unique form of records mining to extract knowledge from event logs recorded with the aid of an information system and the concept of system mining is to find out, reveal, and enhance actual approaches through extracting information from event logs [2–4] and for conformance checking [5] and additionally reading different social networks [3, 4, 6]. Procedure mining is applicable to the occasion records of a wide range of a gadget. Procedure mining appears internal technique at various abstraction tiers (Fig. 1).

Process mining types: Health care processes are highly dynamic, and increasing number of multidisciplinary, complicated, adhoc in nature. In procedure mining specifically three types are exist [1].

(A) **Process Discovery**: Analysis can be started based totally on event logs. Event log can be described as a collection of occasions. An occasion may additionally have the statistics like case id, activity, time stamp, useful resource, quantity, and ordered objects. Process mining area has been used in healthcare approaches for discovering process models from occasion logs [2, 3, 7].

(B) **Conformance Checking**: Conformance checking is used to check compliance checking, auditing, certification, and monitoring at run time. It may be used to find out the quality of models [5].

(C) **Model Enhancement**: Right here model development may be feasible while the non-becoming models can be corrected. The fashions can be corrected by using alignment of the version and the log. Widespread category strategies may be used for model enhancement.

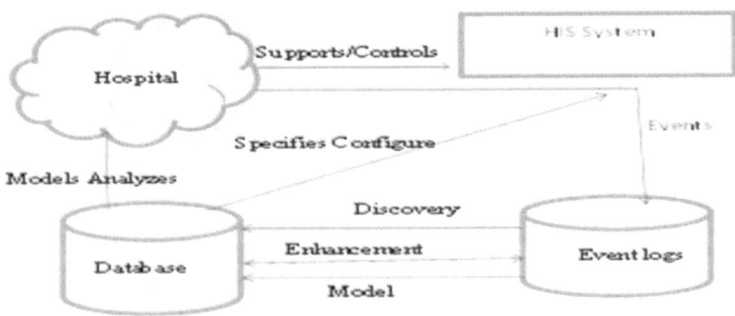

Fig. 1 Process mining in health care

Perspectives of Process Mining:

There is quantity of prospective inside the utility of process mining.

Control flow Perspective: It is one of the maximum critical perspectives of process mining. Based at the event log information new process model can be generated. There are numerous process mining algorithms, that are heuristic mining set of rules, α-mining set of rules and area mining set of rules region mining set of rules [2]. These mining algorithms particularly cope with noise and exceptions and allow customers to cognizance on the main system go with the flow in place of the conduct appearing within the process log.

Performance Perspective: In process mining, there are quantity of performance analysis techniques. In overall performance attitude, it identifies the bottlenecks, synchronization time, idle time primarily based at the study of execution time activities. Performance angle may be carried out.

Conformance Perspective: On this perspective, it allows the detection of system deviations with regard to pre-decided version. Conformance attitude may be applied.

Organizational Perspective:

This perspective mainly based on analyzing the collaboration among sources. There are numerous process mining strategies that address the organizational attitude. Right here collaboration between assets is like collaboration between departments is clinic.

2 Process Mining Tools: ProM

There are variety of tools used for system mining, the most used device in health care is proM (which is short for process mining framework) is an open supply framework for procedure mining algorithms [7]. ProM is straightforward to use and easy to extend offerings for both customers and builders. It offers a platform of the system mining algorithms. Prom has implemented a large range of strategies and algorithms were utilized in extraordinary base papers.

Disco: Disco has been designed to make the information import truly clean by way of automatically detecting timestamps, remembering your configuration settings, and by loading statistics units with excessive velocity. One actually opens a csv or excel record and configures which columns keep the case identification, timestamps, and activity names, in which other attributes ought to be covered inside the evaluation, and the import may begin. Fact sets are imported in a read only mode, so the original files cannot be modified (which is vital, e.g., for auditors). Disco is likewise completely likeminded with the academic tool sets proM 5 and proM 6.

Rapid ProM: Rapid prom affords 37 process mining operators, which includes several process discovery algorithms and filters in addition to importers and exporters from/to specific process-modeling notations.

Techniques or Algorithms: Procedure mining may be applied in distinctive domains like banking, insurance, e-authorities; patron courting control, production, far-flung tracking, and diagnostics [5]. There is extra variety of techniques used in

Table 1 Techniques for process mining

Algorithm	References	Work
Heuristic miner	[5–17]	Discovery algorithm—deals with noise in event logs
Fuzzy miner	[2, 3, 5, 7, 9, 16]	Configurable discovery algorithm—deals with unstructured processes
Trace clustering	[2, 18, 19]	Partitioning of the event logs to generate structured process models
Alpha miner	[6]	Discover process models

process mining. Procedure mining can be used to enhance overall performance while lowering value. Series of more statistics in hospitals is not beneficial, make the most the information to comprehend greater effective care approaches. The main strategies used in system mining are a heuristic miner, fuzzy miner, and trace clustering. Those techniques have already applied and available in open supply environment inclusive of proM [7] (Table 1).

Methodologies: To make the evaluation of statistics, clustering technique may be used that is the primary methodology used in process mining, it is the short method for process diagnosis. First methodology executes a analysis method through system mining technique, it is mainly based totally on the five levels which can be log propagation, log inspection, control flow analysis, overall performance evaluation, and role analysis. a. refuge, d. r. Ferreira, "commercial enterprise method evaluation in health care environments: a method based on system mining". Inside the above research paper methodology, a new step may be brought after log inspection, this step particularly pertains to sequential clustering evaluation. Sequential clustering involves quantity of activities for contending with unstructured technique. These activities helped to find out both standard and infrequent conduct. The second method is life cycle model. The life cycle model may be partitioned the complete system mining assignment into five various tiers which might be plan and justify, extract, create the control model and connect the event log, create the integrated process version and affords operational guide. It is used for steering process mining initiatives, mainly in the evidence primarily based clinical compliance cluster. Methodologies which are carried out above are not area specific which can be applied in other fields except health care.

3 Implementation Strategies

Different types of strategies are used for imposing procedure mining. The primary method is direct implementation, it encompasses information and this is extracted from health information system (his) via applying process mining for building the event log. 17 case studies as an example [7] use this approach and obtain models, diagrams and tables for accomplishing analysis on statistics. It constitutes two chal-

Table 2 Implementation strategies and gaps

Strategy name	Case studies	Gaps
Direct implementation	[2, 4, 7, 9, 21, 22, 23, 24]	Data extraction building the event log, need to understand tools and techniques
Semiautomated	[10]	Building the event logs by using queries, need to understand tools and techniques
Integrated suit	[20, 25]	Fails to provides suitable solutions

lenges that are information extraction and accurate event log introduction and want to apprehend the tools, algorithms, and techniques used in procedure mining for carrying out evaluation. The second approach is semi automated. In this approach the statistics extraction and constructing of the event logs are made by way of custom-made development. It hyperlinks one or extra information sources and extracts the statistics that are required to build up the event log via the use of queries. Procedure mining tool knowledge is required for this strategy. The primary disadvantage by way of the usage of this strategy is to define adhoc way for the extraction of statistics from unique tools. The third approach is the implementation of an incorporated suite, wherein information assets are connected, records are extracted, the occasion log is built and the implemented technique mining strategies may be used. For this strategy, no need to precise information on techniques and algorithms works on information. The main downside of incorporated suit is, it has been advanced for specific precise environments and information resources do not offer portable solutions. Examples for the integrated suits are medtrix process mining studio [20] (Table 2).

4 Conclusion

Process mining in health care explains actual execution from begin to result in health-care domain by the use of process mining types. Discovering process models and checking conformance with medical guidelines, and discovering new development possibilities. This paper is in particular regards the survey about process mining in health care. It consists of perspectives, process mining tools, techniques and algorithms. This survey consisting of range of case research conducted previously on this domain. The future project in phrases of method mining in health care explains its boom and how much it is important in healthcare domain. Future enhancement of system mining in health care is good visualization of process models.

References

1. van der Aalst W (2011) Process mining: discovery, conformance and enhancement of business processes. Springer Science & Business Media
2. Mans RS, Schonenberg H, Song M, van der Aalst WMP, Bakker PJM (2008) Application of process mining in healthcare—A case study in a dutch hospital. In: Fred ALN, Filipe J, Gamboa H (eds) Biomedical engineering systems and technologies, international joint conference, BIOSTEC 2008, Funchal, Madeira, Portugal, 28–31 January 2008
3. Bose RPJC, van der Aalst WMP (2011) Analysis of patient treatment procedures. In: Daniel F, Barkaoui K, Dustdar S (eds) Business process management workshops—BPM 2011
4. Mans R, Reijers HA, van Genuchten M, Wismeijer D (2012) Mining processes in dentistry. In: Luo G, Liu J, Yang CC (eds) ACM international health informatics symposium, IHI'12, Miami, FL, USA, 28–30 January 2012. ACM, pp 379–388
5. Zhou J (2009) Master: process mining: acquiring objective process information for healthcare process management with the CRISP-DM framework. PhD thesis, Master's thesis. Eindhoven University of Technology, Eindhoven
6. Lang M, Burkle T, Laumann S, Prokosch H (2008) Process mining for clinical workflows: challenges and current limitations. In: Andersen SK, Klein GO, Schulz S, Aarts J (eds) eHealth beyond the horizon—get IT there. Proceedings of MIE2008. The XXIst international congress of the European federation for medical informatics, Goteborg, Sweden, 25–28 May 2008. Studies in health technology and informatics, vol 136. IOS Press, pp 229–234
7. Mans R, van der Aalst W, Vanwersch RJ (2015) Process mining in healthcare: evaluating and exploiting operational healthcare processes. Springer
8. Partington A, Wynn M, Suriadi S, Ouyang C, Karnon J (2015) Process mining for clinical processes: a comparative analysis of four australian hospitals. ACM Trans Manag Inform Syst (TMIS) 5(4):1–19
9. Kim E, Kim S, Song M, Kim S, Yoo D, Hwang H, Yoo S (2013) Discovery of outpatient care process of a tertiary university hospital using process mining. Healthc Inform Res 19(1):42–49
10. Cho M, Song M, Yoo S (2014) A systematic methodology for outpatient process analysis based on process mining. In: Asia pacific business process management. Springer, pp 31–42
11. Kaymak U, Mans R, van de Steeg T, Dierks M (2012) On process mining in health care. In: Proceedings of the IEEE international conference on systems, man, and cybernetics, SMC 2012, Seoul, Korea (South), 14–17 October 2012. IEEE, pp 1859–1864
12. Gupta S, Master's thesis: workflow and process mining in healthcare. PhD thesis, Master's Thesis, Technische Universiteit Eindhoven
13. Caron F, Vanthienen J, Vanhaecht K, Van Limbergen E, Deweerdt J, Baesens B et al (2014) A process mining-based investigation of adverse events in care processes. Health Inform Manag J 43(1):16–25
14. Lakshmanan GT, Rozsnyai S, Wang F (2013) Investigating clinical care pathways correlated with outcomes. In: Business process management. Springer, pp 323–338
15. Fei H, Meskens N et al (2008) Discovering patient care process models from event logs. In: 8th international conference of modeling and simulation, MOSIM 2008, vol 10, Citeseer, pp 10–12
16. Suriadi S, Mans RS, Wynn MT, Partington A, Karnon J (2014) Measuring patient flow variations: a cross-organisational process mining approach. In: Asia pacific business process management. Springer, pp 43–58
17. Montani S, Leonardi G, Quaglini S, Cavallini A, Micieli G (2013) Mining and retrieving medical processes to assess the quality of care. In: Case-based reasoning research and development. Springer, pp 233–240
18. Caron F, Vanthienen J, Baesens B (2013) Healthcare analytics: examining the diagnosis–treatment cycle. Proced Technol 9:996–1004
19. Delias P, Doumpos M, Grigoroudis E, Manolitzas P, Matsatsinis N (2015) Supporting healthcare management decisions via robust clustering of event logs. Knowl-Based Syst 84:203–213

20. Rebuge A, Ferreira DR (2012) Business process analysis in health-care environments: a methodology based on process mining. Inform Syst 37(2):99–116
21. Peleg M, Soffer P, Ghattas J (2007) Mining process execution and outcomes—position paper. In: ter Hofstede AHM, Benatallah B, Paik H (eds) Business process management workshops, BPM 2007 international workshops, BPI, BPD, CBP, ProHealth, RefMod, semantics4ws, Brisbane, Australia, 24 September 2007, Revised Selected Papers. Lecture notes in computer science, vol 4928. Springer, pp 395–400
22. Rinner C, Dunkl R, Gall W, Froschl KA, Grossmann W, Rinderle-Ma S, Dorda W, Kittler H, Duftschmid G (2014) Cutaneous Melanoma surveillance by means of process mining. In: European federation for medical informatics and IOS Press
23. Dagliati A, Sacchi L, Cerra C, Leporati P, De Cata P, Chiovato L, Holmes J, Bellazzi R et al (2014) Temporal data mining and process mining techniques to identify cardiovascular risk-associated clinical pathways in type 2 diabetes patients. In: 2014 IEEE-EMBS international conference on biomedical and health informatics (BHI). IEEE, pp 240–243
24. Boere J-J (2013) An analysis and redesign of the ICU weaning process using data analysis and process mining. PhD thesis, Maastricht University Medical Centre
25. Neumuth T, Liebmann P, Wiedemann P, Meixensberger J et al (2012) Surgical workflow management schemata for cataract procedures. Methods Inform Med 51(5):371–382

Named Entity Recognition in Tamil Language Using Recurrent Based Sequence Model

V. Hariharan, M. Anand Kumar and K. P. Soman

Abstract Information extraction is a key task in natural language processing which helps in knowledge discovery by extracting facts from the semi-structured text like natural language. Named entity recognition is one of the subtask under information extraction. In this work, we use recurrent based sequence models called Long Short-Time Memory (LSTM) for named entities recognition in Tamil language and word representation for words is done through a distributed representation of words. For this work, we have created a Tamil named entities recognition corpus by crawling Wikipedia and we have also used openly available FIRE-2018 Information Extractor for Conversational Systems in Indian Languages (IECSIL) shared task corpus.

Keywords Information extraction · Named entity recognition · Recurrent neural network · Long short time memory

1 Introduction

The rise of the internet has seen a huge growth in the amount of information available online in the form of text and as long as we continue to log our daily inferences and other works on the platform such as social media, blogs, email, etc., amount of information will be continuously increasing. Elective methods to process this information on a large scale without the need for human intervention will be an indispensable tool for the understanding this large amount information, this process is called as Information Extraction. One of the crucial subtasks of information extraction is Named

V. Hariharan (✉) · M. Anand Kumar · K. P. Soman
Center for Computational Engineering and Networking (CEN), Amrita School of Engineering,
Amrita Vishwa Vidyapeetham, Coimbatore, India
e-mail: hari03@live.in

M. Anand Kumar
e-mail: m_anandkumar@cb.amrita.edu

K. P. Soman
e-mail: kp_soman@amrita.edu

© Springer Nature Singapore Pte Ltd. 2019
H. S. Saini et al. (eds.), *Innovations in Computer Science
and Engineering*, Lecture Notes in Networks and Systems 74,
https://doi.org/10.1007/978-981-13-7082-3_12

Entity Recognition (NER). This task helps in the classifying the words in the text into predened categories like a person, location, and organization, etc. By knowing the named entities in the text helps us to categorize the text where these named entities occur. This NER subtask helps in creating a structured database from semi-structured text corpus [1]. In this work, supervised learning methods are used for solving the task of named entity recognition. A recurrent based sequence model such called LSTM is trained via supervised learning [2] approach to predict the named entities of the words in the corpus.

2 Literature Survey

Named entity recognition is a key task in ending the factual information from the semi-structured natural language. This has done through unsupervised methods by employing clustering and ranking techniques [3]. In semi-supervised (weakly supervised) methods, techniques like bootstrapping [4] are very popular approach, i.e., having a small sample of good annotation for training. In earlier supervised methods, support vector machines and conditional random fields are used for training [5]. All these earlier methods used features such as co-occurrences, word substrings, punctuation, and handcrafted linguistic patterns. These features are crafted specially according to the domain and the language, so these features cannot easily be scaled. The recent development in the distributed representation proposed by [6] produces as dense vector representation for the word based on the entire corpus this forms a good representation for the word. In this work, we employ a class of neural network architecture called recurrent based sequence models for the learning of the named entities in the text.

2.1 *Distributed Representation*

Word representation is one of the crucial components in any natural language processing system. The distributed representation framework like word2vec, Glove and fastest gives a dense vector representation for each word in the corpus. The dense vector representation for the word is formed by taking softmax for the word over the context window through an iterative process and for updating the information of the word not in the context window, negative sampling is done, i.e. (randomly selecting words that are not in the context window instead of iterating through the all the words in the corpus) [7]. This iteration over the whole corpus is continued until a good analogy between the word vectors is obtained like in the Eq. 1.

$$\text{'king'} - \text{'man'} + \text{'woman'} = \text{'queen'} \tag{1}$$

or a good performance of the system where these word vectors are employed [8]. This dense vector representation formed by the distributed representation framework aims to capture the syntactic and semantics of the word with respect to all words in the corpus. One of the major drawbacks of the word2vec and Glove model is it does not consider the internal structure of the words so it fails the capture the morphology present in the language, fastText overcomes this drawback by considering the different character n-grams of the word along with the word itself to jointly learn the vector representation for word. By considering the character n-grams of words, the morphology of the language is learned to some extent.

2.2 Sequence Based Recurrent Models

All natural language can be best inferred when it is processed as a sequence or with respect to some context. So in order for the neural network to understand the sequential information of the language [9], a recurrent based neural network called recurrent neural network (RNN) which processes each word in the sentence at different time steps, so that the information of the previous timestep is passed on the next timestep. This is done by connecting the hidden state of the previous time step with current timestep. In RNN, the number of time steps depends on the upon the number of words in the sentence given in as input. So RNN can naturally handle variable length sequence.

An RNN unrolled across its time steps is shown in Fig. 1. At each time step $x_{0:t}$, the RNN takes one input in Fig. 1, x_t corresponds to the input at various time steps, h_t corresponds to the output at each time step and A corresponds to the hidden layer.

$$a_t = g(V[x_t : a_{t-1}] + c) \tag{2}$$

$$h_t = Wa_t + b \tag{3}$$

In Eq. 2, the hidden state of each time step a_t is calculated and g is the nonlinear activation function which is applied over the linear transformation between the weight matrix V and the input of the current time step x_t concatenated with a previous hidden

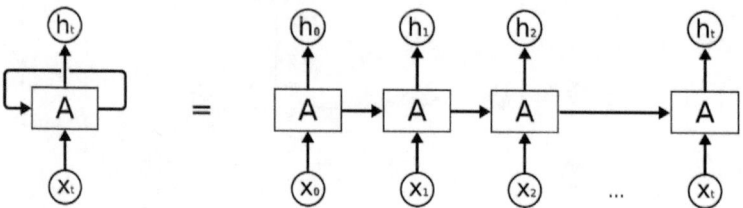

Fig. 1 A single RNN cell versus single RNN cell unrolled across its time steps

94   V. Hariharan et al.

state a_{t-1}. In Eq. 3, linear transformations are done over the hidden state a_t and added with bias term b, to get the output of current time step.

2.3 Long Short Time Memory

Long Short Time Memory (LSTM) is slight modification over the vanilla RNN to overcome exploding and vanishing gradient problem [10]. In LSTM, there is an additional state called cell state, this state is often referred as the memory element of the LSTM, because in the cell state, addition operation is performed with the previous cell state, weighted by input and forget gate. Instead of multiplication with the previous hidden state in the vanilla RNN, this causes the gradient to ow across many time steps with decay. The mathematical equation of LSTM is given below

$$c_n = f_n c_{n-1} + i_n \tanh(V[x_n; h_{n-1}] + b_c) \qquad (4)$$

$$h_n = o_n \tanh(W_h c_n + b_h) \qquad (5)$$

The cell state c_n is calculated in the Eq. 4 and hidden state h_n is calculated in the Eq. 5. Where i_n, f_n and o_n is the input gate, forgot gate, and output gate, respectively. A single time step of LSTM cell is shown in Fig. 2.

The equation of the input gate i_n, forget gate f_n and output gate o_n is given in Eqs. 6, 7, and 8.

$$i_n = \sigma (W_i[x_n; h_{n-1}] + b_i) \qquad (6)$$

$$f_n = \sigma (W_f[x_n; h_{n-1}] + b_f) \qquad (7)$$

Fig. 2 A single time step of LSTM cell

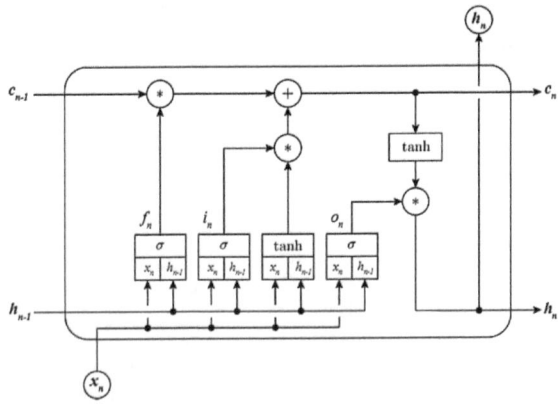

$$o_n = \sigma \ (W_o[x_n; h_{n-1}] + b_o) \qquad (8)$$

The output of these gates is a vector and value of each element in the vector is bounded between 0 and 1 which is due to the sigmoid nonlinearity. The input gate i_n and forget gate f_n grant the LSTM network the ability like how much the network can remember from past timesteps and how much to copy from current timestep. The input gate set to zero is like ignoring the current timestep and forget gate set to zero is like forgetting everything in the cell state (memory) therefore starting fresh. The output gate o_n gives a vector which controls how much amount of information is exposed by the current cell state to the network which is reading from it. If some elements in the output gate vector are zero, then it cancels out those corresponding dimension in the cell state, so it is not visible to the network which is reading from it.

3 Methodology

In this work, supervised learning methods are used for solving the named entity recognition problem. In the approach, a recurrent based neural network called LSTM is used to learn the named entities in the text sequence. LSTM is preferred over the recurrent based neural network since it captures the long-term dependencies in the text sequence without suffering from vanishing and exploding gradient problem. The architecture of the NER model is shown in Fig. 3.

For this work, we have used two corpora for named entity recognition in the Tamil language. The rst corpus is taken from the openly available FIRE-2018 Information Extractor for Conversational Systems in Indian Languages (IEC-SIL) shared task [11], in this corpus, there are about 1,34,030 sentences with 8 entity tag and an other tag. The second corpus is made by crawling Tamil Wikipedia and the annotation for the named entity tag is made by using property tag present in the Wikidata website. In this corpus, there are about 2,12,873 sentences with 20 entity tags and another

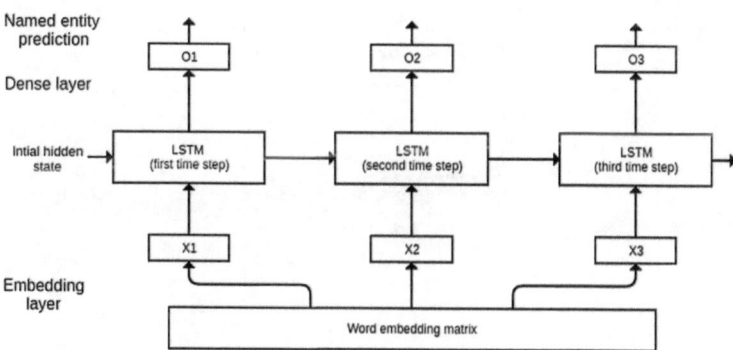

Fig. 3 Named entity recognition model

tag. The corpus statistics are shown in Tables 1 and 2. The word representations are done through distributed representation framework like Glove [12] and fastText [13]. Through distributed representation, each word in the corpus is given a dense vector of certain dimension. These word vectors are fed into the recurrent based neural network model called Long Short Time Memory (LSTM). The recurrent networks can naturally handle variable length sequences. So in this case, the number of words in the sentence determines the time step. To predict the named entity tag for each word in the sequence, dense layer is applied for each time step. The output of the dense layer corresponds to the number of classes, i.e. (number of named entity tags). In addition to initializing of word vectors through distributed word representation framework, we have also used randomly initialized the word embedding and feed into the LSTM network.

Table 1 IECSIL corpus statistics

Sl. no	Entity tag	Freq (%)
1	Datenum	1.03
2	Event	0.34
3	Location	8.99
4	Name	7.90
5	Number	5.18
6	Things	0.41
7	Occupation	1.10
8	Organization	0.67
9	Other	74.34

Table 2 Tamil Wiki corpus statistics

Sl. no	Entity tag	Freq	Sl. no	Entity tag	Freq
1	Date	2.88	12	Geometry	0.05
2	Work of art	6.33	13	Law	0.22
3	Property	0.07	14	Norp	5.95
4	Concept	0.17	15	Money	0.05
5	Time	5.11	16	Event	0.63
6	Org	1.8	17	Quantity	0.97
7	Gpe	10.7	18	Entity	5.9
8	Abstract	0.14	19	Language	2.57
9	Person	6.4	20	Facility	0.17
10	Location	3.38	21	Others	46.01
11	Product	0.38			

4 Result and Discussion

In this work, the task of named entity recognition is solved via supervised learning approach using recurrent based sequence network called Long Short-Time Memory (LSTM). This named entity recognition model is applied on FIRE-2018 IECSIL shared task corpus and Tamil Wiki corpus. On these two corpora, we have trained a distributed word representation network called fastText and Glove to give dense word vectors of different dimensions and word vectors which are randomly initialized is also used for evaluation. These word vectors are fed into the NER model. From the results of the two corpora, it is seen that word embedding obtained by the distributed word representation framework like Glove and fastText performs better than the randomly initialized word vectors. It is also seen that the fastText based distributed representation performs better than the Glove based distributed representation framework this is because in fastText based distributed representation, the word representations are jointly learned along with the different character n-grams of the word and it also forms a better word representation for the unknown words. Since Tamil is a morphologically rich language fastText based word embedding is seen to perform better than other distributed representation methods (Tables 3 and 4).

Table 3 Results for IECSIL@FIRE-2018 corpus

Sl. no	Dimension	Rand emb + LSTM (F1 macro)	Glove + LSTM (F1 macro)	fastText + LSTM (F1 macro)
1	50	85.44	90.26	93.1
2	100	86.03	90.62	93.55
3	150	87.11	92.3	93.73
4	200	87.43	93.03	94.32
5	250	87.01	93.01	94.3
6	300	87.23	93.06	94.54

Table 4 Results for Tamil Wikipedia corpus

Sl. no	Dimension	Rand emb + LSTM (F1 macro)	Glove + LSTM (F1 macro)	fastText + LSTM (F1 macro)
1	50	83.89	86.23	89.1
2	100	83.83	88.82	90.55
3	150	84.11	88.79	90.73
4	200	84.43	89.23	90.32
5	250	84.33	90.31	91.3
6	300	84.03	90.06	91.29

5 Conclusion and Future Work

The named entity recognition is a critical task in information extraction as it helps in identifying the entities like a person, location, etc. This helps in organizing the unstructured text in a structured format. In this paper, for identifying the named entities in the text, a recurrent based neural network called LSTM is applied on IECSIL@FIRE-2018 shared task corpus and Tamil Wiki corpus. The word representation is done by using a distributed word representation framework called fastText and Glove. In this work, word vectors of different dimension are used and it is seen that words vectors obtained through the fastText distributed representation perform better when compared to other embeddings. This is because fastText enriches the word vector with character n-gram information and Tamil is a morphologically rich language, fastText embeddings perform better for the named entity recognition task. In our next work, we are planning to make an end to end neural network for information extraction. Which does named entity recognition and relation extraction in a single stretch.

References

1. Remmiya Devi G, Veena PV, Anand Kumar M, Soman KP (2018) Entity extraction of Hindi–English and Tamil–English code-mixed social media text. In: Forum for information retrieval evaluation. Springer, pp 206–218
2. Lample G, Ballesteros G, Subramanian S, Kawakami K, Dyer C (2016) Neural architectures for named entity recognition. arXiv:1603.01360
3. Remmiya Devi G, Veena PV, Anand Kumar M, Soman KP (2016) Entity extraction for malayalam social media text using structured skip-gram based embedding features from unlabeled data. Proced Comput Sci 93:547–553
4. Mintz M, Bills S, Snow R, Jurafsky D (2009) Distant supervision for relation extraction without labeled data. In: Proceedings of the joint conference of the 47th annual meeting of the ACL and the 4th international joint conference on natural language processing of the AFNLP, vol 2. Association for Computational Linguistics, pp 1003–1011
5. Abinaya N, Anand Kumar M, Soman KP (2015) Randomized kernel approach for named entity recognition in Tamil. Indian J Sci Technol 8(24)
6. Mikolov T, Sutskever I, Chen K, Corrado GS, Dean J (2013) Distributed representations of words and phrases and their compositionality. In: Advances in neural information processing systems, pp 3111–3119
7. Barathi Ganesh HB, Anand Kumar M, Soman KP (2018) From vector space models to vector space models of semantics. In: Forum for information retrieval evaluation. Springer, pp 50–60
8. Anand Kumar M, Soman KP, Barathi Ganesh HB (2016) Distributional semantic representation for text classification and information retrieval. In: CEUR workshop proceedings, vol 1737, pp 126–130
9. Sundermeyer M, Schlüter R, Ney H (2012) LSTM neural networks for language modeling. In: Thirteenth annual conference of the international speech communication association
10. Hochreiter S, Schmidhuber S (1997) Long short-term memory. Neural Comput 9(8):1735–1780

11. Barathi Ganesh HB (2018) Information extractor for conversational systems in Indian languages @ forum for information retrieval evaluation. http://iecsil.arnekt.com/!/home
12. Pennington J, Socher R, Manning C (2014) Glove: global vectors for word representation. In: Proceedings of the 2014 conference on empirical methods in natural language processing (EMNLP), pp 1532–1543
13. Bojanowski P, Grave E, Joulin A, Mikolov T (2016) Enriching word vectors with subword information. arXiv:1607.04606

Gender Identification of Code-Mixed Malayalam–English Data from WhatsApp

Vineetha Rebecca Chacko, M. Anand Kumar and K. P. Soman

Abstract The boom in social media has been a topic of discussion among all generations of this era. It most certainly has its positives, such as real-time communication, and a platform for all to voice their opinions. There are a few shady sides to it too, such as anonymity of those communicating. Such anonymity, especially in mediums of messaging such as WhatsApp, can turn out dangerous. Here, comes the crucial role of author profiling. This paper describes the analysis of code-mixed Malayalam–English data, collected from WhatsApp, and its classification based on the basic demographic, the gender, of the author. The text has been represented as Term Frequency–Inverse Document Frequency (TFIDF) matrix and as Term Document Matrix (TDM). The classifiers used are SVM, Naive Bayes, Logistic Regression, Decision Tree, and Random Forest.

1 Introduction

The basic link between humans, since time immemorial, is communication. With new innovations in technology, and in the field of computer science, the modes of communication have developed over the years. Once, it took days to convey a message to a dear one. But with messaging applications such as WhatsApp, and other social media platforms such as Twitter and Facebook, real-time communication has become a reality. There are a few negative aspects to all this such as cyberstalking and bullying, hacking, spread of fabricated news, etc. What is common in all these cases, is the anonymity of authors. These days, anybody can access the internet and

V. R. Chacko (✉) · M. Anand Kumar · K. P. Soman
Center for Computational Engineering and Networking (CEN), Amrita School of Engineering,
Amrita Vishwa Vidyapeetham, Coimbatore, India
e-mail: vineethachacko7@gmail.com

M. Anand Kumar
e-mail: m_anandkumar@cb.amrita.edu

K. P. Soman
e-mail: kp_soman@amrita.edu

© Springer Nature Singapore Pte Ltd. 2019
H. S. Saini et al. (eds.), *Innovations in Computer Science and Engineering*, Lecture Notes in Networks and Systems 74,
https://doi.org/10.1007/978-981-13-7082-3_13

to create a profile in any of these social media platforms is child's play. Males can project themselves as females, and vice versa. Fake profiles of celebrities, as well as their fans, are available in abundance. Social media has even been used to propagate the wrong ideas about sensitive topics among the masses. Hence, fraudulence is a common issue when it comes to social media. Thus, the field of **Author Profiling** gained its importance in cyber forensics. It involves the analysis of the traits of an author.

Along with the rise of social media, the world is in the process of transitioning into a global community. At the snap of a finger, people from different countries communicate with each other, be that through WhatsApp or Facebook or gaming portals. Hence, the role of multilingual information has also gained its importance. India, being a country having its own diversity in the languages used, has also adopted English as the *lingua franca*. Hence, there is also a huge diversity in the code-mixed languages used in the country. Not much research has been done in analyzing all the code-mixed Indian language varieties, especially when it comes to Dravidian languages, with respect to author profiling. Here, code-mixed Malayalam–English data, from WhatsApp has been collected, and the basic demographics of the author— the gender—is identified, using Machine Learning algorithms. Further, other works related to author profiling, a description of the creation of corpus, the approach used, which involves Machine Learning algorithms and finally, the results analyzed for each corpora have been discussed.

2 Related Work

The systems submitted for PAN@CLEF 2017 Author Profiling shared task, and the creation of corpora for the same, have been referred to the most, to create and analyze the corpora and for the chosen methodology [1]. CLEF or Conference and Labs of the Evaluation Forum aims at research in the field of multilingual data. PAN@CLEF is a shared task, where tasks like author profiling, author obfuscation, author identification, etc., are put up. The PAN 2017 corpora is balanced in the case of number of authors for each gender, as well as for the number of sentences per document (100 Tweets per document). Of the 20 teams that submitted their notebooks for the shared task, the systems that secured the top three positions have been analyzed here. The system that secured the first position, submitted by Basile et al. [2] has used SVM classifier. They have used word unigrams and character trigrams, tetragrams, and pentagrams. Even though they trained with more data, such as the Twitter 14k dataset and the PAN 2016 corpora, the accuracy was not improved. Other aspects analyzed include POS tags, Twitter handles, emojis, etc., but in vain.

The next best system, submitted by Martinc et al. [3], did preprocessing on the data which included checking the spelling of English words, removal of stop words, hashtags, etc. The English words spelled wrong were removed. Here, the classifier used is Logistic Regression which performed well compared to Random Forest and linear SVM. The system that secured the third position, submitted by Tellez et al. [4], uses MicroTC, a framework that works irrespective of the domain and language specifications. Here, binary and trivalent features have been used for preprocessing and text is represented as TFIDF. The classifier used is SVM. Barathi Ganesh et al. [5] submitted a notebook to PAN 2017, which used Vector Space Models for text classification. Chacko et al. [6] has analyzed the PAN 2017 corpora for gender as well as language variety identification. Here, the features used are minimum document frequency, character as well as word n-gram and sublinear term frequency. Basic Machine Learning algorithms such as SVM, Random Forest, Decision Tree, and AdaBoost have been used.

3   Methodology

In this section, it is discussed how the corpora has been created, the challenges faced while creating it and how it has been processed to obtain the final categories of corpora. Then, the methods in which the text has been represented, the parameters used for further feature extraction and how the classification has been done is explained.

3.1   Code-Mixed Malayalam–English Corpora

The code-mixed Malayalam–English corpora have been made after processing 30 WhatsApp conversations, involving 14 males and 20 females. To consider an author's data, he/she must have authored a minimum of 100 sentences in code-mixed Malayalam–English language, each having a minimum of 3 words. The major challenge faced during data collection was the lack of data from males. The collected data had to be further processed to remove the author's names, date and time of the messages, and also those sentences that don't satisfy the above mentioned condition. The purely English script also had to be removed. Some authors send a single sentence as multiple messages
For example, 30/10/2017, 9:44 AM: Njaan
30/10/2017, 9:44 AM: Innu
30/10/2017, 9:44 AM: Pokunnilla.
Such messages have to be combined. Different messages may be conveyed in a single message, which has to be separated.
For example, 27/07/2017: Naale aanu dance program. You will come or njaan potte. Finally, the cleaned data, in text format, was compiled in three ways. The statistics is as shown in Table 1.

The anonymity of the authors is ensured by using randomly generated strings to name the text files containing the data. Data split I has 43 files from male authors and 79 files from female authors, each having 100 sentences. Data split II has 43 files each from male and female authors with the same 100 sentence split. This split was decided upon with reference to the PAN 2017 corpora [1]. Data split III has 215 files

Table 1 Statistics of code-mixed Malayalam–English corpora

Category of data	Data split I	Data split II	Data split III
Number of files of male authors	43	43	215
Number of files of female authors	79	43	215
Sentence split	100	100	20
Number of sentences	12,200	8600	8600

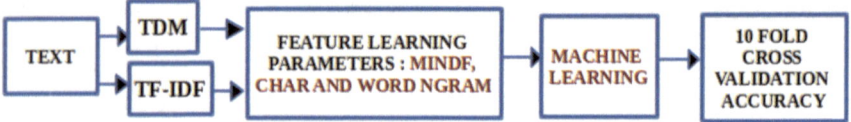

Fig. 1 Methodology for gender identification

each from male and female authors, having 20 sentences in each file. This split was decided upon with reference to the INLI PAN at FIRE 2017 corpora [7].

3.2 Methodology for Gender Identification

The basic approach used for gender identification is as given in Fig. 1. The data, in text format, is represented either as a TDM or as a TFIDF matrix. The aspects of this text are further learned using minimum document frequency, word n-grams, and character n-grams. Then the classification is done using Machine Learning algorithms like SVM, Naive Bayes, Logistic Regression, Decision Tree and Random Forest. A tenfold cross-validation has been done, and the accuracy is recorded and observed.

3.3 Text Representation and Feature Learning

The first step of classification is to represent text in a format on which Machine Learning techniques can be easily applied [8]. Here, TFIDF matrix and TDM have been used to represent the data.

3.3.1 Term Frequency–Inverse Document Frequency (TFIDF)

In TFIDF matrices, the documents are depicted in the rows and columns depict words. It gives more value to unique words by dividing each word's occurrence by the sum of the occurrences of the word in all documents. It can be mathematically expressed as

$$TF\text{-}IDF = \log \frac{N}{n_i} \tag{1}$$

where N is the number of documents under consideration and n_i is the number of times the ith word is occurring in the document.

3.3.2 Term Document Matrix (TDM)

In TDMs also, the rows depict documents whereas the columns depict words. Here, the count of occurrence of words in each document is kept as such. It can be expressed mathematically as

$$TDM = \frac{f_{t,d}}{\sum_{t'd} f_{t',d}} \tag{2}$$

where $f_{t,d}$ is the frequency of the term 't' in document 'd'.

3.3.3 Parameters for Feature Extraction

The parameters utilized for extraction of features are as follows:

1. Minimum Document Frequency (*mindf*)—If mindf is 'm', the terms occurring in at least 'm' documents are considered. The values used for mindf are 1 and 2.
2. N-gram—If ngram is 'n' it means that 'n' continuous occurrences of words or characters form a single term in the matrix of the text. Unigram, bigram, trigram, and tetragram have been used here.
3. Analyzer—By default, word-by-word analysis is done. Character analysis has also been done here.

Depending on how these parameters change, the size of TDM and TFIDF will change. Character unigram analysis will have all characters in the entire corpora as individual terms in the matrix. Word bigrams or trigrams will combine two or three words respectively; hence the size of the matrix will reduce.

3.4 Classifiers for Gender Identification

Here, Machine Learning has been used for gender identification. Even though Deep Learning has scored well in analyzing speech, stock market data, etc., Machine Learning and traditional methods still score well when it comes to text analytics [9]. The algorithms used are SVM, Naive Bayes, Logistic Regression, Decision Tree and Random Forest. Finally, tenfold cross-validation is done on the data, and the accuracy is recorded and analyzed.

4 Experiments and Results

The tenfold cross-validation results have been calculated for Data split I, Data split II, and Data split III. It has been observed that SVM scores well for all the three categories.

4.1 Experimental Results: Data Split I

Table 2 shows the results for the different feature learning parameter values for mindf and character and word n-grams, for the imbalanced dataset of 43 males and 79 females. Here, it can be observed that SVM gives the highest accuracy for both cases 1 and 3. In case 2, SVM closely follows Logistic Regression that scores the highest in accuracy of gender identification.

4.2 Experimental Results: Data Split II

Table 3 shows the results for the different feature learning parameter values for mindf and character and word n-grams, for the balanced dataset of 43 male and female authors each. In case 1, when using word ngram, SVM gives the highest accuracy, whereas in case 2, Logistic Regression gives the highest accuracy.

Table 2 Tenfold cross-validation accuracy of imbalanced dataset: 100 sentence split

Case 1: mindf = 1, word n-gram range = (1, 1)

Text repn	Classifier				
	SVM	NB	LR	DT	RF
TDM	94.15	84.10	94.01	80.35	65.35
TFIDF	94.03	84.10	75.04	78.46	69.20

Case 2: mindf = 2, word n-gram range = (2, 3)

Text repn	Classifier				
	SVM	NB	LR	DT	RF
TDM	85.91	81.78	87.51	74.62	64.52
TFIDF	81.80	81.72	65.29	71.27	64.52

Case 3: mindf = 2, char n-gram range = (2, 3)

Text repn	Classifier				
	SVM	NB	LR	DT	RF
TDM	96.79	82.56	96.02	84.33	72.67
TFIDF	70.09	82.44	64.52	79.36	71.63

Table 3 Tenfold cross-validation accuracy of balanced dataset: 100 sentence split

Case 1: mindf = 1, word n-gram range = (1, 1)

Text repn	Classifier				
	SVM	NB	LR	DT	RF
TDM	92.75	92.75	93.75	61.69	72.91
TFIDF	95.00	93.00	89.00	70.63	73.77

Case 2: mindf = 2, char n-gram range = (2, 3)

Text repn	Classifier				
	SVM	NB	LR	DT	RF
TDM	95.25	85.38	96.50	71.52	71.13
TFIDF	89.13	86.63	89.13	76.02	67.25

Table 4 Tenfold cross-validation accuracy of balanced dataset : 20 sentence split

Case 1: mindf = 1, word n-gram range = (1, 1)

Text repn	Classifier				
	SVM	NB	LR	DT	RF
TDM	93.27	91.09	93.73	76.73	65.35
TFIDF	94.64	88.79	92.32	75.54	64.20

Case 2: mindf = 1, word n-gram range = (1, 2)

Text repn	Classifier				
	SVM	NB	LR	DT	RF
TDM	93.04	92.04	92.57	77.66	59.46
TFIDF	94.65	90.65	90.45	74.22	63.23

Case 3: mindf = 1, char n-gram range = (2, 3)

Text repn	Classifier				
	SVM	NB	LR	DT	RF
TDM	93.27	78.82	93.25	78.75	73.82
TFIDF	94.43	77.44	91.62	75.05	74.65

Case 4: mindf = 1, char n-gram range = (2, 4)

Text repn	Classifier				
	SVM	NB	LR	DT	RF
TDM	93.96	87.39	93.48	79.90	73.92
TFIDF	95.81	86.01	91.85	74.79	68.27

4.3 Experimental Results: Data Split III

Tables 4 shows the results for gender identification of code-mixed Malayalam–English corpora for the different feature learning parameter values for mindf and character and word n-grams, for the balanced dataset of 215 files each of male authors and female authors. Each file contains 20 sentences.

Hence, it has been observed that for all datasets, SVM performs well. SVM, being an algorithm that incorporates optimization, has always scored well in the area of classification of data. Logistic Regression is a close competitor for SVM for this corpora.

5 Conclusion and Future Work

Code-mixed Malayalam–English data from WhatsApp has been collected for gender identification. The main challenges faced during data collection was the lack of data for males, as well as the manual cleaning of the data. The cleaned data has been compiled as Data split I, Data split II, and Data split III. Data split I and split II uses a 100 sentence split per file, whereas Data split III uses a 20 sentence split. Data splits II and III are balanced, the former having 43 files each, and the latter having 215 files each of male and female authors' data. Data split I is imbalanced, with 43 files from male authors and 79 files from female authors. This data is represented as TDM and TFIDF matrix, on which Machine Learning algorithms have been applied for gender classification. The tenfold cross-validation results show that SVM gives the highest accuracy. Further, word embedding can be used, and the data can be classified based on the age group. Socio-linguistic features of the data can also be analyzed [10]. If the corpora size is increased, Deep Learning techniques can also be applied for author profiling.

References

1. Rangel F, Rosso P, Potthast M, Stein B (2017) Overview of the 5th author profiling task at PAN 2017: gender and language variety identification in Twitter. In: CLEF 2017
2. Basile A, Dwyer G, Medvedeva M, Rawee J, Haagsma H, Nissim M (2017) N-GrAM: new groningen author-profiling model, Notebook for PAN at CLEF 2017. In: CLEF 2017
3. Martinc M, Krjanec I, Zupan K, Pollak S (2017) PAN 2017: author profiling-gender and language variety prediction, Notebook for PAN at CLEF 2017. In: CLEF 2017
4. Tellez ES, Miranda-Jimnez S, Grafi M, Moctezuma D (2017) Gender and language-variety identification with MicroTC, Notebook for PAN at CLEF 2017. In: CLEF 2017
5. Barathi Ganesh HB, Anand Kumar M, Soman KP (2017) Vector space model as cognitive space for text classification, Notebook for PAN at CLEF 2017. In: CLEF 2017
6. Chacko VR, Anand Kumar M, Soman KP (2018) Experimental study of gender and language variety identification in social media. In: Proceedings of the second international conference on big data and cloud computing. Advances in intelligent systems and computing (AISC). Springer
7. Anand Kumar M, Barathi Ganesh HB, Singh S, Soman KP, Rosso P (2017) Overview of the INLI PAN at FIRE-2017 track on Indian native language identification. In: CEUR workshop proceedings, vol 2036, pp 99–105
8. Barathi Ganesh HB, Anand Kumar M, Soman KP (2016) Distributional semantic representation for text classification and information retrieval. In: CEUR workshop proceedings, vol 1737, pp 126–130

9. Medvedeva M, Kroon M, Plank B (2017) When sparse traditional models outperform dense neural networks: the curious case of discriminating between similar languages. In: Proceedings of the fourth workshop on NLP for similar languages, varieties and dialects. Association for computational linguistics, pp 156–163
10. Rangel F, Franco-Salvador M, Rosso P (2016) A low dimensionality representation for language variety identification. In: CICLing-computational linguistics and intelligent text processing

Explicit Decision Tree for Predicating Impact of Motivational Parameters on Self Satisfaction of the Students

Aniket Muley, Parag Bhalchandra and Govind Kulkarni

Abstract The decision trees are widely accepted as a novel tool for supervised classification and prediction. We have implemented a decision tree over own educational dataset to get insights for predictive academic analytics, which otherwise are invisible. The educational dataset contains personal and socioeconomical variables related to students. It is known that individuality, lifestyle, and responsiveness-related variables have a close association with motivational aspects which together harshly affect student's performance. The decision tree is deployed to escalate the role of motivational variables on self-satisfaction aspects. Analytics were carried out with R software package.

Keywords Educational analytics · Decision trees · Classification · Prediction · Data mining

1 Introduction

This paper is a preliminary endeavor for applied analytics over student's academic data. This analysis is interesting and can augment the quality of the higher education. Student's raw data regarding course preference, results, further progression, are crucial capital for all higher educational organizations. Data mining can be applied to such databases in order to gain challenging outputs [1]. This is called as Educational Data Mining where we primarily investigate analytics for good insights [2]. These insights are in terms of associations, correlations, clusters, etc. Despite the

A. Muley
School of Mathematical Sciences, SRTM University, Nanded, MS, India
e-mail: aniket.muley@gmail.com

P. Bhalchandra (✉) · G. Kulkarni
School of Computational Sciences, SRTM University, Nanded, MS, India
e-mail: srtmun.parag@gmail.com

G. Kulkarni
e-mail: govindcoolkarni@gmail.com

© Springer Nature Singapore Pte Ltd. 2019
H. S. Saini et al. (eds.), *Innovations in Computer Science and Engineering*, Lecture Notes in Networks and Systems 74,
https://doi.org/10.1007/978-981-13-7082-3_14

percentage of GDP for the year 2013–14 was 1.34% [3], the academic performance of students in India is not improving. There could be many reasons behind it. Self-satisfaction of students regarding the courses they are pursuing can also beone of the reasons. Since these aspects are not visible directly, people have not tackled for the same. This is the right case for data mining explorations. Self-satisfaction is also related to a number of other aspects including academic ambience, easiness in pursuing course, interested areas, family support, etc. These aspects can be categorized as self-related, family-related, motivational, and financial concerns. Since we need to precisely understand the main concern behind self-satisfaction, we have worked out for all these categories. Financially deprived students can secure good ranks. This reflects self-satisfaction of students and also highlights that financial aspects and family support aspects are less significant. Hence, family-related and financial aspects cannot be significantly impactful. Of other remaining aspects, self-related and motivational aspects are more important. The enthusiasm and passion to do something in life is the main motivational catalyst. If we apply analytics for elaboration of this, then it will be very interesting to see the exact role of motivational aspects on self-satisfaction. This work is a multiparty work undertaken together by faculties of educational, computational and statistical sciences. The key idea of this study is to escalate collected datasets through data mining. The minor research objective is to investigate whether motivational parameters significantly contribute to the increase or decrease of self-satisfaction of students? If no, what other variables are related to the same? The below sections brief experimental setup and discussions related to such performance analysis.

2 Research Methodology

Every study from data analytics needs a dataset for implementation of algorithms. This dataset must match with the context of analytical exploration. To do so, we have taken efforts to tailor our own real dataset related to personal, socioeconomical, habitual, and self-related aspects of students of our university. This dataset has 360 records. Every record has 46 fields, each one related to some information/fact about a student. The closed questionnaire is the question with predefined answers, and the method was followed for the creation of this dataset [4]. A standard benchmark illustrated in Pritchard and Wilson [4] was adopted for the same. Some contemporary works [5–8] were also reviewed discover important variables from our dataset/questionnaire. Some preselected students were interacted during finalization of the questionnaire and in total; four trial testing have been made to formulate the concluding edition of the questionnaire. Excluding name details, class details, and roll number details, the concluding questionnaire consisted of 43 variables. The MS Excel 2010 software was used to testimony the dataset. Standard policies for data preparation and cleansing were done [1] as per the requirements of data mining experiments. We cannot study all questions in the questionnaire and test for their association with other questions. This will be exhaustive and divert us from the main rationale [9]. Our applied work must lead to a new dimension [10–12]. That is why, we tried to examine the relationship between student's self-satisfaction and motivational aspects.

All implementations were carried out on R data mining platform. The supervised learning methods called decision trees and their algorithms were implemented using R open source software [13]. The *Rattle* is a free graphical user interface for Data Science, developed using R. The decision tree model exploits a recursive partitioning approach. Its traditional algorithm and default part of *rpart* package of R. This ensemble approach tends to produce complication models than a single decision tree. Classification definitely needs a potential data which can be classified by setting some rules [1]. There are two crucial steps to do so, first learning and second is classification itself. The system is trained first with sample data and once the system learns, test data is supplied to the system to check accuracy. If the accuracy is good enough, the rules can be practical for the new data [1]. This workout can be frequently done via a decision tree or a neural network for the actual workout [1, 2]. A decision tree has branches and nodes [1, 14]. It contains split, which tests the value of an appearance of the attributes. The outcome of the test is labeled on edges. A class label linked with every leaf node [1, 14]. The given entropy as appraise of the impurity is then defined. This calculation is called information gain. The section two below summarizes all these considerations.

3 Experimentations and Discussions

To devise the decision tree, we tried out some set of experiments. Our aim is to evaluate the decision tree induction method. To draw the decision tree, *rattle* uses a command *rpart* from *rpart* package. We obtained the rules associated to the tree and the summary of the Decision Tree model for Classification is given in Table 1. Table 1 represents the root node, level of trees, split of the observations, and loss of frequency at each level split and corresponding probability values. The * denotes terminal node. The total number of population selected was n = 251.

The terminal node variables actually used in the tree construction are: CAREER-DREM, F_T_FRIEND, F_T_STUDY, PLACELVING, REGION, and SELF.LIB. Initially, the root node error is 0.25498 (n = 251). Some of the generated tree rules can be illustrated as follows:

1. Rule number: 3 [PER_SATISF = 1 cover = 17 (7%) prob = 0.94], PLACE-LIVNG >= 3.5
2. Rule number: 19 [PER_SATISF = 1 cover = 25 (10%) prob = 0.92], PLACELV-ING < 3.5, F_T_STUDY >= 2.5, CAREERDREM < 3.5, F_T_FRIEND >= 2.5
3. Rule number: 37 [PER_SATISF = 1 cover = 7 (3%) prob = 0.86], PLACELV-ING < 3.5, F_T_STUDY >= 2.5, CAREERDREM < 3.5, F_T_FRIEND < 2.5, F_T_FRIEND < 1.5

Table 1 Details of classification of nodes

Variable	Nodes	Children gener- ated	Probability
Root	251	64, 1	(0.25498008 0.74501992)
PLACELVING < 3.5	234	63, 1	(0.26923077 0.73076923)
F_T_STUDY >= 2.5	76	27, 1	(0.35526316 0.64473684)
CAREERDREM >= 3.5	16	6, 0	(0.62500000 0.37500000) *
CAREERDREM < 3.5	60	17, 1	(0.28333333 0.71666667)
F_T_FRIEND < 2.5	35	15, 1	(0.42857143 0.57142857)
F_T_FRIEND >= 1.5	28	14, 0	(0.50000000 0.50000000)
SELF.LIB < 0.5	20	8, 0	(0.60000000 0.40000000)
REGION < 1.5	13	4, 0	(0.69230769 0.30769231) *
REGION >= 1.5	7	3, 1	(0.42857143 0.57142857) *
SELF.LIB >= 0.5	8	2, 1	(0.25000000 0.75000000) *
F_T_FRIEND < 1.5	7	1, 1	(0.14285714 0.85714286) *
F_T_FRIEND >= 2.5	25	25, 2, 1	(0.08000000 0.92000000) *
F_T_STUDY < 2.5	158	158, 36, 1	(0.22784810 0.77215190) *
PLACELVING >= 3.5	17	17, 1, 1	(0.05882353 0.94117647) *

4. Rule number: 5 [PER_SATISF = 1 cover = 158 (63%) prob = 0.77], PLACELV-
   ING < 3.5, F_T_STUDY < 2.5
5. Rule number: 73 [PER_SATISF = 1 cover = 8 (3%) prob = 0.75], PLACELV-
   ING < 3.5, F_T_STUDY >= 2.5, CAREERDREM < 3.5, F_T_FRIEND < 2.5,
   F_T_FRIEND >= 1.5, SELF.LIB >= 0.5.

To judge the accuracy of the classifier, we worked out for generation of the error
matrix. This matrix is generated after validating the proportions. It is evident that the
overall error is 22.6% and averaged class error is 41.1% (Table 2).

Table 2 Error matrix

		Predicted		Error
		0	1	
Actual	0	5.7	15.1	72.7
	1	7.5	9.5	9.5

Table 3 Accuracy of results

Train	Test	Validate
0.6894	0.6157	0.5325

The area under the ROC curve for the *rpart* model is computed for 70% training set data, 15% test data and balance 15% validation data. The accuracy of the result was summarized in below Table 3. The results validated simply state that the random selection is responsible for the accuracy level of the data.

The output decision tree is shown in Fig. 1. In order to interpret Fig. 1, we look at the dataset variables and determine crucial and important variables, and then comes up with a node, and so on. The tree is shaped by splitting data up by variables and then counting to see how many are in each bucket after each split.

The decision tree model explores that the satisfaction of student is directly dependent on his/her place of living, and then in the next stage, it depends on free time available for study. Further, it classifies free time available to spend with friends, and then splits in terms of availability of having a self-library with the student. Further splitting is done through regional classification. Finally, there is no classification. This means the final induction of decision tree has been reached. All splitting parameters are playing an important role in the identification of the satisfaction of students. Further, it is suggested that if we can properly focus only on these parameters of individuals, significant improvement in their satisfaction level could be witnessed.

4 Conclusions

It was overtly unspoken that numerous societal, routine and many other aspects are linked with the satisfaction level of students and mere good performance cannot be judged by studious nature alone. In order to escalate these, the study took it as confront and drew a decision tree for selected parameters. Using the interdisciplinary approach, we found that, besides studies, other aspects like the student's satisfaction did depend on self-motivation. This work highlights that the place of living significantly matters related to the satisfaction of the students. Second such variable boosting self-motivation is the quantum of the free time for study. The self-career dream is the third major motivational variable. If prognostication methods are implemented, then change in place of living, motivational canceling can boost student's performance.

Decision Tree d2.csv $ PER_SATISF

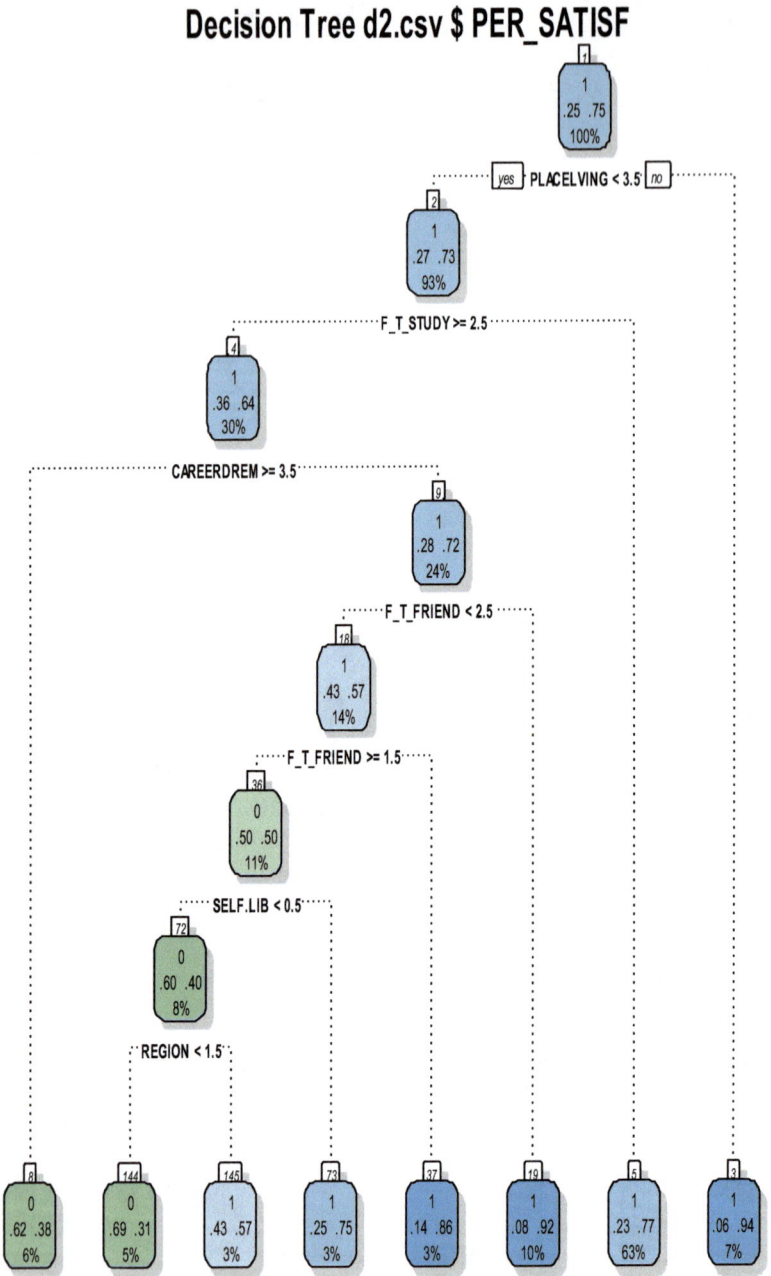

Rattle 2018-Jun-25 18:40:15 hp1

Fig. 1 Final Decision Tree

References

1. Dunham M (2002) Data mining: introductory and advanced topics, by Margaret H. Dunham, Pearson Publications
2. Han J, Kamber M (2006) Data mining: concepts and techniques, 2nd edn. The Morgan Kaufmann Series in Data Management Systems, Jim Gray
3. https://economictimes.indiatimes.com
4. Pritchard ME, Wilson GS (2003) Using emotional and social factors to predict student success. J Coll Stud Dev 44(1):18–28
5. Ali S et al (2013) Factors contributing to the students academic performance: a case study of Islamia University Sub-Campus. Am J Educ Res 1(8):283–289
6. Graetz B (1995) Socio-economic status in education research and policy in John Ainley et al., socio-economic status and school education DEET/ACER Canberra. J Pediatr Psychol 20(2):205–216
7. Considine G, Zappala G (2002) Influence of social and economic disadvantage in the academic performance of school students in Australia. J Sociol 38:129–148
8. Bratti M, Staffolani S (2002) Student time allocation and educational production functions. University of Ancona Department of Economics Working Paper No. 170
9. Field A (2000) Discovering statistics using R for Windows. Sage Publications
10. Joshi M, Bhalchandra P, Muley A, Wasnik P (2016) Analyzing students performance using academic analytics. In: IEEE international conference ICT in business industry and government (ICTBIG), pp 1–4
11. Muley A, Bhalchandra P, Joshi M, Wasnik P (2016) Prognostication of student's performance: factor analysis strategy for educational dataset. Int J 12
12. Muley A, Bhalchandra P, Joshi M, Wasnik P (2018) Academic analytics implemented for students performance in terms of Canonical correlation analysis and chi-square analysis. Inform Commun Technol. Springer, Singapore, pp 269–277
13. Web resource. https://cran.r-project.org/web/packages/rattle/vignettes/rattle.pdf
14. Rokach L, Maimon O (2014) Data mining with decision trees: theory and applications. World Scientific

Improved Clusters in Time Series Data Mining Using Hybridized Adaptive Fuzzy C-Means

J. Mercy Geraldine and S. Jayanthi

Abstract Temporal data is time-based data and real-time applications involve such data. Clustering time-based data improves the efficiency of the frequent itemsets obtained. An existing method for improving the clusters used Ant Colony Optimization (ACO) in FCM. The major drawbacks of ACO are that the search is not fast and it is dependent on many parameters. In this paper, we have proposed the Hybridized Adaptive Fuzzy C-Means (HAFCM) approach, which combines the best features of ACO and Cuckoo Search (CS) in FCM for cluster refinement. This makes search faster and also leads to optimized clusters with better cluster efficiency. The resulting clusters are validated using the Partition Coefficient and Exponential Separation (PCAES) validity index. The proposed method has better performance when applied to real time datasets.

Keywords Temporal data · Hybridized adaptive fuzzy c-means (HAFCM) · Ant colony optimization (ACO) · Cuckoo search (CS) · Cluster refinement · Partition coefficient and exponential separation (PCAES)

1 Introduction

Data mining also known as knowledge discovery extracts information from huge volumes of data [1–3]. Data mining is concerned with the analysis of large volumes of data to discover interesting regularities or relationships in the data. Clustering is the process of assigning data into groups or clusters in such a manner that similar data items are present in the same group and dissimilar data items belong to different groups. Clustering of sequences or time series deals with the grouping of time series

J. M. Geraldine (✉)
Guru Nanak Institute of Technology, Hyderabad, Telangana, India
e-mail: mercygeraldine@gmail.com

S. Jayanthi
Samskruti College of Engineering and Technology, Hyderabad, Telangana, India
e-mail: nigilakash@gmail.com

© Springer Nature Singapore Pte Ltd. 2019
H. S. Saini et al. (eds.), *Innovations in Computer Science and Engineering*, Lecture Notes in Networks and Systems 74,
https://doi.org/10.1007/978-981-13-7082-3_15

119

data or sequences on the basis of similarity [4–7]. The applications where time series clustering is appropriate are activity logs in the web, navigation patterns of user groups, clustering of biological sequences, weather forecasting, clustering of stock exchange data, etc.

A fuzzy clustering algorithm is one of the most widely used clustering algorithms. The clusters formed may not be of 100% quality due to misclustering. Therefore, clusters have to be refined and validity indices have to be used for evaluating the quality of the clusters [8, 9]. In FCM and its variations, it is necessary to previously assume the number 'c' of clusters. In general, the number 'c' is unknown. The problem of finding an optimal 'c' is usually called cluster validity (Bezdek).

In our proposed HAFCM method, initially, the time series database values are clustered using FCM. These clusters are refined using an optimization approach. The clusters formed are validated using the PCAES validity index, which indicates clusters with improved quality. The remaining sections of the paper are as follows: Sect. 2 reviews the literature work related to the proposed technique. Section 3 briefs the proposed technology, Sect. 4 shows the experimental results and Sect. 5 concludes the paper.

2   Review of Related Research Works

Mary and Raja [9] described a cluster refinement method to improve the clusters formed by FCM. These clusters were validated using Xie–Beni validity index. Ants were used to refine the clusters rather than clustering the data points. The clusters formed as a result of applying FCM algorithm to the data were provided as input to the cluster refinement process which used ACO. This method was found to produce better quality clusters than the standard FCM algorithm.

Babukartik et al. [10] proposed an algorithm that used the advantages of ACO and CS for job scheduling. The problem was defined as that there were N jobs and M machines. Each job had a unique order of execution that was to be performed on M machines. The search had to be performed faster in any optimization problem. But ACO was slow in searching because the ant walked through the path where pheromone was deposited, which may also lure the ants. This was overcome by using CS to do the local search.

Wu and Yang [8] described the cluster validity indices that have been used to evaluate the fitness of partitions produced by clustering algorithms. A validity index called PCAES was described, which considered the compactness and separation measure of each cluster to evaluate the cluster quality [11]. When compared with other validity indices, PCAES had the capability of producing a good estimation of the number of clusters and was also applicable for computing cluster validity in noisy environments.

3   Improved Clusters Using Hybridized Adaptive Fuzzy C-Means

The clusters formed using any clustering algorithm may not lead to 100% efficiency. In order to improve the quality of the clusters formed, the proposed HAFCM approach is used for cluster refinement. Before clustering, the time series data are preprocessed and converted into itemset values. These values are clustered using the standard FCM algorithm. The clusters formed are refined using the best features of two optimization algorithms. This leads to efficient association rule mining resulting in optimized frequent itemsets. Figure 1 shows the architecture of the proposed cluster refinement technique.

3.1   Preprocessing of Time Series Data

Consider the database (D), which has the size of $M \times N$ where $M \in 1, 2, \ldots, m$ denotes the number of rows and $N \in 1, 2, \ldots, n$ denotes the number of columns and it has numerical data. The database (D) is converted into itemset database (D') which has itemsets only. From the database D, data $d \in D$ is taken with the size of i x j where $i \in m$ and $j \in n$ and the max (d_i), min(d_i) are computed. The median value is calculated using Eq. (1).

$$Med(d_i) = (max(d_i) + min(d_i))/4. \tag{1}$$

Four values are chosen in order to convert the numerical data into itemset data namely, Very Low, Low, High, and Very High. Consider the numerical data as X.

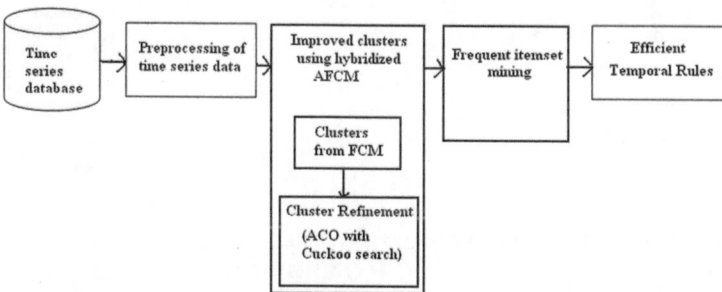

Fig. 1   The architecture of the proposed cluster refinement technique in temporal rule mining

$$\min(d_i) >= X <= \text{Med}(d_i) \quad \text{Very Low (VL)}$$
$$VL > X <= \text{Med}(d_i) \times 2 \quad \text{Low (L)}$$
$$L > X <= \text{Med}(d_i) \times 3 \quad \text{High (H)}$$
$$H > X <= \max(d_i) \quad \text{Very High (VH)}$$

3.2 Cluster Refinement Using ACO and CS

Any fuzzy clustering algorithm is not capable of producing 100% quality clusters. The clusters resulting from the FCM method are given as input to the hybridized method for cluster refinement. This method combines the best features of ACO and CS optimization algorithms. In the proposed methodology, ants are used to refine the clusters. The major problem of ACO is the ant walks through the path where pheromone, which is a chemical substance is deposited. This may be an act of luring the ants. This limitation of ACO is avoided by using CS in ACO. CS is used for performing the local search more efficiently and also it involves only a single parameter besides population size.

CS algorithm is a metaheuristic algorithm, which is inspired by the breeding behavior of the cuckoos [10]. Cuckoos lay their eggs in communal nests by removing the eggs of host birds, thereby increasing the hatching probability of their own eggs. The CS algorithm is given below:

Initialization: Primarily, the population of host nest hn_i, $i \in 1, 2, \ldots, s$ is initiated erratically. Here, s is the size of the population. $s = 2 \times Q$ where Q denotes the number of clusters.

Generating new cuckoo: A cuckoo is picked at random and it generates new solutions using levy flights. After that, the generated cuckoo is evaluated using the objective function for determining the quality of the solutions.

Fitness function: The fitness function (F) of all the nests are computed by using Eq. (2).

$$O = \sum_{i=1}^{N} \sum_{j=1}^{ce} me_{ij}^m \|x_i - Ce_j\|^2 \tag{2}$$

where

me_{ij} membership of the ith data to the jth cluster,
x_i ith data,
Ce_j centroid of the jth cluster,
N number of data points.

The quality of the solution is evaluated and a nest is selected among 's' arbitrarily. If the quality of the new solution in the selected nest is better than the old solutions, it will be replaced by the new solution (Cuckoo). Otherwise, the previous solution is kept as the best solution.

Discard worst nest: In this part, the worst nests are discarded based on their probability p_a values and new ones are built. Then, based on their fitness function the best solutions are ranked. Then the best solutions are identified and marked as optimal solutions.

Stopping Criterian: This process is repeated until the maximum iteration is reached. Finally, the best-clustered dataset is obtained.

Hybridized cluster refinement technique

1. *Obtain the clusters and its centers from FCM.*
2. *Refinement using Hybridized method*

   (i) *Input the FCM clusters and calculate the PCAES index value for all these clusters.*

   (ii) *The ant picks an item one at a time from some cluster by a random walk and drops the item into some other cluster.*
   In this, the choice of clusters for picking the right item and dropping the item, respectively, is performed by a local search by CS algorithm.

   (iii) *Check if the quality has improved or not by calculating PCAES index value after dropping the item.*

   (iv) *Drop the item permanently if the quality improves (that is, if the calculated PCAES index value for that cluster is larger than the previous value for the same cluster) and update the corresponding cluster's PCAES index value. Continue by repeating the steps (ii) to (iv) for a random number of times. Otherwise, repeat the steps (ii) to (iv) for a random number of times with the cluster having unchanged PCAES index value.*

4 Experimental Results

Two time series real datasets have been considered for computing the PCAES index and also for applying the refinement process. The rainfall dataset contains 5-year rainfall data in millimeters recorded on monthly basis from 2004 to 2008 for different districts of Tamil Nadu, India. The dataset has been obtained from the Department of economics and statistics, Government of Tamil Nadu. The temperature dataset contains four years temperature data in degrees recorded on day to day basis for various districts of Tamil Nadu. It has been obtained from the Indian Meteorological Department (IMD) governed by the Ministry of Earth Sciences.

The larger the PCAES index value, the more compact and well separated are the clusters. Table 1 shown above displays the PCAES index values for the clusters resulting from the FCM algorithm, the existing ACO cluster refinement method, and the proposed hybridized cluster refinement method. It is also found that the clusters obtained from the proposed refinement method are of better quality. This is identified by the large PCAES index value. In Table 2, the time taken for refining the clusters

Table 1 PCAES index value for the obtained clusters

Real dataset	No. of clusters	PCAES index value		
		FCM	ACO cluster refinement	Hybridized cluster refinement
Rainfall	4	2.6025	2.8322	3.0025
Temperature	4	2.5955	2.8562	3.0152

Table 2 Refinement time analysis

Real datasets	Refinement time (s)	
	ACO	Hybridized
Rainfall	2.6	1.1
Temperature	2.3	1.0

by the existing ACO cluster refinement method and the proposed hybridized cluster refinement method is shown.

5 Conclusions

The proposed HAFCM method refined the clusters produced by FCM and resulted in improved clusters with a large PCAES index value than the existing ACO refinement method. The PCAES validity index was found to identify highly compact and well-separated clusters than the other cluster validity indices. Also, the refinement time taken by the proposed method was comparatively less than the refinement time taken by the existing method. High-quality clusters lead to efficient frequent itemsets and strong association rules. In future, refining clusters for high-dimensional data can be considered, which will be a research problem of interest to the emerging researchers in the field of data mining.

References

1. Nishi MA, Ahmed CF, Samiullah M, Jeong BS (2013) Effective periodic pattern mining in time series databases. Expert Syst Appl 40(8):3015–3027
2. Mabroukeh NR, Ezeife CI (2010) A taxonomy of sequential pattern mining algorithms. ACM Comput Surv 43(1):1–41
3. Lee YJ, Lee JW, Chai DJ, Hwang BH, Ryu KH (2009) Mining temporal interval relational rules from temporal data. J Syst Softw 82(1):155–167
4. Hong TP, Wang CY, Tseng SS (2011) An incremental mining algorithm for maintaining sequential patterns using pre-large sequences. Expert Syst Appl 38(6):7051–7058
5. Fuchs E, Gruber T, Pree H, Sick B (2010) Temporal data mining using shape space representations of time series. Neuro Comput 74(1):379–393

6. Gharib TF, Nassar H, Taha M, Abraham A (2010) An efficient algorithm for incremental mining of temporal association rules. Data Knowl Eng 69(8):800–815

7. Wu HW, Lee AJT (2010) Mining closed flexible patterns in time-series databases. Expert Syst Appl 37(3):2098–2107

8. Wu KL, Yang MS (2005) A cluster validity index for fuzzy clustering. Pattern Recognit Lett 26:1275–1291

9. Mary CI, Raja SVK (2010) Improved fuzzy C means clusters with ant colony optimization. Int J Comput Sci Emerg Technol 1(4):1–6

10. Babukartik RG, Dhavachelvan P (2012) Hybrid algorithm using the advantage of ACO and cuckoo search for job scheduling. Int J Inform Technol Converg Serv 2(5):51–60

11. Saad MF, Alimi AM (2012) Validity index and number of clusters. Int J Comput Sci Issues 9(3):52–57

Empirical Orthogonal Functions
Analysis of the Regional Indian Rainfall

K. C. Tripathi and Pooja Mishra

Abstract The rainfall over India has attracted the attention of the researchers of a variety of domains such as the meteorology, physics, mathematics, communications, computer science, sociology, geography, and others because of the huge implications involved in all such domains. The rainfall over India is broadly seen at three levels—All India, Regional, and Local. In the present study, the Empirical Orthogonal Functions (EOF) analysis of the rainfall over India at the regional scale has been addressed. The 146-year monthly rainfall records of five regions—North West, West Central, North East, Central North East, and Peninsular India—have been used. It is observed that the five-dimensional data can be effectively reduced in one dimension, on leading EOF, with 80% information retained and in two dimensions, two leading EOFs, with 90% information retained. The implications have been discussed along with future prospects and limitations.

Keywords Principal components · All India rainfall · Prediction · Eigenvectors · Variance

1 Introduction

The rainfall over Indian subcontinent has attracted the attention of researchers since long because of the huge socioeconomical impact it has over a large fraction of the world population. India being an agricultural nation, a major portion of its population is dependent on agricultural products which, in turn, depend on rainfall. Accurate

K. C. Tripathi (✉)
Department of Computer Science and Engineering, Inderprastha Engineering College, Ghaziabad, India
e-mail: Krishna.tripathi@ipec.org.in

P. Mishra
Department of Electronics and Communication Engineering, Inderprastha Engineering College, Ghaziabad, India
e-mail: pooja.mishra@ipec.org.in

© Springer Nature Singapore Pte Ltd. 2019
H. S. Saini et al. (eds.), *Innovations in Computer Science and Engineering*, Lecture Notes in Networks and Systems 74, https://doi.org/10.1007/978-981-13-7082-3_16

and timely forecast of the rainfall can help this population cope up with their losses owing to natural disasters. Further, the timely forecast of flood and drought help the disaster management agencies in advance preparations for rescue operations.

The rainfall in the Indian subcontinent is seen at three levels in the spatial domain—the All India rainfall, the regional rainfall, and the Local rainfall and at three levels in the temporal domain—annual, seasonal, and monthly. The seasonal rainfall is classified in four categories: (i) January and February, (ii) March, April, and May (MAM), (iii) June, July, August, and September (JJAS or the monsoon season), and (iv) October, November, and December (OND) [1]. Studies have been done in the past to analyze the precipitation in these seasons [2–4]. The Indian Summer Monsoon Rainfall and allied phenomena have also been at the center of meteorological and computational studies [5–7]. The prediction of ISMR has been done by statistical [8, 9] as well as dynamical models [10, 11].

The Empirical Orthogonal Functions (EOF) analysis or the Principal Components Analysis (PCA) [12] is used for dimension reduction of multidimensional data. The meteorologists employ this to eliminate the redundant information present in the original data space and retain only those features which carry major information content. Some recent applications include the EOF analysis of sea surface temperatures [13–15] and rainfall [16, 17].

The present study aims to revisit the EOFs of the regional rainfall over India. We have used the data set of the Indian Institute of Tropical Meteorology (IITM) obtained from the official website (http://www.tropmet.res.in/Data%20Archival-51-Page). The data set comprises monthly rainfall with a resolution of up to 1 decimal in mm. The following records are available: (i) 146-year All India rainfall, (ii) rainfall of 30 subdivisions of India during the period 1871–2016, and (iii) rainfall of 5 homogeneous rainfall regions. It is the (iii) set that is being looked upon in the present study. As the above data set is the latest available data set of homogeneous regions, the analysis presents the current state of the affairs. The regions that comprise the data set are: (i) North West (NW), (ii) West Central (WC), (iii) North East (NE), (iv) Central North East (CNE), and (v) Peninsular India (Pen). The regions have been redefined based on the homogeneity of the rainfall. The present division consists of the five regions listed above instead of the erstwhile seven listed in the 1971–2012 data set.

2 The Method of Principal Components

The Principal Components Analysis (PCA) or the Empirical Orthogonal Functions (EOF) Analysis is a feature extraction technique that ultimately reduces the dimension of the data by rotating the actual data space by a "suitable angle" so as to redistribute the variances in the data in a lesser number of dimensions. The mathematical details of the PCA can be found in the references cited above. Here, the mathematical summary pertaining to the regional rainfall is presented.

The actual data set is of 146 years and 5 regions. Each year has 12 data points. Thus, each region has $12 * 146 = 1752$ data points corresponding to 1752 time steps $(t_1 - t_{1752})$. The data space is a five-dimensional space with each axis (attribute) representing one of the five regions. This is called the D^5 space. The matrix of the data set can be thought of as a $1752 * 5$ matrix. At every time step, the vector $r = (r_1, r_2, r_3, r_4, r_5)$ is a point in the five-dimensional space with r_1 representing the coefficient along the first axis and so on. We consider $r_1 = NW, r_2 = WC, r_3 = NE$, $r_4 = CNE$, and $r_5 = Pen$. Thus, each tuple in D^5 is a point in the five-dimensional space.

The objective of PCA here is to transform D^5 to U^5, using a linear transformation

$$\emptyset : D^5 \rightarrow U^5 \tag{1}$$

where U is the transformed space where the axes are linearly independent. Further majority of the information content of D^5 **is retained in a few dimensions in** U^5. The "information" in statistical processing is represented by the amount of "variance", v, in the data. Let v_j be the variance along D^j in D^5, $1 \leq j \leq 5$. Then

$$v = \sum_j v_j \tag{2}$$

We have

$$Var(D^5) = Var(U^5) = v \tag{3}$$

Also, if p_j is the variance along U^j in U^5, $1 \leq j \leq 5$. Then,

$$v = \sum_j p_j \tag{4}$$

The difference between (2) and (4) is that in (2) the variances are of the same order while in (4), the variances are concentrated in a few p_js and the remaining p_js have a small fraction of v so that the axes having these small variances may be discarded. The entire U^5 space can then be represented by a subspace $U^t, t < 5$. Thus, we have effectively transformed D^5 to U^t, $t < 5$. It can be shown that $\emptyset$ defined in (1) is effected by taking the U^5 to be eigenvectors of the covariance matrix of D^5. The eigenvectors corresponding to largest eigenvalues make the space U^t.

3 Results and Discussion

Figure 1 shows the homogeneous regions of the Indian rainfall as defined by the IITM and described in Sect. 2. The regions shown make the coordinate system U^5. These regions account for about 90% of the total Indian rainfall. Figure 2 shows the

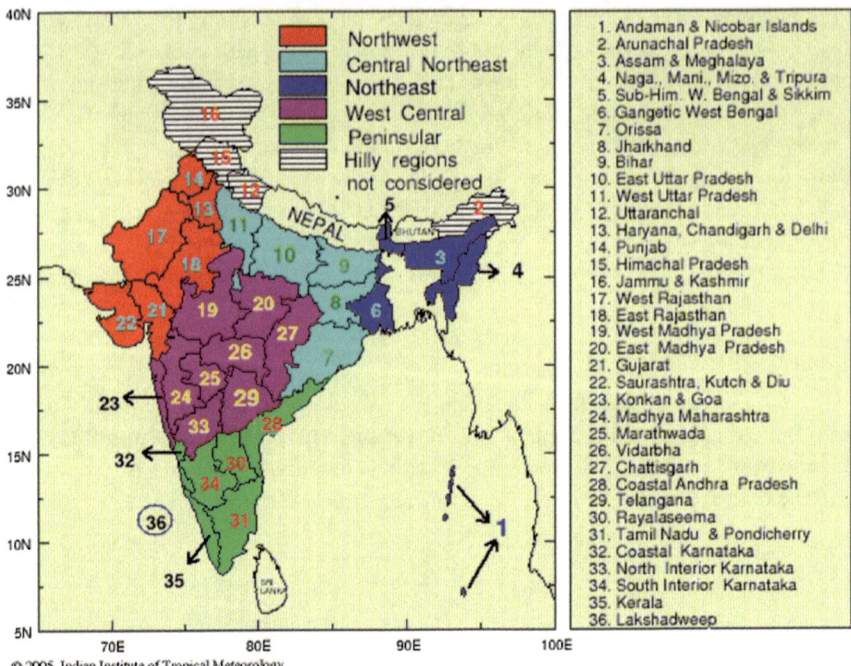

Fig. 1 Regions of homogeneous Indian rainfall

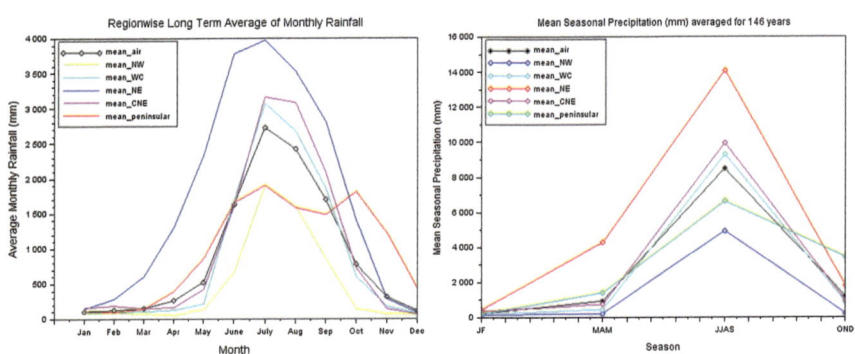

Fig. 2 Climatology of regional rainfall (left) and seasonal rainfall (right)

long-term monthly (left) and seasonal climatology (right) of 146 years of regional and All India rainfall. Bimodal nature of peninsular region rainfall is evident from both the figures. The seasonal variations demonstrate the predominant effect of the monsoon season (JJAS).

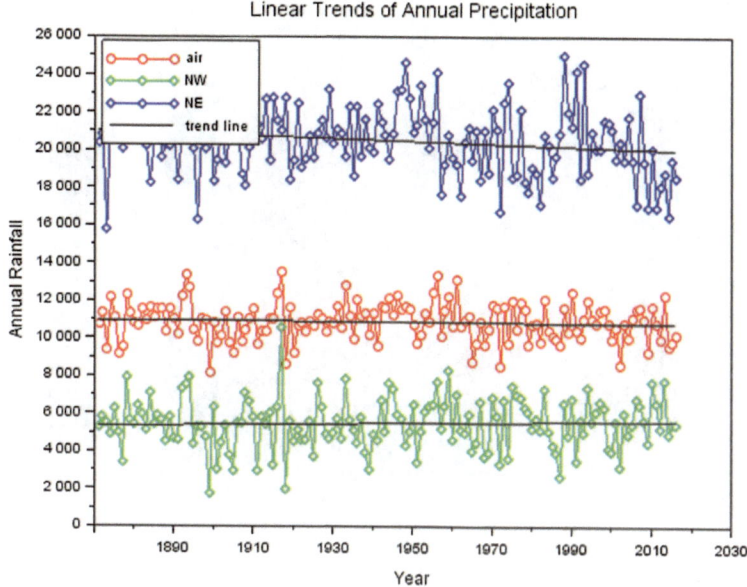

Fig. 3 Trends of the precipitation patterns—All India, NW, and NE

Figure 3 shows the trend of the yearly rainfall at All India and regional level for two dominant regions—NW and NE. It can be seen that the AIR shows a declining trend, which indicates that the amount of precipitation received at the AIR level is slowly decreasing with time.

The EOFs were extracted by creating the correlation matrix and finding the eigenvectors and the eigenvalues thereof. The correlation matrix is plotted in Fig. 4. The abbreviated names of the regions form the axes and the shading forms the correlation coefficient. As can be seen from the figure, the minimum value for the correlation coefficient is 0.55 and the maximum value is 1 (because of autocorrelations). This shows that there is a high degree of association among the rainfalls of this region. Thus, there is a high possibility of transforming these axes to orthogonal system in which the features are not correlated.

Since the original D^5 space is a five-dimensional space, each eigenvector is a point in the D^5 space. Figure 5 (left) shows the eigenvectors and (right) eigenvalues. In the figure (left), the x-axis is the region in space as per the map shown in Fig. 1 and the y-axis represents the strength of contribution by each region. It can be seen that the eigenvector corresponding to largest eigenvalue (i.e., eigenvector 5) has almost the same strength in all the directions, thereby implying that this vector passes through the mean of the D^5 space. The vectors shown in Fig. 5 (left) make the U^5 space. The largest eigenvalue is 4.00 which is corresponding to the fifth eigenvector, EV5, in Fig. 5 (left). This explains 80% of the total variance. Let us call it u_1. The second largest eigenvalue is 0.52 and hence, it accounts for the 52% of the remaining

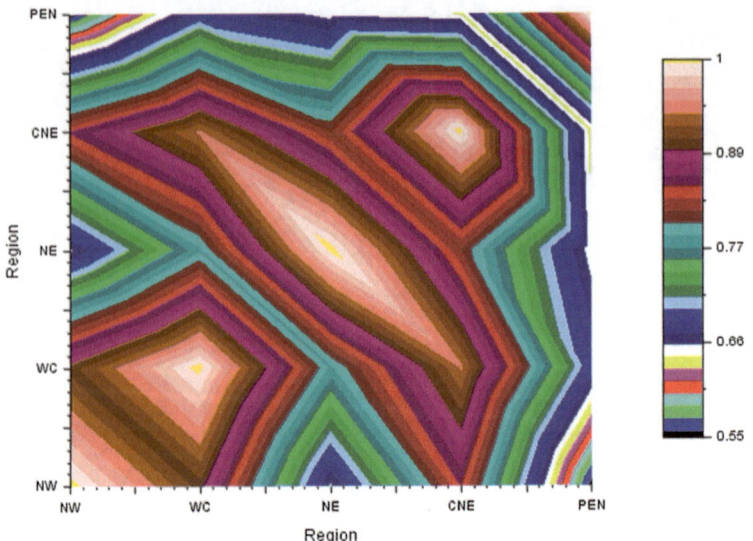

Fig. 4 Correlation among rainfalls of various regions

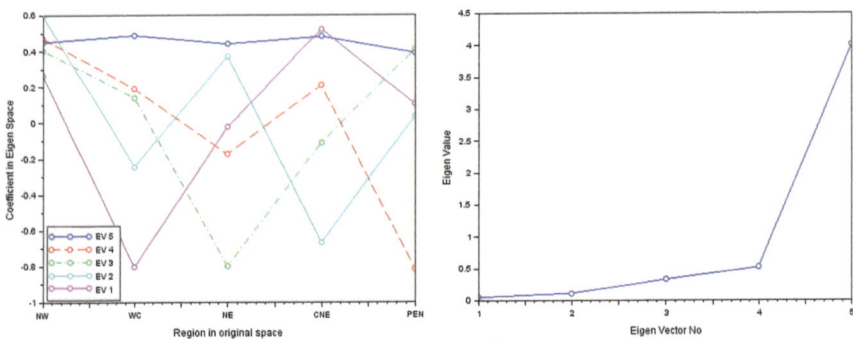

Fig. 5 Eigenvectors (left) and their eigenvalues in increasing order

variance. This vector is EV4 in Fig. 5 (left). Let us call it u_2. Taken together u_1 and u_2 make the space $U^2 = (u_1, u_2)$. This is essentially U^t discussed in Sect. 2 and it explains $(4.00 + 0.52)/5 = 90.4\%$ of the total variance. Thus, the D^5 is transformed to U^2 space reducing the dimension of the data from 5 to 2 and retaining 90.4% of the information.

4 Conclusion and Future Scope

The EOF analysis of five homogenous rainfall regions of India, based on 146-year monthly data, was performed. The basic structure of the data was analyzed and the climatological structure and trend revealed. The EOF analysis was done and it was found that the leading eigenvector accounted for 80% of the total variance. The two leading EOFs together accounted for 90.4% of the total variance of the data. The coefficients of the eigenvectors in the original space revealed that the leading eigenvector actually passes almost through the mean of the data. The fact that two leading eigenvectors explain more than 90% of the data is a significant observation. This can be utilized for making better statistical regional forecast models.

References

1. Frequently Asked Questions, IMD. http://imd.gov.in/section/nhac/wxfaq.pdf
2. Jin Q, Wang C (2017) A revival of Indian summer monsoon rainfall since 2002. Nature 7:587–594. https://doi.org/10.1038/nclimate3348
3. Asharaf S, Ahrens B (2015) Indian summer monsoon rainfall processes in climate change scenarios. J Clim 28:5414–5429
4. Chowdary JS, Harsha HS, Gnanaseelan C, Srinivas G, Parekh A, Pillai P, Naidu CV (2017) Clim Dyn 48:2707. https://doi.org/10.1007/s00382-016-3233-1
5. Gadgil S (2006) The Indian monsoon, GDP and agriculture. Econ Polit Weekly 41:4887–4895
6. Konwar M, Parekh A, Goswami BN (2012) Dynamics of east-west asymmetry of Indian summer monsoon rainfall trends in recent decades. Geophys Res Lett 39:1–6
7. Roxy MK, Ritika K, Terray P, Murtugudde R, Ashok K, Goswami BN (2015) Drying of Indian subcontinent by rapid Indian Ocean warming and a weakening landsea thermal gradient. Nature Commun 6:1–10. https://doi.org/10.1038/ncomms8423
8. Shukla RP, Tripathi KC, Pandey AC, Das IML (2011) Prediction of Indian summer monsoon rainfall using Niño Indices: a neural network approach. Atmos Res 102:99–109. https://doi.org/10.1016/j.atmosres.2011.06.013
9. Sahai AK, Soman MK, Satyan V (2000) All India summer monsoon prediction using an artificial neural network. Clim Dyn 16:291–302
10. Shashikanth K, Sukumar P (2017) Indian monsoon rainfall projections for future using GCM model outputs under climate change. Adv Comput Sci Technol 10:1501–1516
11. Sujata K, Mandke, Shinde M, Sahai AK (2014) Dynamical seasonal prediction of indian summer monsoon using AGCM: weighted ensemble mean approach. Earth Sci India 7:67–72. http://www.earthscienceindia.info/
12. Bishop CM (1995) Neural networks for pattern recognition, Oxford University Press, New Delhi, pp 140–148, 203, 267–268, 372
13. Keiner LE, Yan X-H (1997) Empirical orthogonal function analysis of sea surface temperature patterns in Delaware Bay. IEEE Trans Geosci Remote Sens 35:1299–130. https://doi.org/10.1109/36.628796
14. Sukresno B (2010) Empirical orthogonal functions (eof) analysis of sst variability in indonesian water concerning with enso and iod. Int Arch Photogram Remote Sensing Spat Inf Sci 38:116–121
15. Timothy CG, James JS (1994) An empirical orthogonal function analysis of remotely sensed sea surface temperature variability and its relation to interior oceanic processes off Baja California. Remote Sens Environ 47:375–389. https://doi.org/10.1016/0034-4257(94)90105-8

16. Singh CV (2004) Empirical orthogonal function (EOF) analysis of monsoon rainfall and satellite—observed outgoing long-wave radiation for Indian monsoon. A comparative study. Meteorol Atmos Phys 85:227–234
17. Singh CV (1999) Principal components of monsoon rainfall in normal, flood and drought years over India. Int J Climatol 19:639–652

Novel Semantic Discretization Technique for Type-2 Diabetes Classification Model

Omprakash Chandrakar, Jatinderkumar R. Saini
and Dharmendra G. Bhatti

Abstract Semantic discretization, which is relatively a new concept, can be viewed as the discretization technique that uses the semantics of the data along with its value. The semantics of the data refer to the domain knowledge inherent in the data. The semantics of data is derived from the data value itself. Objective and context of the study also contribute significantly to identifying semantic of the data. Since no explicit ontology is associated with the data in semantic discretization, identifying, interpreting, and exploiting, the semantics of the data is a challenging task. This paper presents a novel algorithm for semantic discretization, in which machine learning techniques such as classification and association rule mining is used to derive semantic knowledge, which is further used for discretization. To show the effectiveness of the proposed semantic discretization algorithm, we applied it on diabetes dataset. Experimental results show 2–15% improvement in classification accuracy on semantically discretized dataset in comparison to the original and statistically discretized dataset.

Keywords Association rule mining · Classification · Clustering · Discretization · Prediction model · Semantic discretization · Type-2 diabetes

1 Introduction

There are certain data mining algorithms that can be used with discrete dataset only. Apriori and Id3 are such algorithms for association rule mining and classification, respectively. But on the contrary, most of the time data are continuous in real-life

O. Chandrakar (✉) · D. G. Bhatti
Uka Tarsadia University, Bardoli, Gujarat, India
e-mail: opchandrakar@utu.ac.in

D. G. Bhatti
e-mail: dgbhatti@utu.ac.in

J. R. Saini
Narmada College of Computer Application, Bharuch, Gujarat, India
e-mail: saini_expert@yahoo.com

© Springer Nature Singapore Pte Ltd. 2019
H. S. Saini et al. (eds.), *Innovations in Computer Science
and Engineering*, Lecture Notes in Networks and Systems 74,
https://doi.org/10.1007/978-981-13-7082-3_17

scenario. The result produced by these algorithms is significantly reduced in such a scenario. To overcome this limitation, various discretization techniques are used, that converts continuous data into discrete data. Several studies have shown that the accuracy of a classification model is better if, the model is built on a discretized dataset rather than a continuous dataset [1]. That is why the discretization is important, and sometimes, essential operation during data preprocessing [2].

2 Literature Review

Several categorizations of discretization methods have been reported in the literature based on the characteristics of discretizer [3–5]. Supervised discretizer takes class information into account while finding an appropriate interval, while unsupervised discretizer do not. Equal Width and Equal Frequency [6], PKID and FFD [7], and MVD [8] are some examples of supervised discretization. Direct methods divide the range of k intervals simultaneously (i.e., equal width), while incremental methods begin with a simple discretization and pass through an improvement process [9]. Static methods do not take other features in the consideration, while, dynamic methods take it into account for discretization [10]. C4.5 is the representative algorithm of static methods. ID3 [11] and ITFP [12] discretizer are representative of dynamic methods. The top-down approach begins with an empty split points list and continuously add a new one to the list by cutting intervals. On the contrary, bottom-up approach begins with the list of all continuous values of the features while the other methods begin with all the continuous values of the feature as split points and remove some of them by merging intervals as the discretization progresses [13]. Chandrakar and Saini have proposed a novel semantic discretization technique [2]. Discretization of the continuous attribute is based on the semantics of data rather than data values. The semantics of data is defined as the domain knowledge inherent in the data.

3 Research Methodology

The objective of this research study is to demonstrate the new technique for semantic discretization using clustering. The underlying idea of semantic discretization is to capture the semantic of data by exploiting intra-attribute dependency. Exploitation of this intra-dependence is hard to achieve because the classical measures of dependence do not apply. To overcome this difficulty, we used clustering, which provides a natural criterion for exploiting this intra-attribute dependency. Clustering is a process in which similar objects tend to group together. It provides us with the intrinsic grouping of the unlabeled data. Thus, the cluster centroid captures the meaningful pattern, or intra-dependencies exist in the data, which leads toward the most suitable split point, which in turn, gives the most natural, distinct, and categorized intervals. The semantics of the data is further used to find the optimal discretization rules. Dataset is discretized using these rules. Classification models are built to predict Diabetes Status (Yes or No), each corresponding to original, semantically discretized, and statistically discretized dataset. The efficiency of semantic discretization is evaluated by comparing the classification accuracy. The step-by-step process for semantic discretization is given in Algorithm 1.

Algorithm 1. SemanticDiscretization
Input: Dataset with IndependentVariable and ClassVariable, NPartition, MinValue, MaxValue
Output: Partition Rules and Semantically Discretized Dataset
1. Start
2. Input no of partitions, NPartition.
3. Set no of cluster, N = NPartition.
4. Set for Cross validation with 10 folds.
5. Apply distance based clustering with MakeDensityBasedClusterer() method with k-mean algorithm on IndependentVariable, with respect to ClassVariable. Get ClusterCentroid[i] and StandardDeviation[i] {for i= 1 to N}
6. Arrange ClusterCentroid[i] and StandardDeviation[i] {for i= 1 to N} in ascending order.
7. Set SplitPoint[i] = ClusterCentroid[i] + StandardDeviation[i] (for i =1 to N-1).
8. Repeat step 9 for i=1 to N

9. Set Partition range
Set Partition[i] = (MinValue, SplitPoint[i])   {for i = 1}
       = (SplitPoint[i-1] + 1, SplitPoint[i]) {for i = 2 to N-1}
       = (SplitPoint[i-1, MaxVal])   {for i = N}

10. Discretized the dataset by replacing the value of continuous variable with their corresponding partition obtained in step 9.
11. Apply association rule mining on the discretized dataset obtained in step 10 using apriori algorithm. Set Delta = 0.05 Set Number of Rules =100 Set Lower Bound Minimum Support = 0.005 Set Metric Type = Confidence Set Minimum Confidence = 0.01
12. Check the Confidence of the rule of following type and assign it to RuleConfidence[i]. Partition[i] → Diabetes = Y
13. Arrange RuleConfidence[i] in ascending order and assign the corresponding partition to Lowest Risk to Highest Risk.
14. Get the Semantically Discretized the dataset by replacing the partition with their corresponding Risk Level in the dataset obtained in step 10.
15. End.

4 Experiment

A comprehensive data collection form [14, 16] is designed after intensive consultation with medical experts and literatures. The primary dataset used in this study contains 844 records of diabetes patients, with 11 attributes. 10 attributes are diabetes risk factors and 1 is Diabetes Status. Except for Gender and Diabetes Status, other 9 attributes are continuous attributes [14]. We applied the proposed semantic discretization algorithm on diabetes dataset that contains nine continuous attributes (Age, Family History, Personal History, BMI, Waist Circumference, Diet, Stress, Physical Activity, and Life Quality). The semantic discretization rules obtained for each attribute is shown in Table 1. All experiments are performed using well-known data mining tool Weka 3.6 [15].

Table 1 Discretization rules for diabetes dataset

1. If age < 41 then diabetes risk = low
2. If age >=41 and age <=61 then diabetes risk = high
3. If age > 61 then diabetes risk = moderate
4. If waist circumference < 36 then diabetes risk = low
5. If waist circumference >=36 and waist circumference <=43 then diabetes risk = moderate
6. If waist circumference > 43 then diabetes risk = high
7. If BMI < 25 then diabetes risk = low
8. If BMI > = 25 and BMI <=30 then diabetes risk = moderate
9. If BMI > 30 then diabetes risk = high
10. If family history < 2 then diabetes risk = low
11. If family history >=2 and family history <=7 then diabetes risk = moderate
12. If family history > 7 then diabetes risk = high
13. If personal history < 3 then diabetes risk = low
14. If personal history >=3 and personal history <=5 then diabetes risk = moderate
15. If personal history > 5 then diabetes risk = high
16. If diet < 3 then diabetes risk = low
17. If diet >=3 and diet <=8 then diabetes risk = moderate
18. If diet > 8 then diabetes risk = high
19. If stress < 14 then diabetes risk = low
20. If stress = 14 then diabetes risk = moderate
21. If stress > 14 then diabetes risk = high
22. If physical activity >=5 then diabetes risk = low
23. If physical activity = 4 then diabetes risk = moderate
24. If physical activity < 4 then diabetes = high
25. If life quality < 5 then diabetes risk = low
26. If life quality >=5 and life quality <=8 then diabetes risk = moderate
27. If life quality > 8 then diabetes risk = high

Table 2 Comparative classification accuracy for diabetes dataset

Classification algorithm	Classification accuracy (%)			Improvement in accuracy (%)	
	Un-discretized dataset (A)	Statistical discretiza-tion (B)	Semantic discretiza-tion (C)	Semantic discretiza-tion over un-discretized dataset (C-A)	Semantic discretiza-tion over statistical discretiza-tion (C-B)
Bayes algorithms					
BayesNet	70.09	73.95	78.77	8.68	4.82
NaiveBayes	73.31	73.95	79.74	6.43	5.79
NaiveBayesUpdatable	73.31	73.95	79.74	6.43	5.79
Function-based algorithms					
Logistic	73.31	73.63	81.99	8.68	8.36
Multilayer Perception	74.27	72.66	76.84	2.57	4.18
SimpleLogistic	72.02	73.95	81.02	9	7.07
SMO	73.31	72.66	79.09	5.78	6.43
VotedPerception	62.37	72.02	77.17	14.8	5.15
Rules-based algorithms					
Jrip	68.81	67.84	72.66	3.85	4.82
OneR	64.63	67.84	69.77	5.14	1.93
PART	68.16	70.09	74.59	6.43	4.5
Trees-based algorithms					
DecisionStump	66.23	62.37	69.77	3.54	7.4
LMT	70.09	73.95	78.45	8.36	4.5
Average	**69.99**	**71.45**	**76.89**	**6.9**	**5.44**

4.1 Building Classification Model

The above discretization rules are used to derive semantically discretized dataset. Classification accuracy is obtained by applying 13 classification algorithms on dataset which is shown in Table 2.

5 Result Analysis and Conclusion

Table 2 and Fig. 1 shows that classification accuracy significantly improved when the dataset is discretized using the proposed semantic discretization algorithm. 1.93–14.80% improvement in classification accuracy is observed. Average clas-

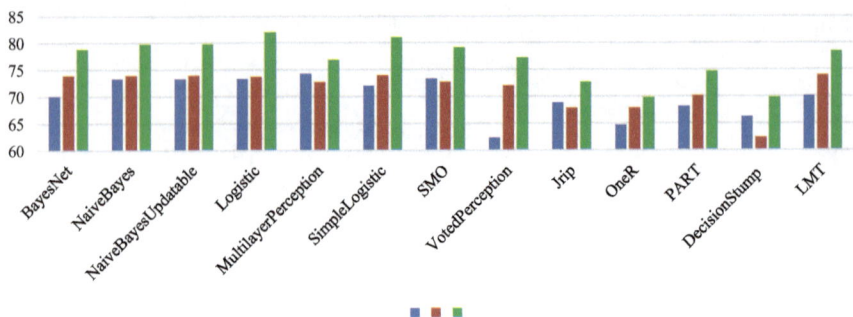

Fig. 1 Improvement in classification accuracy after semantic discretization

sification accuracy for the un-discretized and statistically discretized dataset is around 70 and 71%, while it is around 77% for the proposed semantically discretized dataset. Looking at the significant improvement in classification accuracy, researchers strongly suggest applying the proposed semantic discretization technique as a preprocessing step for Bayes and function-based classification algorithms.

References

1. Yang YW, Wu GI, Maimon X, Oded Rokach L Book section, discretization methods, data mining and knowledge discovery handbook, 2005, Springer US, Boston, MA @ 978-0-387-25465-4
2. Chandrakar O, Saini JR (2017) Knowledge based semantic discretization using data mining techniques. Int J Adv Intell Parad
3. Liu H, Hussain F, Tan CL, Dash M (2002) Discretization: an enabling technique. Data Min Knowl Disc 6(4):393–423
4. Dougherty J, Kohavi R, Sahami M (1995) Supervised and unsupervised discretization of continuous features. In: Proceedings of the twelfth international conference on machine learning (ICML), 1995, pp 194–202
5. Yang Y, Webb GI, Wu X (2010) Discretization methods. In: Data mining and knowledge discovery handbook, pp 101–116
6. Li R-P, Wang Z-O (2002) An entropy-based discretization method for classification rules with inconsistency checking. In: Proceedings of the first international conference on machine learning and cybernetics (ICMLC), pp 243–246
7. Yang Y, Webb GI (2009) Discretization for naive-bayes learning: managing discretization bias and variance. Mach Learn 74(1):39–74
8. Bay SD (2001) Multivariate discretization for set mining. Knowl Inf Syst 3:491–512
9. Cerquides J, Lopez R (1997) Proposal and empirical comparison of a parallelizable distance-based discretization method. In: III international conference on knowledge discovery and data mining (KDDM97). Newport Beach, California, USA, pp 139–142
10. Steck H, Jaakkola T (2004) Predictive discretization during model selection. In: XXVI symposium in pattern recognition (DAGM04). Lecture notes in computer science 3175, Springer, Tbingen, Germany, pp 1–8
11. Quinlan JR (1993) C4.5: programs for machine learning. Morgan Kaufmann Publishers Inc.

12. Au W-H, Chan KCC, Wong AKC (2006) A fuzzy approach to partitioning continuous attributes for classification. IEEE Trans Knowl Data Eng 18(5):715–719

13. Kerber R (1992) ChiMerge: discretization of numeric attributes. X national conference on artificial intelligence American association (AAAI92). USA, pp 123–128

14. Chandrakar O, Saini JR Development of Indian weighted diabetic risk score (IWDRS) using machine learning techniques for type-2 diabetes. In: COMPUTE '16 proceedings of the 9th annual ACM India conference. ACM New York, NY, USA, pp 125–128. ©2016, ISBN: 978-1-4503-4808-9. https://doi.org/10.1145/2998476.2998497

15. Bouckaert RR, Frank E, Hall M, Kirkby R, Reutemann P, Seewald A, Scuse D (2016) WEKA manual for version 3-8-1. University of Waikato, Hamilton, New Zealand

16. Chandrakar O, Saini JR Questionnaire for deriving diabetic risk score for Indian population. Accepted for presentation and publication at international conference on artificial intelligence in health care, ICAIHC-2016

Automatic Selection of Sensitive Attributes in PPDP

V. Uma Rani, M. Sreenivasa Rao and K. Jeevan Suma

Abstract Privacy-preserving data publishing and data analysis conventional respectable work in past age as encouraging methodologies for sharing data as well as preserving privacy of individual. A remedy to this is data scrambling. Numerous algorithms are formulated to replace lawful information by fictitious but that is practical. Yet, nothing was proposed to automate the detection associated with information scrambling which is very much essential. In this paper, we propose a novel method for automatic selection of sensitive attributes for data scrambling. This is done by ranking the attributes based on attribute evaluation measures. And also, an innovative method for privacy-preserving data publishing with automatic selection of sensitive attributes which provide secure release of information for a data-mining task while preserving sensitive information.

Keywords Privacy preservation · Selection of sensitive attributes · Preserving sensitive information

1 Introduction

With the flared use of information production, magnanimous volumes of individual aggregation are frequently gathered and explored. Such data include shopping habits, medical history, criminal records, credit records, etc., but privacy should be protected while using personal or sensitive information. However, it is the most useful data for decision-making processes and to provide social benefit, such as medical research,

V. Uma Rani (✉) · M. Sreenivasa Rao
CSE, School of IT, JNTUH, Hyderabad, Telangana, India
e-mail: umarani@jntuh.ac.in

M. Sreenivasa Rao
e-mail: srmeda@gmail.com

K. Jeevan Suma
SE, School of IT, JNTUH, Hyderabad, Telangana, India
e-mail: sptavisi@gmail.com

© Springer Nature Singapore Pte Ltd. 2019
H. S. Saini et al. (eds.), *Innovations in Computer Science and Engineering*, Lecture Notes in Networks and Systems 74,
https://doi.org/10.1007/978-981-13-7082-3_18

crime reduction, national security, etc. on the other hand. Privacy-preserving data publication has become the most explored topics and becomes a serious concern in the publication of personal data in past period. Thus, a motivating new direction of data-mining research has been emerged, known as privacy-preserving data mining (PPDM). The purpose of these algorithms is the mining of relevant knowledge from a substantial collection of data with privacy preservation of private information. Privacy-preserving is not the same as conventional data security, access control and encryption which prevent information disclosure against unauthorized access, since privacy preserving is to prevent information disclosure due to authorized access to the data. The main goal of privacy-preserving data mining is to protect identity information that is used to identify a person or entity stored in the database and confidential information which is harmful if revealed. We are taking both kinds of information into consideration while scrambling data in a database.

2   Related Work

The sensitive attributes are recognized by computing the threshold measure with the assigning weights to each attribute by Kamakshi et al. [1]. First, the client's query is studied and then sensitive attribute or set of attributes are automatically recognized. The threshold value is set by the data owner. The sensitive attribute or attributes are identified by taking sum of weights of the attributes submitted in the client's task. If the total weight exceeds the limit then the values are qualified using information modification technique in PPDM.

The automatic recognition of the sensitive attributes proposed by du Mouza et al. [2] based on two functionalities: (1) automatic detection of the data to be scrambled and (2) automatic propagation to other semantically linked values. They propose an expert system based on rules which guides the selection of sensitive data. They also presented an inference mechanism with the help of semantic graph for propagation of the confidentiality on near values and the consistency with the other relations.

In 2002, Sweeney [3] proposed the k-anonymity for privacy protection but it cannot prevent attribute disclosure. To address this limitation of k-anonymity, Machanavajjhala et al. [4] introduced l-diversity, which needs the distribution of a sensitive attribute in each equivalence class which has at least l "well represented" values. Li et al. [5] proposed t-closeness, which necessitates that the distribution of a sensitive attribute in any equivalence class is close to the distribution of the attribute in the overall. In addition, several ideologies were introduced, such as (c; k)-safety [6] and δ_presence [7]. In 2006, Xiao and Tao [8] proposed anatomy, which is a data anonymization method that divides one table into two for data publication. In [9–13], similar methods are proposed but nobody addressed the issue of automatic selection of sensitive attributes for PPDP.

3   Selection of Sensitive Attributes

Numerous algorithms for data scrambling focus on smartly changing true data by factious yet practical. However, nothing was formulated to automate the task that is essential for the detection associated with information to be scrambled. In this paper, we propose a novel method for automatic selection of sensitive attributes for PPDP. This is done by ranking the attributes based on attribute evaluation measures First, we computed sensitivity ranks of attributes and selected the attributes whose rank is below the threshold. And also, an innovative method for privacy-preserving data publishing which provides secure release of information for a data-mining task while preserving sensitive information.

3.1   Framework for Selection of Sensitive Attribute

We select the sensitive attribute by the following steps:

Step 1:   Compute the attribute evaluation measures
Find attribute measures like information gain, relief, correlated-based feature selection (CFS), gain ratio, correlation attribute evaluation, one R attribute evaluation, symmetric uncertainty attribute evaluation, symmetrical uncertainty attribute evaluation, wrapper subset evaluation, and principal component analysis for each attribute.

Step 2:   Rank the attributes based on attribute evaluation measures
Take one attribute measure (e.g., gain ratio) and compute score for each attribute. Assign a rank to the attribute based on computed score.
Repeat ranking attributes by taking each of the remaining attribute measures.

Step 3:   Compute absolute sensitivity ranks of attributes
Now assign absolute sensitivity ranks for each attribute which is nothing but the sum of ranks assigned by each attribute measure for that attribute.

Step 4:   Compute relative sensitivity ranks of attributes
Now, the relative rank is assigned based on absolute rank.

Step 5:   Select the attributes whose rank is below the threshold value.

4   Privacy-Preserving Data Publishing

Microdata have been studied expansively in recent years. The microdata are which that contains the information about the individuals publishing or sharing such data, the major threats can expect. The solution is to use anonymization techniques to show data in a less-specific manner.

4.1 Framework for Privacy-Preserving Data Publishing

The proposed privacy-preserving data publishing method has two goals: preserving privacy while revealing useful information for sensitive (i) numerical attributes and (ii) categorical (nonnumerical) attributes and to find a generalized data D′, such that it includes all the attributes of D and an individual tuple from D is non-identifiable in D′.

Step 1: Find the sensitive attributes guided by sensitivity rank
        Use the framework for selection of sensitive attribute proposed in section and
        Find the sensitivity ranks of attributes.
Step 2: Generalize the categorical data based on sensitivity rank and
        Based on sensitivity rank generalize using taxonomy tree.
Step 3: Get the anonymized data.

5 Experiments

Example 1: Adult dataset also known as "Census Income" dataset is taken and computed sensitivity rank which is the last column in Table 1. Here is a summary of the data.

Table 1 Summary of "Census Income" dataset

1 age		2 workclass		3 education		4 education–num		5 marital–status	
Min	17.	Private	22 696	HS-grad	10 501	Min	1.	Married-civ-spouse	14 976
1st Qu	28.	Self-emp-not-inc	2561	Some-college	7291	1st Qu	9.	Never-married	10 683
Median	37.	Local-gov	2093	Bachelors	5355	Median	10.	Divorced	4443
Mean	38.5816	?	1836	Masters	1723	Mean	10.0807	Separated	1025
3rd Qu	48.	State-gov	1298	Assoc-voc	1382	3rd Qu	12.	Widowed	993
Max	90.	Self-emp-inc	1116	11th	1175	Max	16.	Married-spouse-absent	418
		(Other)	981	(Other)	5134			Married-AF-spouse	23
6 occupation		7 relationship		8 race		9 sex		10 capital–gain	
Prof-specialty	4140	Husband	13 193	White	27 816	Male	21 790	1st Qu	0.
Craft-repair	4099	Not-in-family	8305	Black	3124	Female	10 771	3rd Qu	0.
Exec-managerial	4066	Own-child	5068	Asian-Pac-Islander	1039			Median	0.
Adm-clerical	3770	Unmarried	3446	Amer-Indian-Eskimo	311			Min	0.
Sales	3650	Wife	1568	Other	271			Mean	1077.65
Other-service	3295	Other-relative	981					Max	99 999.
(Other)	9541								
11 capital–loss		12 hours–per–week		13 native–country		14 income			
1st Qu	0.	Min	1.	United-States	29 170	<=50K	24 720		
3rd Qu	0.	1st Qu	40.	Mexico	643	>50K	7841		
Median	0.	Median	40.	?	583				
Min	0.	Mean	40.4375	Philippines	198				
Mean	87.3038	3rd Qu	45.	Germany	137				
Max	4356.	Max	99.	Canada	121				
				(Other)	1709				

5.1 Selection of Sensitive Attributes

Step 1: Compute the attribute evaluation measures
Computed attribute measures like information gain, relief, correlated-based feature selection (CFS), gain ratio, correlation attribute evaluation, one R attribute evaluation, symmetric uncertainty attribute evaluation, symmetrical uncertainty attribute evaluation, wrapper subset evaluation, and principal component analysis for each attribute.

Step 2: Rank the attributes based on attribute evaluation measures
Take one attribute measure (e.g., gain ratio) and compute score for each attribute. Assign a rank to the attribute based on computed score.
Repeat ranking attributes by taking each of the remaining attribute measures. The values for age, workclass, etc. are shown in Table 2 under columns 2–7

Step 3: Compute absolute sensitivity ranks of attributes
Now, assign absolute sensitivity ranks for each attribute which is nothing but the sum of ranks assigned by each attribute measure for that attribute.

Step 4: Compute relative sensitivity ranks of attributes
Now, the relative rank is assigned based on absolute rank. Here, in Table 2 marital status, absolute sensitivity rank are min values so its relative sensitivity rank is 1.

Step 5: Select the attributes whose rank is below the threshold value (14 * 1/3)
Here, attributes whose relative rank is 1–4 can be taken as highly sensitive. Attributes 5–8 are medium-sensitive attributes and 9–14 are low-sensitive attributes.

5.2 Privacy-Preserving Data Publishing

Step 1: Find the sensitive attributes guided by sensitivity rank
Use the framework for selection of sensitive attribute proposed in Sect. 3.1 and compute the sensitivity ranks of attributes in Sect. 5.1.

Step 2: Generalize the categorical data based on sensitivity rank
Based on sensitivity rank the attributes are generalized using taxonomy tree. Figure 1 shows the taxonomy tree for attribute education and Table 3 shows "Census Income" dataset with taxonomy tree height for each attribute.

Step 3: Get the anonymized data
Based on sensitivity rank generalize using taxonomy tree. Table 4 shows generalization score for attribute "Education" for each level. Sensitivity score for attribute education $= 1 - 6/14 = 0.57$, which is >0.5 and <0.75. So, attribute education should be generalized to Level 2. If the attribute value is HS-grad, then it should be replaced with secondary.

Table 2 "Census Income" dataset with attribute sensitivity rank

Attribute_name	CFS_Subset_ Evaluator	Correlation ranking filter	Gain ratio feature evaluator	ig	One R feature evaluator	Relief ranking filter	Symmetrical uncertainty ranking filter	Abs- Rank	Rel- Rank
Age	20	5	7	3	11	5	6	277	7
Workclass	20	10	11	11	5	9	11	307	10
fnlwgt	20	14	14	14	14	14	14	428	14
Education	20	9	8	5	4	3	7	247	6
Education- num	1	2	6	6	3	7	5	98	2
Marital status	2	1	3	2	6	4	2	89	1
Occupation	20	11	10	7	7	1	10	285	8
Relationship	3	3	4	1	8	2	3	115	4
Race	20	12	13	13	10	6	13	352	12
Sex	20	7	5	10	9	8	8	294	9
Capital-gain	4	6	1	4	1	11	1	114	3
Capital-loss	5	8	2	9	2	13	4	161	5
Hours-per- week	20	4	9	8	12	10	9	317	11
Native-country	20	13	12	12	13	12	12	396	13

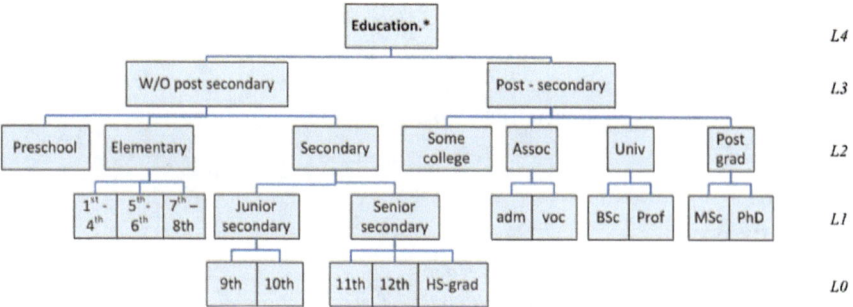

Fig. 1 Taxonomy tree for attribute "Education"

Table 3 "Census Income" dataset with attribute taxonomy tree height

Attribute	Taxonomy tree height
Age	3
Workclass	5
fnlwgt	3
Education	5
Education-num	5
Marital status	4
Occupation	3
Relationship	3
Race	3
Sex	2
Capital-gain	3
Capital-loss	3
Hours-per-week	3
Native-country	5

Table 4 Generalization score for attribute "Education" for each level

Level	Level 0	Level 1	Level 2	Level 3	Level 4
Generalization score	0	0.25	0.5	0.75	1.0

6 Conclusion

Until now, nobody addressed the issue of automatic selection or detection of sensitive attributes. We proposed a novel method for automatic selection of sensitive attributes for PPDP. This is done by ranking the attributes based on attribute evaluation measures. First, we computed sensitivity ranks of attributes and also an innovative method for PPDP for secure release of information for a data-mining task while preserving sensitive information. This work motivates several directions for future research. First, data publishing with secured access and authorization and then the privacy-preserving data mining of datasets with sensitive attributes can also be wise in this magnitude. There is still a need to analyze the privacy and utility metrics.

References

1. Kamakshi P, Vinaya Babu A (2012) Automatic detection of sensitive attribute in PPDM. In: 2012 IEEE international conference on computational intelligence and computing research
2. du Mouza C, Métais E, Lammari N, Akoka J, Aubonnet T, Comyn-Wattiau I, Fadili H, Si-Said Cherfi S (2010) Towards an automatic detection of sensitive information in a database. In: 2010 second international conference on advances in databases, knowledge, and data application
3. Sweeney L (2002) k-anonymity: a model for protecting privacy. Int J Uncertain Fuzziness Knowl Based Syst 10(5):557–570. http://dx.doi.org/10.1142/S0218488502001648
4. Machanavajjhala A, Kifer A, Gehrke J, Venkitasubramaniam M (2007) l-diversity: privacy beyond k-anonymity. ACM Trans Knowl Discov Data 1(1):1–47. http://dx.doi.org/10.1145/1217299.1217302
5. Li N, Li T, Venkatasubramanian S (2007) t-closeness: privacy beyond k-anonymity and l-diversity. In: Proceedings of the IEEE 23rd international conference on data engineering, pp 106–115. http://ieeexplore.ieee.org/stamp/stamp.jsp?tp=&ar-number = 4221659&isnumber = 4221635
6. Martin DJ, Kifer D, Machanavajjhala A, Gehrke J, Halpern JY (2007) Worst-case background knowledge for privacy-preserving data publishing. In: Proceedings of the IEEE 23rd international conference on data engineering, pp 126–135. http://doi.ieeecomputersociety.org/10.1109/ICDE.2007.367858
7. Nergiz ME, Atzori M, Clifton C (2007) Hiding the presence of individuals from shared databases. In: Proceedings of the 2007 ACM international conference on management of data, pp 665–676. https://doi.org/10.1145/1247480.1247554
8. Xiao X, Tao Y (2006) Anatomy: simple and effective privacy preservation. In: Proceedings of the 32nd international conference on very large data bases, pp 139–150. https://dl.acm.org/cita-tion.cfm?id=1164141
9. Sweeney L (2002) Achieving k-anonymity privacy protection using generalization and suppression. Int J Uncertain Fuzziness Knowl Based Syst 10(5):571–588
10. Blanning R (1988) Sensitivity analysis in hierarchical fuzzy logic models. In Proceedings of the international conference on decision support and knowledge-based systems track, pp 471–476
11. Meyerson A, Williams R (2004) On the complexity of optimal K-anonymity. In: Proceedings of the international symposium on principles of database systems (PODS), pp 223–228
12. Bayardo RJ Jr., Agrawal R (2005) Data privacy through optimal k-anonymization. In: Proceedings of the international conference on data engineering (ICDE), pp 217–228; Park H, Shim K (2007) Approximate algorithms for K-anonymity. In: Proceedings of the international conference on management of data (SIGMOD), pp 67–78
13. Xiao X, Tao Y, Koudas, N (2010) Transparent anonymization: Thwarting adversaries who know the algorithm. ACM TODS 35(2)

Investigation of Geometrical Properties of Kernels Belonging to Seeds

M. Aishwarya, Vaidya Srivani, A. Aishwarya and P. Natarajan

Abstract The estimation of the kernel density has been successfully tried on several data mining tasks. In this paper, the geometrical properties of kernels belonging to seeds using Naive Bayesian classification and Hierarchical clustering techniques are analyzed. A method, which has been applied to real data set of grains and analysis of the kernels belonging to three types of seeds namely Canadian, Rosa, and Kama are classified based on the geometrical properties like perimeter, compactness, length of kernel, width of kernel, asymmetric coefficient, and length of kernel groove has been proposed. Also, a comparison between the accuracies obtained after performing Naive Bayes Classification and Hierarchical Clustering has been made to conclude a better data mining technique.

Keywords Naive Bayes · Hierarchical clustering · Kernel · Data mining · Confusion matrix · Dendrogram

M. Aishwarya (✉) · V. Srivani · A. Aishwarya
School of Information Technology and Engineering, Vellore Institute of Technology, Vellore 632014, India
e-mail: aishwaryam286@gmail.com

V. Srivani
e-mail: vaidyasrivani@gmail.com

A. Aishwarya
e-mail: aishu18498@gmail.com

P. Natarajan
School of Computer Science and Engineering, Vellore Institute of Technology, Vellore 632014, India
e-mail: pnatarajan@vit.ac.in

© Springer Nature Singapore Pte Ltd. 2019
H. S. Saini et al. (eds.), *Innovations in Computer Science and Engineering*, Lecture Notes in Networks and Systems 74,
https://doi.org/10.1007/978-981-13-7082-3_19

151

1 Introduction

The area of information science in which the raw data is analyzed to produce significant information by extracting meaningful and functional pattern is data mining [1]. The purpose of classification and clustering algorithms is to extract values from data sets like structured or unstructured data and providing a useful sense of information. The idea behind classification is to target class by analyzing the training dataset. The dataset is trained to get better boundary conditions, which ultimately can be used for determining every target class. The idea behind clustering is grouping the similar kind of things by taking the most satisfying conditions into consideration. It considers all the items that are in the same category and no two different categories of items must not be alike [2].

Naive Bayes is a supervised machine learning technique. It is very simple, but surprisingly powerful used for the predictive model [3]. It works on conditional probability. It is a probability that something will occur, given that some event has already happened. In this analysis, R Language is used to implement Naive Bayes algorithm. It predicts membership probabilities for each and every class such as the probability that the given record belongs to one particular class.

Hierarchical clustering is an unsupervised learning technique [4]. There are two types of hierarchical clustering approaches namely agglomerative and divisive. In this paper, an agglomerative approach [5] is used in which the distance between two cluster points is calculated based on Euclidean distance and subsequently, it is measured using a complete linkage method. In this approach, every data point is considered as a singleton cluster and it is proceeded by integrating all those clusters until all the other points are united into a single cluster [6]. The output of this approach is represented by a Dendrogram or a tree graph.

2 Implementation

2.1 Formulas

I. The following is used to calculate conditional probability [7]:

$$P(h|x) = P(x|h)P(h)/P(x) \tag{1}$$

P(h|x) is the probability of target class given the probability of predictor attribute.
P(h) is the prior probability of the given class.
P(x|h) is the likelihood, which is the probability of predictor class.
P(x) is the prior probability of predictor.

II. Euclidean distance

Consider two points (x1, y1) and (x2, y2) then the distance "d" between two points is [8]

$$d = \text{SQRT}\big((x1 - x2)^2 + (y1 - y2)^2\big) \tag{2}$$

2.2 Methodology

2.2.1 Methodology for Naives Bayes

1. First, the preprocessing of uncertain data [2] is to be done and is saved in .csv format. Then, the data is read.
2. Then the data is segregated in the proportion of 60:40, where 60 is used for training the dataset and remaining 40 is used for testing the data set, which is considered as a thumb rule for classifying the data.
3. For training the dataset, class label is considered, which is usually the last column in the given dataset. On the contrary, class label is not taken into account for testing.
4. The conditional probabilities for each and every attribute are found out using the inbuilt library e1071 and related functions (using Formula (1)).
5. Then class labels namely Canadian, Kama, Rosa are predicted for test data based on the input, that is received from the conditional probabilities of training data.
6. By using the rpart library the confusion matrix is estimated, which helps in calculating the accuracy and errors of the model [9].

2.2.2 Methodology for Hierarchical Clustering

1. The function hclust divides the dataset into the required number of clusters based on the distance (Euclidean distance-using Formula (2)) between the two data points.
2. Then, the Dendrogram [10] is plotted based on the output obtained from the above step.
3. A confusion matrix table is obtained based on the above output.
4. Finally, the function and the library ggplot is used to get the plot of clusters, which demonstrates, how the clusters are scattered based on the clustering technique.

2.3   Program Code

The following inbuilt libraries have been used to extract the desired output.

(a) e1071—This mainly functions for latent class analysis, fuzzy clustering, Naive
    Bayes classifier, etc. [1].
(b) rpart—It is a recursive partitioning for classification, regression, and survival
    trees.
(c) ggplot—It is a package used for creating elegant and complex plots.

2.3.1   Sample Source Code for Naive Bayes Classification

```
trainD<-seeds[data==1,]
testD<-seeds[data==2,]
nrow(trainD)
sd1<-naiveBayes(cls ~.,data=trainD)
pred<-predict(sd1,testD)
mod=rpart(cls ~.,data=trainD)
pred=predict(mod,type="class")
train_control1=trainControl(method="repeatedcv",
number=10,repeats=3)
```

2.3.2   Sample Source Code for Hierarchical Clustering

```
clusters1<-hclust(dist(seeds[,4:5]))
plot(clusters1)
clustcut1<-cutree(clusters1,3)
table(clustcut1,seeds$cls)
```

2.4   Results and Discussions

Here, the model performance is compared by the confusion matrix obtained after
performing classification and clustering techniques. The Confusion Matrix is a table
that is used to represent the performance of a classifier or a cluster model for a given

```
> table(pred,trainD$cls)

pred          canadian kama rosa
    canadian        28    0    0
    kama             3   37    1
    rosa             0    0   47
>
```

Fig. 1 Confusion matrix using Naive Bayes classification

```
> table(clustcut1,seeds1$cls)

clustcut1 canadian kama rosa
        1        0   22   30
        2       65   44    1
        3        0    0   37
```

Fig. 2 Confusion matrix by hierarchical clustering

set of data for which the actual values are known. It is a commonly used technique for summarizing performance. It gives a clear idea of which classification or cluster model is getting right and what kinds of errors it is making.

From the above outputs, it is concluded that Naive Bayes model (Fig. 1) is a better performer, as most of the classes were correctly predicted in comparison to hierarchical clustering (Fig. 2).

A Dendrogram is a kind of tree diagram exhibiting the hierarchical clustering, which tells about the relationships between the datasets that alike. It illustrates the arrangement of the clusters (Fig. 3).

The above plot depicts the formation of clusters each one represented in different colors (Fig. 4). Few clusters, which are overlapped indicates that one particular type of seed is wrongly predicted as another. This can be understood from the confusion matrix that is given above (Figs. 1 and 2).

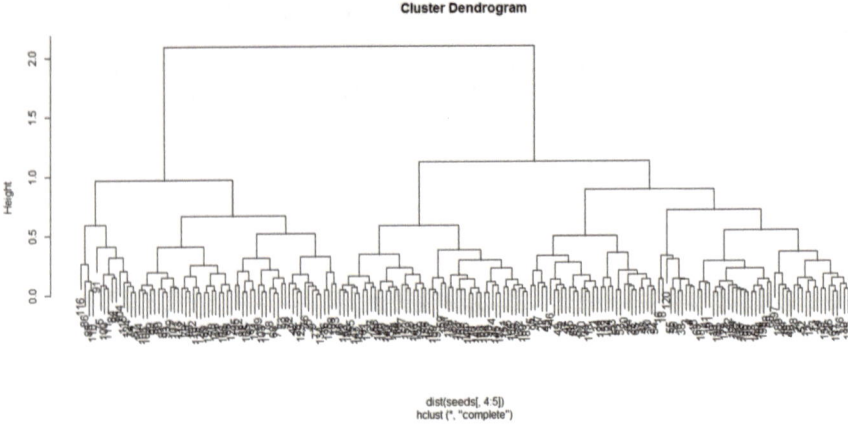

Fig. 3 Dendrogram by hierarchical clustering

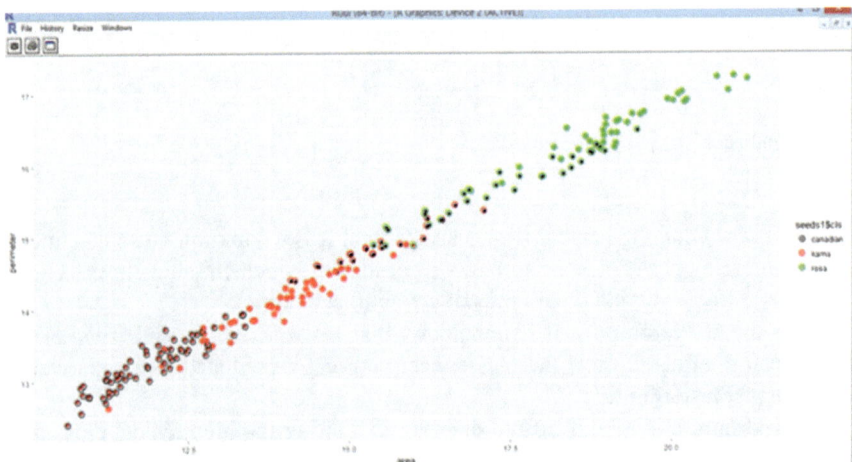

Fig. 4 Formation of clusters using hierarchical clustering

3   Conclusion

Here two effective algorithms are used namely Naive Bayes and Hierarchical clustering, with their implementation and outputs. The better accuracy is achieved in the analysis using Naive Bayes algorithm comparatively to hierarchical clustering. These techniques were implemented in R language. Also, the presented outputs prove that high efficiency and performance can be achieved by the Naive Bayes. This is clearly known from the confusion matrix, where the classes were correctly predicted compared to hierarchical clustering. Therefore, from the analysis made from the above investigation, it is clear that Naive Bayes classification can be applied to many real-time data sets to retrieve valuable information from it. The future work of this project will be to practically test the data with the advanced machine learning techniques, also with Spark and Hadoop, etc. Also, the study will be made to propose and apply an effective algorithm that would work with modern big data analysis and machine learning techniques and test it for challenging and bigger datasets.

References

1. Dimitriadou E, Hornik K, Leisch F, Meyer D, Weingessel A (2006) The e1071 package, Functions of the Department of Statistics (e1071), TU Wien
2. Ren J, Lee SD, Chen X, Kao B, Cheng R, Cheung D (2009) Naive Bayes classification of uncertain data. Department of Computer Science, Sun Yat-Sen University, Guangzhou, China
3. Lowd D, Domingos P (2005) Naive Bayes models for probability estimation. Department of Computer Science and Engineering, University of Washington, Seattle, WA, USA
4. Algorithms for Document Datasets. Department of Computer Science, University of Minnesota, Minneapolis, MN (2002)
5. Balcan M-F, Liang Y, Gupta P (2014) Robust hierarchical clustering. J Mach Learn Res 154011–154051
6. Krishnamurthy A, Balakrishnan S, Xu M, Singh A (2012) Efficient active algorithms for hierarchical clustering. Carnegie Mellon, Pittsburgh, PA
7. Rish I (2001) An empirical study of the Naïve Bayes classifier. T.J. Watson Research Center
8. Ranjini K, Rajalingam N (2011) Performance analysis of hierarchical clustering algorithm. Tirunelveli, India
9. Visa S, Ramsay B, Ralescu A, van der Knaap E (2011) Confusion matrix-based feature selection. Computer Science Department, College of Wooster, Wooster, OH
10. Podani J, Schmera D (2006) On dendrogram-based measures of functional diversity. Oikos 115:179–185

A Tent Map and Logistic Map Based Approach for Chaos-Based Image Encryption and Decryption

Muskaan Kalra, Shradha Katyal and Reena Singh

Abstract A cryptosystem incorporating multiple chaotic maps is proposed. Visual cryptography is necessary in today's world to combat a large number of data thefts. Image encryption using chaotic maps is a common procedure, but it does not ensure absolute security. We use two maps in this paper, to curb the problem of security. The use of two chaotic maps has made the process of image encryption more secure. Image encryption is done using tent map, which is superimposed on another image encrypted using the logistic map. The encryption scheme has a greater keyspace because of two chaotic maps instead of one and thus has better security against statistical attacks.

Keywords Chaos-based encryption · Tent map · Logistic map · Visual cryptography · XOR gates

1 Introduction

Chaos-based systems in cryptography have been used since the past decade and are increasingly secure and safe. A small change in the key value gives rise to new encryption because of which the decryption process becomes difficult and is thus more safe and secure. Chaos uses two main properties—confusion and diffusion. Chaotic systems have the following properties: Butterfly effect, unpredictability, mixing, feedback and fractals [1]. Data encryption transforms data into a secret form that makes it difficult to interpret by intruders. At present, a lot of research has taken

M. Kalra (✉) · S. Katyal · R. Singh
Department of Computer Science, Bharati Vidyapeeth's College of Engineering, New Delhi, Delhi, India
e-mail: muskteer_dps@yahoo.com

S. Katyal
e-mail: skatyal2204@gmail.com

R. Singh
e-mail: reena.singh@bharatividyapeeth.edu

© Springer Nature Singapore Pte Ltd. 2019
H. S. Saini et al. (eds.), *Innovations in Computer Science and Engineering*, Lecture Notes in Networks and Systems 74,
https://doi.org/10.1007/978-981-13-7082-3_20

159

place in the field of cryptosystems but due to the presence of only a single-level encryption in most cases, we come across several breaches in security which lead to the loss of confidentiality of data. Predicting the outcome of a chaotic system is not possible since we never know the exact initial conditions chosen by the user. The image in a chaotic system is protected because it is difficult to get access to the key used for the encryption process. We have proposed a technique that involves the use of two chaotic maps simultaneously for the purpose of image security. The key space for such a procedure is large, thereby making it more secure and resistant to statistical attacks.

Organization of the paper is as follows. Section 2 introduces the literature survey for the paper. Our image encryption and decryption results have been evaluated in Sect. 3 and results: histogram analysis and correlation coefficient analysis tests have been performed in Sect. 4.

2 Literature Survey

In the year 2012, Ye et al. [2] put forward a rather interesting image hiding scheme that employed the 3D skew tent map. A 3D tent map with three parameters and a coupled lattice has been used to encode the necessary image which can then be embedded into one host image.

In 2013, Peng et al. [3] projected a replacement image encoding technique that was digital and was a mix of chaotic maps and deoxyribonucleic acid coding. Analysis of correlation of adjacent pixels is performed that shows coefficients of the encrypted pictures as extremely tiny. This shows that associate degree persona non grata cannot get any valuable data by employing applied mathematics attacks.

Also, Al-shameri and Mahiub [4] analyzed dynamic properties of the tent map like stability and period orbits and used MATLAB to show the results. Bifurcation diagram was also plotted for the tent map. In the same year, Hassani and Eshghi [5] proposed an algorithm consisting of two parts for analysis of the tent map-frequency domain and time domain.

In 2014, Zhang and Cao [6] introduced a method much different from the conventional Logistic and Tent map, that displayed better encoding properties and maximal Lyapunov exponent. This was compared with the standard maps—Logistic Map and Tent Map and was discovered to be simpler and secure.

In 2015, Scheicher et al. [7] projected driving properties of tent map. They started with defining the equation of the tent map followed by the graph. A relation between the tent map's dynamic systems those induced by a beta-expansion has been mentioned.

In 2016, Liu and Miao [8] projected a parameter-varied Logistic Map that resolved all problems and issues of the standard Logistic Map. Various tests and methods were employed to judge its performance.

In 2017, Abdullah and Abdullah [9] proposed an algorithm involving Arnold Cat's Map, Henon and Logistic Maps. The process was done in three ways—confusion, shuffling, and diffusion. Histogram, NCPR, UACI, and correlation are analyzed.

3 Proposed Research Scheme

(i) Encryption Process

Our proposed scheme consists of two processes: application of tent map and application of the logistic map (Fig. 1). The input parameters use key worth $\mu = 1.777$ for tent map and also the original image as shown in Fig. 2a.

The image to be encrypted is split into its constituent red, green, and blue (RGB) parts. The equation of the tent map is applied to every element separately to disarrange the pixels. From this, the constituent values obtained are extremely low. They are, therefore, increased by an element of 10,000 to make computation easier and quicker. The initial value of the pixel is combined with the new value obtained by applying the tent map equation in order to change the pixel. It is then divided by 256 so as to get the value required for cryptography. This may thereby facilitate the coding method. A new image is then chosen as in Fig. 3a for the dual cryptography. This image is then broken down into its RGB parts just like the previous one. The logistic equation is applied to each of its parts to create chaos amongst the pixel values. The input

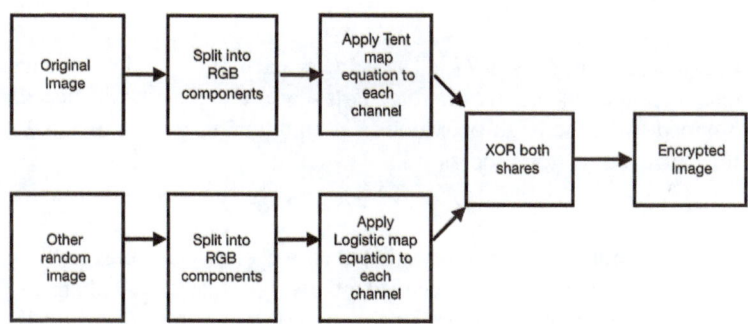

Fig. 1 Block diagram of the encryption process

Fig. 2 **a** Original image to be encrypted, **b** encrypted image share 1 using tent equation

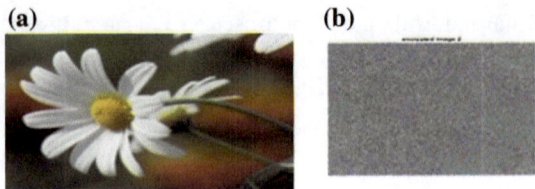

Fig. 3 **a** Randomly selected image, **b** random image encrypted share 2 using logistic equation

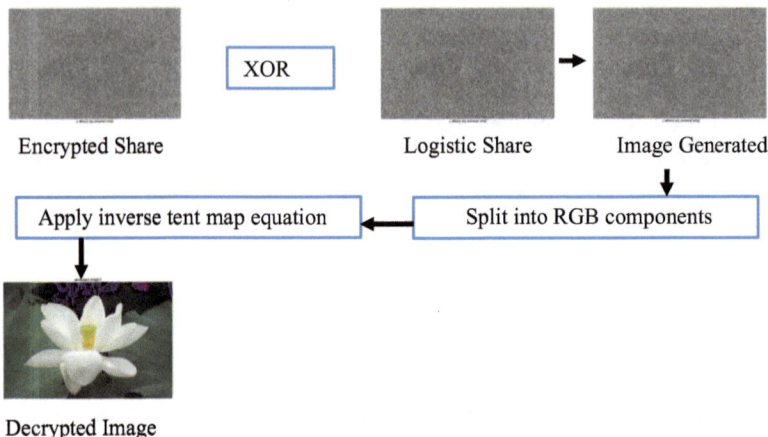

Fig. 4 Decryption algorithm

parameters use key value r = 3.777. The value of each pixel needed for encryption is obtained as in the method used before. Two shares are generated. The shares are then combined using the XOR operation to get the final encrypted image. Figure 1 shows the entire encryption process.

(ii) **Decryption**

Bitwise XOR operation is carried out between the encrypted image and share 2 (logistic image). Divide the image obtained into its consequent RGB channels. Then, apply the inverse tent map equation using the same value $\mu = 1.777$ (Fig. 4).

4 Results

Cryptography schemes need to be checked for safety against statistical attacks. We have tested our method using the following methods:

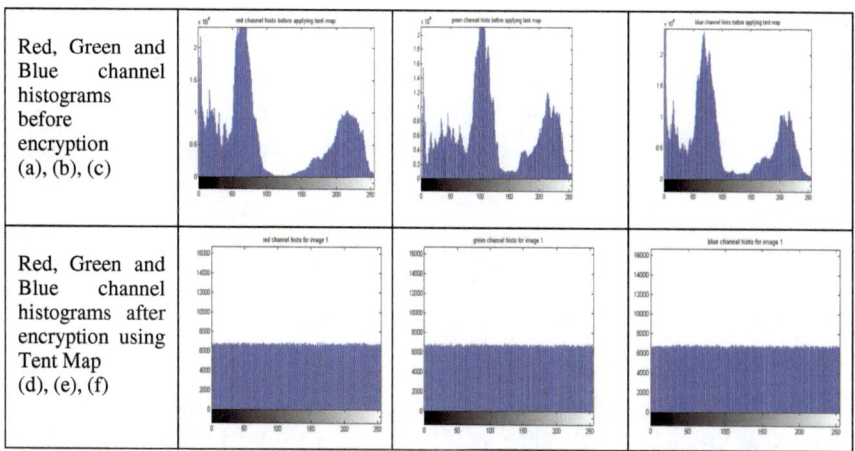

Fig. 5 **a, b, c** Histograms of red, green, and blue channels before encryption, **d, e, f** histograms of red, green, and blue channels after encryption using tent map

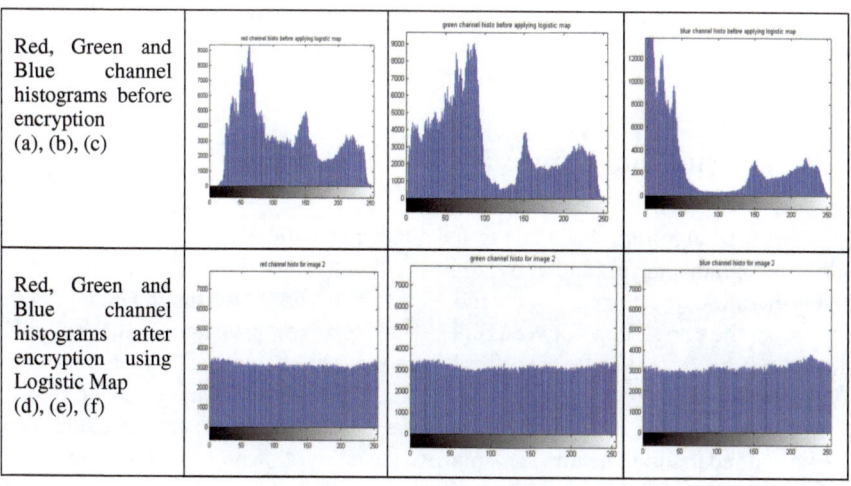

Fig. 6 **a, b, c** Histograms of RGB channels before encryption, **d, e, f** histograms of RGB channels after encryption using logistic map

(i) **Histogram Analysis**

As can be seen from Figs. 5 and 6, both the images were chosen for the proposed encryption technique are safe from statistical attacks because the encrypted image histograms are uniform using chaotic maps.

Table 1 Value of the correlation coefficient for different key values

Image	Key	R	G	B
White lotus	1.666	−0.0013	−0.0032	0.00032
White lotus	1.888	0.0018	0.0015	0.0027
White lotus	1.765	−0.0030	−0.0039	−0.0019
Cat	1.666	−0.0099	−0.0015	−0.0058
Cat	1.888	−0.0017	−0.0022	−0.0019
Cat	1.765	−0.0023	−0.0017	−0.0022

(ii) **Correlation Coefficient Analysis**

The correlation coefficient determines the level of encryption. The lesser the value, the better is the encryption technique. As can be seen from Table 1, all correlation values are very low indicating a safe encryption technique.

Also, the correlation coefficients for two different images have been obtained at three different key values. All values are different which indicates that different key values yield different results. This shows that the key space in the encryption technique is large. Thus, the proposed scheme becomes difficult to crack by intruders.

5 Conclusion and Future Work

The proposed algorithm has been tested against statistical attacks using two methods—histogram analysis and correlation coefficient analysis.

Furthermore, the correlation coefficient analysis has shown that on changing the key value, the correlation between pixel values changes, proving that the method is successful and secure. There are several paths that can be followed after this research. The key selection process can be randomized. The number of shares superimposed can be increased to increase the layers of security. Multiple types of chaotic maps can be applied to the same image to improve the encryption process. The speed of the decryption process can be increased with the use of an additional data structure—the lookup table.

References

1. Fractal Foundation. https://fractalfoundation.org/resources/what-is-chaos-theory/. Accessed 24 Apr 2018
2. Ye R, Zhou W, Zhao H (2012) An image hiding scheme based on 3D skew tent map and discrete wavelet transform. In: 2012 fourth international conference on computational and information sciences, Chongqing, 2012, pp 25–28. https://doi.org/10.1109/iccis.2012.66

3. Peng J, Jin S, Lei L, Han Q (2013) Research on a novel image encryption algorithm based on the hybrid of chaotic maps and DNA encoding. In: 2013 IEEE 12th international conference on cognitive informatics and cognitive computing, New York, NY, 2013, pp 403–408. https://doi.org/10.1109/icci-cc.2013.6622274
4. Al-shameri WFH, Mahiub MA (2013) Some dynamical properties of the family of tent maps. Int J Math Anal (Ruse) 7. https://doi.org/10.12988/ijma.2013.3361
5. Hassani E, Eshghi M (2013) Image encryption based on chaotic tent map in time and frequency domains. ISC Int J Inf Secur 5(1):97–110. https://doi.org/10.22042/isecure.2013.5.1.7
6. Zhang X, Cao Y (2014) A novel chaotic map and an improved chaos-based image encryption scheme. Sci World J 8 p. https://doi.org/10.1155/2014/713541. Article ID 713541
7. Scheicher K, Sirvent VF, Surer P (2016) Dynamical properties of the tent map. J Lond Math Soc. https://doi.org/10.1112/jlms/jdv071
8. Liu L, Miao S (2016) A new image encryption algorithm based on logistic chaotic map with varying parameter. SpringerPlus 5:289. PMC. Web, 14 Nov 2017
9. Abdullah HN, Abdullah HA (2017) Image encryption using hybrid chaotic map. In: 2017 international conference on current research in computer science and information technology (ICCIT), Slemani, 2017, pp 121–125. https://doi.org/10.1109/crcsit.2017.7965545

Two-Way Encryption and Decryption Technique Using Chaotic Maps

Muskaan Kalra, Shradha Katyal, Reena Singh and Narina Thakur

Abstract A cryptosystem incorporating multiple chaotic maps is proposed. The use of chaotic maps for image encryption is a common procedure but this does not ensure absolute security. This paper presents encryption of novel images, a technique that incorporates three different chaotic maps—Tent map, Logistic Map and Gauss Iterative Map for encryption. Image encryption will be a two-step process involving one chaotic map to randomly rearrange the pixels of individual RGB components and other maps to reorganize another image. These two will then be XORed to provide dual security. The technique's strengths have been analysed and presented.

Keywords Encryption technique · Tent map · Logistic map · Gauss Iterative map · XOR gates

1 Introduction

Chaotic systems are secure because they have a wide range of values that can be used to encrypt the images and each value produces a different random image, making the encryption secure and difficult to breach by intruders. Most techniques employ only a single-level security by employing a single chaotic map but in this research, we have employed three different chaotic maps for two-level security which involves performing an XOR operation with another randomized image as well. What makes

M. Kalra (✉) · S. Katyal · R. Singh · N. Thakur
Department of Computer Science, Bharati Vidyapeeth's College of Engineering, New Delhi, Delhi, India
e-mail: muskteer_dps@yahoo.com

S. Katyal
e-mail: skatyal2204@gmail.com

R. Singh
e-mail: reena.singh@bharatividyapeeth.edu

N. Thakur
e-mail: narina.thakur@bharatividyapeeth.edu

© Springer Nature Singapore Pte Ltd. 2019
H. S. Saini et al. (eds.), *Innovations in Computer Science and Engineering*, Lecture Notes in Networks and Systems 74,
https://doi.org/10.1007/978-981-13-7082-3_21

this scheme more secure than the previous ones is that it employed multiple chaotic maps and XOR gate to encrypt an image, thereby, providing an additional layer of security. Each chaotic map is designed in such a way that on changing the values of one parameter, the randomness of the pixels' changes and a new shuffled image is produced. This paper has been proposed to overcome the drawbacks of the paper by Kalra et al. [1] in 2017.

The remainder of this paper is organized as follows. Section 2 introduces the literature survey for the paper. Section 3 explains our submitted image encryption and decryption pattern and its results have been evaluated in Sect. 4 where histogram analysis, and correlation coefficient analysis have been executed to analyse the efficiency of the recommended algorithm.

2 Literature Survey

In 2017, Kalra, Dua and Singh [1] proposed a dual encryption scheme using Logistic Map and noise. This paper explained the encryption technique by employing logistic map on one image and performing an XOR operation with the noise image. This technique proved to be effective but was less secure. Thereby, we have devised a technique that makes use of three chaotic maps in order to increase the key space as well as to enhance the security of the encryption process. In 2007, Gao and Chen [2] presented a modern image encryption strategy that made use of picture absolute rearrangement matrix to reorder the image pixels and then make use a hyper-chaotic system to confuse the link between images, which is plain and the cipher image. In 2010, Liu and Wang [3] produced a stream-cipher strategy based upon single-time keys and powerful chaotic maps. They used a definite chaotic map to generate a pseudo-irregular keystream sequence. NCPR and UACI measures were used to calculate the feasibility of the proposed algorithm. In 2011, Sathishkumar et al. [4] proposed an algorithm in which a combination of subkeys is generated using the logistic map, which is used for image encryption and its transformation leads to diffusion. Then subkeys are produced by four different chaotic maps. In 2013, Cao [5] proposed a new hybrid chaotic map. MD5 is used as part of an initial condition and as a control parameter to alter the path or trajectory to increase security against plaintext-chosen and differential attacks even further. A ciphered image is embedded into several carrier images to reduce suspiciousness and increase robustness. In 2014, Zhang and Cao [6] introduced an altogether new chaotic map and Arnold Cat's Map based image encryption strategy. This was compared with the traditional maps—Logistic Map and Tent Map and was discovered to be more effective and secure. In 2015, Mishra and Prakash [7] used multiple chaotic maps on different sections of the image. This provided a robust m method for encryption of coloured images. In 2016, Gopalakrishnan and Ramakrishnan [8] proposed a beneficial block-wise picture encoding model based on many chaotic maps. The image is broken into four chunks and each is permuted with the Cat Map and its restrictions are controlled by Henon map. A double permutation is produced due to overlapping of the blocks.

3 Proposed Research Scheme

(i) Encryption Process

We propose an encryption technique consisting of three processes. The steps are as follows:

1. The image to be encrypted—Fig. 5a is broken into its Red, Green and Blue (RGB) channels.
2. The red channel is encrypted using the Logistic Map.
3. The blue channel is encrypted using the Tent Map.
4. The green channel is encrypted using the Gauss Iterative Map.
5. These three components are then combined to produce a single encrypted image.
6. A random image is then chosen and broken down similarly into its red, green and blue (RGB) components.
7. Logistic Map is employed on each of these components and a final encrypted image of this random image is also produced.
8. The two final encrypted images thus obtained are then combined using the XOR function to produce a single final encrypted image which is sent on the network.
9. Different key values for each map are used according to the ranges and number of parameters. The block diagram for the operation is displayed in Fig. 1.

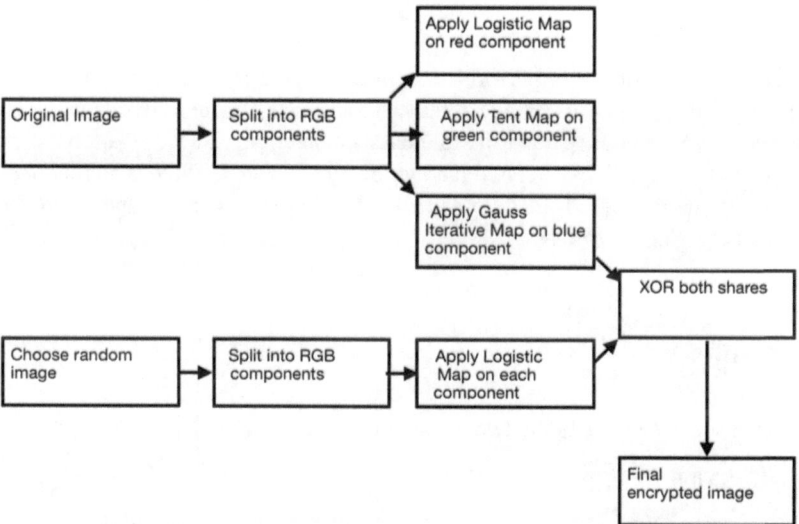

Fig. 1 Block diagram of encryption process

(a)   (b)

Fig. 2 **a** Original image to be encrypted, **b** encrypted image share 1

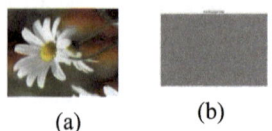

(a)   (b)

Fig. 3 **a** Randomly selected image, **b** random image share 2 after encryption

(a)   (b)   (c)   (d)

Fig. 4 **a** Red channel decrypted for image 1, **b** blue channel decrypted for image 1, **c** green channel decrypted for image 1, **d** final decrypted image

(ii) **Decryption**

Bitwise XOR operation is performed between the encrypted image and share 2 (Logistic image). Divide the image obtained into its consequent RGB channels. Then, apply inverse logistic map equation to the red channel, inverse tent map to the green channel and inverse Gauss Iterated map to the blue channel. There is some information loss in the decrypted image obtained. The technique can thus be improved in this aspect (Figs. 2, 3 and 4).

4   Results

Our proposed scheme is being tested using the following analysis:

(i) **Histogram Analysis**

As can be seen from Figs. 5 and 6, the histograms obtained after applying the chaotic maps are uniform. Thus, both the images were chosen for the proposed encryption technique are safe from statistical attacks.

(ii) **Correlation Coefficient Analysis**

The correlation coefficient determines the level of encryption. The lesser the value, the better is the encryption technique. As can be seen from Table 1, all correlation values are very small indicating a safe encryption technique.

Histograms of Red, Green and Blue channels before encryption (a), (b), (c)	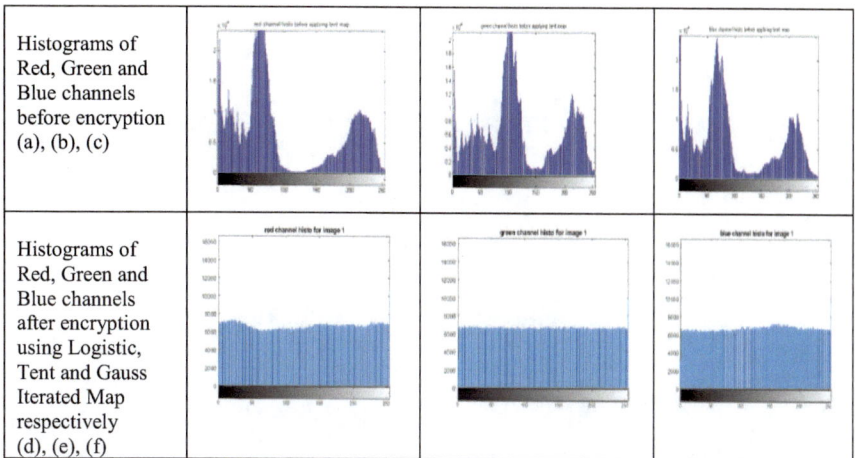
Histograms of Red, Green and Blue channels after encryption using Logistic, Tent and Gauss Iterated Map respectively (d), (e), (f)	

Fig. 5 **a, b, c** Histograms of red, green and blue channels before encryption, **d, e, f** histograms of red, green and blue channels after encryption using logistic map, tent map and Gauss iterated map respectively

Histograms of Red, Green and Blue channels before encryption (a), (b), (c)	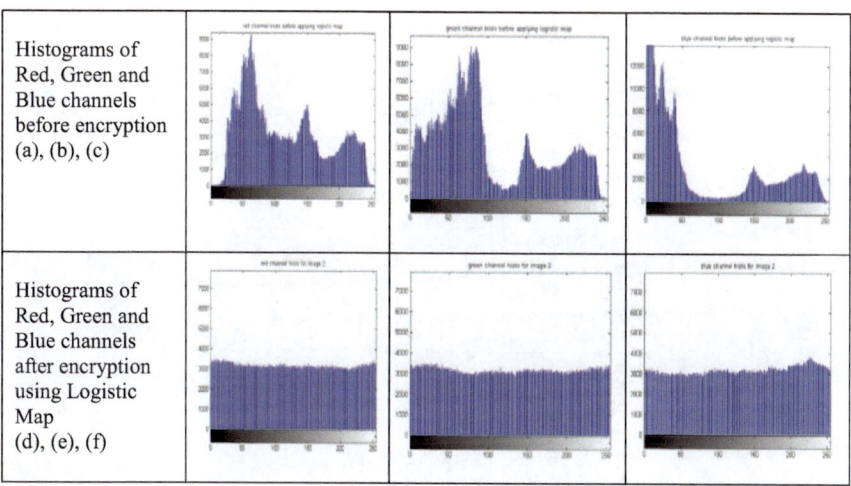
Histograms of Red, Green and Blue channels after encryption using Logistic Map (d), (e), (f)	

Fig. 6 **a, b, c** Histograms of RGB channels before encryption, **d, e, f** histograms of RGB channels after encryption using logistic map

Also, the correlation coefficients for two different images have been obtained at three different key values. All values are different which indicate that different key values yield different results. This shows that the key space in the encryption technique is large. Thus, the proposed scheme becomes difficult to crack by intruders.

Table 1 Value of correlation coefficient for different key values and two different images

Image	Key for red channel (logistic map)	Correlation coefficient of red channel	Key for green channel (tent map)	Correlation coefficient of green channel	Key for blue channel (Gauss Iterated map)		Correlation coefficient of blue channel
					Alpha (α)	Beta (β)	
Cat	3.777	0.0077	1.888	−0.0022	6.20	−0.58	0.0037
Cat	3.585	0.0266	1.765	−0.0017	8.20	−0.58	0.0016
Cat	3.585	0.0266	1.765	−0.0017	8.20	−0.68	−0.0052
					8.20	−0.40	1
Flower	3.777	−0.00054529	1.888	−0.003	6.20	−0.58	0.0038
Flower	3.585	0.0369	1.765	−0.0039	8.20	−0.58	0.0011
Flower	3.585	0.0369	1.765	−0.0039	8.20	−0.68	0.0103
					8.20	−0.40	1

5 Conclusion and Future Work

The algorithm produces good results as can be seen by the histogram analysis that produces uniform graphs. Furthermore, the correlation coefficient analysis has shown that on changing the key value even to a small extent, the correlation between pixel values changes significantly, proving that the method is secure and efficient.

There are several paths that can be followed after this research. The key selection process can be randomized. The number of shares superimposed can be increased to increase the layers of security. Multiple types of chaotic maps can be applied to the same image to improve the encryption process. The speed of the decryption process can be increased with the use of an additional data structure—the lookup table.

References

1. Kalra M, Dua HK, Singh R (2017) Dual image encryption technique: using logistic map and noise. In: 2017 17th international conference on intelligent systems design and applications, New Delhi, 2017, pp 201–208. https://doi.org/10.1007/978-3-319-76348-4_20
2. Gao T, Chen Z (2008) A new image encryption algorithm based on hyper-chaos. Phys Lett A 372(4):394–400. https://doi.org/10.1016/j.physleta.2007.07.040. ISSN 0375-9601
3. Liu HJ, Wang XY (2010) Color image encryption based on one-time keys and robust chaotic maps. Comput Math Appl 59:3320–3327. https://doi.org/10.1016/j.camwa.2010.03.017
4. Sathishkumar GA, Bhoopathy Bagan K, Sriraam N (2011) Image encryption based on diffusion and multiple chaotic maps. Int J Netw Secur Its Appl (IJNSA) 3(2). https://doi.org/10.5121/ijnsa.2011.3214
5. Cao Y (2013) A new hybrid chaotic map and its application on image encryption and hiding. Math Probl Eng 13 p. https://doi.org/10.1155/2013/728375. Article ID 728375
6. Zhang X, Cao Y (2014) A novel chaotic map and an improved chaos-based image encryption scheme. Sci World J 8 p. https://doi.org/10.1155/2014/713541. Article ID 713541
7. Mishra P, Prakash S (2015) Encryption of images by multiple chaotic maps. Int J Adv Found Res Comput (IJAFRC) 2
8. Gopalakrishnan T, Ramakrishnan S. Image encryption in block-wise with multiple chaotic maps for permutation and diffusion. https://doi.org/10.21917/ijivp.2016.0177

Legitimate Privilege Abuse and Data Security in Database

S. Aravindharamanan, Somula Ramasubbareddy and K. Govinda

Abstract Data is fundamental in today's digital life. Therefore, the essential need is to protect data from unauthorized users. Choosing a database application depends on cost. Business database is expensive to a great degree. This paper delineates database security utilizing true blue benefit manhandle and cryptography. There are numerous dangers in the database, which releases the data for the denied purpose. Among the fundamental dangers in database security, this paper will portray the true blue benefit abuse. Among the approaches to secure data from gatecrashers/assailants is by using cryptography techniques, the access-control approach is the primary imperative technique to secure database by using cryptography and row-level security.

Keywords Security · Threats · DBMS · Privilege abuse · Row-level security · Cryptography · Encryption · Content control

1 Introduction

Day by day, protection is fundamental in all viewpoints. Everyone desires to protect his/her advantages, basic information. Protection is the main problem in all associations. For all, associations have their own associated and relevant data. So it should be protected from intruders. So the protection of data is the far-reaching extent to protect information from unauthorized customers. Various associations have to collect or access their data by spending some portion of the money. Therefore, data must be guaranteed and protected. The protection of the database should be in the two spots, either in government zone or in private zone. Wherever there is some essen-

S. Aravindharamanan · S. Ramasubbareddy (✉) · K. Govinda
Department of Computer Science and Engineering, VIT University, Chennai, TamilNadu, India
e-mail: svramasubbareddy1219@gmail.com

S. Aravindharamanan
e-mail: aravindharamanan.s2016@vitstudent.ac.in

K. Govinda
e-mail: kgovinda@vit.ac.in

© Springer Nature Singapore Pte Ltd. 2019
H. S. Saini et al. (eds.), *Innovations in Computer Science
and Engineering*, Lecture Notes in Networks and Systems 74,
https://doi.org/10.1007/978-981-13-7082-3_22

tialness of information is to protect. In addition, the information must be guaranteed. The request for protection shorelines is unapproved data discernment, wrong data adjustment, and unavailability. The protection of data needs three properties, privacy, trustworthiness, and openness. Protection insinuates confirmation from unapproved customers. Reliability keeps the unapproved customers and misguided data change and openness suggests recovery from hardware and programming botches or harmful activity realizing the refusal of data availability [1–8].

2 Literature Survey

Database Threats
The repeat of assaults against these databases has in as manner extended. Database overseers truly can deal with those DDOS attacks [4–9].

Excessive Privilege-Based Abuse
Exactly when database customers are outfitted with getting to benefits that outperform their essential action, these advantages can be misused by intention or unexpectedly. For example, a database executive in budgetary affiliation. If he drops audit trails or makes counterfeit records he can have the ability to trade money beginning with one record then onto the following so mistreating the unnecessary advantage intentionally. Another case is a DBA in the bank, whose action is to change customer contact can access other details. An affiliation is giving a task at home, other option of agents and the laborer takes a fortification of extraordinarily sensitive information to manage from home. This is not only neglects the protection techniques of affiliation, yet what's more may realize data protection break, if a system at home is dealt. So this advantage can be misused incidentally.

Legitimate Privilege-Based Abuse
Customers in a similar manner misuse bona fide database benefits for ill-conceived purposes. Exactly when the affirmed customer mishandles the true blue advantage for an unapproved reason, this is called genuine advantage abuse. Good old fashioned advantage misuse can be as mishandle by database customers, chiefs or a system boss doing any unlawful or deceptive development. It is, however not confined to, any manhandling of sensitive data or unjustified usage of advantages [2]. For example, affiliation laborer with advantages to see particular specialist records by methods for a custom Web application. The structure of the web application normally obliges customers to audit an individual laborer's history. A couple of records cannot be seen in the meantime and electronic duplicates are not good old fashioned. Regardless, the heel laborer may dodge these imperatives by a partner with the database using different customers, for instance, MS Excel and his genuine login qualifications, the laborer may recover and spare every single delegate record.

Platform Vulnerabilities
Vulnerabilities in previous working structures like Windows 2000, UNIX, Linux, etc.,

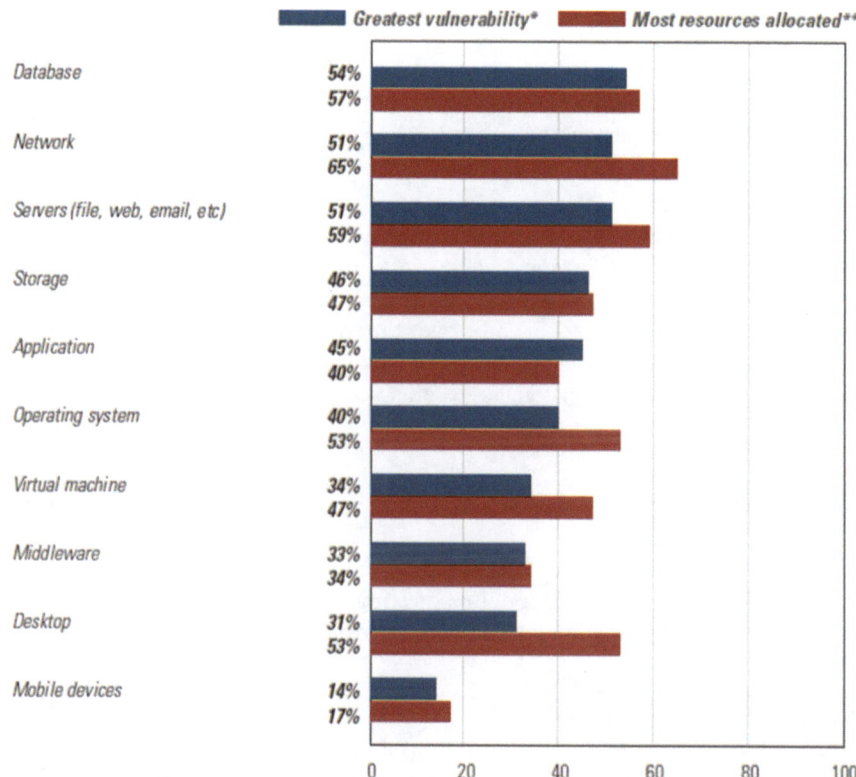

Fig. 1 Vulnerabilities versus priorities

and additional organizations presented on databases may provoke unapproved access to database management or forswearing of administration. For instance, the Blaster Worm exploited in a Windows 2000 helplessness to make dissent of administration situations [2]. In a study report of information security [6]: pioneers versus slouches. The vulnerabilities versus priorities to audit graph is shown in Fig. 1

Stored Procedure

It communicates that in a Database Organization System (DBMS), a putaway methodology is a plan of Structured Query Language (SQL) clarifications with an allotted name that is secured in the database fit as a fiddle so it can be shared by different undertakings. The usage of putaway strategy philosophy can be helpful in controlling access to data (end-customers may enter or change data yet do not create frameworks), protecting data respectability (information is entered consistently), and improving effectiveness (clarifications in a set away method just ought to be formed one time).

3 Proposed Method

Row-level security and data masking are the better methods to be implemented for providing security to a database [8].

Row-Level Security

Row-Level Security for SQL Database is presently large accessible. RLS empowers you to store information for some clients in a solitary database and tables, while in the meantime confining row-level access in view of a client's character, part, or implementation setting. RLS unifies get to rationale inside the database itself, which improves and decreases the danger of blunder in your application code. RLS-Diagram-4 RLS can help customers develop secure applications for a variety of scenarios. For instance,

Confining access to money-related information in the view of a worker's area and part.

Guaranteeing that occupants of a multi-inhabitant application can just access their own particular rows of information. Empowering diverse experts to investigate distinctive subsets of information in light of their position.

Discretionary Access Control

Approval control comprises of scrutiny if given subject/client, activity/benefit name, database object will have permission to continue [3]. So an approval will be seen as subjects, task write, obj definition, which determines that the subject has the privilege to play out an activity of activity compose on a protest. To oversee approvals properly, the DBMS requires significance of subjects/customer, inquiries, and rights or exercises. The system for enforcing discretionary access control in database structure relies upon the Grant and Revoke benefits. Surrender, Revoke clarifications are used to endorse triplets (customer/subject, action/advantage, data challenge) [5–7].

In **UNIVERSITY** database application, consider Student and Professor table.

After Row-Level Security

execute as user = 'moorthy'
go
select * from dbo.detstudent

	name	regno	branch	city	cgpa	emp_id
1	NIKHII	15BCE0589	CSE	ATP	9	1
2	praneeth1	15MIS0081	MIS	kadapa	9	1

Parts of Row-Level Security

a. Security Policy
b. Filter Predicate

You might use Dynamic Data Masking to enable developers to work with production tables without exposing sensitive data or you might use it in help desk or call center scenarios where you want to restrict the data the help desk personnel can see. It is important to understand that Dynamic Data Masking is not encryption. The key difference is that Dynamic Data Masking obfuscates data as it is displayed to the end user. However, it does not change or encrypt the data that is stored on the disk. Dynamic Data masking is implemented when a table is built using the CREATE TABLE statement or you can add it after the fact by using the ALTER COLUMN statement. Implementing Dynamic Data Masking on a column does not prevent authorized users from performing updates to that column. Despite the fact that the end clients may see veiled information when they question the covered segment they can even now refresh, embed, and erase the information on the off chance that they have composed consents. SQL Server 2016's Dynamic Data Masking gives a few implicit functions that you can quickly use without the need to compose any information concealing functions all alone. You can see the inherent information covering capacities in the table beneath.

Default
Full concealing as indicated by the information sorts of the assigned fields.
Conceal WITH (FUNCTION = 'default()') NULL

Email
A concealing technique which uncovered the primary letter of an email address and the steady postfix ".com" as an email address. aXXX@XXXX.com.
Veiled WITH (FUNCTION = 'email()') NULL

Custom String
A veiling strategy which uncovered the first and last letters and includes a custom cushioning string in the center. prefix,[padding],suffix.
Covered WITH (FUNCTION = 'partial (prefix, [padding], suffix)') NULL.

Arbitrary
An arbitrary concealing capacity for use on any numeric sort to cover the first incentive with an irregular incentive inside a predefined extend.
Veiled WITH (FUNCTION = 'random([start range], [end range])'.
Giving the UNMASK consent enables a client to see the information unmasked. You can see cases of utilizing the MASK and UNMASK consents in the accompanying posting.

Adding the UNMASK consent.
Give UNMASK TO User1;
Removing the UNMASK consent.

Disavow UNMASK TO User1;

There are a few considerations you need to be aware of with Dynamic Data Masking. You cannot use it with Always Encrypted columns or the FILESTREAM data type. Although the data is not masked when it is stored in the database if the client executes the SELECT INTO or INSERT INTO to duplicate information from a covered section into another table the outcomes in the objective table will be concealed. You can utilize the sys. masked_columns view to see the sections that have a veiling capacity connected to them. This DMV restores all sections and the is_masked and masking function segments show if a segment is concealed and the veiling capacity that was utilized.

Dynamic Data Masking can be combined with Always Encrypted, Row-Level security, and Transparent Data Encryption (TDE) to help create a comprehensive and layered security strategy.

After Data Masking
execute as user = 'anyuser'
select * from dbo.detstudent.

4   Experimental Analysis

An analysis is made on the two methods, Stored procedure and Row-Level Security based on the execution time in eight attempts as show in Table 1.

Avg. execution time $= 4.8 * 10^{-2}$ s for Row-Level Security
Avg. execution time $= 6.6 * 10^{-2}$ s for Data Masking

So, it is predicted that the stored procedure takes more time for execution. Whereas Row-Level Security takes less time for execution when compared to the stored procedure method (Fig. 2).

Table 1 No. of attempts in executing

No. of attempts in executing	RLS	Stored procedure
1	4.4	6.9
2	4.6	6.8
3	4.8	6
4	5.3	6.6
5	4.6	6.6
6	4.2	8.3
7	4.4	8.4
8	4.6	7.9

Fig. 2 Stored procedure versus row-level security

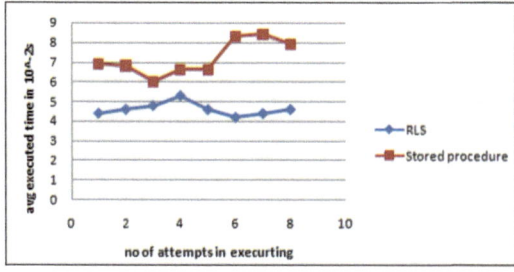

5   Conclusion

This paper discusses database security using Legitimate Privilege Abuse the methods for providing security to the database from assailants. In this paper, a few database dangers have been talked about. This security is possible by various cryptographic techniques. The paper discusses the cryptography in detail. An analysis made on two methods discuss that the the optimized method is Row-Level Security than Stored Procedure. So the security of the database can be provided in two levels, one is at the content level and the next is at access level. The rule destinations of database protection are to guarantee unapproved access to data, guarantee the unapproved change of data to guarantee that data continually open when required. Further usage of this venture will manage the encryption on how the secured content like watchword will be put away in the ambiguous configuration.

References

1. Bertino E, Sandhu R (2005) Database security-concepts, approaches, and challenges. IEEE Trans Dependable Secure Comput 2(1):2–19
2. Asole SS, Mundada MS (2013) A survey on securing databases from unauthorized users. Int J Sci Technol Res 2(4):228–230
3. Basharat I, Azam F, Muzaffar AW (2012) Database security and encryption: a survey study. Int J Comput Appl 47(12)
4. Rohilla S, Mittal PK (2013) Database security: threats and challenges. Int J Adv Res Comput Sci Softw Eng 3(5)
5. Bertino E, Jajodia S, Samarati P (1995) Database security: research and practice, Pergamon. Inf Syst 20(7):537–556
6. Maurer U (2004) The role of cryptography in database security. In: Proceedings of the 2004 ACM SIGMOD international conference on management of data, ACM, pp 5–10, June
7. Sandhu RS, Jajodia S (1993) Data and database security and controls. Handbook of Security Management
8. Le Grand C, Sarel D (2008) Database access, security, and auditing for PCI compliance. EDPAC: EDP Audit, Control, Secur Newsl 37(4–5):6–32
9. Hashmi MJ, Saxena M, Saini R (2012) Classification of DDoS attacks and their defense techniques using intrusion prevention system. Int J Comput Sci & Commun Netw 2(5):607–614

Encryption Using Logistic Map and RSA Algorithm

Krishna Sehgal, Hemant Kumar Dua, Muskaan Kalra, Alind Jain and Vishal Sharma

Abstract Data collected by each individual is increasing day by day and so is the rate of transfer of that data. Hence, there is a need to protect this data while the transfer is being done. One simple solution to this problem of data security is cryptography. Our paper proposes a method to share images after being encrypted by logistic maps and Linear-Feedback Shift Register (LSFR) followed by the RSA Algorithm. Thereby, we provide extra layers of security over the existing encryption technique making it difficult to decrypt the images after data transfer. The image can be decrypted only with the help of a key.

Keywords Logistic maps · RSA algorithm · Linear-Feedback Shift Register (LSFR) · RGB matrix · Encryption

1 Introduction

For many years, the backbone of human existence is information. As technology increases day by day, electronic devices have been advancing and most of the information exchange between devices is done with the help of digital signals. Due to this increasing demand for data transfer via digital signals with the help of elec-

K. Sehgal (✉) · H. K. Dua · M. Kalra · A. Jain · V. Sharma
Department of Computer Science, Bharati Vidyapeeth's College of Engineering,
New Delhi, Delhi, India
e-mail: krishnasehgal2108@gmail.com

H. K. Dua
e-mail: hemant.dua56@gmail.com

M. Kalra
e-mail: muskteer_dps@yahoo.com

A. Jain
e-mail: alindjain11@gmail.com

V. Sharma
e-mail: vishalmtr@gmail.com

© Springer Nature Singapore Pte Ltd. 2019
H. S. Saini et al. (eds.), *Innovations in Computer Science
and Engineering*, Lecture Notes in Networks and Systems 74,
https://doi.org/10.1007/978-981-13-7082-3_23

tronic devices, there is an increasing demand for security of information, which is being transferred as this information may be confidential and an increasing number of cases have come to light where data is being hacked in order to get access to this vital information. This leads to a loss of integrity and authenticity of the information. This problem leads us to a field, which was introduced in 1948 and was termed as cryptography. It involves rules of mathematics and protocols of computer science in order to encrypt the vital data, which can then be decrypted on the other end with the help of some preset rules. Cryptography can be described by its two forms.

1.1 Symmetric

It is the form of cryptography in which there is only one key present which is responsible for both encryption as well as decryption operations at end of sender and receiver.

1.2 Asymmetric

It is the form of cryptography in which there are two types of keys, one is the public key, which can be provided to anyone, and the other is the private key which is to be stored securely by the user.

Images that are accessed by hackers can be easily modified which can lead to loss of integrity and authenticity and the resulting image could be used for the wrong purposes. Areas of military and defence where a high-level of security is desired, if the images are accessed illegally, national security may be under threat. In this paper along with RSA algorithm, we have used chaotic maps in order to provide an additional security layer and remove the discrepancies in encryption using the RSA algorithm. Chaotic maps are based on the principle of chaos, which simply means disorder and are extremely sensitive to their initial states.

2 Literature Survey

Cryptography is an emerging research field and many electronic devices are using encryption before transferring data. Other than logistic maps [1], a lot of chaotic maps [2–4] have been used for encryption, for example, multi-chaotic system [5–14], standard maps [8], baker maps [9] and cat maps [10]. Use of systems based upon multi-chaotic in order to encrypt coloured images is shown in [11]. There are cases where two of these chaotic maps are combined in order to perform encryption in two different stages [13]. In another research, encryption using logistic and pixel table is proposed [14]. Another author used encryption composed of the baked map [15], logistic map and folded map. After achieving better efficiency even, the histogram is non-uniformly distributed for resultant encrypted images.

3 Implementation

The encryption algorithm comprises the following steps:

Step 1 Formation of RGB matrix

In order to present the mathematical part in cryptography with image depth, it is necessary to transform the image in a valid mathematical model.

i. The image is converted to a three-dimensional matrix by the sender. The command in MATLAB for this conversion is
    X = imread ("image.jpg");
    Display (X)
ii. Once the correct RGB three-dimensional matrix is formed by converting the image, the sender has to convert X matrix into another row matrix Y in order to have mathematical operations available. Command in MATLAB to achieve this is
    Matrix Y = X(:); Y = Y';

Step 2 Chaotic sequences of keys generated by logistic map sequence and progression of states of linear feedback shift

The function of the logistic map is given by Eq. (1) which creates chaotic sequences $\{A_n\}$, where A_n lies between 0 and 1.

$$A_{n+1} = x\, A_n(1 - A_n) \tag{1}$$

Here x is a bifurcation parameter that has a range of 0–4 and A_0 is the initial value from 0 to 1. This generates a sequence of elements according to Eq. (1).

By selecting a value of x between 3.75 and 4, a random chaotic sequence is generated, where A_0 is taken as initial value, which can be any value between 0 and 1. This value received is used as the key sequence after being converted to an 8-bit sequence and called as $L_{1,i}$.

$$L_{1,i} = Round(A_n * 255) \tag{2}$$

Simultaneously an m-stage Linear-Feedback Shift Register is created by feedback polynomial of d (2), if the feedback polynomial used is primitive then the sequence of states generated is periodic and is of period (2m − 1).

Here, m = 8 is chosen with a polynomial $x^8 + x^6 + x^5 + x^4 + 1$. Every initial state leaving zero forms sequence states with period $2^8 - 1 = 255$.

Fig. 1 Block diagram of the
proposed scheme

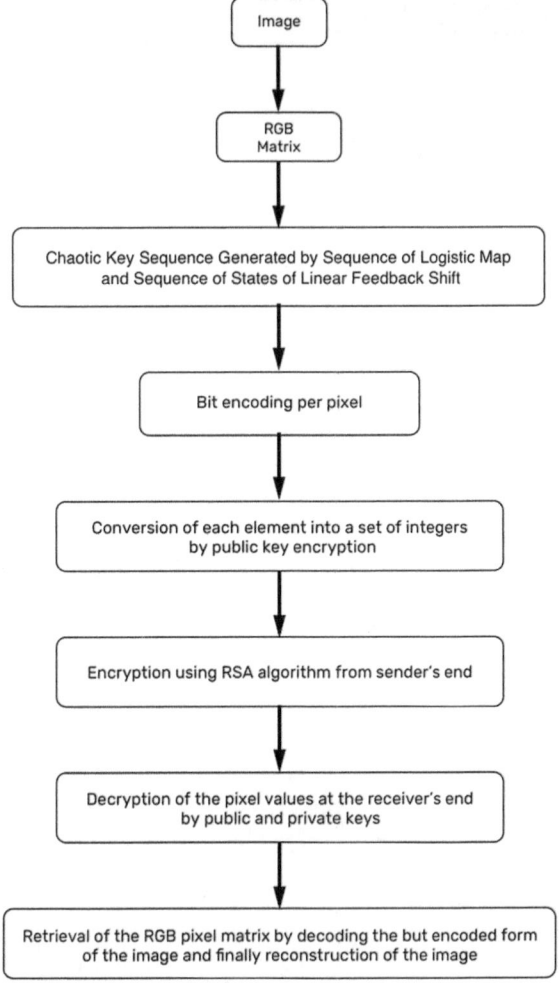

Generated sequences are shifted versions of one another. In the proposed scheme, LFSR of 8 bit is used to create the sequence, which we call $L_{2,i}$. The final sequence is generated by XORing $L_{1,i}$ and $L_{2,i}$ and this sequence is further used in the next step. Figure 1 shows the block diagram of implementation.

Step 3 Non-standard number system is received from pixels

Let pixel 210 be taken as a sample. (They range from 0–255)

i.   First, the binary number of 8 bits is calculated: $(210)_{10} = (1101\ 0010)_2$.
ii. The octal number is calculated by taking 4 bits from the 8-bit binary number: $(1101)_2 = (15)_8$ and $(0010)_2 = (02)_8$. Hence, $(1101\ 0010)_2 = (1502)$.
iii. To increase the security layer, nine's complement is taken: $(1502) = (8497)$.

iv. The result comprises of all the digits present in a decimal system i.e. 0 to 9.
v. The resultant is converted into two parts and every part is used to obtain its octal number, which is then combined to get a number of 6 digits.

$$(84)_{10} = (124)_8 \text{ and } (97)_{10} = (141)_8$$

Hence resultant is $(8497)_{10} = (124141)$

Step 4 Encryption using RSA algorithm at receiver's end

Once we have the matrix form of the image, the RSA algorithm is applied to all the pixels.

i. Two primes are taken by the receiver, such as a $= 17$ and b $= 29$. C $=$ a $\times$ b $= 493$.
ii. Another number has to be generated by the receiver (e), which is given follows:

$$(a - 1) = 16, (b - 1) = 28 \text{ and } (a - 1) * (b - 1) = 448$$

We take e $= 5$, there are many possibilities.
iii. Sender can now encrypt the converted bit pixel with C and e. $(124141)^5 = 2.946645 \times 10^{25}$
iv. The number that the receiver has i.e. 448 is used for decryption. Receiver needs a multiple of 5 which is more than 448's multiple by 1.
Multiples of 448 are 448, 886, 1344, 1792, 2240, 2688

The desired number is 1345 which is greater than 1344 by 1, and $1345 = 5.269$ and d $= 269$.

4 Results

In order to calculate encryption quality, correlation coefficient among adjacent pixels is calculated and histograms are formed as shown in Figs. 2 and 4. Table 1 shows the calculated correlation coefficients that show how much correlated they are with each. Use of three thousand pixels is done in order to calculate the correlation coefficient. Figure 3 shows the RGB components of the image. The equation of correlation is given by Eq. 3. 'r' is the correlation coefficient (Fig. 4).

$$r = n \frac{\left(\sum xy \right) - \left(\sum x \right) \left(\sum y \right)}{\sqrt{\left[n \sum x^2 - \left(\sum x \right)^2 \right]\left[n \sum y^2 - \left(\sum y \right)^2 \right]}} \tag{3}$$

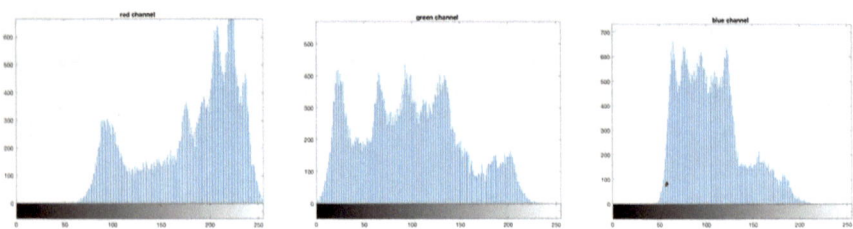

Fig. 2 Histograms before encryption

Table 1 Correlation coefficient between adjacent pixels

	Original image	Cipher Image		
		P=113,q=71, e=6469,d=589	P=47,q=59, e=31,d=1291	P=17,q=23, e=109,d=2661
Lenna				
Corr.	0.9230	0.0064	0.0039	0.0169
Tower				
Corr.	0.9549	0.0030	0.0255	0.0048

Fig. 3 RBG components of the image

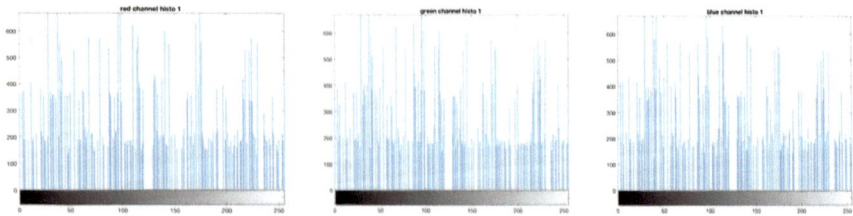

Fig. 4 Histograms after encryption

5 Conclusion

This paper presents a unique way in encrypting images with help of logistic maps, LSFR and RSA algorithm. These lead to the addition of various extra layers of security and preservation to the existing administrative structure of systems. The problem we are facing while using the technique at hand is of space and time complexity, and we have to look at the problem because we have the limit of both. In order to create a system that is hard to penetrate and cannot be hacked easily, a compromise has to be reached for this trade-off. The encryption technique can be used in highly secure projects that are of high value, importance and tremendous military applications.

References

1. Rivest RL, Shamir A, Adleman L (1978) A method for obtaining digital signatures and public-key cyptosystems. Commun ACM 21(2)
2. Kaltz J, Lindell Y (2008) Introduction to modern cryptography: principles and protocols. ISBN: 978-1-58488-551-1
3. Abusukhon A, Talib M (2012) A novel network security algorithm based on private key encryption, IEEE, 2012
4. Yan-Bin Z, Qun D (2011) A new digital chaotic sequence generator based on logistic map, IBICA, pp 175–178, 2011
5. Jemaa ZB, Belghzth S (2002) Correlation properties of binary sequences generated by the logistic map-application to DSCDMA. In: Proceedings of IEEE international conference on systems, man and cybernetics, Hammamet, Tunisia, pp 447–451, 2002
6. Zhang D, Gu Q, Pan Y, Zhang X (2008) Discrete chaotic encryption and decryption of digital images, CSSE, pp 849–852, 2008
7. Di X, Liao X, Wang P (2009) Analysis and improvement of image encryption algorithm. Chaos Solut Fractals 40(5):2191–2199
8. Liu J-m, Qu Q (2010) Cryptanalysis of a substitution–diffusion based image cipher using chaotic standard and logistic map, ISIP, pp 67–69, 2010
9. Salleh M, Ibrahim S, Isnin IF (2003) Enhanced chaotic image encryption algorithm based on Baker's map. In: IEEE conference on circuits and systems, vol. 2, pp 508–511, 2003
10. Wang K, Pei W, Zou L, Song A, He Z (2005) On the security of 3D Cat map based symmetric image encryption scheme. Phys Lett A 343:432–439
11. Hong L, Li C (2008) A novel color image encryption approach based on multi-chaotic system, IEEE, ASID, pp 223–226, 2008
12. Honglei Y, Guang-shou W (2009) The compounded chaotic sequence research in image encryption algorithm. In: Proceedings of WRI global congress on intelligent systems, vol. 3, pp 252–256, 19–21 May 2009
13. Weihua Z, Ying S (2010) Encryption algorithms using chaos and CAT methodology, ASID, pp 20–23, 2010
14. Al-Najjar HM, AL-Najjar AM, Arar KSA (2011) Image encryption algorithm based on logistic map and pixel mapping table, ACIT, 2011
15. "The RSA Algorithm"—Adrian Dudek, PhD, Australian National University

A Dual-Metric Monitoring Scheme in Energy-Secure Routing in Ad Hoc Network

Ugendhar Addagatla and V. Janaki

Abstract Data integrity and range coverage are the two primal issues in the deployment of a ad hoc network. In the development of route in ad hoc network, various constraint such as offered bandwidth, number of hops, link blockage, energy limitation, coverage range, etc., were observed. However, the resource constraints and the need of data preservation are minimizing the reliability of the routing efficiency of ad hoc network. To develop a reliable route in ad hoc network in this paper, a dual-metric monitoring for energy-secured routing is presented. The energy constraint is focused to be minimized by a dynamic scheduling and maximization of aggregate energy level in a route. The proposed approach outcomes with a new routing scheme where the nodes are dynamically scheduled for listen phase on the available request only. This proposed approach illustrated higher energy conservation and less switching overhead as compared to the conventional periodic scheduling scheme.

Keywords Dual-metric monitoring · Energy saving · Scheduling approach · Ad hoc network · Dynamic scheduling

1 Introduction

Evolution of wireless network in data exchange has given opportunity in developing new communication architecture to provide higher service adaptability, long-range communication with the existing wireless communication infrastructure. In the development of such network architecture, ad hoc network has evolved as a new paradigm in data exchange, and gained a lot of attention in the research community. The dynamic property of node adaptability, self-establishing property, and no

U. Addagatla (✉)
Department of CSE, Guru Nanak Institutions Technical Campus, Ranga Reddy, Telangana, India
e-mail: ugendhar2008@gmail.com

V. Janaki
Department of CSE, Vaagdevi Engineering College, Warangal, Telangana, India
e-mail: janakicse@yahoo.com

© Springer Nature Singapore Pte Ltd. 2019
H. S. Saini et al. (eds.), *Innovations in Computer Science and Engineering*, Lecture Notes in Networks and Systems 74,
https://doi.org/10.1007/978-981-13-7082-3_24

pre-infrastructure dependency has made this network more focused on future genera-tion wireless communication architecture. Wherein the advantage of self-developing property makes ad hoc network more focusing, the dynamic nature of node placement and link among them constraints the network from obtaining performance efficiency. The route developed for the data exchange in this network are highly dynamic in nature resulting in low reliability and increases the network overhead with respect to blockage, error rate, power dissipation, etc. The limited power source gives a pos-sibility of faster network collapse under a large distributed network. The nodes, in this case, are scheduled to conserve their power based on the transfer from the wake, sleep, and ideal states. This scheduling is developed to conserve the energy, however, the node enters the scheduled wake period to listen for a request, dissipating power. A scheduling which is dynamically controlled could save this power dissipation. To develop a dynamic schedule in energy conservation a dual-metric monitoring in ad hoc network is proposed. To present the developed approach the rest of the paper is outlined in six sections. Section 2 outlines the past literature developed for energy conservation approaches in ad hoc network. Section 3 outlines the conven-tional energy scheduling protocol for optimal energy routing in ad hoc network. The proposed approach of dual-metric monitoring is outlined in Sect. 4. Section 5 out-lines the simulation results and the comparison for the developed approaches to the conventional method. The concluding remarks for the developed work are outlined in Sect. 6. The last section presents the references made for the development of the proposed work.

2 Literature Survey

To achieve the objective for improving the security of route reliability, various past developments were made by the research community. In [1] a, trust-based routing framework in ad hoc network is presented. The approach defines the routing selec-tion by the correlation of trust driven by multiple interconnected social contexts and builds a reliability reputation for a route to select. The reliability factor is validated by the selection of data integrity in the network. For the secure route establishment, a security management approach is outlined in [2]. The routing approach is based on the geographical location added network routing used for broadcasting and data exchange in the distributed network. The network monitors the data exchange per-formance by means of ending accuracy and data lost in data exchange. In building security during data exchange in [3] an anonymous routing protocol, where the sender and the receiver are made anonymous, and also the intermediate nodes are defined anonymously and simultaneously in the network during routing. In order to adapt the protocol more practical on the Dynamic Network, a password is used instead of a secret key that is shared as a password in the network. In the reception, the recipient receives both the sender and the default public key without primary key information. In [4] a non-passive approach for malicious attack detection is pro-posed. An on-demand route protocol, known as MASK, is used which is used in the data exchange in the MAC layer and network layer communication. This approach uses a dynamic key rather than static MAC and network addresses. MASK computes the sender and receiver credentials as well as sender-reliable credibility to provide reliability. However, the attacks during the data exchange are not being solved in this approach. In [5] a tree-based routing protocol for data exchange over ad hoc network is suggested. The approach derives the source location and develops an approach to maximize the network lifetime based on the dictionary based routing approach. The approach defines the security of routing based on the redundant information exchanged and the lifetime of operation. In [6] a similar objective for route security based on redundant coding is developed. In a multi-attribute approach in [7] a spectrum-energy interpolation is developed. The routing approach derives medium selection based on the radio interference model, and the bandwidth availability. The route in this method is chosen based on the energy interference approach. In [8] a load balancing approach for route optimization is developed. The load balancing in ad hoc network based on Demand Distance Vector (LB-AODV) protocol suggests avoiding network congestion, when considering a non-incorporated transmission bandwidth to achieve a similar level of high data transmissions. A rounding metric has been designed to optimize a path load condition. In [9] addition to power optimization, a route selection at the destination is carried out. It is a routing approach coupled with the user in the route establishment. However, this approach introduces an initial delay in route development process [10]. Suggested channel assignment and routing to cognitive radio networks using an integrated approach such as routing and channel assignment. The Joint Cross-Layer Routing and the Channel Assignment Protocol, based on AODV, which works without a central control system. In this approach, a backup channel is kept for dynamic allocation of data, minimizing the search process. A Local Geocaching Scheme based on novel distance approach for a multi-hop wireless network is outlined in [11]. It does not use the nodes for their geographical information. In this approach, the blind geographical area was derived by flooding focusing on solving the broadcasting effort. In [12], a location for privacy provided for a data sink around the flooding is proposed. An approach has been suggested to provide accurate and reliable tracking of a small mobile device in this network. This approach uses the approach to accurately detect two precise and erroneous reports. The main inspiration for the exploitation of the attack is the harmful sensors which cannot control the physical layer signal-propagation features. In [13] mobility constraint spectrum utilization for ad hoc network is outlined. The spectrum control gives higher network throughput to the system. In [14] the research activity focuses on protecting strategically sensitive areas against location-based service (LBS) providers and unrelated members of the geospatial networks. On the other hand, the method described in [15] uses a threat to define the globally emerging reality of privacy Concern. The threat model specifically guesses global information about the entire network. To achieve security objectives, a hierarchical identity-based signature scheme [16] has been proposed to create a space-based signature to ensure location security and a fake name for unknown authentication scheme used in privacy protection. In this way, two security identities which are not in agreement, increase the attack capabilities. In [17] a tracking challenge is proposed. It specifies a privacy-friendly on-demand location-centralized management routing protocol. By avoiding private friendly harmful to track the attacker and insider node [18]. A reliable and efficient unknown secure routing protocol management scheme is proposed. In this, the communication parties can choose the most reliable means based on reliable relationships between nodes, and then feed the connection experience to keep in the link. In addition, an efficient routing protocol with strong security and high network activity is considered. In [19], open-source software was proposed for security for content exchange on the mobile wireless network. A BitHoc ideal for the Wireless Ad Network environment is suggested. This architecture includes two main components: a membership management service (tracker), and a content sharing service (client). BitHoc tracker agents installed in nodes for linking the Global Distributed Membership Tracking Service linked to each other. To guarantee content sharing, the BitHoc node decides on a distributed manner with routing information of a distributed overlay structure. In [20], the network uses the system parameters of IBC to provide node-specific specifications of the broadcast. Based on system parameters and limited broadcasting band, a secure routing is made. An open survey of routing on open access secure routing in the wireless adhoc network is presented in [21]. The effort in developing the security measure for routing in ad hoc network was made in different directions. However, the optimization of secure routing in terms of reliability and energy conservation simultaneously is not been made. A multi-attribute monitoring of network parameter in securities the route selection is proposed.

3 Energy Scheduling Protocol in Ad Hoc Network

Toward energy conservation, an optimal route is chosen during the setup phase of an ad hoc communication. In the selection process, each of the possible routes is evaluated with the aggregated energy level. The route with the possible energy level having greater than the required energy level is then chosen. The required energy level for transmission E_{tx} is defined by Eq. 1 as

$$E_{tx} = \frac{P_{sz} \times p_{tx}}{B_L} \tag{1}$$

where

P_{sz} is the size of the packet of transmit,

p_{tx} is the required power for each packet transmission, and B_L is the offered bandwidth for data exchange. The transmission power p_{tx} depends on the distance required for the data to exchange. The transmission energy given by Eq. 1 get modified as Eq. 2:

$$E_{tx} = dX\frac{P_{sz} \times p_{tx}}{B_L} \tag{2}$$

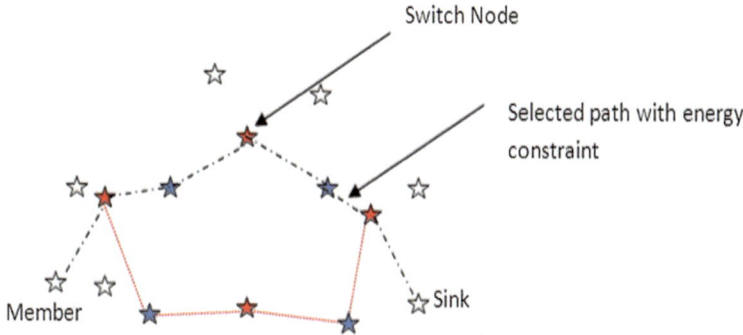

Fig. 1 A energy-efficient routing scheme in topology-driven routing

where

d   is the distance to travel

In the exchange of data packet, the network follows a topology driven communication scheduling protocol, where each of the nodes is operated on a schedule period of listening and sleep mode based on the energy level and the type of node registered. Each of the nodes in the network is registered to its neighbor node for the energy availability, for a node with a higher energy level in the network takes the responsibility of data exchange. This process controls the flow of data in a specified link format, where the fewer power nodes called members are scheduled to exchange data in the listening period and the highest energy link node takes the responsibility of data exchange towards the sink. The topology driven secure routing in ad hoc network is as shown in Fig. 1.

The route scheduling scheme for each node minimizes energy consumption. However, the reliability of this route is not been evaluated with respect to the scheduled period of operation and the link energy utilization.

4   Dual-Metric Monitoring Scheme

In the proposed approach of energy conservation, each of the nodes is a constraint with the registration of the highest energy link node with two add-on parameters of dynamic energy conservation period and maximization of the residual energy level.

The proposed approach select the link route based on the maximization of energy saving period and total residual power. The residual power E_{res} given by Eq. 3 is measured as an energy parameter in the link route selected after each packet exchange. Here, a sub setup period is allocated per communication period, where the node undergoes the residual power consumption to validate the energy consumption and residual energy for the retention of the route. The residual energy of route is given by

$$E_{res} = E_{tx} - E_{avl} \tag{3}$$

where

E_{tx} is the transmission energy; and
E_{avl} is the available energy per node

For each of the communication phase, the node enters into the residual energy consumption when not engaged in data exchange and define the energy as current energy for next packet exchange. The constraint of required energy given by Eq. 2 is then validating to retrain the route. If the aggregated energy defined by Eq. 4 of a route fails in satisfying constraints of Eq. 2, a new route selection process is initializing. The aggregated energy E_{agg} is given by

$$E_{agg} = \sum_{j=1ton} E_{tx,j} \tag{4}$$

where

$E_{tx,j}$ is the transmission energy of all nodes $j = 1ton$

In the communication process in this approach, a route is selected which satisfy the constraint of

$$R_{sel} = \max(E_{agg}) \tag{5}$$

For the selected route given by Eq. 5, a topology governed data exchange is performed where the mode with the highest energy level is given the responsibility of data exchange and all the nodes with lower energy level are set to listen period for a scheduled period. Wherein conventional approach, the member node sends a request to wake after a schedule period, a non-pending request leads to extra power dissipation, wherein in this approach, the registering node sends an 'ON' signal to the registered node when a communication request is observed. Here each of the nodes is set to power save mode until the registering node sends a request to the registered node. The extension of the power save period leads to higher energy conservation hence giving more route lifetime. The illustration of an extended power saving approach is illustrated in Fig. 2.

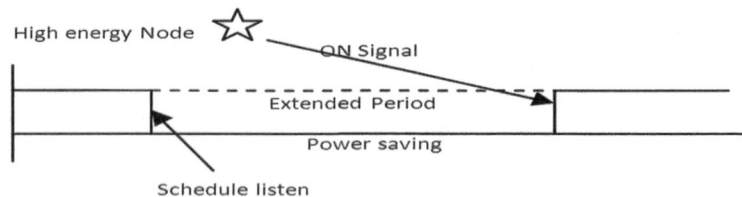

Fig. 2 Extended power saving schedule

The optimal selection of the route with max (E_{agg}) and the extended energy save scheduling leas to higher energy saving and hence leads to a more reliable routing approach in ad hoc network.

5 Simulation Result

In the evaluation of the proposed approach, a simulation Model for a randomly distributed Ad Hoc network with the following network parameter is developed (Table 1).

The simulation result for the developed approaches is validated and the results obtained are as illustrated below.

The random deployment of the node in the network is illustrated in Fig. 3.

The region bound marking for each of the nodes in the registration area is illustrated in Fig. 4.

The selected link route based on the energy optimization approach is illustrated in Fig. 5.

In the communication process during the data exchange, the energy conserved scheme leads to the fixation of the selected route for multiple packet exchange. This

Table 1 Network parameter for simulation

Network parameter	Characteristic
Node density (N_d)	30
Network area	$N_d \times N_d$
Communication range	85 units
Topology	Random
MAC	802.11
Power model	IEEE 802.11-NIC card

Fig. 3 Node deployment in a random manner

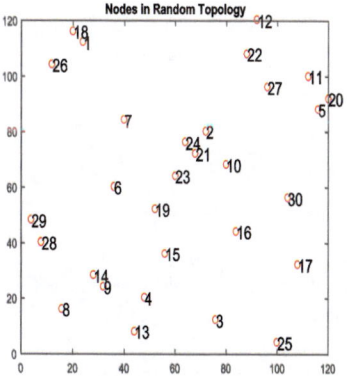

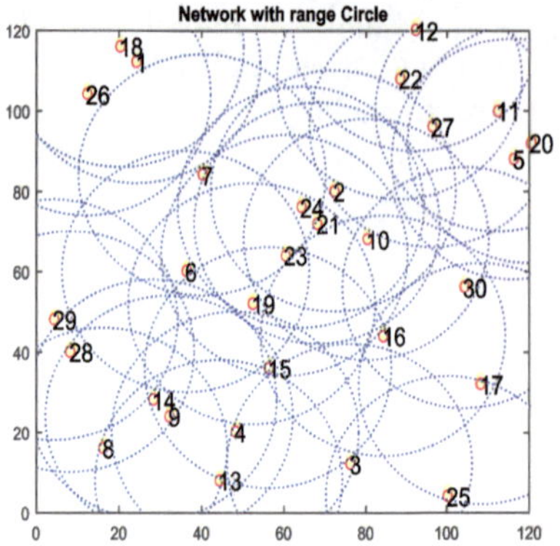

Fig. 4 Region bound for
registration

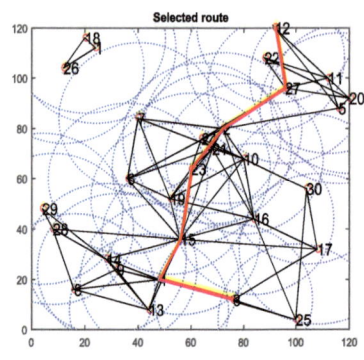

Fig. 5 Selected link node
routes in the network

Fig. 6 Route overhead of
the network

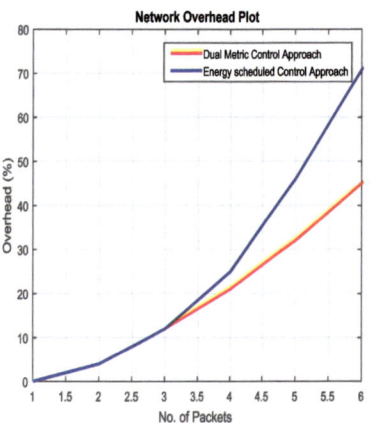

minimizes the route search overhead in the network. The overhead of the two methods
is illustrated in Fig. 6.

The network throughput is observed to be higher due to lower processing overhead,
which leads to higher packet delivery. The approach leads to more packet exchange
from a source to sink with the conservation of energy per node (Fig. 7).

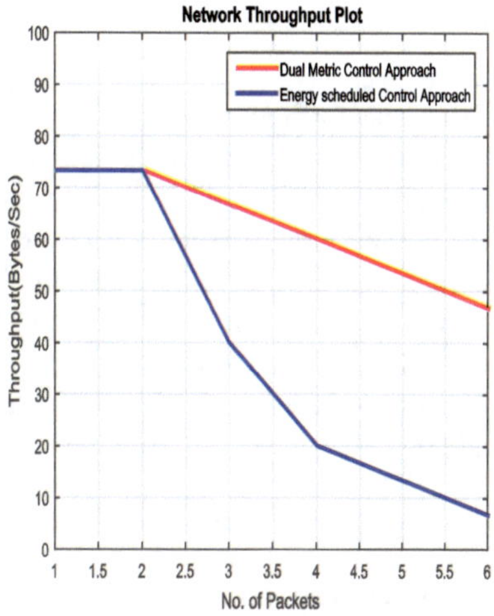

Fig. 7 Throughput with the number of packets exchanged

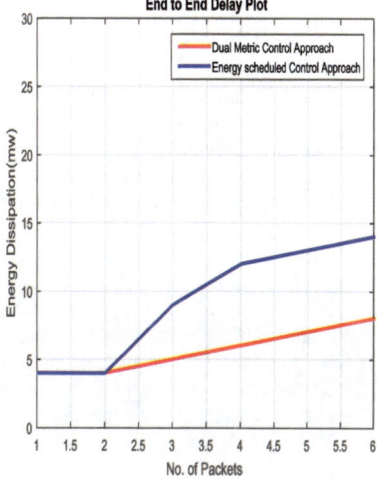

Fig. 8 Energy dissipation for the developed approaches

The energy consumption for the two developed methods is illustrated in Fig. 8. It is observed that the energy consumption increase in the conventional model, due to large intermediate route search overhead, and the enabling of each node after a schedule period to listen for a request. The additional listen period dissipate more amount of power and hence leads to faster energy dissipation.

6 Conclusion

This paper outlines a dual attribute monitoring of providing an energy-secure routing approach in ad hoc network. The suggested approach gives and defines a new energy conservation approach by using a topology based communication approach, where the highest energy node is selected as a junction link node for data exchange. The communication process is scheduled for a listen and communicate period, where the power save operation is extended and the monitoring node controls the operation of node listening. The proposed approach defines energy conservation in ad hoc network which secures the route lifetime, giving higher data exchange reliability in ad hoc network.

References

1. Sun YL, Han Z, Yu W, Ray Liu KJ (2006) A trust evaluation framework in distributed networks: vulnerability analysis and defense against attacks. In: The proceeding of IEEE INFOCOM, 2006
2. Pathak V, Yao D, Iftode L (2008) Securing location aware services over VANET using geographical secure path routing. In: Proceeding of IEEE international on conference vehicular electronics and safety (ICVES), 2008
3. Yuan W (2014) An anonymous routing protocol with authenticated key establishment in wireless ad hoc networks. Int J Distrib Sens Netw
4. Zhang Y, Liu W, Lou W, Fang Y (2006) MASK: anonymous on-demand routing in mobile ad hoc networks. IEEE Trans Wirel Commun 5(9)
5. Long J, Dong M, Ota K, Liu A (2014) Achieving source location privacy and network lifetime maximization through tree-based diversionary routing in wireless sensor networks. IEEE Access, 2014
6. Niu Q (2009) Formal analysis of secure routing protocol for ad hoc networks. IEEE, 2009
7. Kamruzzaman SM, Kim E, Jeong DG, Jeon WS (2012) Energy-aware routing protocol for cognitive radio ad hoc networks. IET Commun
8. Ye H, Tan Z, Xu S, Qiao X (2011) Load balancing routing in cognitive radio adhoc networks. IEEE, 2011
9. Dutta N, Sarma HKD (2013) A routing protocol for cognitive networks in presence of cooperative primary user. IEEE, 2013
10. Zeeshan M, Manzoor MF, Qadir J (2010) Backup channel and cooperative channel switching on-demand routing protocol for multi-hop cognitive radio ad hoc networks (BCCCS). ICET, 2010
11. Chen QJ, Kanhere SS, Hassan M, Rana YK (2007) Distance-based local geocasting in multihop wireless networks. In: Proceedings of IEEE, WCNC 2007
12. Mehta K, Liu D, Wright M (2011) Protecting location privacy in sensor networks against a global eavesdropper. IEEE Trans Mob Comput 11(2)
13. Min AW, Shin KG (2013) Robust tracking of small-scale mobile primary user in cognitive radio networks. IEEE Trans Parallel Distrib Syst 24(4)
14. Damiani ML (2011) Fine-grained cloaking of sensitive positions in location-sharing applications. IEEE, 2011
15. Bicakci K, Bagci IE, Ta B (2011) Lifetime bounds of wireless sensor networks preserving perfect sink unobservability. IEEE Commun Lett 15(2)

16. Park Y, Rhee K-H, Sur C (2011) A secure and location assurance protocol for location-aware services in VANETs. In: International conference on innovative mobile and internet services in ubiquitous computing, 2011
17. Defrawy KE, Tsudik G (2011) Privacy-preserving location-based on-demand routing in MANETs. IEEE J Sel Areas Commun 29(10)
18. Shao M-H, Huang S-J (2009) Lightweight anonymous routing for reliability in mobile ad-hoc networks. J Res Pract Inf Technol 41(2)
19. Krifa A, Sbai MK, Barakat C, Turletti T (2009) BitHoc: a content sharing application for wireless ad hoc networks. IEEE, 2009
20. Zhao S, Kent RD, Aggarwal A (2012) An integrated key management and secure routing framework for mobile ad-hoc networks
21. Cadger F, Curran K, Santos J, Mof S (2013) A survey of geographical routing in wireless ad-hoc networks. IEEE Commun Surv Tutor 5(2)

Automatic Identification of Bird Species from the Image Through the Approaches of Segmentation

M. Surender, K. Chandra Shekar, K. Ravikanth and R. Saidulu

Abstract This paper focus on the automatic identification of bird species from the images captured. Bird monitoring is crucial to perform many tasks, which include evaluating the quality of their living environment, to identify the birds under extinction, to find the migration rate of birds and to monitor the birds which may cause fatal damage to aircrafts near the airports, etc. This paper defines identification of bird species as the task to find the species of the bird from its outlook features. This paper identifies the bird species by implementing techniques named detection and segmentation algorithm which includes Laplacian propagation, histogram of oriented gradients, and support vector machine algorithm. The scope is that the algorithm proposed here is more efficient and is best among the other methods in the corresponding scenarios. The proposed method is also simpler and can be applied to various classes of objects such as birds, flowers, and animals.

Keywords Machine learning · Bird species identification · Image segmentation · Laplacian propagation · Support vector machine

M. Surender · K. Ravikanth
Department of CSE, RGUKT, Basar, India
e-mail: surender.mogilicharla@gmail.com

K. Ravikanth
e-mail: ravikanth27787@gmail.com

K. Chandra Shekar (✉)
Department of CSE, GNITC, Hyderabad, India
e-mail: chandhra2k7@gmail.com

R. Saidulu
Department of ECE, RGUKT, Basar, India
e-mail: saidulurapolu@gmail.com

© Springer Nature Singapore Pte Ltd. 2019
H. S. Saini et al. (eds.), *Innovations in Computer Science and Engineering*, Lecture Notes in Networks and Systems 74,
https://doi.org/10.1007/978-981-13-7082-3_25

1 Introduction

Protecting bird species under extinction is very important for protecting the environment. The problem and resolution are in its principle that itself based on facile question "which bird species are endangered with extinction." The answer involves various interdisciplinary sources that may be expensive, time-consuming expeditions to data analyses. The solution for it is to have a database system which has the capability to collect the condign information about bird species automatically.

The other problems include safeguarding flights in airports. The Aeronautical Centre CENIPA [1], according to its report nearly 1321 aerial accidents occurred in Brazilian airspace, which involved bird collisions which resulted in huge financial losses near about US $3 million [2]. So, knowing in detail information of birds, the authorities can take regulated measurements on specific birds to reduce these sorts of problems.

The above mentioned practical problems justify the importance of identifying the bird species. Here, we emphasize on the identification of bird species automatically using techniques from machine learning. In earlier time, a direct contact of birds is necessary to determine any bird species, but with the evolvement of advanced computational devices and machine learning algorithms [3, 4] just with the image, the bird species can be identified.

Paper Outline: Section 2 demonstrates methodology and process description. Section 3 demonstrates framework which includes Laplacian propagation, Histogram of Oriented Gradients, and SVM classifier. Section 4 demonstrates results and Sect. 5 concludes with future work.

2 Methodology

Bird species identification is a challenging task; even it is very hard for the professional bird watchers to identify the bird species from the image. Even though most of the bird features are alike, they can vary in appearance and shape. The challenges which may include high intraclass variance in lighting, background, and variation in pose that may include birds which are swimming, flying and partially covered by branches.

2.1 Caltech-UCSD Birds Dataset

i. It is an image dataset [5] with photos of 200 bird species (mostly North American). Dataset is a collection of similar species of different groups. For each single image, it contains several variations of that image.
ii. Number of categories: 200.

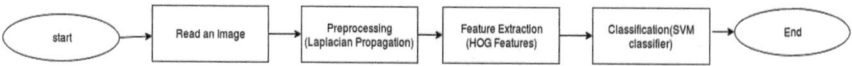

Fig. 1 Process flowchart

iii. Number of images: 6033.
iv. Annotations: Bounding Box, Rough segmentation, Attributes.

2.2 Process Description

The image is given as the input for the system, which has to be preprocessed using Laplacian propagation [6] to remove the noise and the required features are extracted from the image of the bird using Histogram of Oriented Gradients [7] based on which the bird is identified using SVM classifier [8].

The process basically includes four steps as follows (Fig. 1).

3 The Automatic Bird Species Identification Framework

3.1 Laplacian Propagation

Laplacian is a measure of the second spatial derivative of an image. It is very useful in detecting abrupt changes. In edge detection, Gaussian smoothing is done prior to Laplacian to remove the effect of noise. Laplacian propagation is useful to segment the given image to separate the image from the background partially. Here we use Laplacian mathematical formulas to segment the image. In that, initially, we generate one mask matrix by using those formulas then we apply that mask on the image matrix. Then we generate a matrix of 0s and 255s for low intensity and high-intensity regions, respectively.

It takes an image and converts into greyscale. After converting into grayscale, it considers an imaginary line passing through the high-intensity point and line. Then it takes a record of various intensity values across the pixels of the image (Fig. 2).

For white regions, it takes values like 5, 4, 3, and 2. For black regions, it takes values like 1 and 0's.

(3.1.1) The Laplacian is defined as follows:

$$\nabla^2 f = \frac{\partial^2 f}{\partial^2 x} + \frac{\partial^2 f}{\partial^2 y} \tag{1}$$

(3.1.2) here in the x-direction, the partial 1st order derivative is defined as follows:

`5 5 4 3 2 1 0 0 0 6 0 0 0 0 1 3 1 1 0 0 0 0 7 7 7 7`

Fig. 2 Intensity values

Table 1 Mask matrix

$(x - 1, y + 1)$	$(x, y + 1)$	$(x + 1, y + 1)$
$(x - 1, y)$	(x, y)	$(x + 1, y)$
$(x - 1, y - 1)$	$(x, y - 1)$	$(x + 1, y - 1)$

Table 2 Mask matrix

0	1	0
1	-4	1
0	1	0

$$\frac{\partial^2 f}{\partial^2 x} = f(x + 1, y)\, f(x - 1, y) - 2f(x, y) \tag{2}$$

(3.1.3) in the y-direction, it is defined as follows:

$$\frac{\partial^2 f}{\partial^2 y} = f(x, y + 1)\, f(x, y - 1) - 2f(x, y) \tag{3}$$

(3.1.4) Add those x and y derivatives. So, the Laplacian can be given as follows:

$$\nabla^2 f = \left[f(x + 1, y)\, f(x - 1, y) + f(x, y + 1)\, f(x, y - 1) \right] - 4f(x, y) \tag{4}$$

(3.1.5) Replace with coefficient values in mask matrix (Tables 1 and 2).
(3.1.6) Choose matrix representation of one image (Table 3).
(3.1.7) Applying mask on the pixels representation (matrix) of an image, it returns a convolution matrix (Table 4).
(3.1.8) In convolution matrix if the pixel value is >'255' replace it with '255' and if the pixel value is <'0' replace it with '0' (Table 5).

3.2 Histogram of Oriented Gradients

Navneed Dalal and Bill Triggs introduced HOG features. The histogram of oriented gradients (HOG), which is a feature descriptor and is used frequently for object

Table 3 Image matrix

150	150	**10**	150	150	150	150	150	150	150	150
150	150	**10**	150	150	150	150	150	150	150	150
150	150	**10**	150	150	150	150	150	150	150	150
10	**10**	**10**	**10**	**10**	**10**	**10**	**10**	**10**	**10**	**10**
150	150	**10**	150	150	150	150	150	150	150	150
150	150	**10**	150	150	150	150	150	150	150	150
150	150	**10**	150	150	150	150	150	150	150	150
150	150	**10**	150	150	150	150	150	150	150	150

Table 4 Convolution matrix

0	−140	280	−140	0	0	0	0
−140	−280	280	−280	−140	−140	−140	−140
280	280	0	280	280	280	280	280
−140	−280	280	−280	−140	−140	−140	−140
0	−140	280	−140	0	0	0	0
0	−140	280	−140	0	0	0	0

Table 5 Resultant matrix

0	0	255	0	0	0	0	0
0	0	255	0	0	0	0	0
255	255	0	255	255	255	255	255
0	0	255	0	0	0	0	0
0	0	255	0	0	0	0	0
0	0	255	0	0	0	0	0

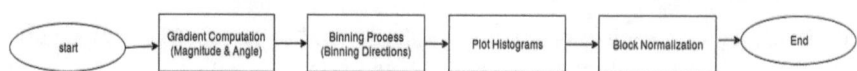

Fig. 3 HOG flow chart

detection in computer vision and also in image processing. In localized portions of the image, the method counts the occurrences of gradient orientations. This method computes on a dense grid of cells which are uniformly spaced and it uses overlapping local contrast normalization to enhance accuracy (Fig. 3).

3.2.1 Gradient Computation

The method usually requires the color or image intensity data with the following filter kernels, Mitchell [3] tested more complex masks which included a 3 × 3 label mask and diagonal masks but their detection of humans in images resulted in poor performance. In the experiment with Gaussian smoothing no improvement was found, and thus derivative mask was applied to find that in practice if removal of any smoothing performed better.

Steps for gradient computation:

i. With no smoothing, compute the centered horizontal and vertical gradients.
ii. Compute the gradient orientation and magnitude.

For an image, for each pixel choose the channel with the greatest gradient magnitude and angle.

So, being given an image I, we obtain the x and y derivatives using a convolution operation:

$$I_x = I(c + 1, r) - I(c - 1, r) \tag{5}$$

$$I = I(c, r - 1) - I(c, r + 1) \tag{6}$$

The magnitude of gradient is

$$|G| = \sqrt{I_x^2 + I_y^2} \tag{7}$$

The orientation of the gradient is given by

$$\Theta = \arctan\left(\frac{I_y}{I_x}\right) \tag{8}$$

3.2.2 Orientation Binning

In cell histograms creation, based on the values found in the gradient computation, each pixel within the cell casts a weighted vote for the orientation based histogram channel. Depending on the gradient ("signed" or "unsigned") the histogram channels are spread evenly from $0°$ to $180°$ or from $0°–360°$ and the cells shape either can be rectangular or radial.

3.2.3 Descriptor Blocks

The cells must be assembled into larger, especially connected blocks to normalize the gradient strengths for accounting the variation in illumination and also in contrast. From all of the block regions, the HOG descriptor is then the integrated vector of the inherent part of the normalized cell histograms. Here each cell contributes several times to the final descriptor. Here, it has two block geometries namely RHOG blocks and CHOG blocks. At some single scale without orientation alignment, RHOG blocks are computed in dense grids, to encode spatial form information RHOG blocks are used.

3.2.4 Block Normalization

There are various methodologies for block normalization. In a given block, let v be the non-normalized vector consisting of all the histograms, $|v_k|$ is its k-norm for $k = \{1, 2\}$ and e be a small constant (whose value has no effect on the results). Then the normalization factor is one among the following:

$$L_2\text{-norm: } f = \frac{v}{\sqrt{|v|_2^2 + e^2}} \tag{9}$$

$$L_1\text{-norm: } f = \frac{v}{\sqrt{|v|_1 + e}} \tag{10}$$

$$L_1\text{-sqrt: } f = \sqrt{\frac{v}{|v|_1 + e}} \tag{11}$$

The scheme computed by taking the L2 norm and clipping the result followed by renormalizing. In the experiment [3], it was found that L2 norm and L1 sqrt schemes provide equal performance, while the L1 norm provides moderately less reliable performance. However, working on non-normalized data all the four methods showed remarkable improvement.

3.3 Support Vector Machine

Classification divides the feature space into several classes. Many techniques are used for classification, in that here we are using Support Vector Machine. It is a supervised machine learning algorithm and it can be used either for classification challenges or for regression challenges. In n-dimensional space (whereas, n defines number of features), here we plot each data item as a point where the value of particular coordinate is defined as the value of each feature.

SVM determine the optimal hyperplane between two classes of data. After getting the normalized feature values for each bird image we can train them into different classes. For two different classes, we classify by using SVM linear classifier. More than two classes we are using SVM multi-classifier. We are classifying species by giving labels as 0's and 1's and predicting the sample by using trained dataset.

4 Results

We have chosen Caltech dataset for our project, which contains 200 varieties of species with 6300 birds. We have taken an image and performed Laplacian propagation for preprocessing. Histograms of oriented gradients for feature extraction and SVM Classifier are used for classifying the species.

4.1 Result of Laplacian Propagation

Laplacian propagation is a technique which is used to separate the bird from the background. Here in our project, we applied Laplacian propagation. In the process of Laplacian propagation, first we took an image and converted to grayscale. Then after with Laplacian formula and taking the high intensity and low-intensity points, we generated a mask matrix. Applying mask matrix over image matrix we got a convolution matrix. In convolution matrix, if the pixel value is >'255', replace it with '255'. And if pixel value is <'0' replace it with '0'. Then we got a final matrix indeed which we could see as a preprocessed image. Final image which we got is not fully separated from background then we have gone for another approach for full object detection i.e. HOG (Fig. 4).

4.2 Result of Histograms of Oriented Gradients

Histograms of oriented gradients descriptors are also feature descriptors and are used for object detection frequently applied in image processing and computer vision. In localized portions of an image, it counts the occurrences of the gradient orientation. In this approach, we first calculated computation of the gradient values horizontally and vertically and we calculated magnitude and angles then we plotted the histograms of 9 bins by dividing 180 degrees into 9 bins. Then we extracted feature values for each cell and by combining all those values we have taken into a feature vector for

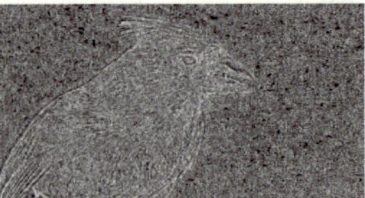

Fig. 4 Result of Laplacian propagation

Fig. 5 Magnitude

Fig. 6 Angle

that respective image. Finally, we got a feature vector which was used further for classification by SVM (Figs. 5 and 6).

4.2.1 Plotting Histograms

i.   Bin the directions, consider the degrees from 0 to 180, and divide as 0–20, 20–40…160–180, as like make 9 bins from 1 to 9.
ii. Divide the image into blocks having 4 cells (2 * 2). Each cell is of 8 * 8 sizes.
iii. Plot histograms for each cell.
iv. Concatenation of four cells histograms forms one block feature results in the following plot (Fig. 7).

4.3 Result of Support Vector Machine

Support vector machine is mainly used to train the data and to predict the test sample by comparing with the trained data. After extracting feature values, we have gone for SVM for classification. As a part of this, we calculated feature vector values for all similar type of bird species and training them as a particular species. Like this, we have trained many species. We also calculated feature values for test samples. Then by comparing these values with trained data values, we classified the test images as a

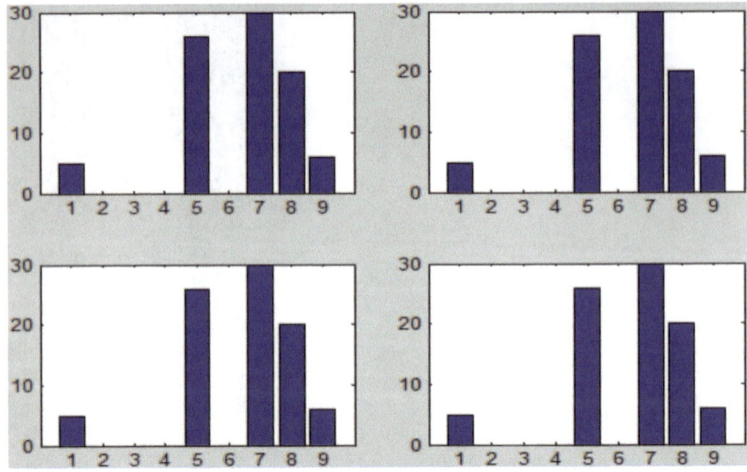

Fig. 7 Plotting histograms (*x-axis: Orientation bins, y-axis: Gradient Values*)

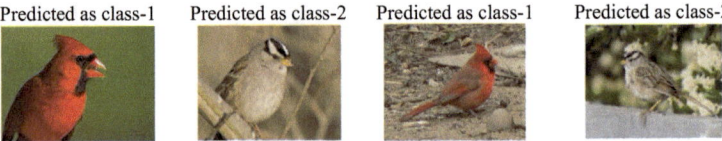

Fig. 8 Test images with SVM

particular species. The training was done on a similar type of birds into classes with corresponding labels 1's and 2's. Then predicted sample test images with SVM are shown in (Figs. 8 and 9).

5 Conclusions and Future Work

It has been proved that in a dense overlapping grid using features from a locally normalized histogram of gradient orientations produces best results for bird detection. We have studied the influence of various descriptor parameters and concluded that for better performance fine orientation binning, fine-scale gradients, and high-quality local contrast normalization in overlapping descriptor blocks are all important. After extracting feature values we have gone for SVM for classification. As a part of this, we calculated feature vector values for all similar types of bird species and training them as a particular species. Like this, we have trained many species. We also calculated feature values for test samples. Then after by comparing these values with trained data values, we classified the test images as a particular species.

Fig. 9 Classification and accuracy

```
>> namesmain
       15

>> classification
          12            41616

misclassified
     'cardinal'

     'cardinal'

     'cardinal'

     'rustybird'

misclassified
misclassified
     'rustybird'

     'herring'

     'herring'

     'herring'

misclassified
     66.6667
```

Future work will include examining the different ways in minimizing the gap among the proposed system to that of ground truth system (assignment of all test samples to the correct subset), which achieves the classification accuracy of 72.6% which is better than that of proposed system whose accuracy is 66.7%, which means that performing furthermore precise assignment of sample to the subset can generate adequate improvement in performance. One of the approaches to get a more precise assignment is to learn visual features that will differentiate subsets instead of all the classes.

References

1. Automatic bird species recognition using periodicity of salient extremities. In: IEEE international conference on robotics and automation (ICRA), Karlsruhe, Germany, 6–10 May 2013
2. Mendonça CAF (2009) The management of the aerial danger in Brazilian airports (in Portuguese). SIPAER report, vol. 1, no. 1, November
3. Mitchell TM (1997) Machine learning. McGraw-Hill
4. Witten IH, Frank E (2005) Data mining: practical machine learning tools and techniques. Morgan Kaufmann, San Francisco

5. http://www.vision.caltech.edu/visipedia/CUB-200-2011.html
6. Angelova A, Zhu S (2013) Pattern recognition by Anelia Angelova and Shenghuo Zhu. CVPR
7. Dalal N, Triggs B (2005) Histograms of oriented gradients for human detection. In: IEEE Computer Society Conference on Computer Vision and Pattern Recognition (CVPR '05), 2005
8. Ge Z, McCool C, Sanderson C, Bewley A, Chen Z, Corke P (2015) Fine-grained bird species recognition via hierarchical subset learning

An Investigation on Existing Protocols in MANET

Munsifa Firdaus Khan and Indrani Das

Abstract Ad hoc networks provides communication without the help of any fixed infrastructure. Due to dynamic topology changes, mobility of nodes and lack of central coordination routing in MANET is a challenging task. In MANET, achieving higher QoS is a hard job, various researchers are working to boost the performance. In this paper, we have done rigorous experiments on existing protocols such as AODV, DSDV, DSR, and AOMDV to expose the performance and QoS using NS2. The parameters that have been considered for experimental analysis are the throughput of the network, delay incurred, and packet delivery ratio (PDR).

Keywords MANET · AODV · AOMDV · DSR · DSDV · QoS · Throughput · Delay and PDR

1 Introduction

MANET (Mobile Ad Hoc Network) is a game changer network technology in the wireless network domain. Therefore, MANET meets huge application domains which include battlefields, emergency operations, search and rescue, and collaborative computing. Thus, human beings are greatly benefited from MANET. Because MANET has the capability to change market tactics in the IT industry. The performance of a MANET depends on the routing protocols. Routing in MANET, itself is a great challenge and numerous researchers are working on MANET to boost up the performance of the existing protocols. Unlike a cellular network, a MANET does not have any fixed infrastructure. It has multiple hops and the network is self-organized [1]. MANET has some unique characteristics that create lots of challenges, namely,

M. F. Khan (✉) · I. Das
Department of Computer Science, Assam University, Silchar 788011, Assam, India
e-mail: munsifa737@gmail.com

I. Das
e-mail: indranidas2000@gmail.com

© Springer Nature Singapore Pte Ltd. 2019
H. S. Saini et al. (eds.), *Innovations in Computer Science and Engineering*, Lecture Notes in Networks and Systems 74,
https://doi.org/10.1007/978-981-13-7082-3_26

dynamic topologies, lack of secure boundaries, lack of centralized management facility, unreliability of wireless links between nodes, etc.

Undoubtedly, it is quite complicated to conclude about a particular routing protocol. We have to select a routing protocol on the basis of suitability of the environment for a particular protocol. MANET provides various routing protocols, namely, proactive routing protocol, reactive routing protocol, and hybrid routing protocol [2]. They are as follows:

- **Proactive routing protocol**: Proactive routing protocols are an extended version of wired routing protocols. It keeps a routing table for exchanging routing information. A few examples of this kind of protocols are -DSDV, WRP, OLSR, CGSR, and STAR.
- **Reactive routing protocol**: Information about routing topology is not maintained by reactive routing protocols. Whenever they receive any routing request for transmission of data packets from one host to the other then only they create a path, i.e., a path is created on demand. A few examples of these types of protocols are DSR, AODV, TORA, LAR, ABR, SSA, and FORP.
- **Hybrid routing protocol**: Hybrid routing protocols are the amalgamation of the above two mentioned protocols. For instance, CEDAR, ZRP, and ZHLS.

In this paper, our objective is

- To do the experiment in NS-2 on the existing protocols, protocols such as AODV, DSR, DSDV, and AOMDV are used for evaluating their performance.
- To investigate the behavior of the existing protocols on the basis of considered parameters.
- To identify protocol suitability for a certain environment.
- To check the QoS provided by each of the routing protocols.

The paper is arranged in this way—In Sect. 2 related work and background are discussed. Section 3 exposes the existing protocols' characteristics. Section 4 discusses the experimentation environment; Sect. 4.2 exploits the performance of existing protocols. The experimental results are presented in the section. Section 5 states the discussion and future works. Finally, Sect. 6 gives the conclusion of the paper.

2   Background and Related Work

The performance of a MANET depends on different routing protocols. Research has been going on for improvement in QoS. The protocols are reviewed based on QoS, which is as follows

Rajeshkumar and Sivakumar [3] evaluate performance on AODV, DSR, and DSDV considering parameters packet delivery ratio, throughput, average end-to-end delay, and control overhead by using the NS2 simulator. Thus, it motivates to compare with another advanced protocol, AOMDV, to check the QoS. Rajeshkumar and Sivakumar considered low to high mobility rates and perform the evaluation accordingly. However, we consider various parameter to evaluates the existing protocols.

Kalavatia et al. [4] performed the comparison in packet drop among reactive routing protocols DSR, AODV and LAR using Qualnet 6.1. It is observed that when the drop probability is lower in LAR, then it provides fewer packet drops in compared to the AODV and DSR. However, when the drop probability is higher, then 100% of packet drops in LAR. It is concluded from the observations that AODV is the best among the three with the large-sized network.

Nayak and Gupta [5] evaluated AODV, DSR, and ZRP based on energy-efficient consumption by considering the parameters PDR, throughput and routing load using NS2. It is observed from their experimentation result that DSR routing technique performs well under different pause time. In a dense medium, the behavior of DSR is much satisfactory than the other existing protocols. The utilization of energy per packet in DSR protocol is less and thus it is concluded that the DSR is the best efficient energy routing protocol.

Agarwal [6] performed a comparison on reactive routing protocols, AODV and DSR, by considering the parameters' average delay, average routing overhead, and packet delivery rate using NS2. The performance is analyzed using varying experimentation time. In AODV, packet delivery ratio increases with decreasing mobility whereas it decreases with increasing mobility, but the packet delivery ratio is good in DSR for all mobility rates. Hence, it is concluded that AODV is preferred over DSR for real-time traffic.

Mishra and Singh [7] performed an examination of traffic load and mobility on AODV, DSR, and DSDV in NS2 considering the parameters throughput and packet dropped. It is noted that DSR is preferable for a normal network with average mobility and traffic. DSDV is preferable with low mobility and less number of nodes, whereas AODV performs better for a robust scenario like high mobility, dense nodes, high traffic, large area and network pattern sustains for a longer period.

3 Existing Methods

We have considered the following existing protocols for performance evaluation.

3.1 Dynamic Source Routing Protocol (DSR)

It is a straightforward and proficient reactive routing algorithm based on source routing. A source node initially checks its routing table with the purpose of sending a packet to its destination. If it finds an entry which has the valid route for the destination than it chooses that path otherwise the source node starts the route discovery mechanism. A node when initiates Route Request Packets (RREQ) are accepted by other nodes that remains within the transition range of it. The RREQ discovers the target node for which the route is demanded. If the route is found successfully then the node that initiated the RREQ receives a Route Reply Packet (RREP). This protocol ensures route maintenance easily. If any transmission problem occurs or a path failure occurs because of mobility of nodes then the corresponding nodes transmits a Route Error Packet (RRER) to the original node. For a small network, this protocol is suitable [8].

3.2 Destination Sequence Distance Vector Routing Protocol (DSDV)

DSDV, based on traditional Bellman–Ford routing algorithm, is a proactive routing protocol. The key objective of this protocol is to resolve the routing loop problem. Each node maintains a next hop table to exchange routing information with its neighbors. One of the advantages of this protocol is that paths are available to all destinations which show that less delay is required. It increases convergence speed, reduces control message overhead, and route looping [9]. For big or dynamic network, this protocol is not t.

3.3 Ad Hoc on Demand Distance Vector Routing Protocol (AODV)

AODV is a reactive routing protocol where an original node transmits routing messages to its closest nodes if it does not have any path to the destination. This protocol is a fusion of DSDV and DSR routing protocol. However, it differs with DSR in routing information. Each packet carries full information in the DSR whereas in AODV each packet carries only destination address which signifies that AODV has less overhead than DSR. All the corresponding nodes update its routing information after receiving an RREQ message from the original node. Presuming that the adjacent nodes have the information about the target node, then it sends an RREP. Otherwise, it again broadcasts the packets to its neighbors. A path is created only when a receives RREP packets. If due to the mobility of nodes, a path is broken, then the related node sends RRER packet to the original node. It has an advantage of flexibility for extremely dynamic environment [10].

3.4   Ad Hoc on Demand Multi-path Distance Vector Routing Protocol (AOMDV)

AOMDV is a multiple path distance vector reactive routing protocol for MANET. This protocol is an expansion of the traditional AODV protocol. In all possible route revelation, AOMDV protocol discovers several routes between one node to the other. Multiple routes that have been computed so far are assured to be looped-free and disarranged [11]. AOMDV has several characteristics that are similar to AODV [12]. Like AODV, it also uses hop-by-hop routing method. Moreover, it also uses route discovery procedure to find routes on demand. In AOMDV, several reverse routes are created both at the midway node and the target node for transmission of RREQ packets from source to the target node. Several RREP packets pass through these reverse routes in order to plan multiple forward routes to the target at the source and midway nodes.

4   Experimentation and Results

4.1   Experimental Environment Setup

We have performed the simulation using NS-2 [13]. The framework for the selected protocol has been given in Table 1.

The output generated by the NS-2 simulator consists of a trace file, named *.tr and animation file, named *.nam. The following metrics have been used for simulation.

Table 1 Framework for NS-2 simulation

Paramaters	Values
Area	500 × 400 m
Number of nodes	20, 40, 60, 80, 100, 120, 140, 160, 180, 200
Simulation time	150 s
Mobility model	Random waypoint
Routing protocol	AODV, DSR, DSDV, AOMDV
Channel	Wireless
Propagation	Two ray ground
MAC	802.11
Interface queue	Queue/droptail/priqueue
Antenna	Omnidirectional

Throughput: This is the moderate rate of transmission of fruitful data packets per unit time. It is expressed in bps, Kbps, Mbps, Gbps. We have simulated throughput in kbps. The greater the throughput the better is the performance.

Packet Delivery Ratio (PDR): This is the ratio of total successful number of data packets transmitted to the target to the total data packets initiated by the original node. The performance of the system increases with increasing PDR values.

Delay: The mean time taken to reach its target by a data packet is defined as a delay. With decreasing delay the performance of the system increases. It is measured in fractions of seconds.

4.2 Experimentation Results

The experimentation is carried out rigorously to check the performance of the AODV, AOMDV, DSR, and DSDV protocols. We have considered all possible parameter for extensive experiments as described in Table 1.

Throughput: With the increasing number of node throughput varies which has been shown in Fig. 1. Table 2 represents the numbers of nodes and various protocols with their corresponding throughput.

Figure 1 represents the throughput. From the above observation, it is shown that the average throughput of DSDV is higher when compared with the other routing protocols whereas AODV has the lowest average throughput. AOMDV has also better throughput. However, it is noted that with the increasing number of node throughput decreases. DSDV gives the highest throughput for 20 nodes, i.e., 153.88 Kbps. AODV gives the least throughput of 48.38 Kbps for 200 nodes. However, DSDV gives the

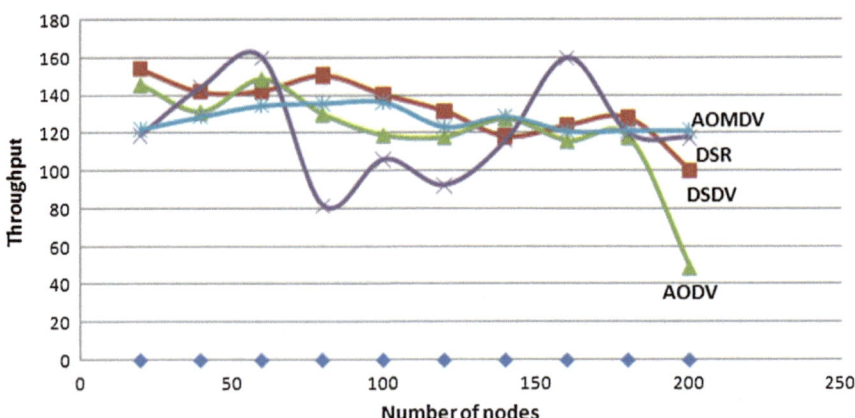

Fig. 1 Variation in throughput with the increasing number of nodes. Y-axis represents throughput in Kbps

lowest throughput of 100.45 Kbps for 200 nodes. In addition, the highest throughputs of AODV, AOMDV, and DSR are 148.18 Kbps in 60 nodes, 136.07 Kbps in 100 nodes, and 159.78 Kbps in 60 nodes, respectively.

Packet Delivery Ratio values have been mentioned in Table 3.

Figure 2 represents the PDR. From the above observation, it is seen that AOMDV has the highest average packet delivery ratio whereas DSR has the lowest. AOMDV shows its highest PDR value of 0.9974 for nodes 60 and 160 and lowest PDR value of 0.9923 for 200 nodes. DSR gives the lowest PDR value for 120 nodes, i.e., 0.9901 and it gives the highest PDR value of 0.9973 for 40 nodes. However, it is noted that highest PDR values of AODV and DSDV are 0.9972 and 0.9976 for nodes 80 and 20, respectively.

Delay: The performance analysis based on the parameter delay has been shown in Table 4 and Fig. 3.

Figure 3 represents the delay. From Fig. 3 and Table 4, it is noticed that DSDV has the minimum delay whereas AOMDV has the maximum. DSDV shows the least

Table 2 Protocols with their corresponding throughput

Number of nodes	AODV	AOMDV	DSR	DSDV
20	145.31	122.16	118.97	153.88
40	131.04	128.69	144.04	141.79
60	148.18	134.42	159.78	142.47
80	129.66	135.54	81.46	150.57
100	118.59	136.07	106.06	140.50
120	117.62	123.37	92.39	131.69
140	127.18	128.54	116.12	118.31
160	115.62	121.02	159.75	124.42
180	117.53	121.29	119.49	128.04
200	48.38	121.34	117.38	100.45

Table 3 Protocols with their corresponding PDR

Number of nodes	AODV	AOMDV	DSR	DSDV
20	0.9961	0.9968	0.9883	0.9966
40	0.9926	0.9955	0.9973	0.9941
60	0.9942	0.9974	0.9955	0.9940
80	0.9972	0.9962	0.6429	0.9962
100	0.9951	0.9970	0.9881	0.9951
120	0.9959	0.9936	0.9901	0.9942
140	0.9964	0.9951	0.9909	0.9927
160	0.9968	0.9974	0.9947	0.9920
180	0.9957	0.9958	0.9967	0.9934
200	0.9954	0.9923	0.9904	0.9925

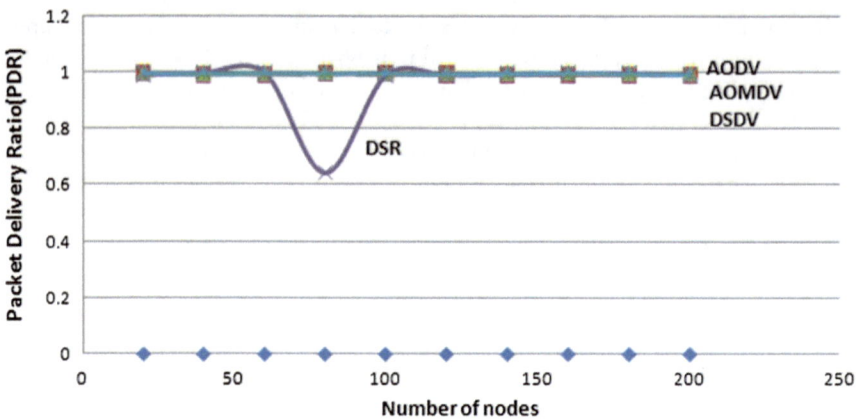

Fig. 2 Variation in PDR with the increasing number of nodes

Table 4 Protocols with their corresponding delay

Number of nodes	AODV	AOMDV	DSR	DSDV
20	261.60	369.95	332.06	235.30
40	268.54	282.61	268.82	179.18
60	213.48	248.29	244.68	214.17
80	265.63	267.17	58.11	182.69
100	259.28	267.27	242.62	160.58
120	268.72	261.46	384.60	212.78
140	249.90	357.64	318.35	226.45
160	196.57	284.05	244.80	216.92
180	216.14	263.61	320.47	292.98
200	118.84	284.09	242.92	247.99

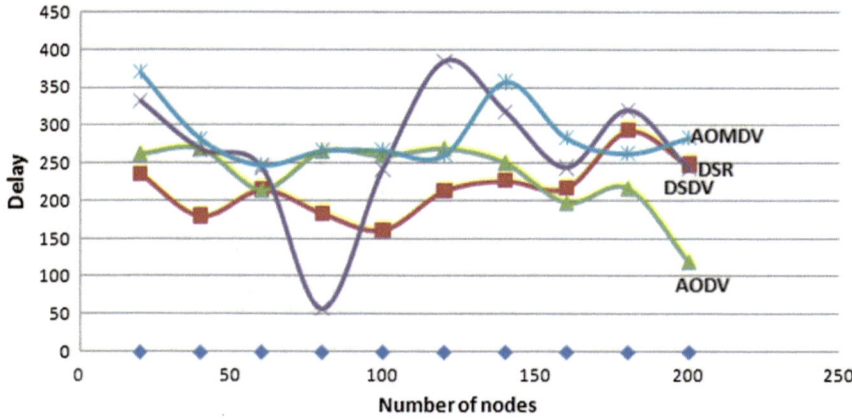

Fig. 3 Variation in delay with increasing number of nodes. Y-axis represents the delay in seconds

delay value of 160.58 s for 100 nodes. AOMDV gives the greater delay value of 369.95 s for 20 nodes. However, it is noted that AODV and DSR also provide the lower delay. The least delay values of AODV and DSR are 118.84 s and 58.11 s for 200 and 80 nodes, respectively.

5 Discussion and Future Work

We have exposed various parameter for understanding the variation in performance of the chosen routing protocols. From the above observation, it is seen that every protocol is not best in terms of every metric. We observed that if we want to improve the quality of services, then we need to improve the throughput and reduce delay. It is achieved if we consider the mobility of nodes in an intelligent way than the delay is reduced by increasing the throughput. Again, another way to reduce delays and increase throughput is by considering the shortest and efficient path mechanism. It has also enough room to work on the route discovery mechanism in order to increase throughput and decrease delays.

6 Conclusion

We have performed a series of experiments using NS-2 on routing protocols, namely AODV, DSDV, DSR, and AOMDV, respectively. Our main objective is to determine which protocol is better in terms of the performance metrics namely throughput, PDR and delay, respectively. From the above simulation result, we have seen that AOMDV has the highest PDR and delay when compared with DSR, DSDV, and AODV, respectively. DSDV has the highest throughput and lowest delay when compared with other routing protocols. We argue that from the above observation AOMDV exceeds with respect to throughput and PDR. It is observed that QoS decreases with a higher number of nodes. However, the QoS is satisfactory in AOMDV with a higher number of nodes too.

References

1. Khan MF, Das I (2017) A study on quality-of-service routing protocols in mobile ad hoc networks. In: 2017 international conference on computing and communication technologies for smart nation (IC3TSN). IEEE, pp 95–98, 2017
2. Rao M, Singh N (2014) Quality of service enhancement in manets with an efficient routing algorithm. In: 2014 IEEE international advance computing conference (IACC), pp 381–384, 2014
3. Rajeshkumar V, Sivakumar P (2013) Comparative study of aodv, dsdv and dsr routing protocols in manet using network simulator ns2. Int J Adv Res Comput Commun Eng 2(9):4564–4569

4. Kalavatia M, Sharma L, Singh RK. Comparative study on a comparison in packet drop among the reactive (on-demand) protocols of mobile ad hoc network (manet) using qualnet
5. Nayak K, Gupta N (2015) Energy efficient consumption based performance of aodv, dsr and zrp routing protocol in manet. Energy 4(11):82–90
6. Agrawal A (2016) Manet: comparion on aodv and dsr. In: For national conference on recent innovations in science, technology and management (NCRISTM). IEEE, pp 77–82, 2016
7. Mishra AK, Singh RR (2015) Performance analysis of traffic load and mobility on aodv, dsr and dsdv routing protocols in manet. Int J 3(9)
8. Ramya ASI (2015) Destination sequenced-distance vector algorithm. Int J Comput Sci Eng 7(11):129
9. Ghuman SS (2016) Dynamic source routing (dsr) protocol in wireless networks
10. Landge P, Nigavekar A (2016) Modified aodv protocol for energy efficient routing in manet. Int J Eng Sci Res Technol 5(3):523–529
11. Marina MK, Das SR (2006) Ad hoc on-demand multipath distance vector routing. Wirel Commun Mob Comput 6(7):969–988
12. Zhou J, Xu H, Qin Z, Peng Y, Lei C (2013) Ad hoc on-demand multipath distance vector routing protocol based on node state. Commun Netw 5(03):408
13. Fall K, Varadhan K et al (2005) The ns manual (formerly ns notes and documentation). The VINT project 47

Intelligent Fatigue Detection by Using ACS and by Avoiding False Alarms of Fatigue Detection

A. Swathi and Shilpa Rani

Abstract One cannot determine one's state of being awake when driving the car for long hours in lengthier journeys. The most dangerous situation comes when the driver is feeling fatigue (sleepy). There are lot of works already been done on the fatigue detection of the driver. This paper acknowledges the past work and presents the improved version of fatigue detection by eliminating false possibilities of the system. Unlike the previous fatigue detection systems, the proposed system discusses the accident prevention by using adaptive cruise control, by which it is very easy to predict whether the vehicle is maintaining the proper speed or not. However, it is very tough to determine the driver's state. This is a very dangerous situation for a driver to immediately take left or prevent an accident. This paper discusses the false fatigue detection and the method to avoid such confusions. When the fatigue detection is accurate, automatically ACS has to be turned on and vehicle-halting measures will be taken.

Keywords Fatigue · Yawn detection · PERCLOS · Adaptive cruise control system · Edge detection · CHT

1 Introduction

Fatigue is extreme tiredness caused by not getting enough rest over a period of time, whether from mental or physical exertion or illness. Drivers with such fatigue can impair reaction time and decision-making when behind the wheel which increases the risk of being involved in an accident. If a driver falls asleep for just four seconds while traveling at a speed of 100 km/h by that time car will have gone 111 m without

A. Swathi · S. Rani (✉)
Department of Computer Science, Sreyas Institute of Engineering and Technology, Hyderabad, Telangana, India
e-mail: shilpachoudhary@sreyas.ac.in

A. Swathi
e-mail: swathigowroju@sreyas.ac.in

© Springer Nature Singapore Pte Ltd. 2019
H. S. Saini et al. (eds.), *Innovations in Computer Science and Engineering*, Lecture Notes in Networks and Systems 74,
https://doi.org/10.1007/978-981-13-7082-3_27

a driver in control. At such a high speed, a crash is likely to happen with a high risk of death or severe injury. Fatigue is a major cause of crashes resulting in some 50 deaths and approximately 300 serious injuries each year. In order to prevent such a damage, this paper introduces an advanced new system to prevent road accidents when fatigue is detected.

Detecting drowsiness is a major challenge to researchers as it depends on various physiological and probabilistic models [1], which can still lead to a false prediction [2]. The major analysis can be categorized based on a driver's facial and body motions. Our proposed system deals with facial properties to detect fatigue. The probabilistic measures of eyelid movement by tracking the pupil [3] is considered as the main property. The other body motion methods of following ECG and EEG signals are bit uncomfortable as the driver have to wear a special shirt constituting sensors [4], always may not be possible.

Fatigue detection merely detects whether the eyes are open or closed. However, there are other factors where the detected closure of the eye may give a false prediction, which is discussed further in the paper. Our approach eliminates the false alarm detection in the system.

To prevent accident the proposed system is using ACS (Adaptive Cruise control System), which is embedded in the vehicle [5] developed for traffic simulation in real-time environment, to prevent the accident.

2 Literature Survey

Existing studies [2, 6] impose the detection into one of the two techniques called intrusive and nonintrusive, by which fatigue can be detected based on physiological activities and nonphysiological behavior of the person. The problem existing with the first approach is the limitation of usage of certain tools to predict the behavior inside the car. Because of this reason maximum research [7–10] is being done on the second approach, which needs a very few parameters in consideration and detects the exact fatigue of the driver. The main equipment used in such techniques can be employed simply by using a camera mounted on the dashboard of the car and few image-processing techniques. There is another universally accepted method, which detects the fatigue based on the steering behavior and vehicle moment on the road. But this system fails in the exact prediction based on the fact that the vehicle moment and sharp steering moments would be obvious based on the road and traffic conditions while driving, rather than driver's sleepiness.

The obvious prediction method that this paper discusses is the second approach, i.e., by using a nonintrusive method [2]. Here, the facial movements of the driver are been predicted in order to predict the sleepiness of the driver. The main parameters considered in the current research is eye position and mouth positioning to predict the sleepiness. Eyes and mouth play a very big role in detecting fatigue because when a person is feeling sleepy his eye movements will become slow, his eyelid movement will be heavy, his gaze will change, and his openness of eye will change. Along with

Fig. 1 Yawning with eyes closed while driving

Fig. 2 Sneezing while driving

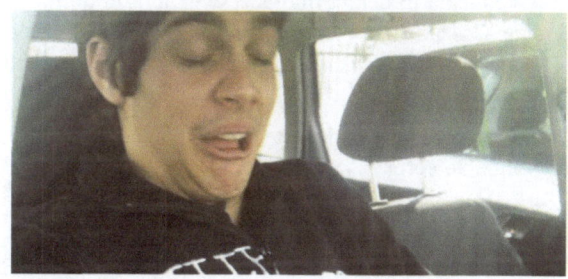

this, his frequent yawning will give an added advantage to detect the fatigue of the driver.

The proposed system also had few challenges to compete with. There may be some false predictions in the system, for example, when a person is speaking on the phone there is a tendency of closing the eye partially in Fig. 1 [2] and for some time he may close the eye. The same thing might happen when a person wants to take a break from driving because of his restlessness of eyes. It may occur even if a person sneezes while driving (Fig. 2 [6]). The other false prediction such as, the person in the driving seat may be singing and driving or speaking [7] in phone and driving or keeping his mouth open for some time and then driving which gives a false alarm of fatigue detection.

Considering all the challenges, the proposed system gives the solution to detect the fatigue of the driver accurately by eliminating false predictions and false alarms.

3 System Working Principle

In order to build this system, the first video should be processed and continuous frames should be generated. From extracted images, the face should be detected by using Haar-like feature based method [8], because of its high accuracy and faster execution. By using iris center localization, if the driver's fatigue is detected. Once the detection is done, the alarm will be sounded. The next continuous frames are checked for fatigue, if the fatigue is detected, then the control is transmitted to the cruise control system, which is received by the receiver.

In this paper, we propose the algorithmic view of a proposed method in [9]. The proposed system has two phases.

Phase 1: Track the face from video
Phase 2: Detecting fatigue of driver
Phase 3: Turn on the cruise control system (Fig. 3).

The next discussion in this paper elaborates about the implementation details.

Phase 1: Detecting face: The traditional model cascade object detector proposed in [9] is used to detect the face, eye, and mouth positions and to track them in the image. Cascade detector algorithm uses Viola–Jones algorithm to detect the objects. The algorithm creates a rectangular box of the tracked face.

```
% Create a cascade detector object.
faceDetector = vision.CascadeObjectDetector();

% Read a video frame and run the detector.
videoFileReader = vision.VideoFileReader('visionface.avi');
videoFrame     = step(videoFileReader);
bbox           = step(faceDetector, videoFrame);

% Draw the bounding box around the detected face.
videoOut = insertObjectAnnotation(videoFrame,'rectangle',bbox,'Face');
figure, imshow(videoOut), title('Detected face');
```

After detecting the face on the video, the tracking algorithm is imposed on it. To track the face, the proposed system is using features like shape, texture, and color of the object to track it. The proposed system is using skin color to track it. Skin tone information can be taken by the following function,
[hueChannel,~,~] = rgb2hsv(video frame);
Then, according to this color, a box is drawn on the screen to identify the face.
figure, imshow(hueChannel), title('Hue channel data');
rectangle('Position', bbox(1,:), 'LineWidth', 2, 'EdgeColor', [1 1 0])
After tracking the face, algorithm movies into second phase.

Phase 2: Detecting fatigue of the driver is a very essential part. To analyze this part of the image, we need to process a few sets of algorithms. First, it has to detect face and then locate the eye and mouthparts in the image. After positioning these parts, fatigue detection algorithm must be run.

There are two broad ways of detecting fatigue.

1. Detecting PERCLOS (Percentage of Closure [10]) of an eye.
2. Detecting yawning position of the mouth.

Fig. 3 System flowchart

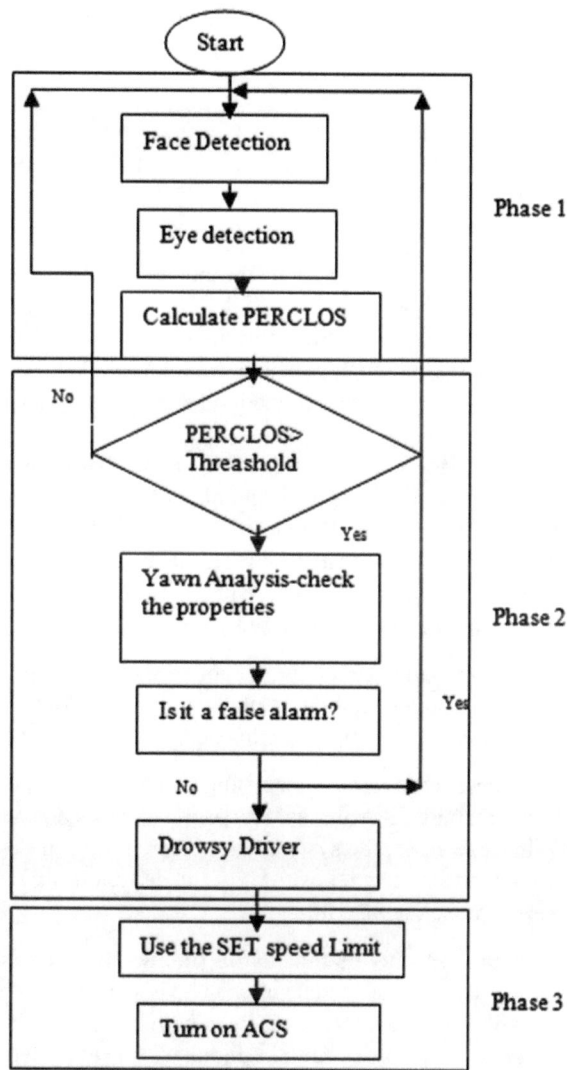

First, eyes are detected on the streaming rectangular box of the face using Haar-based classifier, which is a machine learning based approach and uses a cascade function that is trained from many positive and negative images.

Second, determining the PERCLOS (the percentage of the duration of a closed-eye state in a specific time interval of time) gives the exact measure of neurophysiologic fatigue level. Two states can be determined by this level, open eye or closed eye.

The duration of PERCLOS can be determined by the following formula:

$$PERCLOS = (t/30) * 100\%$$

where t is the duration that the eye remained closed. The threshold value for the proposed system is taken from [11] as 4.8 If it exceeds, it is determined that the driver is sleepy.

Yawn Detection: Mouth can be located from the face and yawning can be detected from the ratio of mouth height and width. The following process has to be done to detect the yawn.

1. Detect facial edge using a gradient edge detector algorithm.
2. Project the lower part of the image by using the horizontal image projection technique [12].
3. Perform edge detection algorithm to find the edge of the mouth.
4. Once the mouth edge has been detected [13], perform circular hough transformation [14] in order to recognize the mouth wide open circle.

Fatigue Detection by Eliminating Other Probabilities

Threshold is calculated by the number of frequent mouth wide opens followed by the PERCLOS threshold. If both are detected, Phase 3 will be activated by an automatic switch provided to switch on the cruise control system. The other parameters will definitely come into the picture here before the system decides that it is fatigue of a driver. Parameters are described below.

1. A person may be talking on phone while driving.
2. Person may be talking in phone with eyes closed.
3. Person may be stretching his body with yawning and eyes closed position.

In these three cases, the existing system may sound alarm and turn on to Phase 3. To avoid these false alarms, this paper suggests an analysis of true alarm.

In three cases, the system has to search the consecutive frames to find accuracy. After analysis several positions of the driver during the parameters specified above, one can conclude like this,

1. **A person may be talking on phone while driving**: His mouth open diameter will continuously change when he is speaking on phone. The continuous frames will show different diameters of mouth with which we can ignore this false alarm.
2. **Person may be talking in phone with eyes closed**: The first parameter detection will also hold here. Along with that, once the person opens his eyes PERCLOS value threshold will not meet, as the person will be awake totally after the position described in this parameter.
3. **Person may be stretching his body with yawning and eyes closed position**: The first and second parameter detection will hold even here. Along with that once he is done with stretching, his body will reach to normal position, and again PERCLOS value will differ to detect the fatigue. Because the person is not feeling sleepy.

Phase 3: Adaptive cruise control system (ACC) is one of the precrash safety systems, which keeps the car speed in a constant user-defined speed rate.

ACC is configured [15] on four properties. The proposed system uses the second property that is Deceleration Control. According to this property, if any vehicle is traveling ahead of the driving vehicle with the slow speed the proposed system understands that there is a high scope of getting crashed with the fatigued driver. Then the system uses the throttle to decelerate driver's vehicle. If the deceleration is insufficient, the system uses a break to stop the vehicle.

4 Results

After repeated trials, we were able to achieve false (Fig. 4) [Author *2] and right predictions (Fig. 5) [Author *1] with live video samples. Further, we can predict that if any right prediction is followed by false prediction and if the false prediction is followed by right PERCLOS value then we can determine that the false prediction is real false prediction only. Otherwise, we can shift on to the alarm and ACC phase to avoid further risk of an accident.

5 Conclusion

This proposed system for monitoring driver fatigue can be implemented, which detects the fatigued state of the driver continuously through monitoring the eyes of the driver and closely watches for driver safety. Though there are approaches to find

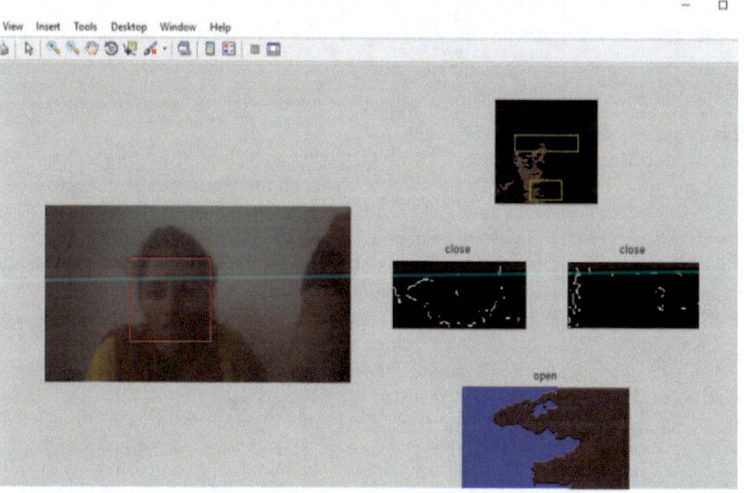

Fig. 4 False prediction

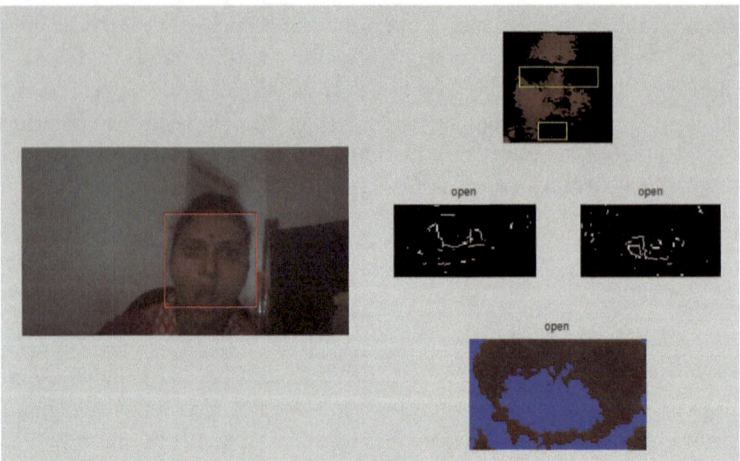

Fig. 5 Right prediction

the fatigue of the driver, the proposed system concentrated mainly on eliminating the false predictions. As this system is not able to process 24 frames per second, the real implementation of this system will not give better results. This system needs to be improved on its image capturing speed, which is otherwise of no use. The proposed system is working well in the less and bright light conditions near the camera. This system also ensures the prevention of false alarms and provides a clear way of detecting fatigue of the driver.

References

1. Ji Q, Zhu Z, Lan P (2004) Real-time nonintrusive monitoring and prediction of driver fatigue. IEEE Trans Veh Technol 53:1052–1069
2. Karchani M, Mazloumi A, Saraji GN, Gharagozlou F, Nahvi A, Haghighi KS, Abadi BM, Foroshani AR (2015) Presenting a model for dynamic facial expression changes in detecting drivers' drowsiness. Electron Physician 7(2):1073–1077
3. Kumar KSC, Bhowmick B (2009) An application for driver drowsiness identification based on pupil detection using IR camera. In: Proceedings of first international conference on intelligent human computer interaction, Allahabad, India, pp 73–82, 20–23 Jan 2009
4. Yang CM, Young M, Wu CC, Chou CM, Yang TL (2009) Vehicle driver's ecg and sitting posture monitoring system. In: Proceedings of 9th international conference on information technology and applications in biomedicine 2009, Larnaca, Cyprus, pp 1–4, 5–7 Nov 2009
5. Pauwelussen J, Feenstra PJ (2010) Driver behavior analysis during ACC activation and deactivation in a real traffic environment. IEEE Trans Intell Transport Syst 11:329–338
6. http://carinsurance.arrivealive.co.za/taxi-driver-in-fatal-accident-found-not-guilty-after-sneezing-fit-caused-accident.php
7. Bhatt PP, Trivedi JA (2017) Various methods for driver drowsiness detection: an overview. Int J Comput Sci Eng (IJCSE) 9:70–74

8. Li Z, Li SE, Li R, Cheng B, Shi J (2017) Online detection of driver fatigue using steering wheel angles for real driving conditions. Sens 17:495. https://doi.org/10.3390/s17030495
9. Swathi A (2018) Driver fatigue detection and accident preventing system. IJARIIT 4(3)
10. Ullah A, Ahmed S, Siddiqui L, Faisal N (2015) Real time driver's drowsiness detection system based on eye conditions. Int J Sci Eng Res 6(3)
11. Zhu W, Chen Q, Wei C, Li Z (2014) A segmentation algorithm based on image projection for complex text layout. Int J Veh Technol. Article ID 678786
12. Duda RO, Hart PE (1972) Use of the Hough transformation to detect lines and curves in pictures. Commun ACM 15(1):11–15
13. Viola PA, Jones MJ (2001) Rapid object detection using a boosted cascade of simple features. IEEE, CVPR, 2001
14. Wu J, Zhou ZH (2003) Efficient face candidates selector for face detection. Pattern Recogn 36:1175–1186
15. Miyata S (2010) Improvement of adaptive cruise control performance. EURASIP J Adv Signal Process. Article ID 295016

Motion Detection in Video Retrieval Using Content-Based Video Retrieval

Sudhakar Putheti, M. N. Sri Harsha and A. Vishnuvardhan

Abstract The most important challenge in content-based video retrieval is detection of moving objects accurately. The videos are stored in the database and their features are extracted. The features of the input query clip were also extracted and compared with the database features. If the features are matched up to the threshold values then the similar videos are identified. This paper proposes a technique to detect the moving objects in the videos and in order to detect the objects an abstract difference technique is used. This technique is followed by noise filtering.

Keywords Conten-based video retrieval · Moving object · Histogram difference · Query clip

1 Introduction

Content Based Video Retrieval (CBVR) is an application that retrieves targeted videos from large databases of digital videos. This application works on the search algorithms that analyzes the content of the videos in question. Features such as colour, shape, size, texture, etc., constitute the content in this context. Especially in a video when a moving object has to be identified, it has to be ascertained that whether an object exists in motion and recognize the position of the targeted object. This paper focuses on improving the efficiency of moving object tracking.

Moving object detection is the process of identifying the physical movement of an object in video content. Applications such as video surveillance, anomaly detection,

S. Putheti (✉) · M. N. Sri Harsha · A. Vishnuvardhan
Computer Science and Engineering, Vasireddy Venkatadri Institute of Technology, Guntur, Andhra Pradesh, India
e-mail: sudhakarp0101@gmail.com

M. N. Sri Harsha
e-mail: sriharsha1103@gmail.com

A. Vishnuvardhan
e-mail: vishnuvardhan.299@gmail.com

© Springer Nature Singapore Pte Ltd. 2019
H. S. Saini et al. (eds.), *Innovations in Computer Science and Engineering*, Lecture Notes in Networks and Systems 74,
https://doi.org/10.1007/978-981-13-7082-3_28

235

video conferencing, traffic analysis, and security become the focus area of CBVR. Two distinct steps of content retrieval from video and object tracking in a video are combined in this proposed method with the aim of extracting targeted videos.

A video retrieval application behaves like a search engine. In CBVR, a video clip or image is sent as a query. After processing the query, the features of the objects are extracted and matched with the videos stored in the database. Depending on the extent of features matched, the videos are displayed to the user. The main advantage of this method over the traditional concept was it needs less memory to store the features. The retrieval time in this method is lower than what it takes in the traditional method.

Video segmentation, identification of keyframes, features extraction, and lastly comparison of the extracted features with the videos stored in the database and detecting the moving objects in the videos are the five stages involved in the process of CBVR.

Moreover, object detection and its applications are open for further exploration. This paper aims at providing a technique for efficiently identifying moving objects in the key video frames along with the noise filtering methods.

The rest of this paper is organized as follows: Section 2 briefly describes the work done in the field of object detection. Section 3 focuses on the proposed method. Section 4 illustrates the results obtained through the proposed method. Section 5 presents the conclusion arrived at after the work.

2   Related Work

A significant amount of work has been done in identifying different techniques for detecting moving objects in the videos. In the method proposed by N Prakash , A Rathis, M Suresh, video retrieval is done in five stages: shot segmentation, identification of keyframes, object detection, extraction of features, and final comparison of features. They identified the moving objects present in the keyframe of the video and also this method eliminates the noise in the frames by using noise-filtering methods.

Object Detection, Object Tracking, and Behavior Recognition are the three phases in the method proposed by Xiaowei Zhou. First, the object is detected by identifying and segmenting the important objects in the videos. This phase is implemented by using background subtraction technique. In the next step, object tracking is implemented by tracking every frame of the video and finally the behavior of the object is obtained [1].

In the method proposed by Jyoti and Jadhav, the Reference Background Subtraction (RBS) technique is used for detecting and tracking the moving object. In this method, a static camera is used to record video and the first frame of video is considered as a reference frame. The current frame is subtracted from the reference frame and the difference in the pixel values is obtained. The difference in the pixel values obtained is compared with the threshold value to identify the moving object. This method is suitable for only simple environment where there are small changes. It may not be suitable for a complex environment with lighting changes. To obtain moving objects, dynamic optimization threshold method is used. This method eliminates the impact of light changes [2].

Yizhong Yang, Qiang Zhang, Pengfei Wang, Xionglou Huand Nengju Wu provide a method that detects the moving objects that are present in the videos containing a dynamic background. This is a nonparametric method containing both spatial and temporal features and detects the dynamic background. This method is more robust and effective in an environment, where the backgrounds are dynamic than the other existing methods [3].

The method proposed by Arghavan Keivani, Jules Raymond Tapamo, and Farzad Ghayoor detects multiple moving objects in the videos. Good Features to Track algorithm is used to obtain feature vector from each frame. Motion-based information identifies features of moving objects and based on this information the number of moving objects are identified and clustered using k-means algorithm. The method is accurate and efficient in determining the number of moving objects and tracking them in a scene [4].

3 Methodologies

The detection of moving objects also enables monitoring their displacement from frame to frame, which is used to identify the behavior of objects by Moving Object Detection (Fig. 1).

First, the video is segmented into shots. There are different techniques for performing video segmentation such as color histograms of consecutive frames. The keyframes are identified in each shot and they are used to identify the moving objects.

CBVR is implemented as follows: The features of the videos are extracted and stored in the databases. When a query video is supplied, the target video is retrieved based on the features stored in the database using similarity measures. First, a user uploads a video clip as input query to the proposed system. The system will divide the video into frames and select the keyframes. After frame segmentation and selection, extract the color, text, and edge detector on selected frames and at last extract the features. The same process is done on videos stored in the database and the features are saved in an SQL database. The CBVR system compares the features and produces relevant videos to the users.

3.1 Motion Detection

The first frame of the visual is taken as a reference frame and the frames of a shot are converted into HSV (Hue, Saturation, Value) format from the RGB (Red, Green, Blue) format. The HSV format can separate the color information from the intensity and thus resolves the intensity related issues. This can also be used for removing shadows.

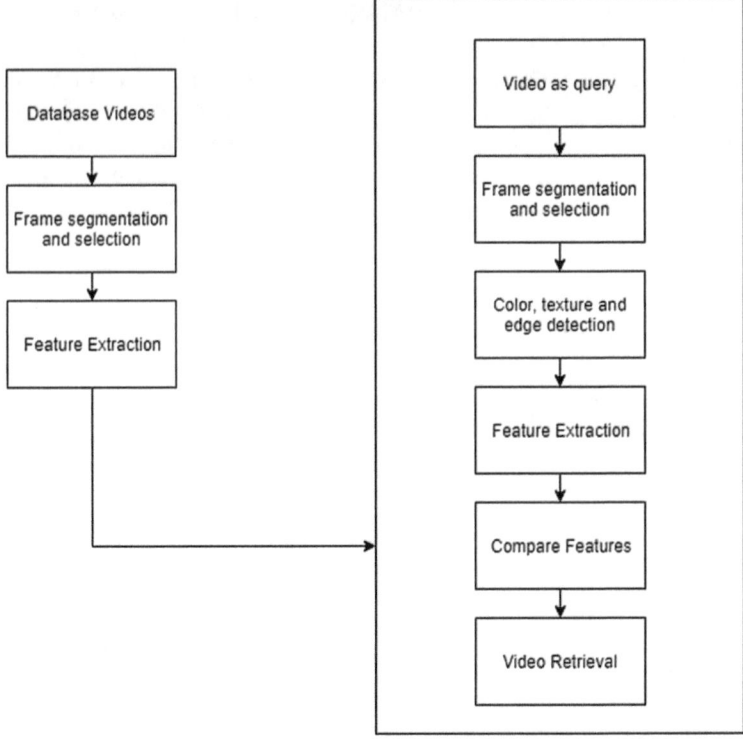

Fig. 1 Architecture of content-based video retrieval

In this method, the absolute difference between the reference frame and the keyframe is obtained. The resulting image contains the objects, and may contain noise. The difference indicates the displacement of the object from the original position. Now the resulting image is transformed into a binary image by converting each pixel to either black or white color based on the threshold value.

The method is repeated for the keyframes of all shots present in the videos. The group of binary images consisting of moving objects will be obtained finally. The steps in this process are represented in Fig. 2.

3.2 Noise Filtering

The binary image obtained may consist of the noise and the detected moving objects may contain holes. A hole is the cluster of background pixels covering the foreground pixels. The methods used for noise filtering are mean filter, median filter, Bitonal algorithm etc. However except median filter, all other methods blur the image to remove noise and some of the edge pixels may be lost. Therefore, the median filter

Fig. 2 Motion detection

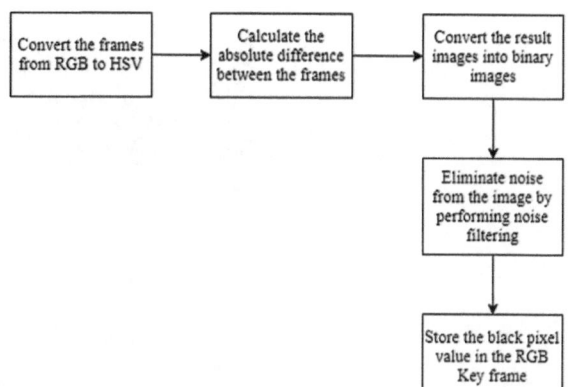

is suggested as an alternative to filter the noise in the image. The median filter functions by passing through the image pixel by pixel. In this method, each value is replaced with the median value of surrounding pixels. In this process, the pattern of the neighboring pixels is called "window". This window is slid over pixel by pixel across the entire image. Then the median is calculated by first sorting all the pixel values from the window into ascending order. Then the central pixel is replaced with the median pixel value. After removing noise, the image is compared with the RGB frame image. By repeating the process for all binary images, the moving objects are detected and the features such as color, shape, size, and edge can be determined.

4 Experimental Results

The moving object detection for CBVR is mainly used in applications such as Traffic surveillance, medicine, satellite surveillance, etc. Whenever a new object enters into the shot, it is taken as a reference frame of the shot.

4.1 Detection of Objects in Motion

Figures 3 and 4 represent the first frame and keyframe of the shot. The absolute difference between the frames is calculated. Figure 5 represents the background frame and current frame of the shot and the noise filtering method such as the median filter is applied to remove the noise. Figure 7 shows the tracking of moving objects from the video. The objects are identified based on the threshold value set by the user. This follows the dynamic threshold for tracking the moving objects in the videos (Fig. 6).

Fig. 3 Reference frame in
the shot

Fig. 4 Current frame

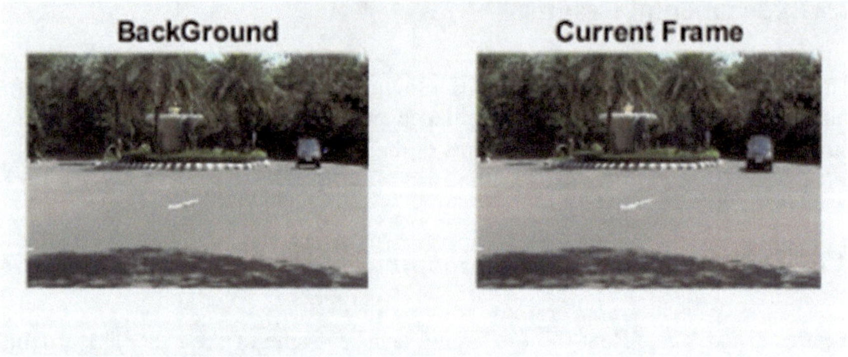

Fig. 5 Background and current frame in the shot

Fig. 6 Blob detected in the frame

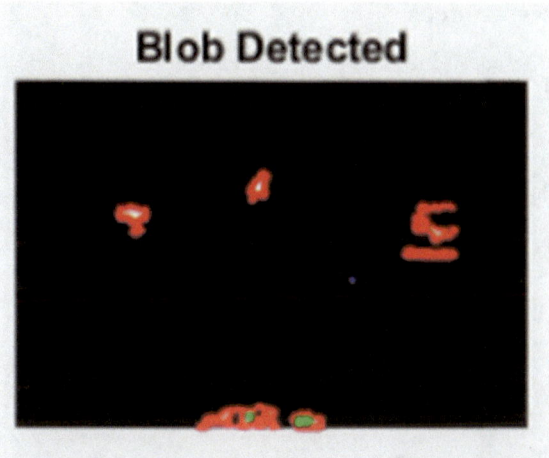

Fig. 7 Tracking of moving objects based on the dynamic threshold value

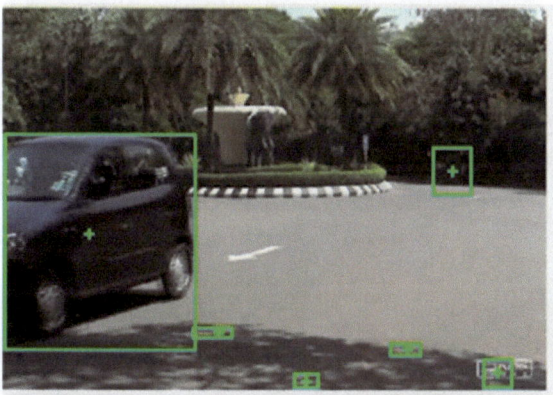

5 Conclusion

This paper aims at providing a method that detects the objects in motion and displays them to the user. The features used for retrieving the videos from the database are color, shape, and size. If only those features are considered for retrieval, there is a possibility that different objects may possess the same features and there is a probability of treating those different objects as a single object. The method is limited to videos in which the background is static and the foreground is dynamic. The method uses dynamic thresholding, while tracking the objects so that the user can see the threshold value based on the user requirement.

References

1. Raviprakash N, Suresh M, Rathis A, Devarla D, Yadav A, Nagaraja GS Moving object detection for CBVR
2. Zhou X, Yang C, Yu W (2013) Moving object detection by detecting contiguous outliers in the low-rank representation. IEEE Trans Pattern Anal Mach Intell 35:1–30
3. Cohen I (1999) Detecting and tracking moving objects for video surveillance. IEEE Proc Comput Vis Pattern Recognit 2:23–25
4. Jadhav MJJ (2014) Moving object detection and tracking for video surveillance. IJERGS 2(4):372–378

Improved Clustering Technique Using Metadata for Text Mining

S. Tejasree and Shaik Naseera

Abstract Metadata delivers a lot of data for the bunching reason, creating groups without shifting the side or extra data it can result to give an awful nature of groups. In content mining application, the content alongside the side data or metadata is introduced in every single report. The way toward applying bunching method to the records containing content is called as the content grouping. The consistent development in unstructured data makes the content mining applications critical to accomplish the quality data as an objective. Content mining basically changes over unstructured information to organized information. Content mining is the way toward getting brilliant data from unstructured information. Content mining is otherwise called content information mining or content investigation. Unstructured information contains an abundance of data that is exceptionally helpful for cutting-edge recognition of dangers why in light of the fact that prepared security investigator feel hard to examine tremendous volume of information. Multifaceted nature is expanding to dissect those information. Our investigation investigates the using the methods for content mining, content grouping, common dialect handling, machine figuring out how to distinguish security dangers by mining the applicable data from unstructured log messages. Enhanced the bunching procedures comes about a solid potential for expanding the execution by expanding the span of datasets and separating the more highlights from the unstructured log messages. In this paper, it primarily center around the bunching strategies and grouping techniques that are utilized as a part of text mining.

Keywords Text Mining · Text classification · Co-clustering · Security information · Natural language processing · Machine learning

S. Tejasree (✉) · S. Naseera
School of Computer Science and Engineering (SCOPE), VIT University, Vellore,
Tamil Nadu, India
e-mail: samakoti.tejasree2016@vitstudent.ac.in

S. Naseera
e-mail: naseerakareem@gmail.com

© Springer Nature Singapore Pte Ltd. 2019
H. S. Saini et al. (eds.), *Innovations in Computer Science and Engineering*, Lecture Notes in Networks and Systems 74,
https://doi.org/10.1007/978-981-13-7082-3_29

243

1 Introduction

Metadata contain the side information, and the side information contains information which is having noise. Form the clusters before filtering the side information, it makes the clustering not a good quality. Efficient selection method is used to perform the mining process to select the side information [1], and it is very useful for clustering. Two-mode clustering isone type of text mining technique that allows the groups by clustering both text and side information [2]. The main purpose is to use side information along with text context in a good way to improve the process of text clustering [3]. Co-clustering with the help of TFIDF [4] and Gini index eliminate the noisy data present inside information, and it is provided similarity by using k-means clustering technique [5]. Co-clustering can be defined as merging of two or more types of data based on their features [6]. In between different types of data, there exists a close relationship in between them. Traditional clustering techniques like k-means clustering, hierarchical clustering find it difficult to analyze and utilize the relationship information [7] Text mining is very useful to make large volumes of unstructured data that can be accessible and very useful.

2 Clustering Techniques

Text mining purpose and their disadvantages can be listed below in Table 1.

Table 1 Different clustering algorithms used in text mining

Clustering algorithm	Purpose	Disadvantages
K-means [29, 25]	Nearest-neighbour classifier to cluster	Accuracy of the cluster is susceptible
Agglomerative hierarchical [25, 25]	Documents of similar class in the same cluster	Mistakes cannot be fixed once they happen
Spatial text clustering [30]	Discovery of interesting relationships and features	Costly and impractical for users
CLARANS [30, 31]	Works well for spatial data clustering	Cannot be applied for normal data and also the side information cannot be presented
BRICH [31]	It is very efficient for huge data sets	It does not consider the side information, it does not provide any semantic relationships
FCDS CLUSTERING [31]	It uses word net ontology it provides semantic relationship	It does not consider the side information

(continued)

Table 1 (continued)

Clustering algorithm	Purpose	Disadvantages
COATES [32]	Text clustering with side information	It performs alternative iterations
Co-clustering with TFIDF [30, 5]	Merging of two or more data types	Fail to detect anticorrelations it implies functional similarity

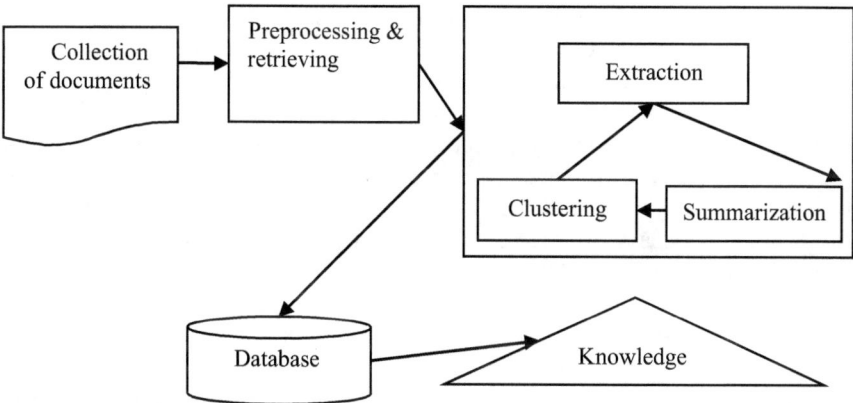

Fig. 1 Text mining architecture

3 Text Mining Architecture

Text mining tool would retrieve the particular document and preprocess it by format checking [8]. Text mining has more importance because of the availability of an increasing number of documents from a variety of sources [9] (Fig. 1).

4 Text Mining Techniques

Search engines are the most well-known information retrieval systems. The main example for the search engine is Google, which retrieves the available documents from the web by using Google search engine. The search engine searches for relevant documents [10]. Search engines are used to extract information, and the information is disordered. So, it provides the order for the information. Information retrieval is the task of getting relevant information; the retrieved information may contain many sources as webpages, images, videos, articles, and pdfs [11]. The process of identifying and searching the information, the information is needed for the user is called as retrieve process [12]. Information retrieval is used to reduce the information overload by using automated information. Information retrieval is used by many places that are universities, libraries, access to journal, articles, and books [13, 14]. Cross-language information retrieval is a subfield of information retrieval, and it is used software engineering to retrieve the information [14, 15]. Information extraction is the process of extracting or mining the useful or relevant information from unstructured or semi-structured document [16]. It recognizes each and every extraction entities, and it provides the relationships between all the entities. The valuable information extracted is without understanding the text [17]. The valuable information is stored in the database like patterns and after that, the information is used for further use [18]. The information collected after retrieving is structured information [1]. The complexity of the information extraction method is very high because it uses natural language process. In each and every document, more information is presented, and the document is a combination of both information and expressions [19]. So, reducing the information from the document is necessary to reduce the information from the document with a computer program that places the main important points in the original document. Summarization can be based on the single-source document and multisource document. Google is a search engine that uses summarization technology. Summarizationt uses multisource document called as the multi-document summarization system [20].

Text Categorization: Original data can be divided into groups of entities [21, 22] ,and for dividing the group of entities categorization process isused. The main goal of text categorization is to assign a category to a new document. The main advantageof using text categorization is to reduce the load on memory, and it provides efficient storage and retrieval of information [23, 24]. Clustering is the process of creating groups of similar objects from a given set of inputs [25]. Objects of the same cluster are similar, and objects of the different clusters are different. First, the idea of clustering comes from statistics. In statistics, clustering is used on numeric data. After that, the clustering technique is used for all data types that are text or multimedia [26]. Unsupervised process classifies the objects into groups. Here, the main problem is unlabeled collection of information without any prior information [27, 28]. Detecting malicious activities using unstructured log messages can be seen as a major problem in text classification [9]. It can be used on different datasets those are tabulated in below (Table 2).

5 Conclusion

Text mining is a method of extracting meaningful information from unstructured documents. Mining the text data with the use of side information or meta information. Text databases contain large volumes of data that contains the side information, and it is used to improve the clustering process. The recently used clustering technique is

Table 2 Classification techniques used for text mining

Author	Dataset	Algorithm used	Purpose
Alsudais et al. [33]	Data of Twitter	Rapid Forest	Identify the type of location users are tweeting from
Pang et al. [34]	Movie Review Data	Machine Learning Methods	Categorize documents according to its subjects
Gabrilovich and Markovitch [35]	Data set is based on Web	Support Vector Machine	Analyze a large collection of data sets
Korde and Mahender [36], Wei et al. [37]	Data set is not specified	K-Nearest Neighbour, Naïve Bayes, Decision tree, SVM, Neural Network	Introduces text classification
Nigam et al. [38]	New group set data	Naïve Bayes Classification, Expectation Maximization	Improvement in the accuracy of learned text classifiers, classification error is reduced
Joachims [39]	Two data sets are used	Support Vector Machine is used	SVM do not require any tuning of the parameters
Fu et al. [40]	Time series dengue data	Support Vector machine and genetic algorithm	Climate factors and dengue incidences
Boyd et al. [41]	Sample of public tweets	Principal Components Analysis (PCA)	Built a reductive rewet model
Delen et al. [42]	Data was collected from Behavioral Risk Surveillance	Artificial Neural Networks and decision trees are used	Provides more efficiency and accuracy
Godin et al. [43]	Twitter API	Expectation Maximization and Naïve Bayes method are used, LDA model is used	Recommended hashtags for tweets in a fully unsupervised manner
Yang et al. [44]	Twitter data	Information filtering Approaches, Topic Modeling approaches	Implementing the system and satisfying the quality requirements
Ha et al. [45]	Public resource sites	Systematic Literature Review method	Detailed analysis by the proposed SLR

co-clustering technique with TFIDF. Gini index for using to eliminate noise present in the information k-means algorithms is used to create clusters. By using the side information in order to improve the quality of the cluster. By using better classification algorithms for getting correct information, it provides security for threat detection.

References

1. Aggarwal CC, Zhao Y, Yu P (2014) On the use of side information for mining text data. IEEE Trans Knowl Data Eng 26(6):1415–1429
2. Thomas RE, Khan SS (2016) Improved clustering technique using metadata for clustering. In: 2016 IEEE conference on communication and electronics systems (ICCES)
3. Divya P, Kumar GSN (2015) Effective feature selection for mining text data with side-information. Int J Trends Eng Technol 4
4. Wang X, Cao J, Liu Y, Gao S (2012) Text clustering based on the improved TFIDF by the iterative algorithm. In: IEEE symposium on electrical and electronics engineering (EEESYM)
5. Thomas RE, Khan SS (2016) Co-clustering with side information for text mining. In: International conference on data mining and advanced computing (SAPIENCE)-IEEE 2016 (in press)
6. Li Y, Abdulmoheson, Albathan M, Shen Y, Bijaksana MA (2007) Relevance feature discovery for text mining. IEEE Trans Knowl Data Eng 6(1)
7. Suh-Lee C, Jo J-Y, Kim Y (2016) Text mining for security threat detection. In: 2016 IEEE conference on communications and network security (CNS). 978-1-5090-3065-1/16
8. Gupta V, Lehal GS (2009) A survey of text mining techniques and applications. J Emerg Technol Web Intell 1(1)
9. Bhumika PSSS, Nayyar PA (2013) A review paper on algorithms used for text classification. Int J Appl Innov Eng Manag 2(3)
10. Mondal AK, Maji DK (2013) Improved algorithms for keyword extraction and headline generation from unstructured text. First Journal publication from SIMPLE groups. CLEAR J
11. Ahmad PH, Dang S (2014) A comparative study on text mining techniques. Int J Sci Res 3(12):2222–2226. ISSN: 2319-7064
12. Saranya S, Munieswari R (2014) A survey on improving the clustering performance in text mining for efficient information retrieval. Int J Eng Trends Technol (IJETT) 08
13. Vidhya KA, Aghila G (2010) Text mining process, techniques and tools: an overview. Int J Inf Technol Knowl Manag 2(2):613–622
14. Singhal A (2001) Modern information retrieval: a brief overview. Bulletin of the IEEE Computer Society Technical Committee on Data Engineering, pp 1–9
15. Subarani D (2012) Concept based information retrieval from text documents. IOSR J Comput Eng (IOSRJCE) 2(4):38–48. ISSN: 2278-0661
16. Gupta V, Lehal GS (2010) A survey of text summarization extractive techniques. J Emerg Technol Web Intell 2(3):258–268
17. Pande VC, Khandelwal AS (2014) A survey of different text mining techniques. IBMRD's J Manag Res 3(1):125–133. ISSN: 2348-5922
18. Zhong N, Li Y, Wu S-T (2010) Effective pattern discovery for text mining. IEEE Trans Knowl Data Eng. © Copyright 2010 IEEE
19. Ghosh S, Roy S, Bandyopadhyay SK (2012) A tutorial review on text mining algorithms. Int J Adv Res Comput Commun Eng 1(4):223–233
20. Foong OM, Oxley A, Sulaiman S (2010) Challenges and trends of automatic text summarization. Int J Inf Telecommun Technol IJITT 1(1):34–39. ISSN: 0976–5972
21. Sharma S, Srivasta SK (2016) Review on text mining algorithms. Int J Comput Appl 134(8)

22. Mooney RJ, Nahm UY (2003) Text mining with information extraction. In: Multilingualism and electronic language management: proceedings of the 4th international MIDP colloquium, pp 141–160
23. Dang S, Ahmad PH (2015) A review of text mining techniques associated with various application areas. Int J Sci Res. ISSN:2319-7064
24. Mahalakshmi B, Duraiswamy K (2012) An overview of categorization technique. Int J Mod Eng Res (IJMER) 2(5):3131–3137. ISSN: 2249-6645
25. Steinbach M, Karypis G, Kumar V (2000) A comparison of document clustering techniques. In: Proceedings of text mining workshop KDD, pp 109–110
26. Mehta N, Dang S (2011) A review of clustering techniques in various applications for effective data mining. Int J Res IT Manag 1(2):50–66. ISSN 2231-4334
27. Patel R, Sharma G (2014) A survey on text mining techniques. Int J Eng Comput Sci 3(5):5621–5625. ISSN: 2319-7242
28. Nasa D (2012) Text mining techniques—a survey. Int J Adv Res Comput Sci Softw Eng 2(4):50–54. ISSN: 2277 128X
29. Sihag VK (2013) Graph based text document clustering by detecting initial centroids for k-means. Int J Comput Appl 62(19):1–4
30. Ng RT, Han J (1994) Efficient and effective clustering methods for spatial data mining. In: Proceedings of VLDB conference, San Francisco, CA, USA, pp 144–155
31. Zhang T, Ramakrishnan R, Livny M (1996) BIRCH: an efficient data clustering method for very large databases. In: Proceedings of ACM SIGMOD conference, New York, NY, USA, pp 103–114
32. Aggarwal C, Zhao Y, Yu P (2014) On the use of side information for mining text data. IEEE Trans Knowl Data Eng 26(6):1415–1429
33. Alsudais A, Leroy G, Corso A (2014) We know where are you tweeting from: assigning a type of place to tweets using natural language processing. In: IEEE International congress on big data
34. Pang B, Lee L, Vaithyanathan S (2002) Thumb up sentiments classification using machine learning techniques. In: Proceedings of EMNLP
35. Gabrilovich E, Markovitch S (2004) Text categorization with many redundant features: using aggressive feature selection to make SVMs competitive with C4.5. ICML
36. Korde V, Mahender CN (2012) Text classification and classifiers: a survey. Int J Artif Intell Appl (IJAIA) 3(2)
37. Wei L, Wei B, Wang B (2012) Text classification using support vector machine with mixture of kernel. A J Softw Eng Appl 5:55–58. https://doi.org/10.4236/jsea.2012.512b012. Published Online Dec 2012
38. Nigam K, Mccallum AK, Thrun S, Mitchell T (2000) Text classification from labeled and unlabeled documents using EM. Kluwer Academic Publishers, Boston. Manufactured in Netherlands
39. Joachims T (1998) Text categorization with support vector machines: learning with many relevant features
40. Fu X, Liew C, Soh H, Lee G, Hung T, Ng L-C (2007) Time series infectious disease data analysis using SVM and genetic algorithm. IEEE
41. Boyd D, Golder S, Lotan G (2010) Tweet, tweet, retweet: conversational aspects of retweeting on twitter. IEEE
42. Delen D, Fuller C, McCann C, Ray D (2007) Analysis of healthcare coverage: a data mining approach. Expert Syst Appl
43. Godin F, Slavkovikj V, De Neve W (2013) Using topic models for twitter hashtag recommendation. In: International World Wide Web conference committee (IC3W2)
44. Yang S, Kolcz A, Schlaikjer A, Gupta P Large-scale high-precision topic modeling on twitter. In: Proceedings of 20th ACM SIGKDD International conference on knowledge discovery and data
45. Ha I, Park H, Kim C (2014) Analysis of twitter research trends based on SLR. In: 16th International conference advanced communication technology (ICACT)

46. Kapugama KDCG, Loresuhewa SAS, Kalyan MAL (2016) Enhancing wikipedia search results using text mining. In: 2016 International conference on advances in ICT for engineering regions (ICTer), pp 168–175
47. Agrawal R, Batra M (2013) A detailed study on text mining techniques. IJSCE 2(6). ISSN: 2231-2307. Paper ID: SUB151800 2465. Int J Sci Res (IJSR). ISSN (Online): 2319-7064. Index Copernicus Value (2013): 6.14 | Impact Factor (2013): 4.438, 4(2), February 2015. www.ijsr. net. Licensed Under Creative Commons Attribution CC BY
48. Niharika S, SnehaLatha V, Lavanya DR (2012) A survey on text categorization. Int J Comput Trends Technol 3(1):39–45. ISSN: 2231-2803
49. Steinbach M, Karypis G, Kumar V (2000) A comparison of document clustering techniques. Comput Sci Eng 1(3):1–20

E-Commerce Security by Quantum Digital Signature-Based Group Key Management

Udayabhanu N. P. G. Raju and R. Vivekanandam

Abstract The Internet development is high over the past decades and there is a lot of cloud computing technology is used in business and other purposes. This leads to the major tread to the security in the cloud and secure communication is much required. Quantum Key Distribution (QKD) is one of the most common methods used to provide better encryption. The main focus of this research is that provide secure communication from the group keying. The Quantum Digital Signature-based Group Key management (QDSGKM) combines QKD with the Fibonacci, Lucas, and Fibonacci–Lucas matrices, which provides the quantum digital signature checks. This method gives the technique to verify the integration of information received by the participants, to authenticate the identity of the participants, and to improve the verification of the signing verification. The experimental result shows that the proposed system provides the security with less delay when compared to the QKD system.

Keywords Encryption technology · Quantum cryptography · Quantum digital signature-based key management · Quantum key distribution · Secure communication · Fibonacci–Lucas matrices

1 Introduction

The fast development of the internet and cloud computing improves the significance of the research of the quantum cryptography [1]. The cloud computing has a significant impact on both scientific and business in information technology. In order to prevent the message sent in the group communication from other members to read, the encryption method is needed [2]. The applications are able to communicate with

U. N. P. G. Raju · R. Vivekanandam (✉)
Sri Satya Sai University of Technology and Medical Sciences, Sehore, Madhya Pradesh, India
e-mail: svkmbbs@gmail.com

U. N. P. G. Raju
e-mail: unpgangadharraju@gmail.com

© Springer Nature Singapore Pte Ltd. 2019
H. S. Saini et al. (eds.), *Innovations in Computer Science
and Engineering*, Lecture Notes in Networks and Systems 74,
https://doi.org/10.1007/978-981-13-7082-3_30

each other in a secure way from the help of a shared key in a secure manner. The important aspect that considers is that quantum key distribution is perfectly safe for the quantum key distribution [3]. The data owners store the encrypted data into the cloud, in which that user wants to access. The decryption key and digital certificates can be distributed to the user by the data owners if the access control policy permits the user [4]. Then, the user shows the certificate to the cloud and can get access to the encrypted data. It is well known that once the large quantum computer is built, the existing encryption system is efficiently broken [5, 6]. Cloud computing grows fast and is quite expensive in the application, for instance, e-commerce, which is the new type of transaction that brings the consumer, logistic, and enterprise into a compressive network era. Theoretically, quantum computer is unbreakable and this is highly needed in quantum computer era. Classical communication is the submission of data in the unsecure connection, for example, Internet [7].

Several algorithms are presented to prevent the leak in the communication and also to withstand some attacks [8]. These algorithms are considered as secure until the quantum computer is publicly available. On the other hand, quantum cryptography provides the security for the large information of the shared data and communication, even after the quantum computer is found [9, 10]. Then, the quantum attack of natural noise gives the major challenge for the quantum encryption. This challenge is prevented by using a robust secret key builder that is made by group key. The QKD is the mechanism of two users initiate the secret shared key (SSK). This key helps to encrypt and decrypt the data in the secure connection. The various QKD protocol was proposed like KMB09 protocol, Coherent One-Way (COW) protocol, EPR protocol, SARG04 protocol, and B92 protocol. The payment service is the most important link in the whole transaction and the mobile payment is highly used. In this method, the QKD is improved with Fibonacci, Lucas, and Fibonacci–Lucas matrices. The quantum signature is verified from the Charlie and is authenticated to the user, and then the information is provided. The proposed method is compared with the QKD method and it has a low delay.

2   Literature Review

Mohajer and Eslami [11] all participants retrieve the final key simultaneously and there is no guarantee that all the participants received the final key. Mogos [12] the error rate of the method is high. Qiu [13] is still vulnerable for man-in-the-middle attack. Tanizawa et al. [14] method still has the delay, so it is applicable for only some applications. Metwaly et al. [15] need to have quantum key distribution and hashing-based authentication helps to implemented on the multicast network.

3 Proposed Method

The matrices are integrated into the QKD to provide a secure connection in the cloud. The Charlie is used to verify the signature of the user and provides the information if the signature matches the Alice's signature. The matrices used to combine the QKD is Fibonacci, Lucas, and Fibonacci–Lucas matrices.

3.1 Quantum Key Distribution

The quantum key distribution protocol of Simon et al. shows that how protocols can be improved in coding efficiency of entangled states, while using the Fibonacci, Lucas, or Fibonacci–Lucas matrices.

3.1.1 Fibonacci Matrices

Fibonacci number F_n is an infinite sequence of integers [16] defined in the following recursion:

$$F_n = F_{n-1} + F_{n-2}, n \geq 2, \tag{1}$$

where the first two elements of the sequences are $F_0 = 0$ and $F_1 = 1$. Taking the first three integers F_0, F_1, F_2 of the Fibonacci sequence, we construct a 2×2 Fibonacci matrix:

$$Q_1 = \begin{pmatrix} F_0 & F_1 \\ F_1 & F_2 \end{pmatrix} = \begin{pmatrix} 0 & 1 \\ 1 & 1 \end{pmatrix}, \tag{2}$$

where $\det(Q_1) = F_0 F_2 - F_1^2 = -1$.

Construction of Q_p. The new class of Fibonacci matrices Q_p, where $p = 2, 3, \dots$ and Q_1 is given in Eq. (2).

The class satisfies the following relation:

$$Q_p = \begin{bmatrix} Q_1 & Q_1 & \dots & Q_1 & Q_1 \\ O & I & O & \dots & O \\ O & O & I & \ddots & O \\ \vdots & \vdots & \ddots & \ddots & O \\ O & O & \dots & O & I \end{bmatrix} \tag{3}$$

where O is the 2×2 matrix that has zero entries and I is 2×2 identity matrix. It is easy to prove that matrices Q_p^n satisfy the following properties in terms of (3):

$$\det\left(Q_p^n\right) = \left(\det\left(Q_p\right)\right)^n = (-1)^{pn}. \tag{4}$$

Q_p^n where $p = 1, 2, 3, \ldots$ is invertible and its inverse can be calculated, and is shown as follows:

$$Q_1^{-n} = \begin{bmatrix} Q_1^{-n} & -I & \ldots & -I & -I \\ 0 & I & 0 & \ldots & 0 \\ 0 & 0 & I & \ddots & 0 \\ \vdots & \vdots & & \ddots & \ddots & 0 \\ 0 & 0 & \ldots & 0 & I \end{bmatrix} \quad \text{where } p = 1, 2, 3\ldots \tag{5}$$

3.1.2 Lucas Matrices

Lucas numbers L_n are an infinite sequence of integers [17] defined by the following recursion:

$$L_n = L_{n-1} + L_{n-2}, n \geq 2, \tag{6}$$

where the integers $L_0 = 2$ and $L_1 = 1$ start the sequences and $n = 1, 2, \ldots$ Lucas and Fibonacci numbers share the following conjugate relation:

$$L_n = F_{n+1} + F_{n-1} \tag{7}$$

Let us define a 2×2 matrix R_1 as

$$R_1 = \begin{pmatrix} 2 & 1 \\ 1 & 3 \end{pmatrix} \tag{8}$$

According to (1) and (3), the nth defines the power of R_1 as

$$R_1^n = \begin{pmatrix} L_{n-1} & L_n \\ L_n & L_{n+1} \end{pmatrix} = Q_1^n \times \begin{pmatrix} -1 & 2 \\ 2 & 1 \end{pmatrix}, \tag{9}$$

$$\det\left(R_1^n\right) = \det\left(Q_1^n \times \begin{pmatrix} -1 & 2 \\ 2 & 1 \end{pmatrix}\right) = 5 \times (-1)^{n+1} \tag{10}$$

This implies that R_1^n is invertible and its inverse matrix R_1^{-n} can also be derived using the properties of Lucas sequences. They are

$$R_1^{-2k} = \begin{pmatrix} \frac{L_{2k+1}}{5} & -\frac{L_{2k}}{5} \\ -\frac{L_{2k}}{5} & \frac{L_{2k-1}}{5} \end{pmatrix}, \quad \text{for } n = 2k, \tag{11}$$

The matrix R_1 is used to build the new class of Lucas matrices R_p.

3.1.3   Fibonacci–Lucas Matrices

Fibonacci and Lucas sequences can be jointly to create a new class of matrices, which we call them Fibonacci–Lucas matrices [18]. They are consecutive power of T_1 and are defined according to the following recursion:

$$T_1^n = \begin{pmatrix} F_{n-1} & F_n \\ L_{n-2} & L_{n-1} \end{pmatrix} \tag{12}$$

where the first Fibonacci–Lucas matrix T_1 is

$$T_1 = \begin{pmatrix} F_1 & F_2 \\ L_0 & L_1 \end{pmatrix} = \begin{pmatrix} 1 & 1 \\ 1 & 1 \end{pmatrix} \tag{13}$$

Lucas and Fibonacci numbers satisfy the relation $L_{n-1} = F_n + F_{n-2}$, thus T_1^n can be written as

$$T_1^n = \begin{pmatrix} F_{n-1} & F_n \\ F_{n-1} + F_{n-3} & F_n + F_{n-1} \end{pmatrix} \tag{14}$$

The determinant of Fibonacci–Lucas matrices.

3.1.4   Matrix Encryption

Consider a message that is a sequence of integers $\{m_i\}_{i=1,2,.....}$. Integers of the message can be packed into a square $l \times p$ matrix M. The arrangements of message in M can be to some extent arbitrary as integers can be determined by selecting odd or even number of digits. Given a matrix K matrix encryption can be defined as follows:

$$E = M \times K \tag{15}$$

The decryption can be done using the inverse matrix K^{-1}

$$M = E \times K^{-1} \tag{16}$$

where K can be either Q_p^n or R_p^n or T_p^n.

The symmetric encryption is given in the encryption method and symmetric cryptography needs a secure channel to distribute secret keys between two communicating parties. This is the linear encryption and this can be easily breakable in a chosen plaintext attack. The Fibonacci sequences is used to to prepare entangled states and two communicating parties can detect the Fibonacci values for the entangled states with the designated sorters. More importantly, Q_n^p or R_p^n or T_p^n can be used just one time and their order is determined by quantum random generators in Alice's, Bob's, and Charlie's side. Considering these, the quantum matrix encryption is secure.

3.1.5 Re-encryption QKD

The ReEnc-QKD is constructed by using meta proxy re-encryption scheme and the scheme is shown to be group key indistinguishable.

3.2 Simon's et al.'s QKD Protocol

The major idea behind Simon et al.'s QKD protocol [19] is the use of a Vogel spiral and this allows either Alice or Bob to prepare a source of entangled Fibonacci-valued orbital angular momentum (OAM) states. The spiral is left after the Fibonacci-valued entangled pairs and enters the down-conversion crystal. The down-conversion breaks each Fibonacci value into two lower OAM values and in both Alice's and Bob's laboratories, there is a beam splitter directing some regular proportion of the beam to two different types of OAM sorters L and D. The entangled photons randomly transmit the beam splitters to either the L or D sorter. The L sorter allows Fibonacci-valued entangled photons to reach at the single photon array detectors only. The D sorter allows "diagonal" superposition in the form $\frac{1}{\sqrt{2}}(|F_n \times |F_{n+2})$ and filters out any non-Fibonacci entangled photons.

3.3 Proposed Quantum Digital Signature

The proposed Quantum Digital signature includes five steps: setup, key distribution, message blinding, signing, and verification. Consider that there are authenticated classic channels and insecure quantum channels in the Alice, Bob, and Charlie. Every pair of parties have the different quantum key matrices K_{AB}, K_{AC} and K_{BC}, respectively, is shown in Fig. 1. The key matrices K_{AB}, K_{Ac} and K_{BC} are generated from the Simon et al.'s QKD algorithm, which is of the form Q_{np} or R_{pm} or T_{pn}.

3.3.1 Setup

From this method, it consists of three participants: (1) the owner of the message, Alice transfer the message into an n-square matrix ($n = 2, 3, \ldots$) and blinds the matrix, (2) Bob the signer who signs blind messages, and (3) Charlie the verifier who checks if a signature matches a message.

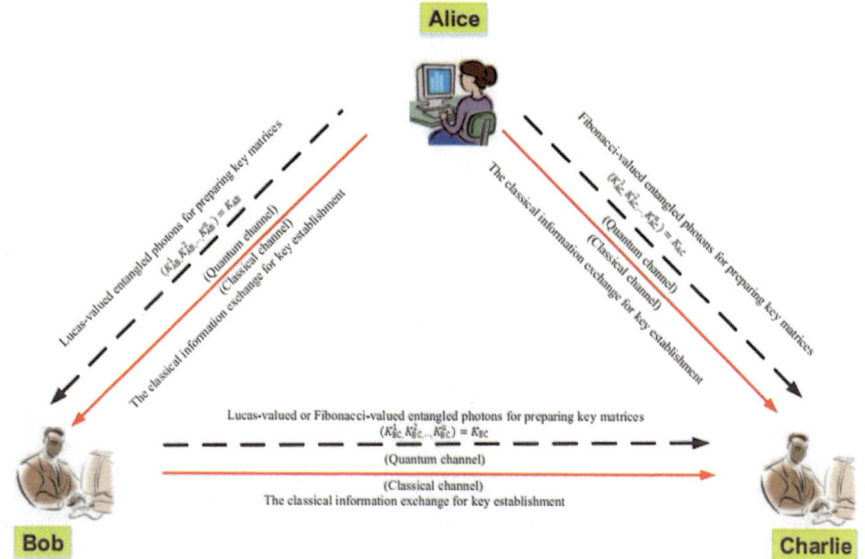

Fig. 1 Quantum key distribution of QDS system

3.3.2 Key Distribution

Alice and Charlie, Alice and Bob, and Bob and Charlie have the pairwise quantum key matrices K_{AB}, K_{AC} and K_{BC}, respectively. The QKD protocol is used by parties and generate their pairwise key matrices $\{K_{AB}^1, K_{AB}^2, \ldots, K_{AB}^\alpha\} = K_{AB}$ between Alice and Bob $\{K_{AC}^1, K_{AC}^2, \ldots, K_{AC}^\alpha\} = K_{AC}$; between Alice and Charlie; and $\{K_{BC}^1, K_{BC}^2, \ldots, K_{BC}^\alpha\} = K_{BC}$ between Bob and Charlie. The key generation order is determined by Alice's, Bob's, and Charlie's quantum random generators.

3.3.3 Message Blinding

Alice message is transformed into matrices $(M_1, M_2, \ldots, M_\alpha) = M$, where $M_k = (m_{tj})_{m \times n}$, $K \in \{1, 2, \ldots, \alpha\}$, $t, j \in \{1, 2, \ldots, n\}$. Then, the key blinds the message matrix M using the key K_{AC} and obtains the blind message

$$M_k' = M_k \times K_{AC}^k, \quad k \in \{1, 2, \ldots, \alpha\}. \tag{17}$$

Then, the blind message M' is encrypted in the Alice side with the key K_{AB} as follows:

$$M_k'' = M_k' \times K_{AB}^k, \quad k \in \{1, 2, \ldots, \alpha\}, \tag{18}$$

and M_k, $M_{k'}$ and M'_k. Finally, Alice sends $(M''_k, \det(M'_k))$ to Bob, and $\det(M_k)$ to Charlie.

3.3.4 Signing

Bob signs message M blindly by creating a signature for the message M' and this means that Bob does not know the contents of M. The execution of the following steps to receive the message:

(1) It checks the authenticity of $(M''_k, \det(M'_k))$. First, the message M''_k with the key K^k_{AB} and obtains

$$M'_k = M''_k \times (K^k_{AB})^{-1}, \tag{19}$$

where $(K^k_{AB})^{-1}$ denotes the inverse matrix of K^k_{AB}. If the determinant of M'_k recovered by Bob is not equal to the value of the determinant obtained from Alice, Bob aborts this communication. Otherwise, he performs the next step.

(2) Bob signs the blind message M'_k using K^k_{BC}. The signature is

$$S^k = M'_k \times k^k_{BC} \tag{20}$$

(3) Sends the signature $S = \{S^1, S^2, \ldots, S^\alpha\}$ to Charlie.

3.3.5 Verification

Charlie verifies the signature provided by Bob and he uses the key K_{AC} and the determinant $\det(M)$. It executes the following steps.

(1) Having received the signature S, Charlie decrypts it using K_{BC} and obtains the blind message M' with K_{AC} and obtains M.

(2) Charlie checks if the determinant of M recovered from the signature is the same as $\det(M)$ obtained from Alice. If the check holds, he verifies the following equations:

$$\det(S^k) = \det(M'_k K^k_{BC}) = \det(M'_k) \times \det(T^n_p)$$
$$= (-1)^n \det(M'_k) = (-1)^{2n} \det(M_k) \tag{21}$$

If the verification holds as well, Charlie accepts S^k. Otherwise, he aborts this communication.

4 Experimental Result

In this section, the evaluation of the proposed system in the cloud computing along with the existing QKD method. The specification of the computer used to evaluate the performance is Intel i5 processor with 4 GB DDR3 RAM of 500 GB storage space and have the network of 5x Gigabit Internet. The tool used to measure the performance of this method in this paper are NetBeans

There are three methods are compared to evaluate the performance of the QDSKM and the establishment of the delay. The two method used to compare the functions are: QKD and QDSKM. The QKD method uses the QKD-QKD-AES256-SHA cipher suite with three different sizes of the global key. The delay of the method is calculated and compared with each other as shown in Table 1. This clearly shows that the QDSKM performs well when it is compared with the QKD method. The delay in this method is less than other methods. In the quantum cryptography method, this QDSKM gives a better result than the QKD. This clearly shows that this method can be applied to the cloud communication, which provides secure communication.

This value is evaluated by using the different keys and delay is calculated. The delay of QKD is more than the QDSKM. The high delay is recorded in the QKD while generating the QKD512 byte as 1115 ms delay. There are two types of delay is calculating handshake delay and overhead delay.

The performance of the QDSKM is calculated in the range of time taken for the methods for the directory access and checking the key amount as shown in Table 2.

Table 1 Comparison of different methods in delay

Method	Condition	Overhead delay	Handshake delay	Total delay
QKD	QKD 160 byte	240.4	13.7	254.1
	QKD 16 byte	443.7	13.6	457.3
	QKD 512 byte	1101.2	13.8	1115
	RSA 1024 bit	7.1	2	9.1
QDSKM	QKD 160 byte	220.2	12.6	232.8
	QKD 16 byte	404.5	12.7	417.2
	QKD 512 byte	946.2	12.4	958.6
	RSA 1024 bit	7	2	9

Table 2 Time taken for accessing the directory and checking key amount delay

Condition	1. Directory access (ms)				2. Checking key amount (ms)			
	QKD 160 byte	QKD 16 byte	QKD 512 byte	RSA 1024 bit	QKD 160 byte	QKD 16 byte	QKD 512 byte	RSA 1024 bit
QKD	2.2	2.3	2.3	2.2	4.2	4.2	4.1	4.1
QDSKM	2.1	2.1	2.2	2	4.1	4	3.8	3.7

Table 3 Time taken for the method selection and key synchronization

Condition	3. Method selection				4. Key synchronization				Total overhead delay			
	QKD 160 byte	QKD 16 byte	QKD 512 byte	RSA 1024 bit	QKD 160 byte	QKD 16 byte	QKD 512 byte	RSA 1024 bit	QKD 160 byte	QKD 16 byte	QKD 512 byte	RSA 1024 bit
QKD	0.7	0.7	0.7	0.7	233.2	436.5	1094	–	240.3	443.7	1101.1	7
QDSKM	0.7	0.7	0.7	0.7	210.4	384.3	942	–	217.3	391.1	948.7	6.4

This only requires less time for the process and both methods are taken the less time to check the key amount. The performance of the two systems is measured for the method selection and key synchronization. The time taken for selecting the method, key synchronization, and total overhead delay are given in Table 3. This is taken for the four different conditions and QDSKM provide low delay compared to the QKD. The QKD 512 byte have more delay in the two process and QDSKM have 948.7 ms delay. The delay of the QKD 160 byte is 240.3 in QKD and 217.3 for the proposed method. The experimental result clearly shows that the QDSKM provides better result compared to the QKD method. This method can be applied in the e-commerce application with more efficiency and with the secure transaction.

5 Conclusion

Quantum cryptography provides better security in the cloud for the shared data and communication. The shared key helps in sharing the information between the user without data leaking. In this method, the technique is used to provide security in the communication from the QKD in the cloud. The QKD is taken from the Simon et al. research, and Fibonacci, Lucas, and Fibonacci–Lucas matrices are used to improve the security. Charlies method, which checks the signature of the Bob and provides the information. This helps to provide security in the cloud and this also improves the performance. This method can be used in the e-commerce application like cloud storage, mobile networks, online transaction, etc. The result shows that it has low delay, while compared to the other methods. The future work of this method is that some attacks have to be applied to this method and check its security performance.

References

1. Molotkov SN (2017) Quantum entanglement and composite keys in quantum cryptography. JETP Lett 105(12):801–805
2. Müller-Quade J (2006) Quantum cryptography beyond key exchange. Informatik-Forschung und Entwicklung 21(1–2):39–54
3. McKague M, Sheridan M (2013) Insider-proof encryption with applications for quantum key distribution. In: International conference on information theoretic security. Springer, Cham, pp 122–141
4. Wu C, Yang L (2016) Bit-oriented quantum public-key encryption based on quantum perfect encryption. Quantum Inf Process 15(8):3285–3300
5. Zhou L, Wang Q, Sun X, Kulicki P, Castiglione A (2018) Quantum technique for access control in cloud computing II: Encryption and key distribution. J Netw Comput Appl 103:178–184
6. Cho K, Miyano T (2015) Chaotic cryptography using augmented Lorenz equations aided by quantum key distribution. IEEE Trans Circuits Syst I Regul Pap 62(2):478–487
7. Nagy M, Nagy N (2013) Quantum secret communication without an encryption key. In: International conference on theory and practice of natural computing. Springer, Berlin, pp 181–192
8. Molotkov SN (2008) Cryptographic robustness of practical quantum cryptography: BB84 key distribution protocol. J Exp Theor Phys 107(1):28–48

9. Kronberg DA, Molotkov SN (2009) Robustness of quantum cryptography: SARG04 key-distribution protocol. Laser Phys 19(4):884–893
10. Molotkov SN (2016) On the complexity of search for keys in quantum cryptography. JETP Lett 103(5):345–349
11. Mohajer R, Eslami Z (2017) Cryptanalysis of a multiparty quantum key agreement protocol based on commutative encryption. Quantum Inf Process 16(8):197
12. Mogos G (2015) Quantum key distribution protocol with four-state systems-software implementation. Proc Comput Sci 54:65–72
13. Qiu L, Sun X, Xu J (2017) Categorical quantum cryptography for access control in cloud computing. In: Soft computing, pp 1–8
14. Tanizawa Y, Takahashi R, Sato H, Dixon AR (2017) An approach to integrate quantum key distribution technology into standard secure communication applications. In: 2017 ninth international conference on ubiquitous and future networks (ICUFN). IEEE, pp 880–886
15. Metwaly AF, Rashad MZ, Omara FA, Megahed AA (2014) Architecture of multicast centralized key management scheme using quantum key distribution and classical symmetric encryption. Eur Phys J Spec Top 223(8):1711–1728
16. Esmaeili M, Moosavi M, Aaron Gulliver T (2017) A new class of Fibonacci sequence based error correcting codes. Cryptogr Commun 9(3):379–396
17. Vajda S (1989) Fibonacci and Lucas numbers, and the golden section: theory and applications. Courier Corporation
18. Mishra M, Mishra P, Adhikary MC, Kumar S (2012) Image encryption using Fibonacci-Lucas transformation. arXiv:1210.5912
19. Simon DS, Lawrence N, Trevino J, Dal Negro L, Sergienko AV (2013) High-capacity quantum Fibonacci coding for key distribution. Phys Rev A 87(3):032312

Enhanced Trust-Based Cluster Head Selection in Wireless Sensor Networks

B. Deena Narayan, P. Vineetha and B. K. S. P. Kumar Raju Alluri

Abstract Advancement in Wireless Sensor Networks (WSN) in recent years led to many new protocols. Wireless sensor networks are formed by the sensor nodes, that are used very often in various fields. As a result, security is the most crucial task to be accomplished. Since the wireless sensor networks are distributed in nature, the sensor nodes are easily prone to security attacks. To address this problem, we avoid the election of a malicious or compromised node as a cluster head. We proposed an approach to elect a trustworthy cluster head. This approach also reduces the overhead in the calculation of trust for cluster head election using overhearing trust mechanism. Simulations results obtained from our approach are effective.

Keywords Wireless Sensor Networks · Cluster head election · Malicious node · Overhearing trust · Direct trust · Indirect trust · Overhear node · Trustworthy cluster head

1 Introduction

Wireless Sensor Networks (WSN) are used in a variety of real-world applications in the fields of health care, military, environmental, and earth sensing. WSN is a collection of distributed sensor nodes [1]. The sensor nodes in WSN senses the information and propagate that information to the base station via a single or multi-hop mechanism. Energy and power are limited resources for the sensor nodes. These sensor nodes are more likely to be failed.

B. Deena Narayan (✉) · P. Vineetha · B. K. S. P. Kumar Raju Alluri
Department of CSE, NIT Andhra Pradesh, Tadepalligudam, Andhra Pradesh, India
e-mail: deenaboddapati@gmail.com

P. Vineetha
e-mail: vineethapuritipati@gmail.com

B. K. S. P. Kumar Raju Alluri
e-mail: pavan0712@gmail.com

© Springer Nature Singapore Pte Ltd. 2019
H. S. Saini et al. (eds.), *Innovations in Computer Science and Engineering*, Lecture Notes in Networks and Systems 74,
https://doi.org/10.1007/978-981-13-7082-3_31

The most critical issue in wireless sensor networks is security. As nodes are distributed in nature, the individual nodes are likely to be attacked by its opponents [2]. If any malicious node becomes cluster head, it welcomes many attacks such as HELLO flood attack, Dos, Sybil attack, and Selective forwarding. So, improving security is an important issue in cluster-based WSN. Malicious nodes may perform some activities like Packet dropping, tampering the information, and disturbing the regular operations of wireless sensor networks.

LEACH is a clustering-based hierarchical routing protocol. It is used in WSN to increment the lifetime of the network, minimize energy dissipation of nodes. This protocol elects cluster head randomly. The main purpose of the cluster head is to collect data from the cluster members and send aggregate of the information to the base station. This leach protocol works in two phases. Setup phase where the cluster head is elected randomly based on probability function, which is not elected as cluster head before. Steady-state phase is where data transmission begins. In this phase, the cluster head aggregates the data sent by the member nodes of the cluster and pass it to the base station [3].

In WSN, trust is defined in different ways. Trust is the amount of confidence of one node for another node to perform any specific task within some time. It can be calculated based on past communications from one node to other, by a direct or indirect study. As this trust value changes frequently, it also takes recommendations from neighboring nodes which can be trusted. Hence, we need a technique which determines trust, based on a mathematical model which is used for taking various decisions according to the trust factor of a given node.

In order to improve security in wireless sensor networks, the trust factor of a node is computed. This trust is calculated both by the node itself and the neighbour nodes, which is combined and used for cluster head election. Acquiring trust value of nodes from all its neighbours is time consuming and more overhead is involved in this. To reduce the overhead in trust calculation, we proposed an effective approach.

1.1 Problem Definition

In this paper, we identified the following challenges in the context of cluster head selection in WSN.

- To enhance the security of wireless sensor networks by ensuring the legitimate node is elected as cluster head.
- To reduce the delay involved in identifying the appropriate cluster head in the considered topology.

1.2 Proposed Approach for Cluster Head Election

Instead of calculating indirect trust from all neighbours of the node, we considered only the overhearing node trust (node through which more packets have been forwarded). This overhear trust combined with the direct trust of each node is used in cluster head election, and this avoids malicious node to become as cluster head.

The organization of the paper in the next sections is as follows. In Sect. 2, we discuss the related work. The proposed approach for cluster head election is discussed in Sect. 3. We validate the proposed approach through extensive simulations and they are described in Sect. 4. Finally, we conclude our contributions in Sect. 5.

2 Related Work

The design of several other protocols is inspired from LEACH (Low-Energy Adaptive Clustering Hierarchy), which has enhanced the Cluster Head (CH) selection by considering the residual energy and improving the lifetime of the network. However, there are more critical issues like security which are to be addressed by the LEACH protocol.

In [4], the authors proposed an algorithm which increases the security of LEACH in WSN. In this, the authors merged two techniques: pre-shared key pairs and low-power CH selection algorithm. This use of pre-shared key pairs increases the security. In [5], the authors used random key pre-distribution for LEACH protocol. In [6], the authors used the technique of sharing pairwise keys between the cluster head and nodes of its cluster. However, cryptographic techniques do not provide sufficient security due to the presence of compromised nodes in the network.

The authors in [7] proposed a technique, where security is ensured by an authentication algorithm. Here, an authentication function is proposed for nodes, which are likely to connect to the cluster. In [8], a random ID is assigned to each node. This ID is verified by sender and receiver to safely transmit the data between them. These are the traditional security measures, which can handle only some type of attacks on the network. By trust management, we can handle some other attacks on the network.

In [9], the authors proposed a trust framework which reduces the probability of electing cluster head as malicious or compromised node. Here, the authors used promiscuous mode which means, if node A wants to send data to node B through node C, node A can hear if node C has forwarded that data or not. Taking this concept, it constructs the trust tables of all neighbouring nodes for a specific node. This creates an overhead as the node needs to maintain all the trust tables of the neighbouring nodes.

The authors in [10] proposed an algorithm which builds trust between nodes which are one hop away. For calculation of direct trust of a specific node, it takes the recommendations from neighbour nodes. Based on the trust calculated, it takes the decision for that node whether it can be trusted or not. These trust values are managed by the local mobile agent of the node.

In [11], the authors proposed a model which represents the node as three variables like node ID, set of attributes, and trust value. This trust value ranges from 0 to 1 and this trust value can be increased, if it has high probability of packet forwarding and encryption level is high. If there is any malicious node, it can be found by observing attributes and then can assign its trust value.

The authors in [12] proposed a better energy-efficient function with distance and count of neighbour nodes. In order to increase the network performance, power amplification is used for elected CH.

3   Proposed Approach for Cluster Head Election in WSN

Our main objective is to build an effective trust-based model, which reduces overhead and also delay involved in the election of cluster head in WSN.

3.1   Architecture of the Proposed Approach

We proposed an architecture for cluster head election and the same is shown in Fig. 1. The details are described below.

Overhearing node The overhearing node of a given node is considered as the nearest node from which most of the packets are forwarded to the destination.

Fig. 1 Proposed architecture for the cluster head election

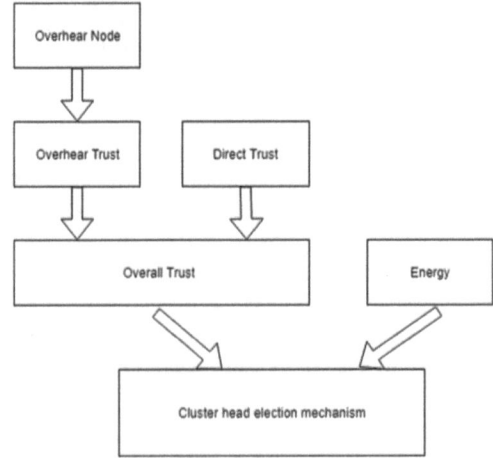

Overhearing trust calculation As we cannot believe the trust value of node given by itself, we ask the overhearing node to calculate its trust. This trust is named as overhearing trust. It is calculated based on old statistics and present overhearing trust value as shown in Eq. (3).

Direct trust The node by itself calculates trust value based on the number of packets forwarded and the number of packets received. The calculation of direct trust is shown in Eq. (2).

Trust evaluation We cannot solely depend on direct trust or indirect trust because there may be a chance of node itself being a malicious node and can tamper its trust value. We should consider overhearing trust along with direct trust for overall trust calculation in Eq. (6).

Trust value In general, energy is the main criteria for cluster head election. There may be a node having more trust value and less energy. If that node is elected as cluster head then the node may die early. In order to avoid this case, we consider trust value along with the residual energy for the cluster head election process.

4 Cluster Head Election Mechanism

In the proposed approach, the cluster head is elected considering the trust value and residual energy. This effective trust value is based on the threshold for trust value and residual energy. If only trust value is considered there may be a chance of having a node having high trust value and low energy, which is an inappropriate choice for cluster head.

The trust value is of two types: Indirect trust and Direct trust.

- Direct trust is confidence that a node has on itself. It is calculated by using the number of packets correctly forwarded by that node and the number of packets received by the same node.
- Indirect trust is confidence that the neighbor nodes in the network have on this node. It is calculated based on the number of packets correctly forwarded from node to destination via neighboring node and the number of packets actually received by the same neighbor node.

These values of direct and indirect trust change frequently. These malicious nodes may enter the network at any possible time. If there is any drastic change in indirect trust or direct trust, there may be a chance of compromised node in the network.

Generally, this indirect trust value is collected from all the neighbors of a node. However, requesting trust value from all its neighbours is time consuming. If trust of a node is requested and one of its neighbour is busy in some other operation, the node has to wait until the operation of neighbour has completed. If this node is to be elected as a cluster head, then it adds a delay in the network. To overcome these problems, we introduce overhearing node trust as indirect trust. Here, the request is sent only to the overhear node instead of all its neighbours. This reduces the consumption of energy and delay in the network. This is applicable only for multi-hop communication of wireless sensor networks. The Enhanced Trust-based Cluster Head election mechanism (ETCHS) is proposed.

$$ETCHS = \frac{S(i).E * ReqTrust(i)}{\sum_{i=1}^{n} S(i).E} \tag{1}$$

where,

S(i).E   is the energy of ith node in the network.
ReqTrust(i)   gives the overall trust of the ith node.

Algorithm 1 Proposed Cluster head election Procedure

1: **function** CLUSTERHEADELECTION(S, C)
2:   **for** $i = 1 : n$ **do**
3:     **if** $S(i).E > 0$ **then**
4:       $p(i) \leftarrow p$
5:       **if** $ETCHS >= (E_{thresh} * Tr_{Thresh})$ **then**
6:         $C(cluster).d \leftarrow d_i$
7:         $C(cluster).id \leftarrow i$
8:         $S(i).type \leftarrow' C'$
9:         $d_i \leftarrow sqrt((S(i).xd - (S(n+1).xd))^2 + (S(i).yd - (S(n+1).yd))^2)$
10:        $cluster \leftarrow cluster + 1$
11:        **if** $d_i > do$ **then**
12:          $S(i).E \leftarrow S(i).E - ((E_{TX} + E_{DA}) * (4000) + Emp * 4000 * (d_i * d_i * d_i * d_i))$
13:        **end if**
14:        **if** $d_i <= do$ **then**
15:          $S(i).E \leftarrow S(i).E - ((E_{TX} + E_{DA}) * (4000) + E_{fs} * 4000 * (d_i * d_i))$
16:        **end if**
17:      **end if**
18:    **end if**
19: **end for**
20: **end function**

Illustrative example of cluster head election We will explain our approach by considering the topology in Fig. 2.

The node (cluster member) has to send packets to the base station. Then, it will send data to the cluster head which aggregate and send to the base station. Here, the mechanism of cluster member sending data to the cluster head is single hop communication, whereas data transfer from cluster head to the base station is a multi-hop communication. If the cluster members are near to cluster head, single hop communication is preferred, else we perform multi-hop mechanism. If we opt for single hop communication then energy usage is more, when compared with multi-hop communication.

Overhear node is considered as the nearest node from cluster head to the base station, which is the next hop from where the maximum number of packets are being transferred. This overhearing node is selected by considering several packet transfers.

Fig. 2 Topology for the considered example

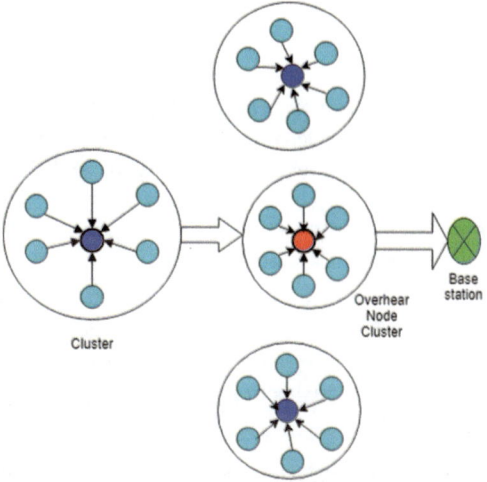

Cluster

Overhear Node Cluster

Base station

Algorithm 1 explains the way to elect a trustworthy cluster head. This algorithm takes overall trust value of overhearing node to calculate ETCHS of a node as shown in Eq. (1). This ETCHS also requires the energy of a node to check the energy threshold. By using this ETCHS, we can verify whether a node meets the threshold requirement of both energy and trust values to become a cluster head. We again classify them into single hop and multi-hop communication based on the distance from the cluster head. If the overhearing node is at a multi-hop distance from cluster head then ETCHS is calculated using trust value from Algorithm 2, where in case of single hop distance we calculate the trust value from all the neighbours. In Algorithm 2 for each node we find the overhear node and calculate the indirect trust from the packets received by overhear node and sent by the node via overhear node. This indirect trust aggregated with direct trust of each node gives overall trust used in calculation of ETCHS of a node in Eq. (1).

$$DT = \frac{P_{df}}{P_{dr}} \tag{2}$$

$$Oh_t = \frac{P_{or}}{P_{os}} \tag{3}$$

$$T_d = W_{old} * DT' + W_{new} * DT \tag{4}$$

$$T_o = W_{old} * Oh_t' + W_{new} * Oh_t \tag{5}$$

$$Overall\ trust = W_i * T_d + W_j * T_o \tag{6}$$

where

P_{df}	Total number of packets forwarded directly.
P_{dr}	Total number of packets received directly.
P_{or}	Number of packets received by overhear node from node.
P_{os}	Number of packets actually sent by index node to base station via overhear node.
DT, DT'	Trust calculated by the node itself based on present and past observations, respectively.
Oh_t, Oh_t'	Trust calculated by the overhear node, based on present and past observations respectively.
W_i, W_j	Weight given for direct and overhear trust respectively.
T_d, T_o	are the overall direct and indirect trust values respectively.
W_{old}, W_{new}	are the weights given to the overall direct and indirect trust values, respectively.

Algorithm 2 Request Trust value from Overhear node

1: **function** TRUST(*index*)
2:   **for** $i = 1 : n$ **do**
3:     **if** $S(i).Overhear = index$ **then**
4:       $Oh_t(index) = P_{or}/P_{os}$
5:       $T_d = W_{old} * DT'(index) + W_{new} * DT(index)$
6:       $T_o = W_{old} * Oh_t'(index) + W_{new} * Oh_t(index)$
7:       $S(index).Trust = W_1 * T_d + W_2 * T_o$
8:     **end if**
9:   **end for**
10:   **return** $S(index).Trust$
11: **end function**

Algorithm 2 is used in the case to find the trust value for a node based on overhear node. This Overhear node is elected on basis of more number of packets transferred from given source to base station via overhear node. These statistics are taken from past transmissions. Here, the above cases discuss only when there may be a chance of either elected node being cluster head or whether overhear node may go into the compromised mode. If both are malicious we consider the trust values from all their neighbour nodes. Overhear node is decided based on their statistics from all past transmissions. If there is any drastic change in behaviour node, we should consider trust from all their neighbours. Using the proposed approach, we can also reduce the overhead involved for cluster head election.

Advantages

– In the proposed approach, we are reducing the overhead of trust calculation in cluster head election. For a node to be elected as cluster head we need, its trust value from all its neighbouring nodes. If some of the neighbouring nodes are busy, there may be a delay in cluster head election mechanism. To reduce this waiting time, the node asks the most contributed node in cluster head election. This approach reduces delay.

– Waiting time for cluster head election decreases.
– The node with more trust and more Residual Energy is elected as the cluster head.

5 Experiments and Results

In this section, we compare the results of the proposed approach with the traditional LEACH protocol. These simulations are implemented with respect to fixed-topology in Matlab. These are the network parameters used in the proposed approach, shown in Table 1.

First, making some of the considerations clear, we considered a topology of 90 nodes for the simulation. These nodes are distributed in a region of 300 m × 300 m randomly. The base station is positioned at the center of the network with coordinates (150, 150). All the sensor nodes have the same initial energy. We consider an assumption, once the topology is created the nodes positions are fixed and it also works for multiple topologies.

Figure 3 shows the network lifetime versus number of rounds for the proposed approach and the LEACH protocol. It shows that in LEACH protocol all the nodes die by 1500 rounds, whereas in TRUSTED LEACH all nodes die after 2000 rounds. This implies the improvement in lifetime of the proposed approach in comparison with LEACH protocol. This is because we have taken residual energy into consideration while electing cluster head (Fig. 4).

Figure 5 shows the number of dead nodes versus the number of rounds. where all the nodes in the network die after 1500 rounds in LEACH and after 2100 rounds in TRUSTED LEACH. We can observe that the nodes die rapidly one after the other in LEACH, but in TRUSTED LEACH the nodes die less frequently preserving the lifetime of the network. So, by the proposed approach we can also preserve the lifetime of network.

Table 1 Simulation environment parameters

Simulation environment parameters	
Parameter	Value
Network size	300 * 300
Location of BS	150 * 150
Number of nodes	90
Cluster head probability	0.1
Initial energy	0.5 J
E_{TX}	50 n J/bit
E_{RX}	50 n J/bit
E_{DA}	50 n J/bit
Data packet size (bits)	4000

Fig. 3 Comparison of
network lifetime between the
proposed approach and
LEACH

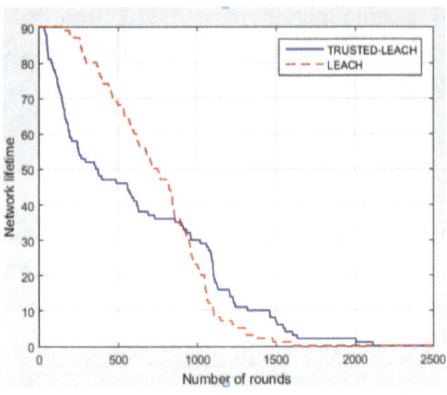

Fig. 4 Comparison of
average energy between the
proposed approach and
LEACH

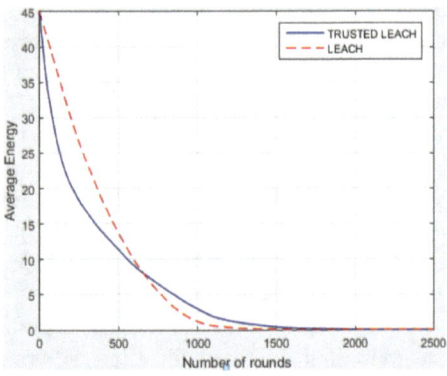

Fig. 5 Comparison of dead
nodes between the proposed
approach and LEACH

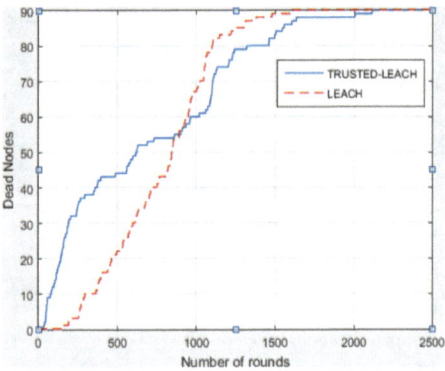

Fig. 6 Comparison of
packets to base station
between the proposed
approach and LEACH

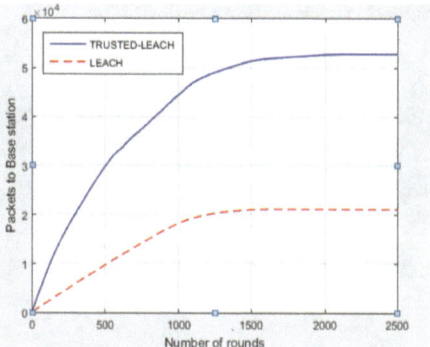

A comparison was made between LEACH protocol and proposed for the results
of packets sent to base station approach. The LEACH protocol does not consider the
malicious nodes. The packets may be dropped before reaching base station, by these
malicious nodes. In the proposed approach, we overcome this problem by electing a
trustworthy cluster head.

We observe that the packets sent to the base station for the proposed approach is
more than the LEACH protocol and the same is shown in Fig. 6.

We compared the delay involved in cluster head election process between
TRUSTED LEACH protocol with neighbours and TRUSTED LEACH with overhearing node approach. It is evident from Fig. 7 that the delay is reduced for the

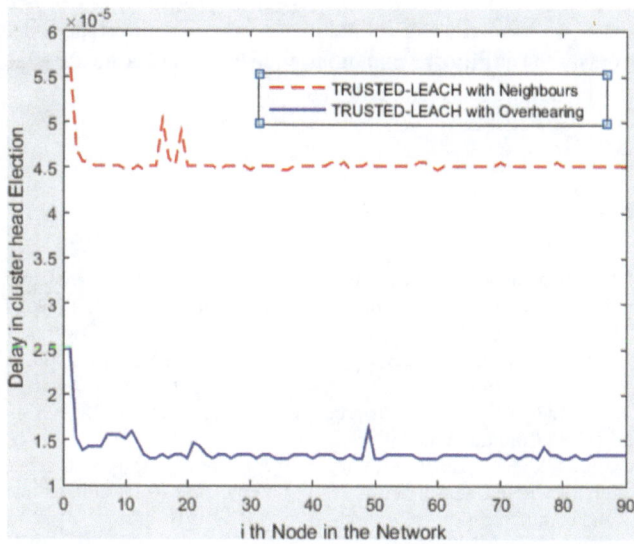

Fig. 7 Comparison of delay between the proposed approach and TRUSTED LEACH with
neighbours

Table 2 Delay in cluster head election

Delay in cluster head election		
Number of nodes	TRUSTED LEACH (s)	Leach with neighbours (s)
500	1.6674×10^{-5}	7.4939×10^{-4}
1000	3.5287×10^{-5}	0.0044
1500	3.8792×10^{-5}	0.0082
2000	3.9757×10^{-5}	0.0152
2500	3.9161×10^{-5}	0.0182
3000	4.3047×10^{-5}	0.0316

proposed approach when the concept of overhearing node is used. We can observe better results in the delay of cluster head election, when network scalability is improved as shown in Table 2.

In comparison with the traditional LEACH protocol, the proposed approach performed better as it improved the network lifetime, decreased the overall delay and increased the throughput.

6 Conclusion

In this paper, we proposed an effective trust calculation of a node and it works for multiple topologies. We also avoid the chance of a malicious node to become the cluster head. An efficient and reliable trust model is developed. The proposed approach decreases the ability of a malicious node to become a cluster head. Overhead involved in trust calculation is also reduced.

References

1. Karthik N, Sarma Dhulipala VR (2011) Trust calculation in wireless sensor networks. In: 2011 3rd International conference on electronics computer technology (ICECT), vol 4. IEEE
2. Safavi-Naini R (2008) Secure key distribution for wireless sensor networks. In: 2nd International conference on signal processing and communication systems, 2008. ICSPCS 2008. IEEE
3. Manohar K, Darvadiya AI (2015) Clustering based routing protocol (LEACH protocol). Int J Innov Res Comput Commun Eng 3(4)
4. Wang J et al (2007) Secure LEACH routing protocol based on low-power cluster-head selection algorithm for wireless sensor networks. In: International symposium on intelligent signal processing and communication systems, ISPACS 2007. IEEE
5. Oliveira Leonardo B et al (2006) SecLEACH—a random key distribution solution for securing clustered sensor networks. In: Fifth IEEE international symposium on network computing and applications, NCA 2006. IEEE

6. Alshowkan M, Elleithy K, AlHassan H (2013) LS-LEACH: a new secure and energy efficient routing protocol for wireless sensor networks. In: 2013 IEEE/ACM 17th international symposium on distributed simulation and real time applications (DS-RT). IEEE
7. Qiang T, Bingwen W, Zhicheng D (2009) MS-Leach: a routing protocol combining multi-hop transmissions and single-hop transmissions. In: Pacific-Asia conference on circuits, communications and systems, 2009. PACCS'09. IEEE
8. Sun Y, Tang M (2014) An enhanced protocol for LEACH based wireless sensor networks. In: 2014 International symposium on computer, consumer and control (IS3C). IEEE
9. Crosby Garth V, Niki P, James G (2006) A framework for trust-based cluster head election in wireless sensor networks. In: Second IEEE workshop on dependability and security in sensor networks and systems, DSSNS 2006. IEEE
10. Ramalingam L, Audithan S (2014) Trust based cluster head selection algorithm for wireless sensor network. In: 2014 2nd international conference on current trends in engineering and technology (ICCTET). IEEE
11. Kodali RK, Soratkal SR (2015) Trust model for WSN. In: 2015 International conference on applied and theoretical computing and communication technology (iCATccT). IEEE
12. Krishnakumar A, Anuratha V (2017) An energy-efficient cluster head selection of LEACH protocol for wireless sensor networks. In: 2017 International conference on Nextgen electronic technologies: silicon to software (ICNETS2). IEEE
13. Zahariadis T et al (2010) Implementing a trust-aware routing protocol in wireless sensor nodes. In: Developments in E-systems engineering (DESE), 2010. IEEE

Homomorphic Encryption Scheme to Enhance the Security of Data

C. K. Deepa and S. Ramani

Abstract Data generated in digital devices need to be secured for its use by the authorized users. Different security algorithms such as symmetric and asymmetric cryptosystems have been developed to secure the data while in transmission as well as at rest. However, data is not secure while processing. This paper explores the concept of Homomorphic Encryption (HE) based solution for secured functional evaluation of data. The proposed FHE-AES/DH cloud security model executes encryption on the data and encrypted data is transmitted to the cloud server. The functional evaluation is carried out on uploaded data by the third party without knowing the contents. These models hide the content of data in order to protect security and privacy. A banking data set has been used for the case study. The FHE-AES model takes less time for functional evaluation compared to FHE-DH model.

Keywords Data security · Cryptosystems · Homomorphic encryption · AES · Diffie-Hellman model

1 Introduction

Nowadays, the business model that is gaining popularity is an aggressive collection and perpetual storage of data. The extensive use of ubiquitous devices like security cameras, wireless sensors and data collected from such devices is massive and ever increasing. Social data is continuously collected by business organizations. Another motivation toward the collection of data is to generate business analytics, which helps to improve the business. Hence, there is an exponential rise in data collection and storage of that data to provide value to the business processes.

C. K. Deepa (✉) · S. Ramani
Department of Computer Science and Engineering, Ramaiah Institute of Technology,
Bengaluru 560054, India
e-mail: deepakiran23@gmail.com

© Springer Nature Singapore Pte Ltd. 2019
H. S. Saini et al. (eds.), *Innovations in Computer Science
and Engineering*, Lecture Notes in Networks and Systems 74,
https://doi.org/10.1007/978-981-13-7082-3_32

277

The cloud computing technology is based on pay-per-use model hence, it becomes a more popular technology in the community. The data storage on the cloud comes with several advantages. On the other hand, cloud computing has disadvantages. Privacy and security of data stored remotely is a major concern. The organizations are outsourcing the business processes [1]. Consider the operations in banks/financial institutions, there are many products and portfolios offered to customers/clients. These institutions are outsourcing some of its bulk/one-time/periodical operational processes like calculating the loan interest, installment payment at the end of every month to the third party. The process is considered to be successful when the third party is able to calculate the interest amount on the loan taken by performing the required operation on the customer data. If the data is provided to the third party in plain text, then the data is exposed to privacy and security violations due to threats and vulnerabilities. "Homomorphic Encryption" (HE) is one of the solutions to handle the security and privacy related vulnerabilities by allowing the user to compute functions on encrypted data directly.

Homomorphic Encryption executes complex numerical operations on encrypted information without decrypting the ciphertext. In other words, let P1 and P2 are the plaintexts and C1 and C2 are ciphertexts of P1 and P2, correspondingly, HE estimates P1 $\ominus$ P2 from C1 and C2 without decrypting P1 or P2. On the basis of operation, the function "$\ominus$" can be multiplicative if multiplication is used or additive if operation used.

Multiple studies have been carried out since 2009 after Gentry et al. [2] showed that it is possible to compute functions on encrypted data directly. This helps the user not to expose the data while processing. In this paper, a Secure Functional Evaluation (SFE) model has been developed using the encryption algorithm that satisfies the HE property. Section 2 discusses the objectives of this work. The related work is discussed in Sect. 3. Section 4 describes the proposed cryptosystems for banking transaction applications. The implementation details are discussed in Sect. 5. Results obtained are discussed in Sect. 6. The conclusions are summarized in Sect. 7.

2 Objective

The objective is to propose a model, which accepts encrypted inputs and then perform blind processing to satisfy the user query without being aware of its content, whereby the retrieved encrypted data can only be decrypted by the user who initiates the request. This allows owners to rely on the services offered by remote applications/third party service providers without the knowledge of the contents of data/information thereby protecting privacy.

3 Related Works

Togan and Pleşca [3] derived a new secure searchable encryption model aided by multi-key fully homomorphic encryption (MFHE) to be used for the SMCE application environment. Kumar and Sharma [4] has performed a survey on several existing encryption algorithm and explored the HE property of the algorithm.

Bharati and Tamane [5] discussed to provide computing on a cloud in a secure way using the HE scheme using the Diffie-Hellman algorithm to have secure communication between the third party and organization. A similar effort was done in [6] where FHE was applied for secure data storage and access in cloud storage. Chen [7] dealt with mining on encrypted data using FHE cryptography. Similarly, it has been found that FHE can enable almost $20\times$ speedup in a sample application such as medical data communication and remote monitoring [8]. Even FHE was found noise free to provide optimal computation [9] on applications.

Based on data significances, Rangasami and Vagdevi [10] applied FHE for cloud computing environment. However, before FHE numerous efforts were made by developing Symmetric key and Asymmetric key encryption algorithms [11]; however, Yassein et al. [11] recommended using more efficient cryptosystem to adopt dynamism in the cloud computing system. Aggarwal et al. [12] developed a new SHE model without employing bootstrapping. Their method applied symmetric keys and found it better than the classical encryption systems.

Ishimaki et al. [13] FHE assisted cryptosystem to assist a client to search on a genome sequence database without revealing his/her query to the server. Ishimaki et al. [14] proposed a private substring search protocol over encrypted data by adopting FHE assessing its feasibility. Cheon and Kim [15] designed a semi-autonomic cloud auditing architecture weaved in privacy optimization measure by applying the public key of the SHE Gong et al. [16] presented a high security encrypted data by using SHE and FHE methods to provide security over encrypted data computations without decrypt the ciphertext and advances the performances of the cloud services.

Chen et al. [17] introduced a Hybrid-HE that combines public-key encryption (PKE) and SHE to provide the improvement over polynomial of arbitrary extent without bootstrapping. Lapotre and Lagadec [18] concentrated on three factors such as efficient modular design the security requirements for ring-LWE encryption and SHE based versatile pipelined architecture to achieve a high-speed polynomial multiplier and polynomial multiplications for different lengths n and moduli p. and contributed to this is threefold. The proposed architecture supports polynomial multiplications for different lengths n and moduli p. Li et al. [19] introduced the FHE model to achieve a noise resilient cryptosystem for real-time applications. Weili et al. [20] proposed the identity-based leveled FHE (IBFHE) scheme over ideal lattice to assure the security of selective-identity aligned to chosen-plaintext attacks in the standard model. Kalpana et al. [21] proposed a modified RSA algorithm to uphold the FHE over cloud computing system. This novel telemedicine system conveys medical data of patient's personal encrypted data using AES to assure medical data security.

Fig. 1 Proposed cloud
application (banking)
security system

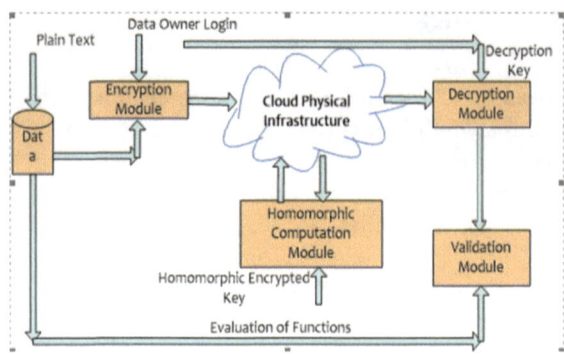

4 System Design

The proposed model uses FHE-AES/DH cloud security model where the data owner at first executes encryption on the data to be transmitted, which is then followed by the upload of the data to the cloud server. The uploaded data can be accessible by the third party provided with the suitable access credentials. Here, the third party recovers the data from cloud and performs necessary action on the encrypted data with the approval of the data owner. The results are then stored back on the cloud for the owner to process further. The owner compares the result stored by the third party with a flag set for decrypting and the result obtained on performing the same functional evaluation on the plain data. If the data matches, then it is termed as success in using homomorphic encryption algorithm. If the data does not match, then it is termed as failure. A snippet of the overall mechanism is depicted in Fig. 1.

5 System Implementation

In this paper, the proposed model exhibits encryption on the banking uploaded data, where the proposed FHE-AES and FHE-DH are applied to perform encryption distinctly. The encryption is performed with FHE-AES and FHE-DH algorithms to fetch the encrypted data to the cloud (Fig. 2).

As depicted in Fig. 1, in the proposed cloud application (banking) security system there are two key actors, the person defined as a bank employee ("user1") and the third party ("user2"). Here, bank employee hires out the evaluation of bank account balance on fund transfer as well as interest on a mortgage if used to a third party.

The details of the clients registered with the respective bank are held by the user1. Account balance, whether loan is taken and such other private information of each client are included in detail. In the proposed model, the information is kept in a database built on SqlYog. The dataset contains the data related to direct marketing campaigns of a Portuguese banking institution. The attributes account ID, balance,

Fig. 2 Homomorphic
computation module

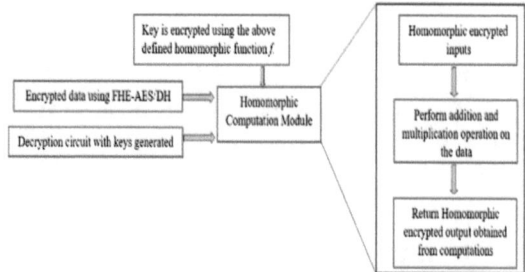

loan attributes are only used for our study. The proposed model has been implemented on the Java platform using Eclipse Oxygen 2.0 IDE.

5.1 FHE-AES Implementation

Java has a built-in guidance for AES in the "javax.crypto.*" packages, that come with any latest adaptations of JDK or JRE. The "AESHelper.java" class in the project wraps the "javax.crypto.Cipher" type to offer a simple to use user interface for encryption as well as decryption operations. The "AESKeystore.java" class in the project is actually the place where encryption key is produced and stored. The "AESKeystore" class is used, to create and produce the encryption element as well as the initialization vector (IV). The "create()" will make use of them, or else it will produce them making use of the "java.security.SecureRandom" class, any "random byte arrays" as the key and the IV. The scale of the IV matches which AES encryption size of the block that is usually 128 bits/16 bytes. For encryption as well as decryption operations, the "AESHelper" class is wrapped in the "javax.crypto.Cipher" class to offer a simple to use user interface. To encrypt as well as decrypt byte arrays, method "encryptBytes" and "decryptBytes" is utilized. To encrypt as well as decrypt strings, methods "encryptString" and "decryptString" is utilized; The algorithm is hard-coded to "AES/CBC/PKCS5PADDING" that is AES algorithm with "CBC" (Cipher Block Chaining) blocking method.

5.2 FHE-Diffie-Hellman Implementation

Java support for "Diffie–Hellman Algorithm" Implementation "KeyPairGenerator" class. To produce pairs of private and public keys, the class "KeyPairGenerator" is used. The factory methods "getInstance" is used to build the key pair generator. "Factory methods" refer to methods that are static and an instance of the class is returned by the method. The initialization of the object is the sole distinction between the methods. Every provider should supply the initialization as default if the client does not explicitly initialize the "KeyPairGenerator" to call the initialization method. This class is actually abstract and extends the interface "KeyPairGeneratorSpi". "Diffie Hellman" is algorithm name for generating key pairs for Diffie–Hellman algorithm. Homomorphic computation module involves several steps. A detailed description is provided in [22].

6 Results and Discussions

The proposed FHE-AES and FHE-DH cryptomodels have been applied for a banking application for assessment under multiparty computation environment. The SFE model developed has been successfully tested with bank dataset. The results obtained in both the modes namely "Data in Plaintext Form" and "Data in Encrypted Form" have been given in the following Table 1. The results demonstrate the time taken for functional evaluation in each mode for one account holder data taken from the dataset. This indicates that time taken in encrypted form is more than in plaintext form which is expected. However, the difference is very marginal. This can be accommodated within the processing budget of the functional evaluation. Hence, it is feasible for a third party to carry out the functional evaluation in an encrypted form. The privacy and security of the data are protected while at rest as well as during processing. Since in the banking sector time efficiency is of utmost significance, we have examined the performance of the FHE-AES and FHE-DH models in terms of computational time. The total time consumed by FHE-AES and FHE-DH has been obtained. To perform encryption of the data in FHE-AES takes 436 ms while for the same activity FHE-DH takes almost 483 ms. It signifies that the first (i.e., FHE-AES) exhibits almost 9.73% faster computational than the FHE-DH encryption algorithm. Similarly, to perform an activity of account balance transfer FHE-AES cryptosystem took 365 ms while for the same activity FHE-DH took 403 ms. It shows that the FHE-AES model exhibits 9.43% faster processing than the counterpart FHE-DH. A snippet of the time efficacy by these cryptosystems is given in Table 1.

7 Conclusion

In this paper, Fully Homomorphic Encryption (FHE) algorithm has been selected for secured functional evaluation of data. Two methods viz., classical symmetric cryptosystem (i.e., AES) and public key exchange cryptosystem (i.e., DH) have been chosen for implementation. The implemented models have been validated with banking application. The results reveal that the third party evaluation of functions like account balance and interest calculation carried out on encrypted data is matching with the expected results performed on plain data. Hence, the privacy and security of data are protected using this method. Also, the results obtained revealed that the proposed FHE-AES cryptosystem exhibits 9.73% better performance than the

Table 1 Results

Function evaluation	Evaluated data	Encryption algorithm	Execution time (ms)	Plaintext balance (before computation)	Encrypted balance (before computation)	Encrypted balance (after computation)	Updated balance
Account balance transfer	Plaintext	–	365	10,000	–	–	7432
	Encrypted	FHE-AES	436	–	7ACB84E6297 35E6F100B93BC 78D4644	B021BC4E484CE 4019682636F8A 5CC5E	–
Deduction of interest rate from the balance	Plaintext	–	113	7432	–	–	7997
	Encrypted	FHE-AES	146	–	B021BC4E484CE 4019682636F8A 5CC5E	FFF14C304F7789 5F7BB9A8EE726 36D59	–
Account balance transfer	Plaintext	–	403	10,000	–	–	9345
	Encrypted	FHE-DH	483	–	434B2B7845618C C2	3D42A5DBC1729 4D8	–
Deduction of interest rate from the balance	Plaintext	–	132	9345	–	–	8690
	Encrypted	FHE-DH	166	–	3D42A5DBC1729 4D8	D336D97BACC52 132	–

FHE-DH model. It signifies that the proposed FHE-AES model can be applied for real-time cloud security or secure data communication purposes. The computation time taken in encrypted data is marginally more than the time taken on plain data. Hence, this approach is feasible for implementation in real life applications. In future, the proposed model can be compared with other cryptosystems and in addition, increasing payload or block size the efficacy could be examined.

References

1. Sha P, Zhu Z (2016) The modification of RSA algorithm to adapt fully homomorphic encryption algorithm in cloud computing. In: 2016 4th international conference on cloud computing and intelligence systems, Beijing, pp 388–392
2. Gentry C, Groth J, Ishai Y, Peikert C, Sahai A, Smith A (2014) Using fully homomorphic hybrid encryption to minimize non-iterative zero-knowledge proofs. J Cryptol 28(4):820–843
3. Togan M, Pleşca C (2014) Comparison-based computations over fully homomorphic encrypted data. In: 2014 10th international conference on communications, Bucharest, pp 1–6
4. Kumar N, Sharma S (2013) Study of intrusion detection system for DDoS attacks in cloud computing. In: 2013 tenth international conference on wireless and optical communications networks (WOCN), Bhopal, pp 1–5
5. Bharati M, Tamane S (2017) Intrusion detection systems (IDS) & future challenges in cloud based environment. In: 2017 1st international conference on intelligent systems and information management (ICISIM), Aurangabad, pp 240–250
6. Liu J, Han JL, Wang ZL (2016) Searchable encryption scheme on the cloud via fully homomorphic encryption. In: 2016 sixth international conference on instrumentation & measurement, computer, communication and control, Harbin, pp 108–111
7. Chen J (2016) Cloud storage third-party data security scheme based on fully homomorphic encryption. In: 2016 international conference on network and information systems for computers (ICNISC), Wuhan, pp 155–159
8. Togan M (2016) A FHE-based evaluation for searching on encrypted data. In: 2016 international conference on communications (COMM), Bucharest, pp 291–296
9. Chen B, Zhao N (2014) Fully homomorphic encryption application in cloud computing. In: 2014 11th international computer conference on wavelet active media technology and information processing, Chengdu, pp 471–474
10. Rangasami K, Vagdevi S (2017) Comparative study of homomorphic encryption methods for secured data operations in cloud computing. In: 2017 international conference on electrical, electronics, communication, computer, and optimization techniques (ICEECCOT), Mysuru, pp 1–6
11. Yassein MB, Aljawarneh S, Qawasmeh E, Mardini W, Khamayseh Y (2017) Comprehensive study of symmetric key and asymmetric key encryption algorithms. In: 2017 international conference on engineering and technology (ICET), Antalya, pp 1–7
12. Aggarwal N, Gupta C, Sharma I (2014) Fully homomorphic symmetric scheme without bootstrapping. In: Proceedings of 2014 international conference on cloud computing and internet of things, Changchun, pp 14–17
13. Ishimaki Y, Imabayashi H, Shimizu K, Yamana H (2016) Privacy-preserving string search for genome sequences with FHE bootstrapping optimization. In: 2016 IEEE international conference on big data (big data), Washington, DC, pp 3989–3991
14. Ishimaki Y, Imabayashi H, Yamana H (2017) Private substring search on homomorphically encrypted data. In: 2017 IEEE international conference on smart computing (SMARTCOMP), Hong Kong, pp 1–6

15. Cheon JH, Kim J (2015) A hybrid scheme of public-key encryption and somewhat homomorphic encryption. IEEE Trans Inf Forensics Secur 10(5):1052–1063
16. Gong Z, Xiao Y, Long Y, Yang Y (2017) Research on database ciphertext retrieval based on homomorphic encryption. In: 2017 7th IEEE international conference on electronics information and emergency communication (ICEIEC), Macau, pp 149–152
17. Chen DD et al (2015) High-speed polynomial multiplication architecture for ring-LWE and SHE cryptosystems. IEEE Trans Circuits Syst I Regul Pap 62(1):157–166
18. Lapotre FV, Lagadec L (2018) Fast evaluation of homomorphic encryption schemes based on ring-LWE. In: 2018 9th IFIP international conference on new technologies, mobility and security (NTMS), Paris, pp 1–5
19. Li J, Teng M, Zhang Y, Yu Q (2016) A leakage-resilient cca-secure identity-based encryption scheme. Comput J 59(7):1066–1075
20. Weili W, Bin H, Xiufeng Z (2017) Identity-based leveled fully homomorphic encryption over ideal lattices. In: 2017 IEEE 2nd international conference on big data analysis (ICBDA), Beijing, pp 377–381
21. Kalpana G, Kumar PV, Krishnaiah RV (2016) Secured cloud computing using user classification and bilinear Diffie-Hellman schema. In: 2016 IEEE 6th international conference on advanced computing (IACC), Bhimavaram, pp 563–568
22. Deepa CK, Ramani S (2018) Homomorphic encryption scheme to enhance the security of the data. M. Tech. thesis, Department of computer science and engineering, Ramaiah Institute of technology, Bangalore, June 2018

An Effective Data Transmission Approach Through Cascading of Clusters in Path Identification for WSN Routing

P. Sachidhanandam and M. Sakthivel

Abstract Wireless Sensor Networks (WSNs) are widely used in sensing applications. The data sensed by the sensor should reach the processing center within a short duration to initiate necessary actions in real-time systems. The efficiency of the WSNs relies on the lifetime of the sensor nodes and it directly depends on the residual energy. Since the Wireless Sensor nodes are operated with battery power, the WSNs are not offered to lose energy within a short interval of time. This paper proposes a novel data transmission method in which the clusters are formed with the nodes lying on the same diagonal that are cascaded in a ring routing path. The energy consumed by the transmitter unit of the sensor nodes is reduced with the multi-hop transmission. K-Nearest Node (KNN) algorithm is used to form the clusters along the diagonal of the ring. This multi-hop transmission ensures the even distribution of energy consumption and increases the lifetime of the entire sensor network.

Keywords Cluster formation · Energy efficient · K-Nearest node · Distribution energy

1 Introduction

Wireless Sensor Networks (WSNs) are a sensitive technology for the recent decade. There is tremendous growth in technology with autonomous schemes to develop and evaluate the performance measuring on data analysis about the sensing node for a transmission from one place to another. Wireless Sensor Network creates a new

P. Sachidhanandam (✉)
Department of Computer Science and Engineering, Knowledge Institute of Technology, Salem 637504, Tamil Nadu, India
e-mail: pscse@kiot.ac.in

M. Sakthivel
Department of Computer Science and Engineering, United Institute of Technology, Coimbatore 641020, Tamil Nadu, India
e-mail: sakthiplacement@gmail.com

© Springer Nature Singapore Pte Ltd. 2019
H. S. Saini et al. (eds.), *Innovations in Computer Science and Engineering*, Lecture Notes in Networks and Systems 74,
https://doi.org/10.1007/978-981-13-7082-3_33

infrastructure based on the design of nodes, installed on the network connectivity in real time, where the node is being placed in the network. Major focus areas of the network are in the deployment of the sensing node, monitor and measure the environmental circumstances either physically or logically created. Lifetime of the network depends on the sensing node routing technique adopted by the network architecture and balances the energy consumed by individual nodes, thus ensuring the uniform consumption of energy by different nodes in the cluster.

The construction of wireless sensor network needs the following knowledge such as analysis of various parameters such as sensing the environmental information, location placed, power consumed, operating time, execution of signal processing techniques, connectivity systems are required. Ring routing techniques are the best methodology for energy-efficient routing protocol for many applications to avoid the delay and quick response in the transmission process of the network. Issues of various researches point out to the improvement of the lifetime of the sensor and make efficient data transmission which depends on the environment. When considering the power factor, the battery lifetime of the sensor is prolonged when distributed equally.

1.1 Ring Topology

The nodes are connected in a ring structure. Every node is connected to two other nodes either to the left or right side, and intermediate nodes between the sources and designation. The reliability of ring topology is better than bus topology. When the connection in the network breaks at the time of data transmission, it may find the alternate path automatically and it makes the data transmission reliable in the system [1]. The object of intermediate nodes can put forward from one neighbor to the other and make a cluster formulation; ultimately it reaches the destination of the target position of the network transmission is as shown in Fig. 1.

Fig. 1 Cluster formation of nodes

To acquire the sink position from the ring with minimal overhead whenever needed. Zhang et al. [2] have given the analysis and experimented about the IPv6 distance vector of routing for a network to minimize the energy consumption. The nodes are divided into different level for optimal direction angle to form interring domain communication to save energy during transmission of data packets as well as clustering algorithm were proposed to create the heterogeneous cluster to find out the residual energy of the balanced node with related position of cluster and avoid energy consumption in a single node for a longer duration. Effusively, based on the energy-efficient algorithm to construct with probability threshold to identify the layer circle of the ring, direction angle, and node residual energy and hop difference. Energy-Efficient Heterogeneous Ring Clustering (E2HRC) Routing Protocol for Wireless Sensor Networks routing protocol to improve lifetime of low power transmission network.

2 Finding Optimal Routing Path Through Cascading of the Ring

Ring routing is an energy-efficient, reliable routing protocol that provides fast data delivery. Ring domain communication topology can effectively use 360° signed transudation while also reducing message collision during the transmission process. Signal intensity decrease during the wireless signal transmission process as transmission distance increases. Where the Distance = speed × Time, Time = distance/Speed and Speed = distance/Time (Fig. 2).

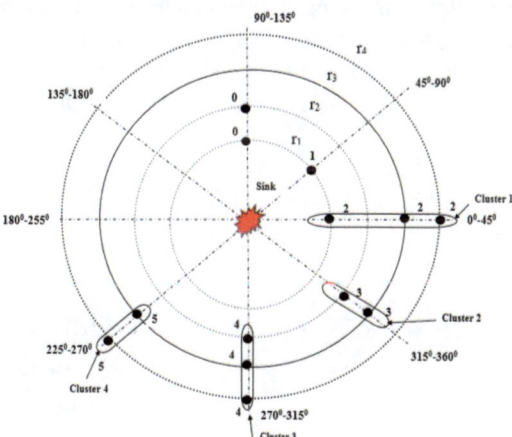

Fig. 2 Clusters in the ring routing path

2.1 KNN Algorithm—Cluster Formation

The cluster formation is done with K-Nearest Node (kNN) algorithm. The KNN uses the information from the K number of nearest neighbors to form the cluster. The cluster formation is based on the distance of each node from the sink. The energy consumed for transmission of data from the sensor node is directly proportional to the distance of the sensor node from the sink.

KNN algorithm:

> For every node
>> Transmit sample bit
>> Wait for response
>> If response received
>>> Increment the counter
> Record the k number of nearest nodes for each node

Table 1 shows the distance between the sensor node and the sink and the energy consumed by the transmitter unit in a single-hop transmission.

Hop by hop transmission, energy revised for 1-bit transmission is n/u joules.

$$\text{Total energy} = n/4 + n/4 + n/4 + n/4$$
$$= 4/4\ n$$
$$\text{Total energy consumed} = n \text{ Joules.} \tag{1}$$

The graph in Fig. 3 shows the variation of the energy consumption by the sensor node in accordance with the distance. As the distance increases the node nearer to the sink has to increase its transmission range to receive the data from the node in the outer ring. Thus, the increase in distance shoots up the energy consumption by the transmitter unit of the sensor nodes in the innermost rings.

Table 1 Energy consumption in single-hop transmission

S. no	Distance between the sensor node and sink	Coverage range of sensor node	Energy consumed by transmitter unit in joules
1	1	1.5	200.125
2	2	2.75	151.0
3	3	4.0	100.50
4	4	5.25	50.25

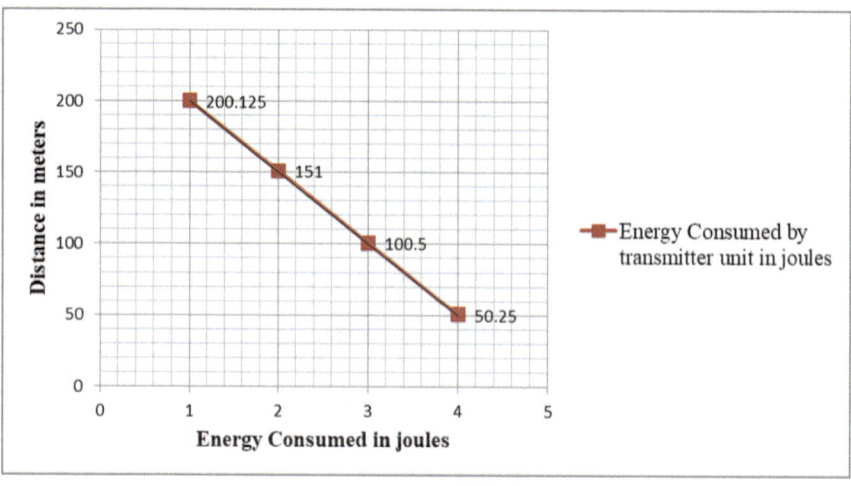

Fig. 3 Energy consumption in single-hop transmission

2.2 *Energy Consumption in Multi-hop Transmission*

The data from the transmitter may reach the sink in multi-hop fashion, which reduces the total energy consumed by the individual sensor nodes. Each node can transmit the data to the next neighbor in one direction towards the sink [3]. Table 2 shows the energy consumed by the transmitter unit and the distance between the neighbors.

The graph in Fig. 4 shows the energy consumed by the sensor nodes during the transmission of data by the sensor unit. The energy consumption is higher around 200 J at a shorter distance. As the distance increases there is a steep decrease and at around 50 J the energy consumed remains at a constant rate [4]. Even the distance increase the energy consumed is constant.

Table 3 shows the comparison of energy consumed by the transmitter unit of the sensor nodes in single-hop and multi-hop transmission [5]. The single-hop transmission involves the energy consumed in transmitting the data from the sensor node directly to the sink and the multi-hop transmission involves the energy consumed in transmission of data to the nearest neighbor in the direction of the sink.

Table 2 Energy consumption in multi-hop transmission

S. no	Distance between the sensor node and sink	Coverage range of sensor node	Energy consumed by transmitter unit in joules
1	1	1.5	50.25
2	2	1.5	50.25
3	3	1.5	50.25
4	4	1.5	50.25

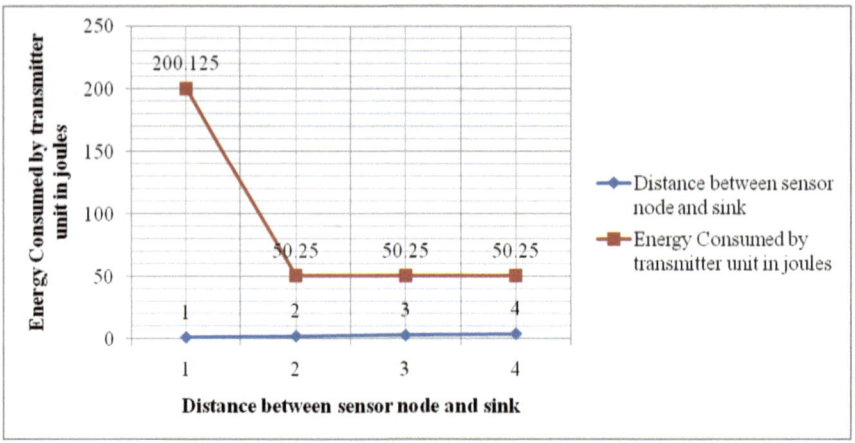

Fig. 4 Energy consumption in multi-hop transmission

Table 3 Comparison of energy consumption in single-hop and multi-hop nodes

S. no	Distance between the sensor node and sink	Coverage range of sensor node		Energy consumed by transmitter unit in joules	
		Single hop	Multi-hop	Single hop	Multi-hop
1	1	1.5	1.5	200.125	50.25
2	2	2.75	1.5	151.0	50.25
3	3	4.0	1.5	100.50	50.25
4	4	5.25	1.5	50.25	50.25

The graph in Fig. 5 shows the comparison of energy consumption by the sensor nodes in single-hop and multi-hop transmission. The comparison gives a clear view of the fall in energy consumption in the multi-hop node with an increase in distance.

3 Conclusion

The energy consumption by the transmitter unit of the sensor node decides the lifetime of the sensor network. Each sensor node when transmitting the data directly to the sink consumes more energy than the sensor nodes transmitting the data indirectly using multi-hop transmission. The multi-hop transmission ensures that every individual node spends only a minimum of energy to transmit the data to the sink. When the clusters are formed with the cascading of the ring, each node transmits its own data and also forwards the data from one neighbor to the other neighbor and

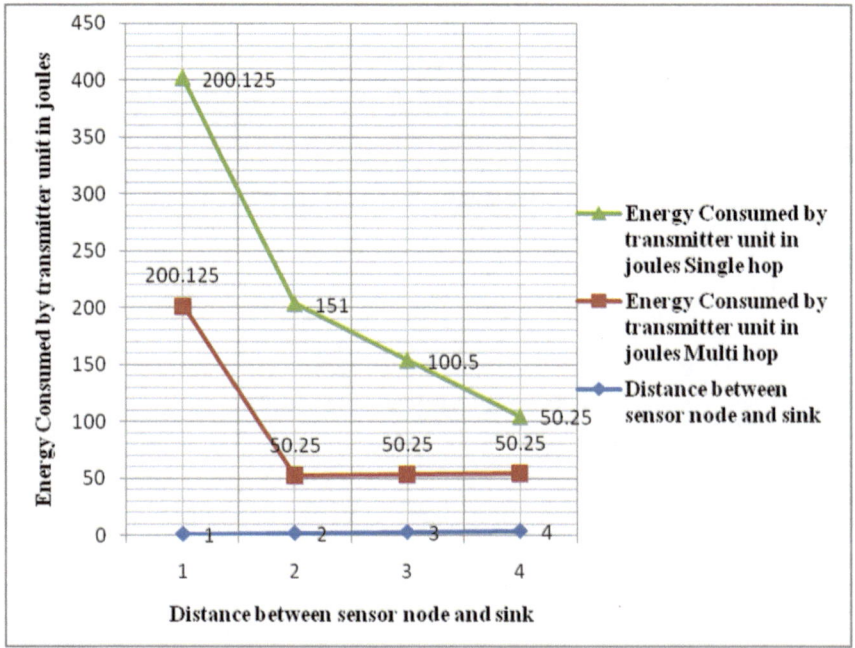

Fig. 5 Comparison of energy consumption in single-hop and multi-hop node

eventually reaching the sink. Thus, the total energy consumed by the sensor network is distributed evenly among the sensor nodes in the network and increases the lifetime of the Wireless Sensor Network.

References

1. Tunca C, Isık S, Donmez MY, Ersoy C (2014) Ring routing: an energy-efficient routing protocol for wireless sensor networks with a mobile sink. IEEE Trans Mob Comput 1–15
2. Zhang W, Li L, Han G, Zhang L (2017) E2HRC: an energy-efficient heterogeneous ring clustering routing protocol for wireless sensor networks. In: IEEE access on special section on future networks: architectures, protocols, and applications, Mar 2017
3. Yu CM, Hsu TW (2017) Determining the optimal configuration of the multi-ring tree for bluetooth multi-hop networks. Energ J 10:1–16
4. Kaur G, Singh J (2017) Power efficient ring and tree based routing protocol. Int Res J Eng Technol (IRJET) 4(4):1914–1922
5. Benaddy M, El Habil B, El Meslouhi O, Salah-ddine (2017) A mutlipath routing algorithm for wireless sensor networks under distance and energy consumption constraints for reliable data transmission. Int J Sensors Sensor Netw 5(1):32–35

A Survey of Various Cryptographic Techniques: From Traditional Cryptography to Fully Homomorphic Encryption

Rashmi R. Salavi, Mallikarjun M. Math and U. P. Kulkarni

Abstract With the advent in Internet and networking applications, security is a major concern in the current era of Information Technology. The huge amount of information exchanged over the Internet is vulnerable to security threats and attacks. Cryptography provides secure exchange of encrypted data by sharing a key. The major concern with this approach is data privacy as anybody with the key can access the data. Moreover, user loses control over data once it is uploaded to the cloud and must rely on cloud service provider. User must share a key with cloud service provider to perform any operations like searching, sorting, etc., or need to download and decrypt the data and then perform the operation. These approaches lead to privacy issue and repeated encryption decryption, even for small computation. These concerns are addressed by Homomorphic Encryption (HE), which enables the cloud service provider to carry out computations on ciphered data without decrypting it. With the advent of HE scheme, in 2009, the Fully Homomorphic Encryption (FHE) scheme was invented by C. Gentry, which allows any computational function to operate on encrypted data. But the practical implementation of FHE is still in research. This survey focuses on various traditional and modern cryptography techniques along with the different schemes of HE and FHE.

Keywords HE · SWHE · FHE

R. R. Salavi (✉) · M. M. Math
Gogte Institute of Technology, Belagavi, Karnataka, India
e-mail: rashmisalvi@gmail.com

M. M. Math
e-mail: mmmath@git.edu

U. P. Kulkarni
SDM College of Engineering and Technology, Dharwad, Karnataka, India
e-mail: upkulkarni@yahoo.com

© Springer Nature Singapore Pte Ltd. 2019
H. S. Saini et al. (eds.), *Innovations in Computer Science and Engineering*, Lecture Notes in Networks and Systems 74,
https://doi.org/10.1007/978-981-13-7082-3_34

1   Introduction

Cryptography is an area of studying various encryption techniques to provide information security. Encryption is a process, which converts the real message, called as
plaintext into the coded message, called as ciphertext. The reverse process, known
as decryption restores plaintext from the ciphertext. The plaintext along with a key is
given as input to encryption algorithm produces the ciphertext. The intended receiver
with key transforms ciphertext back to plaintext. The main problem here is to maintain
the secrecy of the key. The traditional cryptography is based on private key encryption techniques and provides only confidentiality of the information. It operates on
characters (letters and digits). Modern cryptography is a foundation of computer and
communication security. It operates on bits instead of characters and is based on
various mathematical concepts like number theory and probability. Basically, cryptography is defined as two different mechanisms: Symmetric- and Asymmetric-key
cryptography.

2   Symmetric-Key Cryptography [1–4]

When two parties want to converse with each other, they share secret information
known as the key. The sender encrypts the message before it is sent, and the receiver
decrypts the message using the same secret key. The symmetry is achieved as a common secret key is shared by both the communicating parties. The original message is
known as plaintext and the coded message sent by the sender is known as ciphertext.
Figure 1 shows the working of private key encryption.

In this scheme, the same small size keys are used for both encryption and decryption of data. The communicating parties must share a common key before the
exchange of information. The communication between any two parties among n
persons requires n(n − 1)/2 key exchanges.

The different methods used in Symmetric-Key Encryption are: Digital Encryption Standard (DES), Triple-DES (3DES), Advanced Encryption Standard (AES),
International Data Encryption Algorithm (IDEA), and Blowfish. These symmetric
key schemes are compared in Table 1.

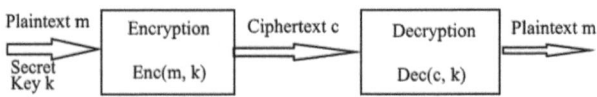

Fig. 1 Private key encryption

Table 1 Comparison of different symmetric-key cryptography algorithms

	DES [2, 3]	Triple-DES [2, 3]	AES [2, 3]	IDEA [2, 3]	Blowfish [2, 3]
Key size	56 bit	112 or 168 bits	128, 192 or 256 bits	128 bits	32-448 bits
Block size	64 bits	64 bits	128 bits	64 bits	64 bits
Rounds	16	48	10, 12, 14	8.5	16
Structure of the algorithm	Balanced Feistel	Feistel	Substitution, permutation	Lai-Massey scheme	Feistel
Invented by	IBM, 1975	IBM, 1978	J. Daemen, V. Rijmen, 2000	J. Massey, X Lai, 1991	B. Schneier, 1993
Analysis	Little modification in plain text results in a huge change in ciphertext, small key size, exhaustive key search	Secure than DES, overcome the drawback of DES, much slower than DES	Operates on bytes instead of bits, six times faster than Triple-DES, adopted and supported in both hardware and software	Weak keys	Faster except when changing key

3 Asymmetric-Key Cryptography [1–4]

Public key techniques allow users to communicate securely without agreeing on any secret key in advance. In this scheme, a key pair (p_k, s_k) is generated by the receiving party. The message is encrypted by the sender using public key p_k and is decrypted by the receiver using private key s_k. Figure 2 shows the working of public key encryption.

This scheme uses a public and private key pair, (pk, sk). It does not require the sharing of any information or key in advance. It addresses the key distribution problem as communicating parties need not share a key in advance. It is extremely secure but is based on complex mathematics and slower than private key encryption.

The different methods used in Asymmetric-key Encryption are RSA, ElGamal, and Elliptic Curve Cryptography (ECC) compared in Table 2.

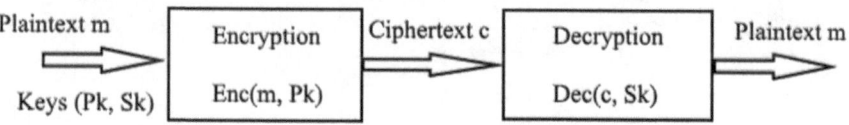

Fig. 2 Public key encryption

Table 2 Comparison of different asymmetric-key cryptography algorithms

	RSA	ElGamal	ECC
Key size	1024 bits	1024 bits	160 bits
Structure of algorithm	Integer factorization	Discrete logarithmic	Elliptical curve
Invented by	Rivest, Shamir, and Adleman, 1977	Taher ElGamal, 1985	Neal Koblitz and Victor S. Miller, 1985
Analysis	Uses two dissimilar keys: public and private key for encryption and decryption, respectively, based on factoring of large prime numbers, finding p and q for given mod n is easy but the reverse is difficult, vulnerable to small integers	Based on discrete logarithm problem, difficult to find the discrete logarithm of a given number in particular time frame, generates different ciphertext for a given plaintext, slow and generates long ciphertext	Based on the elliptic curve, apply discrete logarithm to points on the elliptic curve, small key size, less computing power, expensive

The cryptography schemes discussed above can store data securely but cannot allow any operation on encrypted data.

4 Homomorphic Encryption [5]

The traditional cryptographic algorithms allow users to encrypt their data, but to perform any operation on it, one need to decrypt it, perform computation and further encrypts the result. Homomorphic encryption allows to carry out some computations on encrypted data and generates encrypted results, which when decrypted gives the same result if the same operation is performed on the plaintext. Figure 3 shows the working of HE.

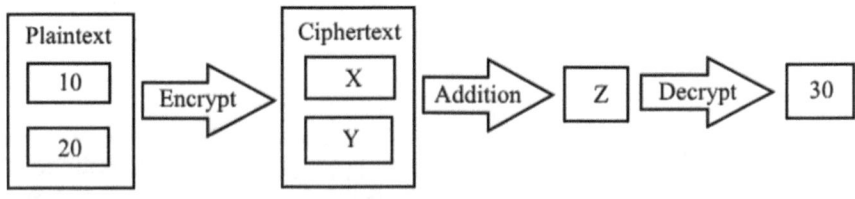

Fig. 3 Working of homomorphic encryption

4.1 Objective

- To store and exchange data securely.
- To carry out a few computations on encrypted data.
- To enhance data confidentiality.

4.2 Unpadded RSA [2–4]

The first public key encryption with homomorphic property is RSA. Let us assume that $(M, \bullet)$ and $(C, \bullet)$ are groups formed with modular multiplication operation $\bullet$ and plaintext, M and ciphertext C, respectively. Let p_k is a public key, where $p_k = (e, n)$ where e and n are large integers.

Let m_1 and m_2 are two plaintexts in M, then

$$E(m_1, p_k) = m_1^e (mod\ n) \tag{1}$$

$$E(m_2, p_k) = m_2^e (mod\ n) \tag{2}$$

$$\begin{aligned} E(m_1, p_k) \cdot E(m_2, p_k) &= m_1^e \cdot m_2^e (mod\ n) \\ &= m_1^e \cdot m_2^e (mod\ n) \\ &= E(m_1 \cdot m_2, p_k) \end{aligned} \tag{3}$$

Thus, RSA holds homomorphic property.

But unpadded RSA is insecure; hence random bits are padded to message before encryption. This leads to loss of homomorphic property. In the literature, several public key encryption schemes have been proposed with the homomorphic property.

4.3 Goldwasser–Micali (GM) Encryption Scheme [6]

In 1982, Goldwasser–Micali developed a first public key encryption algorithm based on probability, known as Goldwasser–Micali (GM) Encryption scheme. This scheme basically constitutes key generation, encryption, and decryption algorithms. The public and private key pair is generated by the key generation algorithm. The probabilistic encryption algorithm encrypts plaintext P randomly to ciphertext C, whereas deterministic decryption algorithm decrypts ciphertext C back to plaintext P. The encryption algorithm encrypts the same plaintext to different ciphertexts if it is encrypted several times. This scheme relies on intractable problem of quadratic residue modulo n, to decide that, given factors (p, q) of n, will give a value of y is square mod n or not.

Remark:

- Due to randomness in an encryption algorithm, it generates separate ciphertext for the same message if encrypted several times.
- It is semantically secure as it is based on the intractable problem, quadratic residue modulo n.
- The ciphertext size is large, as to encrypt every single bit it requires several hundred bits in n.

4.4   ElGmal Encryption Scheme [7]

In 1985, Taher ElGamal invented a Diffie–Hellman key exchange based public key encryption algorithm known as ElGamal encryption scheme. The discrete logarithmic problem is a basis for this scheme. It is defined over a cyclic group G and its security depends upon the problems in G related to the difficulty of computing discrete logarithm. This scheme is used to reduce the size of key.

Remark:

- This scheme depends on the group properties and padding scheme on messages.
- It is insecure under the chosen ciphertext attack as it is vulnerable.
- It is used in the hybrid cryptosystem.

4.5   Pallier Encryption Scheme [8, 9]

The Paillier encryption scheme is a probabilistic public key algorithm and is based on the assumption of the decisional composite residuosity. The problem of finding the nth residue class is difficult. This scheme performs both homomorphic addition and multiplication.

Remark:

- This scheme is based on the decisional composite residue problem.
- It is malleable, hence does not guard against adaptively chosen ciphertext attack.

4.6   Boneh–Goh–Nissim (BGN) Encryption Scheme [10]

The BGN encryption scheme is the first scheme with the constant size ciphertext. It is based on pairing for the elliptical curve, which allows multiplication and addition

Table 3 Comparison of different homomorphic encryption schemes

	GM	ElGamal	Paillier	BGN
Homomorphic property	Additive	Multiplicative	Additive and multiplicative	Additive and multiplicative
Structure of the algorithm	Probabilistic encryption and deterministic decryption, based on the intractability of quadratic residue modulo N problem	Based on Diffie–Hellman key exchange algorithm, defined over cyclic group G	Probabilistic public key encryption	Based on pairing for elliptical curve
Invented by	Shafi Goldwasser and Silvio Micali in 1982	Taher ElGamal in 1985	Pascal Paillier in 1999	Boneh, Goh, Nissim in 2005
Analysis	Randomness in encryption algorithm generates different ciphertext for same plaintext, larger ciphertext	Based on the computational problem within a cyclic group G, used in the hybrid cryptosystem, slower than symmetric	Based on decisional composite residue	Constant size ciphertext

both on cipher text. This scheme is secured on the basis of the subgroup decision problem.

Remark:

- This scheme is based on pairing-based cryptography.
- It allows both addition and multiplication operation on ciphertext.
- It achieves semantic security on the basis of the subgroup decision problem. The different Homomorphic Encryption techniques are compared in Table 3.

5 Fully Homomorphic Encryption [11]

Homomorphic Encryption techniques either perform addition or multiplication operation on encrypted data. It is beneficial if one can able to perform both of these operations simultaneously on the encrypted data. The Somewhat Homomorphic Encryption (SWHE) schemes perform limited operations on ciphertext. The decryption function fails to recover proper ciphertext after certain threshold, as the noise is getting added to the ciphertext. Gentry proposed ideal lattice-based first Fully Homomorphic Encryption (FHE) Scheme. Fully Homomorphic Encryption enables to perform arbitrary operations on ciphertext so that any number of additions and multiplications can be performed to manipulate encrypted data.

5.1 Objective

- To carry out arbitrary operations on coded data.
- To enhance security in Cloud Computing.

In 2009, Gentry [12] proposed the first fully homomorphic encryption scheme based on two atomic operations namely addition and multiplication on the ideal lattice. These two operations can be used to build and evaluate any circuit. This scheme is very promising in the field of FHE but due to some hard mathematical concepts, it is hard to implement and had high computational cost for real-life applications. So many new schemes and optimization have followed this scheme to overcome the above-mentioned limitations.

5.2 FHE Using Ideal Lattice [12]

Gentry proposed ideal lattice-based first Fully Homomorphic Encryption scheme. Initially, Gentry used ideals and rings without lattice to design the homomorphic encryption scheme. Ideals are property preserving subset of the rings. Each ideal is represented by lattice in this scheme. As this scheme is an extension of SWHE, after certain threshold, the ciphertext generated become noisy. The noise must be reduced to convert noisy ciphertext to proper ciphertext. Gentry used squashing and bootstrapping to generate proper ciphertext so that it enables to perform homomorphic operations on encrypted data. These processes are performed repeatedly so that it enables to perform arbitrary operation to make Fully Homomorphic scheme.

Remark:

- This scheme is based on ideal lattice.
- It uses squashing and bootstrapping to reduce noise in the generated ciphertext.

5.3 FHE Over Integers [13]

Van Dijk proposed another Fully Homomorphic scheme based on Gentry's bootstrapping method. This scheme is based on the problem of Approximate Greatest Common Divisor (AGCD). AGCD problem [14] try to recover an integer p from two near multiples m_1 and m_2, where $m_i = q_i p + r_i$, q_i are integers smaller than m_i and r_i are unknown error terms. The homomorphic addition and multiplication can be carried out on encrypted data. The result of homomorphic operation holds the homomorphic property and preserves the format of the original ciphertext. If the noise is less than half of the private key then encrypted data can be decrypted. The noise grows exponentially with multiplication operation.

Remark:

- This scheme is based on Approximate Greatest Common Divisor problem.
- This scheme is simple and based on symmetric HE.
- As noise increases exponentially with multiplication operation, it puts more restrictions on multiplication than addition.

5.4 FHE Based on LWE [15]

This scheme is the first step towards practical implementation of FHE scheme. Brakerski and Vaikuntanathan proposed a new SWHE scheme based on Ring Learning with Error (RLWE). Learning with Errors (LWE) problem for post-quantum algorithms is one of the hardest problems as it is not solved in practical time. If any algorithm can solve LWE problem in efficient time, then it can also solve the shortest vector problem (SVP) in an efficient time. This can be found promising for post-quantum cryptology with relatively small ciphertext size.

Remark:

- It is based on Learning with Error.
- It is the first step toward practical implementation of FHE.
- It can solve a complex lattice problem in efficient time.

The different Fully Homomorphic Encryption schemes are compared in Table 4.

Table 4 Comparison of different fully homomorphic encryption schemes

	FHE using ideal lattice	FHE over integers	FHE based on LWE
Based on the concept	Ideals and rings	Approximate greatest common divisor	Ring learning with errors
Invented by	Gentry in 2009	Van Dijk, C. Gentry, Halevi, and Vaikuntanathan 2010	Brakerski and Vaikuntanathan 2011
Analysis	Represents lattice using ideals and rings. Used squashing and bootstrapping to minimize noise	Is simple and is based on symmetric HE. Able to decrypt ciphertext, if noise is less than half of the private key. With multiplication operation noise increases exponentially	First practical implementation of FHE, solves complex lattice problems

6 Conclusion

In the current era of information technology, the information is heavily exchanged over the Internet. The various encryption techniques used to encrypt data and keys are shared with service providers or third-party operators to achieve data privacy. The modern cryptography techniques based on public and private key cryptography are analyzed with their limitations. Data privacy is a major concern in cryptography as due to shared keys, user's data is accessible to the untrusted service provider or third party operator. One promising direction to preserve the privacy of data is Homomorphic Encryption (HE) scheme, which allows the service provider to operate on encrypted data without sharing keys. HE schemes allow either multiplication or addition operation on encrypted data. Hence, only a few computations are performed on encrypted data. However, Gentry's Fully Homomorphic Encryption (FHE) schemes enable to perform any function on encrypted data, but its implementation is practically expensive for real-life applications. Here, we surveyed the basic idea, different implementations along with their limitations of HE and FHE schemes.

References

1. Kats J, Lindell Y (2014) Introduction to modern cryptography. In: Cryptography and network security. Chapman & Hall/CRC. ISBN:- 13: 978-1-58488-551-1
2. Stallings W Cryptography and network security: principles and practice, 5th edn, Pearson
3. Forouzan BA Cryptography and network security. Tata McGraw Hill
4. Katz J, Lindell Y Introduction to modern cryptography. In: Cryptography and network security. Chapman & Hall/CRC
5. Yi X, Paulet R, Bertino E (2014) Homomorphic encryption: chapter-2. In: Homomorphic Encryption and Applications. Springer briefs in computer science, pp 27–46
6. Goldwasser S, Micali S (1984) Probabilistic encryption. J Comput Syst Sci 28:270–299

7. ElGamal T (1985) A public-key cryptosystem and a signature scheme based on discrete logarithms. IEEE Trans Inf Theory 31(4):469–472
8. Paillier P (1999) Public key cryptosystems based on composite degree residue classes. In: Proceedings of advances in cryptology, EUROCRYPT'99, pp 223–238
9. Okamoto T, Uchiyama S (1998) A new public-key cryptosystem as secure as factoring. In: Proceedings of advances in cryptology, EUROCRYPT'98, pp 308–318
10. Boneh D, Goh E, Nissim K (2005) Evaluating 2-DNF formulas on ciphertexts. In: Proceedings of theory of cryptography, TCC'05, pp 325–341
11. Acar A, Aksu H, Uluagac AS (2017) A survey on homomorphic encryption schemes: theory and implementation. ACM CSUR
12. Gentry C (2009) Fully homomorphic encryption using ideal lattices. In: Proceedings of STOC'09, pp 169–178
13. Van Dijk M, Gentry C, Halevi S, Vaikuntanathan V (2010) Fully homomorphic encryption over the integers. In: Advances in cryptology–EUROCRYPT, pp 24–43. Springer
14. Galbraith SD, Gebregiyorgis SW, Murphy S (2016) Algorithms for the approximate common divisor problem. LMS J Comput Math 19:58–72
15. Regev O (2009) On lattices, learning with errors, random linear codes, and cryptography. J ACM (JACM) 56(6):34

Segmentation of Cotton Leaves Blade Based on Global Threshold and Morphological Operation

Janwale Asaram Pandurang, S. Lomte Santosh and Kale Suhash Babasaheb

Abstract In this article, we proposed a method of segmentation using a global threshold to segment a leaf blade from the dark black background. RGB images of leaf captured using a digital camera get converted into HSI model. The threshold value, t, is computed automatically for separate bands of HSI using Otsu's methods. By combining only hue and saturation images where both are true, we got the colored object mask. Finally, by concatenating these images we got segmented leaf image from the background. In the second step, we applied a sequence of morphological operations to remove petiole from the leaf blade. Experiments have been carried on 90 leaves and we got a success rate of 93.33%.

Keywords Image segmentation · Cotton leaf · Petiole segmentation · Global thresholding · Morphological operations · HSI · RGB

1 Introduction

In India, Cotton is an important crop and India has the largest cotton growing area in the world. In many cases, the health of cotton plant can be determined by leaf's characteristics. A leaf contains different parts like leaf blade, petioles, and stalk. Petioles are an important component of leaf used in photosynthesis and transpiration for green plants, which can affect plants growth and yield. Image processing techniques evolved, can play an important role in leaves image analysis. Many appli-

J. A. Pandurang (✉) · K. S. Babasaheb
Balbhim College Beed, Beed, India
e-mail: janwale26@gmail.com

K. S. Babasaheb
e-mail: ctekbeed@gmail.com

S. Lomte Santosh
Matoshri Pratishthan Group of Institutions, Nanded, India
e-mail: santoshlomte@gmail.com

© Springer Nature Singapore Pte Ltd. 2019
H. S. Saini et al. (eds.), *Innovations in Computer Science and Engineering*, Lecture Notes in Networks and Systems 74,
https://doi.org/10.1007/978-981-13-7082-3_35

Fig. 1 Image of leaf parts

cations need to perform an important step of image segmentation in digital image processing such as plant identification, estimation of Nitrogen, identification of leaf diseases, and so on. The digital image processing techniques are able to do non-destructive measurements for leaves, but the petiole will be a major obstacle on extracting the information accurately. Manual lab technique like Kjeldahl technique used for estimation of nitrogen from plant leaves need to remove petiole from leaf blade before chemical analysis, so in case of automation of this, it needs an automatic method to segment leaf blade and removes petiole from image.

Over several years number of different segmentation techniques proposed for gray or color images. There are still no efficient and simple segmentation techniques for leaf segmentation from background and petiole removal from the stalk. Our interest in this study is to segment and removal of petiole from cotton leaf images (Fig. 1).

The purpose of segmentation is to present image into simple form or parts that are more useful and easily analyzed. Image segmentation will produce results consisting a set of parts that include the entire image, where all pixels from a region are same in case of few characteristic or computed features, for examples, color, intensity, or texture. The threshold is one of the simplest methods of segmentation. This method based on threshold value is calculated by converting the image to binary image. Threshold methods include global and local thresholding, having methods like Otsu's global thresholding and adaptive local thresholding.

The main contribution of this article is to provide a simple method for segmentation of the leaf blade from the background using the global threshold technique and removing petiole from the blade by applying morphological operations.

2 Related Work

In addition to this paper [1], they projected an algorithmic program that features fuzzy threshold and clustering segmentation for varied plant analyses. The result they got from this system is predicated on a fuzzy threshold and bunch techniques for identification of most similar components in plant leaf pictures.

In this paper [2], first by comparing the distribution of multi-model histogram in HSV color model, and then selected the appropriate band and threshold value to process the image. Here, they used the method of mathematical morphology with the element of multiple structure elements and multiple scales to smooth the leaves border and remove the segmented petiole in the image.

The segmentation algorithms [3] are developed that is capable to perform segmentation method victimizing the leaf pictures that were exposed to totally different illuminations. They tried Otsu's international threshold so that the best segmentation supported the tested pictures.

In this paper [4], they presented an algorithm which combined PCNN model and HSU color model for leaf petiole detection. The results show the successful removal of petiole from the leaf blade.

In this paper [5], automatic marker-controlled watershed segmentation methodology was sorted with pre-segmentation and the mathematical morphological operation was to segment leaf pictures with advanced background support the form information. Experimental results on some plant leaves showed that planned classification framework worked compatibly whereas classifying leaf pictures with a difficult background.

In this paper [6], they projected a way for image segmentation of cucumber leaf pictures. First, the color image is analyzed. At that time, a sort of color feature is employed to get the characteristics map, which mixes the RGB model and HSI model. Finally, the morphological methodology is employed to induce the image segmentation.

In this paper [7] for segmentation, they planned the algorithmic rule that consists of two steps. Beginning supported peak of histogram identification victimization Otsu's methodology. Within the second step, the proof theory was accustomed to merge totally different pictures drawn in numerous color areas, to urge correct segmentation result.

In this paper [8], in this analysis, a way for automatic identification and segmentation of leaves has been developed supported the depth image captured by the Kinect detector. Pixels of the depth image is reborn into 3D points and registered during a commonplace coordinate system during ground standardization.

According to the literature survey, there is no single method to segment cotton leaf from background and removal of petiole automatically. It gives the scope for our work to improve and provide a method for segmentation and petiole removal combined technique.

3 Proposed Segmentation Method

A. Image Acquisition

Created own dataset of Cotton leaf images, which are collected from fields. Single leaf placed against black background and images were taken using Sony Cyber-Shot W830 camera of 20 megapixels with CCD sensor. To reduce the illumination problem [9], images are taken in the morning and evening session. Samples of images are

Fig. 2 The image of cotton leaf captured by Sony camera

taken from 15 cm height with the black background in natural light and resized to 1764 * 768 pixels (Fig. 2).

B. Color Model

Digital images are made of Red, Green, and Blue (RGB) color model. Conversion to the grayscale of RGB image will convert each R, G, and B components to level 0–255, which increase space and time complexities. Therefore, instead of working directly on the specific color, we converted images from RGB to HIS color space. The HSI model is perfect for color image processing based on color features, which are natural to humans. In the HSI model, H factor contains information about color, S factor represents dominance of that color, and V component represents intensity. Therefore, we can detect all color shades, which is a very important step in finding nutrients deficiency projects.

C. Leaf blade segmentation

Black background was used to take images using digital camera, which reduces illumination effect and simplifies the segmentation process. Images are taken and stored in jpg format. Images were separated into its components such as Hue, Saturation, and Value images and are shown in Fig. 3 and plot the histogram of H, S, I components to process.

By using Otsu's method, appropriate threshold values are computed. Otsu's thresholding method is used to compute threshold values by considering bimodal histogram. Three bands of HSI image get separated into a different band of H, S, and I. Threshold value t is computed automatically for separate bands of HSI using Otsu's methods. We calculated and assigned a range of the low and high thresholds for each color band. Separate mask for hue, saturation, and value are created by

(a)   **(b)**   **(c)**

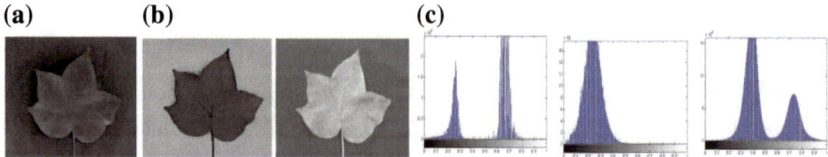

Fig. 3 Shows **a** Hue image. **b** Saturation image. **c** Intensity image and their histograms

assigning low and high threshold values to each color bands. In HSI model, color is represented by only hue and saturation components, so we combined only the hue and saturation images where both are true, we got the colored object mask (Fig. 4).

Masked image contains some noise and edges are also not continuous. Therefore, we applied morphological operations to smooth the border used morphological closing operation then small white particles are removed and borders smoothed (Figs. 5 and 6).

To mask out the colored-only portions of the RGB image use the colored object mask by multiplying it with RGB components using formulas 1, 2, and 3.

$$\mu IP = \chi M. * \chi I(:, :, 1); \tag{1}$$

$$\mu I\Gamma = \chi M. * \chi I(:, :, 2); \tag{2}$$

$$\mu IB = \chi M. * \chi I(:, :, 3); \tag{3}$$

Fig. 4 Image of the specified color

Fig. 5 Morphological operations on the image

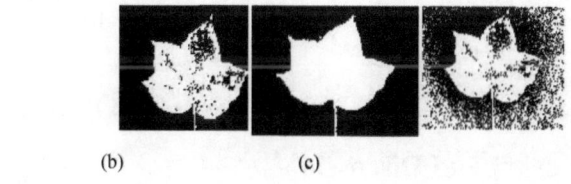

(a)   (b)   (c)

Fig. 6 **a** Colored object mask image. **b** Hue mask. **c** Saturation mask

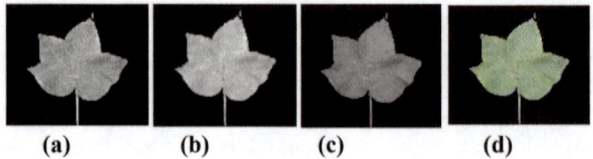

Fig. 7 **a** Masked red image. **b** Masked green image. **c** Masked blue image. **d** Segmented image

Fig. 8 Petiole removed image

where χM denotes colored object mask, χI denotes color image and μIP, μIT, and μIB are masked images of respective color of red, green, and blue. Finally, by concatenating these images, we got segmented leaf image from the background as in Fig. 7d.

D. Leaf petiole elimination

The output image produced by the global threshold method still contains petiole with leaf blade, which can be removed by using mathematical morphological operations. Morphology operations are used in image processing for a task like segmenting touching parts in the image, elimination of undesired image regions and noise and categories objects by their size. Morphological operations use structuring part to an input image, to provide an output image of equivalent size. The basic operations are opening, closing, fill, smooth, thin, and so on. The shapes of structural elements are also different like disk, line, diamond, and round. In our study, we experimented on different structural elements and sequence of morphological operations. After a number of tests, we find that a combination of structural elements and repetition of opening and closing will remove petiole from leaf blade. Therefore, we applied the line structural element with 90 degrees along with the disk-shaped element and performed opening and closing operation to remove petiole from leaf blade successfully as shown in Fig. 8.

4 Conclusion and Discussion

In this paper, we presented the implementation of a fully automatic method, which segments the leaf blade from the dark black background and also eliminates petiole from it. This method includes two steps: first by using global threshold method leaf blade is segmented from the background and second by applying a sequence of morphological operations petiole removed from leaf blade as shown in Fig. 8. Experiments have been carried out on 90 leaves and we got a success rate of 93.33%. Compared with other methods, the proposed method here is simple and suitable for leaf blade segmentation. These results have an important role in nitrogen deficiency detection and other applications were we need leaf blade image.

References

1. Valliammal N (2012) A novel approach for plant leaf image segmentation using fuzzy clustering. Int J Comput Appl 44(13):0975–8887
2. Xingli WU et al (2015) Petiole segmentation method based on multi-structure elements morphology. Appl Mech Mater. Trans Tech Publications, Switzerland. https://doi.org/10.4028/www.scientific.net/AMM.734.581
3. Hairuddin MA et al (2013) Elaeis guineensis leaf image segmentation: a comparative study and analysis. In: 2013 IEEE 3rd international conference on system engineering and technology, Shah Alam, Malaysia, 19–20 Aug 2013
4. Wang Z et al (2015) A new petiole detection algorithm based on leaf image. In: Proceeding of the IEEE 28th Canadian conference on electrical and computer engineering Halifax, Canada, 3–6 May 2015
5. Wang X-F et al (2008) Classification of plant leaf images with complicated background. Appl Math Comput 205. https://doi.org/10.1016/j.amc.2008.05.108
6. Wang L, Yang T, Tian Y (2007) Crop disease leaf image segmentation method based on color features
7. Harrabi R, Ben Braiek E (2012) Color image segmentation using multi-level thresholding approach and data fusion techniques: application in the breast cancer cells images. EURASIP J Image Video Process 2012:11
8. Shao X, Shi Y, Wu W, Yang P (2014) Leaf recognition and segmentation by using depth image
9. Vibhute A, Bodhe SK (2013) Color characteristics in RBG space under outdoor illumination in tropical region. IJCA 62(6):0975–8887
10. Gocławski J et al (2012) Neural network segmentation of images from stained Cucurbits leaves with colour symptoms of biotic and abiotic stresses. Int J Appl Math Comput Sci 22(3):669–684. https://doi.org/10.2478/v10006-012-0050-5
11. Hsiao Y-T et al (2005) A mathematical morphological method to thin edge detection in dark region. In: Conference paper, Jan 2005. IEEE Xplore. https://doi.org/10.1109/isspit.2004.1433746
12. Dowlatabadi M, Shirazi J (2011) Improvements in edge detection based on mathematical morphology and wavelet transform using fuzzy rules. World Acad Sci Eng Technol Int J Electr Comput Eng 5(10)
13. Kaiyan L et al (2014) Measurement of plant leaf area based on computer vision. 978-1-4799-3434-8/14 $31.00 2014© IEEE. https://doi.org/10.1109/icmtma.2014.99
14. Janwale AP, Lomte SS (2017) Plant leaves image segmentation techniques: a review. Int J Comput Sci Eng Rev Paper 5(5). E-ISSN 2347-2693

Electronic Guitar MIDI Controller for Various Musical Instruments Using Charlieplexing Method

Robinson Devasia, Aman Gupta, Sapna Sharma, Saurav Singh
and Neeru Rathee

Abstract Music is nowadays, a way to learn for kids, passion for youth, and a mode of meditation for adults. The urge to be able to use different varieties of sounds for a creative composition of music is the biggest challenge for musicians, which is quite complicated as well as time-consuming task. To resolve this issue, this paper presents an approach to generate sounds of various musical instruments corresponding to the particular notes of a guitar. The fundamental principle behind the proposed work is that, though all the musical instruments play the same notes but have different timbre characteristics. The electronic guitar is designed in such a way that it enables a guitarist to play all kinds of instruments without physically learning them. The guitar is connected to a computer through a microcontroller. It has multiple input ports, a part of which acts as frets and remaining as strings. The number of input pins is far less than the actual number of frets and strings. The technique called Charlieplexing achieves it. The instrument to be played can be selected from the instruments listed in the software. After selecting a particular instrument, a guitarist can play it just like a conventional guitar. Computer connectivity also allows the guitarist to practice music with earphones making it a soundless device for others.

R. Devasia (✉) · S. Sharma
Indraprastha Institute of Information Technology Delhi, Okhla Industrial Estate, Phase III, Near
Govind Puri Metro Station, New Delhi 110020, Delhi, India
e-mail: robinson18173@iiitd.ac.in

S. Sharma
e-mail: sapna18222@iiitd.ac.in

A. Gupta
Continental Pvt. Ltd., Bengaluru, Karnataka, India
e-mail: amanguptamsit@gmail.com

S. Singh
Rochester Institute of Technology, New York 14623, USA
e-mail: ss3337@rit.edu

N. Rathee
Maharaja Surajmal Institute of Technology, Janakpuri, New Delhi 110058, India
e-mail: neeru1rathee@gmail.com

© Springer Nature Singapore Pte Ltd. 2019
H. S. Saini et al. (eds.), *Innovations in Computer Science
and Engineering*, Lecture Notes in Networks and Systems 74,
https://doi.org/10.1007/978-981-13-7082-3_36

Keywords Embedded systems · MIDI generator · Electronic guitar · Music synthesis · MIDI controller · Instruments

1 Introduction

Music is a tool, which connects souls and this is the reason it has been used in various therapies [1], control, and guidance. Creating music is an art. The notes, rhythm, pitch, and texture have to be in perfect proportion otherwise it would be nothing more than noise. Music is also a part of our culture, which we have seen evolving from traditional folk music to rock bands [2]. It has become a way to express oneself, and the passion is increasing day by day.

The world has witnessed musicians like Bob Marley, The Beatles, Tansen, Hari Prasad Chaurasia, Zakir Hussain, Ravi Shankar, and many more. Each of them was/is an expert in their field. The Beatles is known for their rock music. Zakir Hussain is known for his tabla playing art, Ravi Shankar for Sitar, Hari Prasad Chaurasia for flute, and Raghu Dixit for Guitar.

Musicians play different musical instruments to generate a different type of music, but the theory behind each instrument is the same and share the same literature of notes and chords. Despite this fact, one has to put efforts to learn to play different instruments. An expert musician of one instrument will not be an expert of another instrument if he/she has not mastered it, for e.g., an expert Sitar player may not be able to play Guitar or an expert Guitar player may not be able to play flute without learning. To learn all the instruments is very time-consuming, tedious, and challenging task in the music industry. To address the above issue, researchers have put their best efforts to synthesize musical sound artificially, and they achieved great success in this field.

Many researchers have worked in this field and come up with solutions like Orb Composer-A music composing software using artificial intelligence, OrchExtra-A sound enhancement system run from one's laptop, etc. Digital music refers to an art of music that is described, created, spread, and stored via digital technology by using the computer and the Internet, and where sound streams are processed in other processing modes. It is a product of the electronic and computer age [3, 4]. The success achieved in synthesizing various musical instruments is remarkable and appreciable.

All the work done to date toward music synthesis or control is dedicated to a single instrument. The work toward multiple instrument control using a single instrument guitar as a MIDI controller is missing. In the presented work, an attempt has been made to control numerous instruments using a single instrument MIDI controller. The basics behind the proposed work are based on the fact that the music literature of every instrument is same and hence can be used to play all kinds of the instrument through single design (Table 1).

Table 1 MIDI signal instruction

Value (Decimal)	Value (HEX)	Command	Data types
128–143	80–8F	Note off	2 (note, velocity)
144–159	90–9F	Note on	2 (note, velocity)
160–175	A0–AF	Key pressure	2 (note, key pressure)
176–191	B0–BF	Control change	2 (controller no., value)
192–207	C0–CF	Program change	1 (program no.)
208–223	D0–DF	Channel pressure	1 (pressure)
224–239	E0–EF	Pitch bend	2 (least significant byte, most significant byte)

Electronic Guitar is a MIDI Controller that replaces the existing controllers with a cheap and efficient controller, which can be played just like existing guitars. This MIDI controller looks and can be performed exactly like a guitar. However, it does not directly produce sound; it generates MIDI signals, which are sent to the computer to do all the processing for the production of the sound. This input/output system eliminates the use of expensive guitars and MIDI interface as it acts as both at the same time. The conventional strings are not used in this guitar. Hence, it eliminates the handedness of the instrument and the effects on tensed strings due to changes in temperature. This provides some advantage over existing guitars.

The organization of the presented paper is as follows. Section 2 describes the conventional guitar basics; Sect. 3 gives a brief of MIDI controller. In Sect. 4, the Serial peripheral interface is explained in detail. Sections 5 and 6 focus on Sequence generator and Guitar circuit, respectively. The algorithm of the proposed work is explained in Sect. 7. Finally, the conclusion is presented in Sect. 8.

2 Conventional Guitar

2.1 Working Principle

The fundamental principle behind the conventional guitar is resonance. Resonance here refers to the amplification of vibration by a particular design/structure. Any vibrating string generates sound. The frequency of vibration is inversely proportional to the length of the string and its thickness, thus shorter string results in higher vibration frequency. In acoustic guitars, "all of the sound energy that is produced by the body originally comes from energy put into the string by the guitarist's finger."

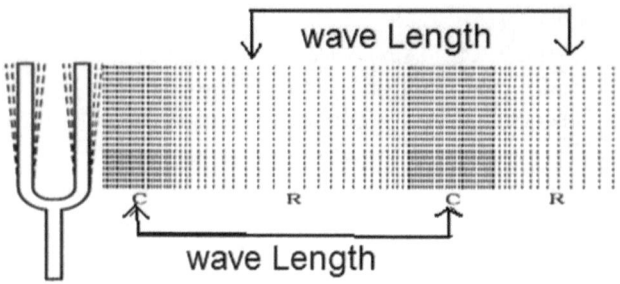

Fig. 1 Wavelength

In the electric guitar, the sound box is not as useful as in an acoustic guitar though the strings vibrate on the same frequencies as an acoustic guitar. The sound pickups of electric guitar act as a transducer for converting the change in air pressure to electrical energy that is then sent to an amplifier (Fig. 1).

2.2 Parts of Conventional Guitar

2.2.1 Strings

Strings of varying lengths and diameters are used in different instruments. The materials used for guitar strings are Nylon, Nickel, Bronze, Silver, etc.

2.2.2 Pickups

Pickups are transducers that may be piezo, magnetic, single coil or of any other type. It converts the string vibrations into electrical signals. These signals are fed directly to amplifiers or devices adding features like distortion etc. to these signals.

3 MIDI Controller

MIDI is a technical standard that describes a protocol, digital interface, and connectors, which allows a variety of device to connect and communicate with each other [5] (Fig. 2).

Fig. 2 MIDI

MIDI carries mainly three components in a message namely notation, pitch, and velocity. The MIDI interface operates at 31.25 k Baud rate using an asynchronous serial data byte comprising 1 Start bit, 8 Data bits (0–7), and 1 stop bit. This makes a total of 10 bits per serial byte with a period of 320 μs [6].

MIDI bytes are divided into two types command bytes and data bytes. Status byte is in range 0x80 to 0xFF. The data byte is in the range 0x00 to 0x7F in hex [6]. Commands include things such as note on, note off, pitch bend, and so forth. Data bytes include things like the pitch of the note to play, the velocity, or loudness of the note, amount of pitch bend and so forth [6, 7]. MIDI messages comprise a STATUS byte (bit 7 = 1) followed by DATA bytes (Bit 7 = 0). Messages are divided into two main categories: Channel and System. Channel messages contain a four-bit channel number encoded into the Status byte, which addresses the message specifically to 1 of the 16 channels. System messages are not encoded with channel numbers and are divided into three main types: System common, system real-time, and system exclusive [6–8].

A. *Abbreviations and Acronyms*
   MIDI: Musical Instrument Digital Interface
B. *Units*
   Baud Rate: It is a unit for symbol rate or modulation rate in symbols per second. *Unit symbol* **Bd**.

4 SPI

The **Serial Peripheral Interface (SPI)** buses are asynchronous serial communication interface that are used for communication between an integrated circuit such as a microcontroller and a set of relatively slow peripherals. The interface was developed by Motorola in mid-1980s and has become a de facto standard [9].

The SPI module allows a duplex, synchronous, serial communication between the MCU and peripheral devices.

During an SPI transmission, data is transmitted (shifted out serially) and received (shifted in serially) simultaneously. The serial clock (SCK) synchronizes shifting and sampling of the information on the two serial data lines. A slave select line allows selection of an individual slave SPI device, slave devices that are not selected do not interfere with SPI bus activities. Optionally, on a master SPI device, the slave select line can be used to indicate multiple master bus contention as described in SPI Block Guide V03.06 released by Motorola Inc. (2003, February).

5 Sequence Generator

24-bit output driven with a single output pin is implemented with the help of SPI protocol which sends 8 bits of data all together serially to the three shift registers connected in the series to each other. The data, which is to be sent to the third register, will be transmitted to the first so that it can be latched to the end. A common clock is given to all these three registers so that when the pulse is provided by the SPI for next 8 bits, the data in register automatically moves to the next register, so the separate clock is not needed for sending the data to the next register.

Once the data is latched on the shift registers the data is then sent to the storage register by sending a clock to the storage register by this method the data will reflect at the output altogether without delay.

As given in Fig. 3 U3 shift register acts like the strings of the guitar. U2 and U4 act as the frets of the guitar. The sequence is generated so that a fret is checked for each string of the guitar, whether it is pressed or not. The LEDs are named D18, D19, etc.

6 Guitar Circuit

In this guitar, the concept of plucking and pressing the fret string is peculiar because the strings for fret and the plucking are entirely independent. In the circuit, J1 and J2 are the fret of the guitar, which is completed, with the help of the conducting wound wire tied over the fret. If the particular fret is pressed, a 5 V signal reaches the collector of a specific transistor through J3, depending on which string of the fret is pressed, J4 acts as the string of the guitar (Fig. 4).

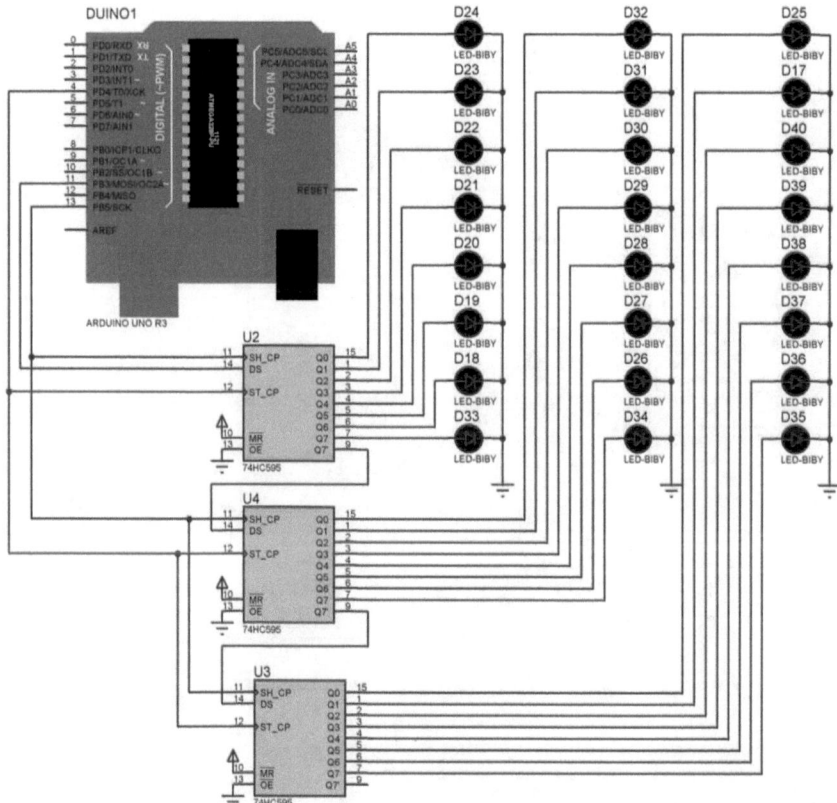

Fig. 3 Sequence generator circuit

The five V signal will remain at collector until a particular string is plucked, i.e., emitter side is plucked with the help of a plectrum. Plectrum is attached to the digital pin of the controller that will detect the voltage at the collector of the particular transistor.

The sequence running in the shift register is very fast compared to the human response it seems that every pin has the voltage at the same time, but every time the sequence inside the shift registers is different. As soon as a voltage is detected at the digital pin, the algorithm looks in the table for the particular sequence, which is inside the shift register at that specific instant. For every sequence, a note is assigned which is sent to the computer as soon as it is detected in the form of the MIDI signal.

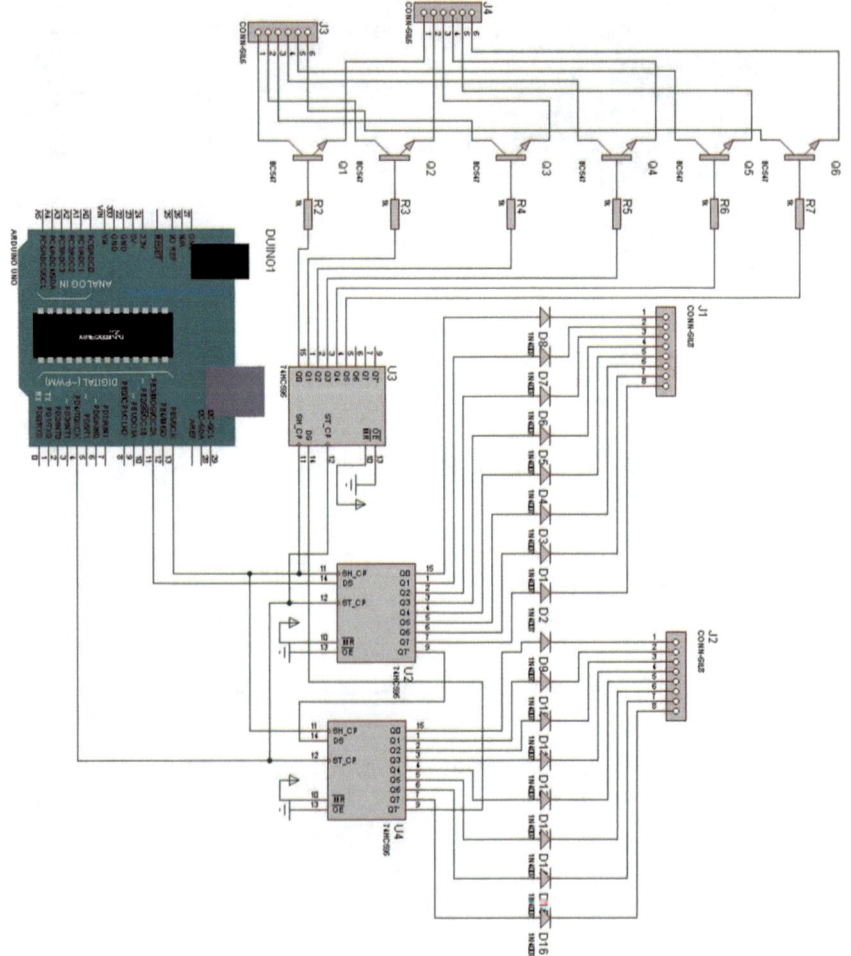

Fig. 4 Guitar circuit

6.1 *Components*

6.1.1 The Processing Unit (*Arduino*)

The processing unit used here, for converting Input Digital Signals into corresponding MIDI Signals, is Arduino. Arduino is an open-source electronics platform based on easy-to-use hardware and software working at the 16 MHz clock frequency. SPI can

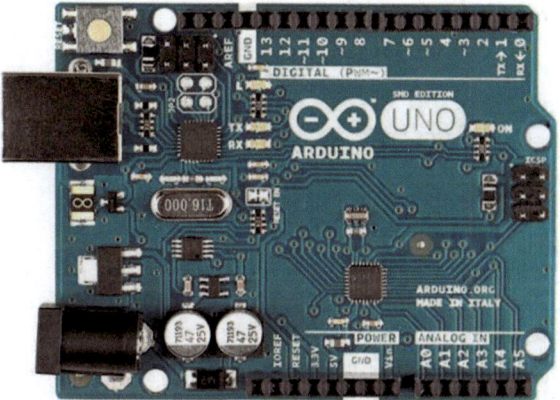

Fig. 5 Arduino UNO

generate and transfer a sequence of 8 bits at 8 MHz frequency, which is fast enough for human perception. The Arduino board used in electronic guitar is Arduino UNO [10] (Fig. 5).

6.1.2 Shift Registers (*74HC595*)

As per the data sheet released by Texas Instruments Incorporated Dallas, Texas 75265, (March 17, 2017), the 74HC595 device contains an 8-bit serial-in, a parallel-out shift register that feeds an 8-bit D-type storage register. The storage register has parallel 3-state outputs. Separate clocks are provided for both the shift and storage register. The shift register has a direct overriding clear (SRCLR) input, serial (SER) input, and serial outputs for cascading. When the output-enable (OE) input is high, the outputs are in the high impedance state. Both the shift register clock (SRCLK) and storage register clock (RCLK) is positive-edge-triggered. If both clocks are connected, the shift register is always one clock pulse ahead of the storage register.

6.1.3 Transistor

It is a semiconductor device, which can be used as an amplifier or a switch. In this system, NPN transistors are used for better switching application. It is used as a switch so it is operated in cutoff and forward saturation region.

7 Algorithm

```
MIDI_message:-protocol to send MIDI signal which consists
of three parameters which are the command, note, and
velocity.
SPI:- Serial Peripheral Interface Bus is a protocol to
send 8 bits of data from a single bus.
1. Loop: 24-bit Sequence is generated.
2. SPI.transfer after breaking it into three parts
3. Latch the sequence data to the output register through
the clock
4. If(digitalRead(input_pin))
    {
      MIDI_message to the computer
    }
   Else
    {
      GOTO Loop;
    }
5. Computer receives the signal and process it.
6. The intermediate software sends the data to the sound
   processing software.
7. Sound processing software assigns different sounds to
   the received signal.
8. The sound is played through the loudspeaker.
```

8 Results and Conclusion

Electronic Guitar is a single remedy for all the problems of the guitar world. It is an embedded system that eliminates the dependency of sound on the acoustics of the guitar or the pickups used. Instead, the sound is generated using the Digital audio software. Hence, the same instrument can sound like a Nylon string guitar, an electric guitar, piano, or even a drum set without compromising with the sound quality. Since the produced sound does not depend on the timber, size of the instrument, type, and quality of strings, etc., thus the electronic guitar is very cost efficient and portable.

Its conventional design preserves the feel of real guitars with steel strings, but it is independent of the tension in the strings. Being a digital device, it excludes the need to tune the guitar, and it eradicates the problems of tensed strings like bending of guitar, breaking of strings, degradation of strings, etc. This functionality of the guitar also makes it "Soundless". Signals are fed directly to a computer and can be heard on headphones.

Through further research, the latency of the sound produced can be further reduced. Further improvements can be made by eliminating the requirement of audio software and thus making the guitar self-sufficient. Added functionality on the guitar can include programmable buttons, thus providing the functionality of pedals, loopers, etc.

References

1. Inoue D, Suzuki M, Matsumoto T (2003) Detection resistant steganography for standard MIDI files. IEICE Trans Fundam Electron Commun Comput Sci E86A(8):2099–2016
2. Zhang XP, Wang SZ (2006) Dynamical running coding in digital steganography. IEEE Signal Process Lett 13(6):1651–1668
3. Oudjida AK, Berrandjia ML, Liacha A, Tiar R, Tahraoui K, Alhoumays YN (2010) Design and test of general-purpose SPI master/slave IPs on OPB bus. In: 7th international multi-conference microelectronics and nanotechnology division centre, systems signals and devices (SSD), Algeria, pp 1–6
4. The MIDI Specification. https://docs.isy.liu.se/pub/VanHeden/DataSheets/The_MIDI_ Specification.pdf
5. MIDI Protocol Guide (2014). http://hinton-instruments.co.uk/reference/midi/protocol/index. htm
6. MIDI Protocol Guide. http://www.slaveksamal.com/files/midicode.htm
7. Bunt L, Stige B (2014) Music therapy an art beyond words, 2nd edn. Routledge, p 266
8. Rose GJ (2014) Between Couch and Piano psychoanalysis, music art and neuroscience. Routledge, p 189
9. Portelli R (2015) France. https://www.orb-composer.com/about-orb-composer/
10. Badamasi YA (2014) The working principle of an Arduino. In: 11th international conference on electronics, computer and computation (ICECCO), Nigerian Turkish Nile University, Abuja, Nigeria, pp 1–4

Music Generation Using Deep Learning Techniques

Somula Ramasubbareddy, D. Saidulu, V. Devasekhar, V. Swathi,
Sahaj Singh Maini and K. Govinda

Abstract Deep learning techniques for generating music that has melody and harmony and is similar to music compositions by human beings is something that has fascinated researchers in the field of artificial intelligence. Nowadays, deep learning is being used for solving various problems in numerous artistic fields. There has been a new trend of using deep learning models for various applications in the field of music that has attracted much attention, and automated music generation has been an active area of research that lies in the cross section of artificial intelligence and audio synthesis. Previously, the work in automated music generation was solely focused on generating music, which consisted of a single melody, which is also known as monophonic music. More recently, research work related to the automated generation of polyphonic music, music, which consists of multiple melodies, has met partial success with the help of estimation of time series probability density. In this paper, we use Restricted Boltzmann Machine (RBM) and Recurrent Neural Network Restricted Boltzmann Machine for music generation by training it on a collection of Musical Instrument Digital Interface (MIDI) files.

S. Ramasubbareddy (✉) · S. S. Maini · K. Govinda
Department of Computer Science and Engineering, VIT University,
Vellore, Tamilnadu, India
e-mail: svramasubbareddy1219@gmail.com

S. S. Maini
e-mail: sahajsingh.maini2014@vit.ac.in

K. Govinda
e-mail: kgovinda@vit.ac.in

D. Saidulu · V. Devasekhar · V. Swathi
Department of Computer Science and Engineering, Guru Nanak Institutions
Technical Campus, Hyderabad, Telangana, India
e-mail: fly2.sai@gmail.com

V. Devasekhar
e-mail: devasekharv@gmail.com

V. Swathi
e-mail: swathivelugoti@gmail.com

© Springer Nature Singapore Pte Ltd. 2019   327
H. S. Saini et al. (eds.), *Innovations in Computer Science
and Engineering*, Lecture Notes in Networks and Systems 74,
https://doi.org/10.1007/978-981-13-7082-3_37

Keywords Deep learning · Restricted Boltzmann Machines · Recurrent Neural
Network Restricted Boltzmann Machine · Musical Instrument Digital Interface

1 Introduction

Deep learning techniques in the domain of automated music generation are an intrigu-
ing topic for research that has not been studied in depth. Initially, the technique for
automated music generation was based upon forming a set of rules for the computer
to generate music accordingly. This technique generated substantial results. However, a defined set of rules to generate music does not produce the ideal results due to
the complexity of the music. It is a popular notion among many individuals in various
music communities that human beings can only produce music compositions and that
machine intelligence cannot produce pleasing music compositions. However, there
are many researchers who disagree with this notion and believe that there is information related to repetition and structure in musical compositions that can be learned
from the existing compositions by using various deep learning techniques for composing pleasing melodic music. Music can be defined as the science of organization of
various sounds and tones in temporal relationships, combination, and succession that
generate a composition, which has continuity and integrity. In this paper, the problem
of learning sequences of notes, which are melodic in nature, is considered. In this
paper, we focus on the pitch and the duration of the notes in the songs for generating
pleasing music. In order to achieve this goal, we use unsupervised generative models
known as Restricted Boltzmann Machines and Recurrent Neural Network Restricted
Boltzmann Machines. Restricted Boltzmann Machines (RBM) is a generative model
and a stochastic neural network, which is artificial in nature and is capable of learning
a distribution of probabilities over a given set of input data [1]. Restricted Boltzmann
Machines (RBM) is a deep learning model, which can be used for collaborative filtering, topic modeling, regression, dimensionality reduction, classification, and feature
learning and depending upon the given task Restricted Boltzmann Machines (RBM)
can be trained in an unsupervised or supervised way. The Recurrent Neural Network
Restricted Boltzmann Machines (RNN-RBM) is a series of Restricted Boltzmann
Machines (RBM) where the parameters are determined by a Recurrent Neural Network (RNN) [2]. The data for training the models in this paper will be in the Musical
Instrument Digital Interface (MIDI) file format. Data in Musical Instrument Digital Interface (MIDI) file format are used for communication between devices like
samplers, synthesizers, and computers that are used for controlling and generating
sound. The rest of the paper is organized as Sect. 2 describes the literature review,
Sect. 3 provide the proposed method, in Sect. 4 described the results followed by
conclusion and references.

2 Related Work

A music research project using deep learning, which is an open-source project by Google known as Magenta [3] aims at the use of machine learning for the generation of music, which is compelling. The organization made this project open source in June month of the year 2016 and the project currently implements a regular recurrent neural network and two long short-term memory networks. Ji-Sung-Kim from Princeton University developed a project called DeepJazz [4]. Theano and Keras are used in this project for the generation of Jazz music. In this project, two-layered long short-term memory networks are built and the learning is done using a MIDI file, which is given for training. BachBot [4] is a research project, which was introduced in Cambridge University by Feynman Liang. This project uses a long short-term memory network, which is trained on Johann Sebastian Bach's chorales. This research project aims at generating and harmonizing the chorales in a manner that they cannot be distinguished from Bach's work. The project generates wandering chorales if at least one voice among the voices is not fixed. This project can be used in order to add chorales to a melody, which is already generated. Vincent Degroote et al. [1] are working toward a project called Flow Machines, which can be used in order to keep artists in their creative flow. This project can be used for generation of lead sheets on the basis of the composer's style using a database of about 13,000 lead sheets. A neural network technique, which is known as Markov constraints, is used in this project. Google researchers at DeepMind have built a project called Wavenet [2]. In this project, Convolutional Neural Networks are used which are predominantly used for image generation and image classification. One of the most promising purposes of this project is enhancing the text to speech applications by the generation of a refined flow in vocal sounds produced. This project can be used for music generation as the output and the input for the model in order to generate music is raw audio. Aran Nayebi et al. [5] at Stanford introduced a research project which, like Wavenet [2], uses audio waveforms as the form of input to the model, where the model uses long short-term memory networks and GRU's rather than convolution neural networks. The demonstrations provided by the researchers of this project seem over-fitted on a particular song due to the small size of the training corpus and a large number of layers of the neural networks.

3 Methodology

The objective of the proposed models for music generation is to learn a structure from a large number of different song examples and generate pleasing music that contains melodies. The data used for learning is in the Musical Instrument Digital Interface (MIDI) file format. This format is an easy way to have access to a lot of free music data. The other advantage of the Musical Instrument Digital Interface (MIDI) file format is that it provides ease of retrieving information from an audio sample. A Musical Instrument Digital Interface (MIDI) file is represented as an arranged set of objects. The pattern object remains at the top which is the header chunk and consists of data pertaining to the overall file, which is followed by one or more chunks of tracks, where a track is a list of Musical Instrument Digital Interface (MIDI) events. Every song consists of melodies that are comprised of various notes, which are accordingly represented in a Musical Instrument Digital Interface (MIDI) file. In a Musical Instrument Digital Interface (MIDI) track, a Musical Instrument Digital Interface (MIDI) tick describes the lowest level of resolution in the track and tempo is equivalent to quarter notes per minute (QPM) which is the same as beats per minute (BPM). Pulse per quarter note is the same as ticks per quarter, which is equal to the number of ticks that make up a quarter. A saved Musical Instrument Digital Interface (MIDI) file first encodes a tempo and initial resolution and these values are then used for initializing the sequencer timer where sequencer is a program that combines all the MIDI events to form a song. The resolution in the track is to be considered static to the track and the sequencer. During the Musical Instrument Digital Interface (MIDI) playback, the file may have timed tempo change, events which are used for modulating the tempo depending on the specified time encoded in it while the resolution remains the same as the initial value in the duration of the playback. The tempo in a Musical Instrument Digital Interface (MIDI) file is represented in microseconds per beat which means one tick is a representation of half a millisecond or 0.0005 s and on increasing the resolution there is a decrease in this number and vice versa. In order to generate music from the proposed models, information pertaining to the notes played in the sample song is required. It is required to know which note is played, when the note is pressed and when it is released in order to describe a sample song. Therefore, the songs are encoded into a binary matrix using the information related to the arrangement of notes in that song. The initial n columns represent the note on events whereas the later n columns represent the note off event. The song is divided into a number of time steps, where it represents the number of columns of the matrix (Fig. 1).

Generative models are models, where a probability distribution is specified over a dataset made up of input vectors. These models are usually probabilistic in nature and specify joint probabilistic distribution over target values and observations. Bayes rule can be used to form a conditional distribution by a generative model. In this paper, generative models are used for the creation of synthetic data through directly sampling from modeled probability distributions [6–10].

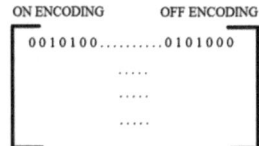

Fig. 1 Data in the form of matrix

Fig. 2 Multi-layer
perceptron model

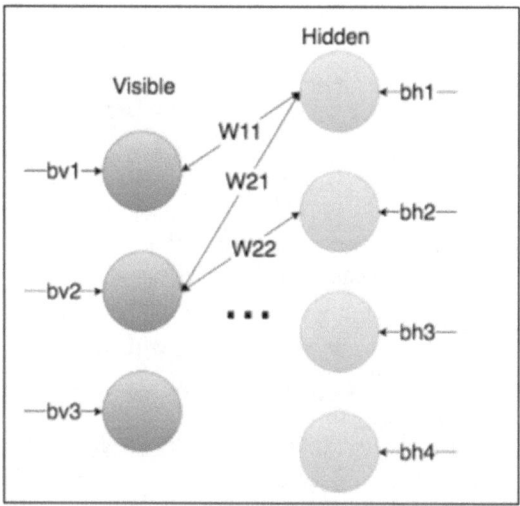

In this paper, two generative models called Restricted Boltzmann Machines (RBM) and Recurrent Neural Network Restricted Boltzmann Machines (RNN-RBM) are used in order to generate music using collected data. Restricted Boltzmann Machines (RBM) is a 2-layered neural network, which consists of a visible layer and a hidden layer where each node in the visible layer is connected to each node in the hidden layer, and vice versa, however, there are no two nodes in the same layer are connected to each other. In Restricted Boltzmann Machines (RBM), the parameters are the bias vectors bv and the matrix W which represents the weights and is also known as the weight matrix (Fig. 2).

In the above diagram, each circle is a representation of a neuron-like unit, which is called a node, and nodes represent where required calculations take place.

In Restricted Boltzmann Machines (RBM), the restriction refers to the fact that there is no intra-layer communication. Nodes are where the computation occurs, inputs are processed, and stochastic decisions (decisions which are randomly determined) are made as to whether the input is to be transmitted or not.

Restricted Boltzmann Machines (RBM) are defined as symmetrical bipartite graphs as all the inputs in a Restricted Boltzmann Machine (RBM) are passed to all the hidden layers of a Restricted Boltzmann Machine (RBM). Boltzmann Machines are a unique kind of log-linear Markov Random Field (MRF) in which energy function is linear with respect to its free parameters. In order to make Boltzmann Machines enough powerful for it to represent complex distributions, it is required to consider that few variables are never observed, these variables are also known as hidden variables. A depiction of a Restricted Boltzmann Machine (RBM) is given below.

For Restricted Boltzmann Machines, S contains a set of hidden units and visible units and as they are conditionally independent, block Gibbs sampling can be performed. In this, given the hidden units' fixed value, visible units are sampled concurrently. In a similar manner, given the fixed values of visible units, hidden units are sampled simultaneously. This can be shown as follows:

$$h^{(n+1)} = sigm(W'v^{(n)} + c) \tag{1}$$

$$v^{(n+1)} = sigm(Wh^{(n+1)} + b) \tag{2}$$

Here, $h^{(n)}$ presents the set of all hidden units which are at the nth step of the Markov chain which means that $h_i^{(n+1)}$ is chosen randomly either to be one (against zero) with the probability of $Sigma(W_i'v^{(n)} + c)$ and in a similar manner $v_j^{(n+1)}$ is chosen randomly to be one (against zero) with the probability of $Sigma(W_j'h^{(n+1)} + b)$. As $t \to \infty$, $(V^{(t+1)}, h^{(t)})$ become accurate samples of $p(v, h)$.

Theoretically, every single parameter update in the process of learning would require a similar kind of chain of convergence to be run. Running such a chain of convergence, needless to say, would be prohibitively expensive. Several such algorithms have been developed in order to sample efficiently from $p(v, h)$ in the duration of the learning process.

The weight matrix shared by all the Restricted Boltzmann Machines (RBMs) across the given model is the same and the visible bias vectors and the hidden bias vectors are determined using the outputs of u_t. The role of this weight matrix is to specify a consistent priority across all of the Restricted Boltzmann Machine distributions, and the required temporal information regarding a song is communicated using the bias vectors. In order to train the Recurrent Neural Network Restricted Boltzmann Machine (RNN-RBM) we use the following steps:

1. During the first step, the Recurrent Neural Network's knowledge about the current state of the song is transmitted to the Restricted Boltzmann Machine. The weight matrix, state of Recurrent Neural Network hidden unit $u_{(t-1)}$, and bias values are used in order to determine b_u^t and b_v^t for RBM_t

$$b_h^t = b_h + u_{t-1} + W_{uh} \tag{3}$$

$$b_v^t = b_v + u_{t-1}W_{uv} \tag{4}$$

2. In this step, a few notes of music are created using the Restricted Boltzmann Machine. RBM_t is initialized with v_t and a single iteration of Gibbs sampling is performed in order to sample v_t^* from RBM_t. The following equations define this procedure:

$$h_j = \sigma(W^t v_j + b_h^t)_i \tag{5}$$

$$v_i^* = \sigma(W_h v_t + b_v^t)_i \tag{6}$$

3. In the third step, the musical notes that are generated by the model are compared with the actual notes in the song at time t. In order to do this, the contrastive divergence estimation for negative log likelihood of v_t is computed with respect to RBM_t. Now, this loss is backpropagated along the network in order to compute the gradients of the network parameters and to update the network biases and network weights. The loss function is computed by taking the difference between free energies of v_t^* and v_t.

$$loss = F(vt) - F(v_t^*) \cong -\log(p(v_t)) \tag{7}$$

4. This step deals with using new information in order to update the Recurrent Neural Network's internal representation of the song's state. The state of Recurrent Neural Network's hidden unit u_{t-1} and the bias and weight matrices of the Recurrent Neural Network is used for determining the state of Recurrent Neural Network's hidden unit u_t. The following equation demonstrates this step:

$$u_t = \sigma(b_u + v_t W_{uv} + u_{t-1} W_{uu}) \tag{8}$$

For generating a sequence using the trained network, an empty array is initialized with an empty array and the same steps as the ones that have been mentioned above are repeated with a few simple changes, we replace step 2 and step 3 with:

4 Experimental Results

The following graph represents the value of cost function with respect to the number of epochs for Recurrent Neural Network Restricted Boltzmann Machine (RNN-RBM) where the cost is represented along the vertical axis and the number of epochs is represented along the horizontal axis (Fig. 3):

Fig. 3 Cost function versus no of epochs

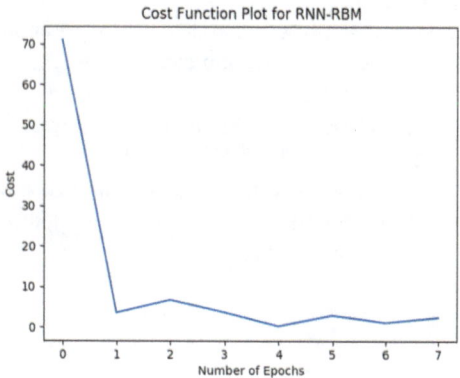

Fig. 4 Cost function versus
no of epochs in RBM

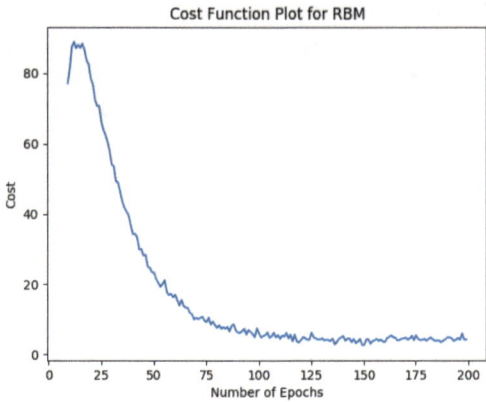

The following graph represents the value of cost function with respect to the number of epochs for Restricted Boltzmann Machine (RBM) where the cost is represented along the vertical axis and the number of epochs is represented along the horizontal axis (Fig. 4):

5   Conclusion

In this paper, we demonstrate the fact that algorithmic music generation using deep learning models is possible with audio waveforms as input by using Restricted Boltzmann Machines (RBM) and Recurrent Neural Network Restricted Boltzmann Machine (RNN-RBM). Investigation of the effect of adding layers to the Recurrent Neural Network in Recurrent Neural Network Restricted Boltzmann Machine (RNN RBM) in order to discover the impact of additional layers on the performance of the model and the melodic structure of the generated samples of music can be considered an interesting future direction. Deep learning for generating music allows an individual to sample music according to his/her preference or taste in art. Deep learning models in the future may become an integral part in the art of music production due to the extensively available musical tracks in various genres that have been accumulated over years and were developed by various artists, who portrayed unique features in their work of art. As the music generated from the above deep learning models depend upon the kind of music that is fed to the models, choosing the kind of music to be modeled and the understanding of how the changes in deep learning models will influence the music that is generated, and will become an important aspect to be considered while producing samples of music using deep learning models.

References

1. Chen H, Murray AF (2003) Continuous restricted Boltzmann machine with an implementable training algorithm. IEEE Proc Vis Image Signal Process 150(3):153–158
2. Medsker LR, Jain LC (2001) Recurrent neural networks. Des Appl 5
3. Engel J, Resnick C, Roberts A, Dieleman S, Eck D, Simonyan K, Norouzi M (2017) Neural audio synthesis of musical notes with wavenet autoencoders. arXiv preprint arXiv:1704.01279
4. Sutskever I, Hinton GE, Taylor GW (2009) The recurrent temporal restricted boltzmann machine. In: Advances in neural information processing systems, pp 1601–1608
5. Abadi M, Barham P, Chen J, Chen Z, Davis A, Dean J, Kudlur M (2016) TensorFlow: a system for large-scale machine learning. In: OSDI, vol 16, pp 265–283
6. Liang F (2016) BachBot: Automatic composition in the style of Bach chorales. Doctoral dissertation, Masters thesis, University of Cambridge)
7. Liang F, Gotham M, Johnson M, Shotton J (2017) Automatic stylistic composition of bach chorales with deep LSTM
8. Van Den Oord A, Dieleman S, Zen H, Simonyan K, Vinyals O, Graves A, Kavukcuoglu K (2016) Wavenet: a generative model for raw audio. arXiv preprint arXiv:1609.03499
9. Bakshi BR, Stephanopoulos G (1993) Wave-net: a multiresolution, hierarchical neural network with localized learning. AIChE J 39(1):57–81
10. Nayebi A, Vitelli M (2015) Gruv: algorithmic music generation using recurrent neural networks. Course CS224D: Deep Learn Nat Lang Process (Stanford)

Development and Investigation of Data Stream Classifier Using Lagrangian Interpolation Method

S. Jayanthi and J. Mercy Geraldine

Abstract Analyzing data streams floated over digitized organizations have become a notable challenge as it is often being obstructed by the incidence of concept drift and concept evolution. Conventional classification algorithms are neither accurate nor reliable on classifying dynamic data streams. For instance, Support Vector Machine (SVM) is a prominent classifier for performing supervised classification on static data; however, it becomes flabby when it is directly applied for data stream classification due to its intrinsic nature. In a bid to avoid this issue, this research work expounds a novel SVM-based Parallel Genetic Ensemble Model, which makes a series of optimization on SVM by synergizing it with Lagrangian interpolation method, fuzzy logic, and parallel genetic algorithm. The proposed model effectively classifies real-time data streams amid of concept drift and concept evolution. The exhaustive experiments probed on real-time data streams assert the efficacy of the proposed system in terms of various metrics.

Keywords Data stream classification · Support vector machine · Fuzzy logic · Parallel genetic algorithms · Parallel genetic ensemble model

1 Introduction

Data stream classification has become an inevitable task for digitized organizations as they are massively invaded by data streams. Despite the issues such as concept drift and concept evolution popped up on data stream classification have imminently fostered the development of an acute solution, as it hinders the efficiency of data stream classifier. Concept drift is a phenomenon which occurs when there is an unforeseen

S. Jayanthi (✉)
Samskruti College of Engineering and Technology, Hyderabad, Telangana, India
e-mail: dean@samskruti.ac.in; nigilakash@gmail.com

J. Mercy Geraldine
Guru Nanak Institute of Technology, Hyderabad, Telangana, India
e-mail: mercygeraldine@gmail.com

© Springer Nature Singapore Pte Ltd. 2019
H. S. Saini et al. (eds.), *Innovations in Computer Science and Engineering*, Lecture Notes in Networks and Systems 74,
https://doi.org/10.1007/978-981-13-7082-3_38

337

change in the features or context of data streams, whereas concept evolution occurs at the emergence of novel features or context in data streams [1].

Many of the existing data stream classification methods explored in the literature are found futile on classifying dynamic data streams, as they do not regard concept drift and concept evolution issues at the time of its inception [2, 3]. They also assumed the classes and instances of data streams as fixed. Some of the existing methods which address one of the issues either concept drift or concept evolution on dynamic data streams are not efficient on both [4, 5].

In light of addressing the scantiness of the existing methods, the proposed system is developed to classify and predict the data streams evolving over a real-time video server dedicated to the proposed research work. Here, the server performs twofold analysis over the attributes of data streams to cater to the dynamic interest of viewers. At first, the server analyzes the attributes of a video such as video type, file size, duration, video resolution, video bit rate, repeat view status, number of audio channels, audio bit rate and category of the video viewed by the users. Second, the server analyzes the attributes of browsing history of users such as protocol, user name, IP address, source and destination data packets, geography class, log-in status, number of failed connection attempts, and service counts.

Approaches proposed on data stream classification in the literature can broadly be categorized into three, namely ensemble classification, incremental classification, and block-based classification [6–8]. Among these, the approaches categorized under ensemble classification are the most prominent for its correctness as they carry out classification by consolidating the results of more than one classifier. The proposed research is also focused on developing a novel ensemble-based classifier, which better suits for data stream classification.

Accordingly, the SVM-based Parallel Genetic Ensemble classifier Model is formulated for better classification of data streams. However, SVM is widely preferred for its accuracy, its latency makes it unfit for data stream classification. In a bid to resolve this deformity of SVM, the proposed research work investigates it and how to perform a series of optimizations on SVM by synergizing it with fuzzy logic, parallel genetic algorithm and Lagrangian interpolation method so as to comply with data stream classification.

2 Proposed System

Support Vector Machine performs classification by generating support vectors and finding a hyperplane separating one class from another. It transforms input space into a high-dimensional feature space. However, SVM works well on linear data streams, it slows down in classifying nonlinear data streams. In general, tuning a classifier through optimization is one of the tactics in research, the performance of SVM is tuned by synergizing it with fuzzy logic, so as to have a good fit for data stream classification. The tuning process includes generating fewer support vectors with fuzzy logic, Lagrange interpolation method, and parallel genetic algorithm.

In the proposed research work, the performance of SVM is tuned up by synergizing it with fuzzy logic and Lagrangian interpolation method so as to generate fewer support vectors and to have a good fit for data stream classification. The cost function of SVM, which supports in optimizing the hyperplane for the linear case is given as follows:

$$\min C \sum_{0}^{m} [y^i \cos t_1(\theta^T x^i) + (1 - y^i)\cos t_0(\theta^T x^i) + 1/2 \sum_{j=1}^{n} \theta^2 j. \tag{1}$$

Here, "C", "θ", "n", and "m" specify the cost parameter, weight parameter, the number of attributes, and number of training instances, respectively, in SVM-based classification task. x(i) and y(i) specify "i"th attribute and "i"th example of the classification label, respectively.

Kernel function plays a notable role in the SVM-based classification process. In general, the Radial Basis Function is chosen as the Kernel function to perform classification on nonlinear dynamic data. However, the features of the SVM with the RBF kernel function achieve high accuracy in nonlinear data, it entails a time delay and complexity in finding the hyperplanes that separate the classes of data streams. To cope up with this concern, it is planned to convert the data streams into a high-dimensional feature space via a fuzzy system, and then SVM is applied to map data into a higher feature space and thereby construct the hyperplanes between the different categories of data streams.

2.1 Lagrangian Interpolation Method

Lagrangian interpolation method is used to compute the value of any function f(x) at different discrete points in time. The Lagrangian interpolation is a desirable approach in finding approximation function from the unknown function at some intermediate points. The SVM is intended to find a hyperplane based on the data points on support vectors.

- By applying the Lagrangian optimization formula, parameters for a hyperplane can be found which will be split two different classes. There can also be a margin parameter, which will maximize the space between two different classes.
- Hence, the Lagrangian interpolation method has been chosen as the component in the proposed classifier to optimize the parameters of the SVM prediction model and in fixing the initial points.

2.2 PGHDSC—Phenomenon on Data Stream Classification

Eighteen attributes of a different category of videos are given as the input to PGHDSC. SVM finds the optimal class boundary between the different categories efficiently as SVM is optimized with fuzzified genetic algorithm and lagrangians interpolation algorithm. The architecture model of **Parallel Genetic Hybrid Data Stream Classifi**cation is shown in Fig. 1.

The proposed system has been deployed in a real-time online data stream, where the possibility for the incidence of concept drift and concept evolution is high. The proposed system analyzes the attributes in two dimensions.

First in the view of the users, browsing history such as Protocol, User Name, IP Address, Source Data Packets to Initialize Connection, Destination Data Packets, Geography Class, Log-in Status, Number of Failed connection attempts, and Service counts.

The second view is based on the attributes of the videos such as Video Type, File Size in Bytes, Duration (HH:MM:SS:mmm format), Video Resolution Width x Height, Video bit rate, Repeat view status, Number of Audio Channels, Audio bit rate, and Category. The Attributes of Data Stream Classification taken for the experiment if slated in Table 1.

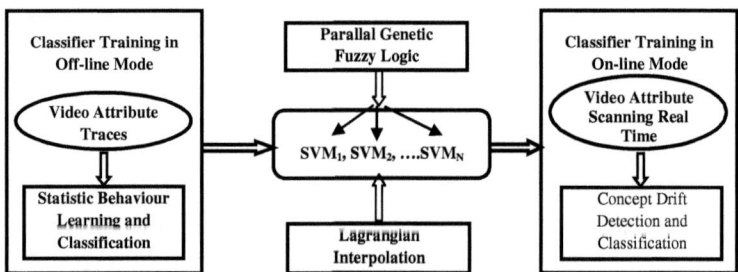

Fig. 1 Parallel Genetic Hybrid Data Stream classification

Table 1 Attributes Data Stream and its priority

S. no	Parameters	Weight	Value = (preset value * 2 ^ weight)
1	Category	11	1 * 2 ^ 11: 2048
2	Video resolution	10	12 * 2 ^ 10: 12,288
3	Video type	9	4 * 2 ^ 9 = 2048
4	Duration	8	2115 * 2 ^ 8 = 541,440
5	Video bit rate	7	450 * ^ 7 = 57,600
6	Number of audio channels	6	6 * 2 ^ 6 = 384
7	Audio bit rate	5	48 * 2 ^ 5 = 1536
8	User name	4	u * 2 ^ 4 = u * 16

(continued)

Table 1 (continued)

S. no	Parameters	Weight	Value = (preset value * 2 ^ weight)
9	Protocol	3	$1 * (2 \wedge 3) = 8$
10	IP address	2	$z * 2 \wedge 2 = z * 4$
11	Geography class	1	$1 * 2 \wedge 1 = 2$

3 Processing of the Sample Record

Each sample record is processed by assessing its value on 18 attributes such as, SFTP, User18949, 144.94.37.237, 4709, 4410, land, Not Logged in, 5, 119, MP4, 1973548800035, 0:35:15:683, 1920 × 1080, 450, Not Repeated, 6, 48, Do It Yourself is given below.

However, all these 18 attributes are analyzed, the contribution of 11 attributes listed in Table 1 is high over other attributes to cater to the interest of users. Each essential parameter is assigned a weight factor in the power of 2 and is listed down with proper ranking. The weight factor in the power of 2 of each essential parameter is multiplied with the preset value. Classification matrix is formulated by combining the resultant value of each parameter. Preset value for each attribute is assigned based on previous track records. Source and Destination Packets of the users are not mandatory, so discarded. Log-in status, Number of failed connections attempts, service counts are also discarded since they do not contribute to predicting the interest of users on videos.

Classification Matrix:

$$\begin{Bmatrix} 8 & U*16 & Z*4 & 2 & 2048 & 541{,}440 \\ 12{,}288 & 57{,}600 & 0 & 384 & 1536 & 2048 \end{Bmatrix}$$

The classification probability p(x) represented by is further sent to further classification rules.

$$p(x) = \sum_{i=0}^{k} \pi_i N\left(\frac{x}{\mu_i}, \sum i\right) \tag{2}$$

where $N(\frac{x}{\mu_i}, \Sigma i)$ is the distribution factor with mean and covariance (Σ).

The average performance of PGHDSC in terms of Accuracy, Precision, Recall, F1-Measure, Processing time, Memory, Average Response Time is measured, on classifying 10 different varying sizes of data chunks, in Table 2.

Table 2 Average performance report of PGHDSC

Chunk #	Accuracy (%)	Precision (%)	Recall (%)	F1-measure %	Processing time (mS)	Memory (B)	Avg. Resp. time (mS)
1	88.09	88.21	88.12	88.165	1259	6,328,477	706
2	88.18	88.41	88.33	88.37	1428	6,494,026	703
3	88.51	88.37	88.39	88.38	1583	6,488,056	703
4	88.44	88.72	88.53	88.6249	1725	6,490,252	726
5	88.47	88.61	88.48	88.545	1880	6,499,419	734
6	88.61	88.74	88.63	88.685	2017	6,496,089	758
7	88.81	88.79	88.72	88.755	2184	6,492,355	777
8	88.86	88.82	88.89	88.855	2335	6,491,728	786
9	88.82	89.01	88.98	88.995	2474	6,495,016	785
10	88.88	89.06	89.03	89.045	2625	6,500,410	836

Performance evaluation of the proposed system comprises twofold analysis: First, to assess the efficacy of applying PGHDSC in the varying size of data chunks of data streams; Second, to assert and compare the efficacy of the proposed system with other systems. Processing time and average response time are measured in microseconds.

4 Conclusion

Parallel Genetic Hybrid Data Stream Classifier proposed in this research work deals with analyzing the dynamic interest of users over various categories of videos by analyzing 18 attributes, which effectively discriminates the type of videos and also recognize the newly emerging videos in real time. The interest of users are categorized into 12 genres namely, Do It Yourself, Technology, Sport, Movie, Drama, Funny, 3D-Animation, Entertainment, 2D Animation, Short Film, Tutorials, and Vector. However, the proposed approach of this research work shall be adopted as the promising approach to perform data stream classification on any other data streams that incur concept drift and concept evolution. Since the proposed approach involves the right mix of and keenly interfaced optimization algorithms, the efficacy of the proposed system is high over other comparative algorithms. However, it also has faltered rarely in identifying the interest of users due to the complex features of some genres of videos. Future work shall be extended to experiment with additional features of videos that support better discrimination of more genres of videos.

References

1. Tahayna B et al (2010) Optimizing support vector machine based classification and retrieval of semantic video events with genetic algorithms. In: Proceedings of 2010 IEEE 17th international conference, pp 1485–1488
2. Charu Aggarwal C, Wang J (2006) A framework for on-demand classification of evolving data streams. IEEE Trans Knowl Data Eng 18(5):577–589
3. Bifet A, Holmes G, Pfahringer B, Gavalda R (2009) Improving adaptive bagging methods for evolving data streams. In: ACM, Proceeding ACML, pp 23–37
4. Tsai CJ, Lee CI, Yang WP (2009) Mining decision rules on data streams in the presence of concept drifts. Expert Syst Appl 36:1164–1178
5. Liang C, Zhang Y, Shi P, Hu Z (2012) Learning very fast decision tree from uncertain data streams with positive and unlabeled samples. ACM J Inf Sci Int J 213:50–67
6. Wang H, Fan W, Yu PS, Han J (2003) Mining concept-drifting data streams using ensemble classifiers. In: Proceedings of the 9th ACM SIGKDD, pp 226–235
7. Brzezinski D, Stefanowski J (2014) Combining block-based and online methods in learning ensembles from concept drifting data streams. Inf Sci 265:50–67
8. Tsai C-J, Yang W-P (2008) An efficient and sensitive decision tree approach to mining concept-drifting data streams. J Inf 19(1):135–156

Quron: Basic Representation and Functionality

B. Venkat Raman, Nagaratna P. Hedge, Dudimetla Mallesh and Bairi Anjaneyulu

Abstract The paper presents an approach to represent a quantum bit as a neuron, which uses a threshold function. The functionality of the perceptron model of a neuron is modeled using qubit, which can be used as a building block for the quantum neural network. Several approaches for building and training a Quantum Neural Network have been proposed, namely step function as measurement, quantum dots, quantum associative memory, and perceptron models. But, the functional model of the elementary quantum neuron is not described in much detail. We attempt to build such quantum neuron using qubits and perform its function using qubit rotation operation.

Keywords Quantum Neural Networks · Qubit · Neuron · Superposition · Quantum rotation

1 Introduction

Quantum Neural Networks is a field, which attempts to combine the concepts of quantum mechanics, super fast computing, cognitive science, neurocomputing, and machine learning. It is basically the realization of an Artificial Neural Network using quantum gates and circuits that would mimic the working of a natural brain.

B. Venkat Raman (✉)
Osmania University, Hyderabad, India
e-mail: venkat521@yahoo.co.in

N. P. Hedge
Vasavi College of Engineering, Hyderabad, India
e-mail: nagaratnaph@gmail.com

B. Venkat Raman · D. Mallesh · B. Anjaneyulu
Department of CSE, RGUKT, Basar, Hyderabad, India
e-mail: mallesh.dmy.98@gmail.com

B. Anjaneyulu
e-mail: anji.bairi98@gmail.com

© Springer Nature Singapore Pte Ltd. 2019
H. S. Saini et al. (eds.), *Innovations in Computer Science
and Engineering*, Lecture Notes in Networks and Systems 74,
https://doi.org/10.1007/978-981-13-7082-3_39

345

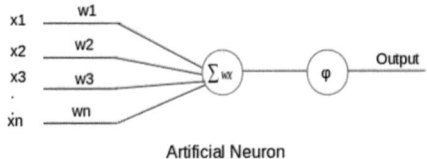

Fig. 1 Model of an artificial neuron

Artificial Neural Networks (ANN) are computational models, which follow biological (real) neurons structure and functions. In ANN, neurons take its input from all
other artificial neurons, perform some simple operations, and pass its output onto all
other neurons. Quantum Neural Networks (QNN) are systems that are a mixture of
the properties of neural networks with the features of the quantum theory. Basically,
neural networks are interlinked units having the features of biological neurons supply
signals from one another.

Figure 1 represents the basic representation of an Artificial Neuron, where x_1, x_2,
and x_n are input values and w_1, w_2, and w_n are weights of respective inputs, $\sum wx$
represents summation function and φ represents Activation function. Summation
function produces sum of the products of the inputs with corresponding weights.
Activation function produces final output based on the summation function.

2   Literature Review

The following may be given as the requirements [1] to state that the given structure
and functionalities of a neural network to be termed as Quantum Neural Networks

(1) A Quron yield input value as any binary string having length 'n' and it will give
    static output as one of the states from 2^m possible output string combinations,
    which have length 'm' and it is near to the input value by any of the distance
    measurement methods.
(2) Quantum Neural Network should reflect basic neural computing mechanisms.
(3) The evolution of Quantum Neural Network must be based on quantum effects.

Few approaches use step function as a measurement. Kak [2] published first
quantum neural network approach, which is based on "Quantum neural computation"
by introducing the stipulation for static state $X^0 = (X_1^0, \dots X_N^0)$ a 'Hopfield-like'
network derived as Eq. (1) with weight matrix W and with entries W_{ij}

$$sgm(WX_0) = X_0 \tag{1}$$

Representation of a quantum system eigenvalue equation is $\hat{w}|x_0\rangle = \lambda|x_0\rangle$.
Where w is weight matrix with an operator and eigenvector $|x_0\rangle$ and eigenvalue
$\lambda = 1$.

Menneer and Narayanan introduced an approach called "Quantum inspired neural networks" [3]. This approach based on the "Many-Universe interpretation".

Some other approaches use Quantum dots. Elizabeth Behrman proposed this approach and this approach received higher attention among all [4]. He reformulated Green function by Feynman path integral

$$|\psi i\rangle T, \varphi 0 = ZiD\varphi(t) \exp\{\bar{h}Zm\, 2\dot{\varphi} - V(\varphi, t)dt\} |\psi i\, 0, \varphi 0 \qquad (2)$$

where V is potential and m stands for mass of the quantum system. The network realization can be done by the propagation of only one Quron at a time, instead of different Qurons.

Faber and Giraldi [4] discussed this approach and raised an incompatibility problem between quantum computing and neural computing. Faber and Giraldi raised a question "how the neural network's nonlinear dynamics can be simulated by a quantum system?".

Another distinct approach is called Quantum Associative Memory (QAM), in this model a quantum algorithm that imitates the behavior of 'Associative memory'. QAM circuit performs an operation in the following way. First, it takes input pattern and produces an output pattern, which is near to the input pattern by using the distance of Hamming method.

Another approach is quantum neural circuits. Using this approach, each quantum operations is computed by a unitary operator $\hat{U}$, and by dissipative operator $\hat{D}$ [5]. If it exceeds threshold δ, these dissipative gate maps the state amplitude onto $c \in C$ or 0.

Quantum perceptron is another approach in which Altaisky's [6] has first introduced the concept of a Quantum Perceptron model, this is modeled by using the function of quantum update

$$|y(t)\rangle = \hat{F} \sum_{i=1}^{m} \hat{w}_{iy}(t)|x_i\rangle \qquad (3)$$

Here, $\hat{F}$ stands for arbitrary quantum gate operator, m is input qubits, $\hat{w}_{iy}$ represents synaptic weights. Using this model, some authors show that quantum perceptron can implement quantum gates like Hadamard [7] and C-NOT [8] gates.

Siomau [9] introduced another quantum perceptron model, which represents its powerful advantage of computing non-separable problems. From all of the above approaches, we considered step function as measurement and proposed a method to build a Quron and to maintain the functionalities of Quron.

3   Proposed Method

A neuron may take several inputs with different significance due to the distribution of weights of corresponding inputs. The sum of the weighted input is taken and applied to the activation function.

$$\emptyset = W_1 X_1 + W_2 X_2 + \cdots + W_n X_n + bias$$

$$\Phi(\emptyset) = \begin{matrix} -1 \ if \ \emptyset <= 0 \\ +1 \ if \ \emptyset > 0 \end{matrix} \tag{4}$$

where Φ represents the activation function.

In our approach, we considered a qubit with the superposition of all inputs along with bias as the quantum neuron(Quron). Equation (5) represents such Qubit mathematically.

$$|X\rangle = |X_1 X_2 X_3 \ \ldots \ X_n\rangle \tag{5}$$

where Xi represents ith input.

The superposition Quron then would be undergoing rotation operation to make the complete weighted input. The rotation operations represent the summation function of weighted inputs.

Rotation operation may be defined as

$$Ry(\emptyset) = \begin{pmatrix} \cos \emptyset/2 & -\sin \emptyset/2 \\ \sin \emptyset/2 & \cos \emptyset/2 \end{pmatrix} \tag{6}$$

Rotation on negative angle is defined as

$$Ry(-\emptyset) = \begin{pmatrix} \cos \emptyset/2 & \sin \emptyset/2 \\ -\sin \emptyset/2 & \cos \emptyset/2 \end{pmatrix} \tag{7}$$

Equations (6) and (7) represent positive and negative rotations of a Quron, respectively, where $\emptyset$ is the angle of rotation. These rotations move toward qubit $|1\rangle$ state for positive rotation, while they move the qubit toward $|0\rangle$ state for negative rotation. Initially, we apply Hadamard gate [7] to the superposition of qubit, the Quron to neutralize the $|0\rangle$ and $|1\rangle$ state.

$$H|0\rangle = 1/\sqrt{2}|0\rangle + 1/\sqrt{2}|1\rangle \tag{8}$$

Equation (8) represents the operation of Hadamard gate [7], when it is applied to single Quron at $|0\rangle$ state, it produces a superposition of $|0\rangle$ and $|1\rangle$ both with an equal amplitude of $1/\sqrt{2}$.

Fig. 2 Neutralization—application of Hadamard gate to $|0\rangle$ state qubit

Fig. 3 a Quantum-negative rotation of a qubit. **b** Quantum-positive rotation of a qubit

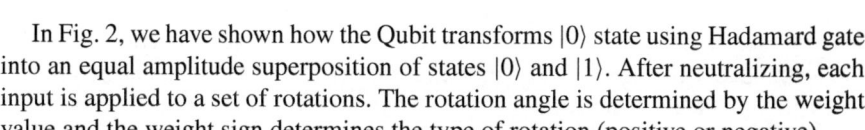

In Fig. 2, we have shown how the Qubit transforms $|0\rangle$ state using Hadamard gate into an equal amplitude superposition of states $|0\rangle$ and $|1\rangle$. After neutralizing, each input is applied to a set of rotations. The rotation angle is determined by the weight value and the weight sign determines the type of rotation (positive or negative).

Figure 3a represents the state of the Qubit after negative rotation, whereas Fig. 3b represents positive rotation. After rotations are applied the amplitude of the output qubit gives the result of the activation function. It results in $|1\rangle$ state, if the amplitude of $|1\rangle$ is greater the amplitude $|0\rangle$. This is assumed as '1' output of the activation function. Vice versa, if the amplitude of $|0\rangle$ is greater the output may be assumed as '0' for the activation function.

4 Results

We have implemented binary Quron, that will take random weights, binary input values and bias. If we take the 2-bit Quron as input and if we apply the Hadamard gate then the output value is neutralized between the two states $|0\rangle$ and $|1\rangle$ with the amplitude of $1/\sqrt{2}$. Using a binary Quron, with two inputs and a bias, we have realized two input digital logic gates. Randomly, weights are assigned to the Quron inputs with any integer. Using backpropogation algorithm, the weights are learned to realize the expected outputs of the gate. The weights of the inputs would act as the angle of rotation to the qubit neuron. A positive weight would make the qubit rotate positively with the weight magnitude, whereas a negative weight would make a negative rotation. The overall effective angle of rotation would be treated as summed input and will be given as input to the activation function. If the effective angle of rotation is neutral or toward positive side, then the output would be 1 else 0 as per threshold activation function.

5 Future Scope

We have implemented Quron for 1-bit representation, if the rotation of Quron is growing greater then the 180° then the adjacent Quron bit value is set to 1 and the next bit is reset to 0, this feature helps us to implement deep learning concepts.

6 Conclusion

Our proposed method main contribution is a simple implementation for the creation of a Quron. A superposition of a qubit along with its rotation operation mimics the functionality of a neuron in quantum perspective. We have implemented binary Quron to implement basic gates.

References

1. Schuld M, Sinayskiy I, Petruccione F (2014) The quest for a quantum neural network. arXiv preprint arXiv:1408.7005v1 [quant-ph]
2. Kak SC (1995) Advances in imaging and electron physics, vol 94, p 259
3. Menneer T, Narayanan A (1995) Department of computer science, University of Exeter, United Kingdom, Technical Report 329, 1995
4. Faber J, Giraldi GA (2002) Electronically available. http://arquivosweb.lncc.br/pdfs/QNN-Review.Pdf
5. Gupta S, Zia R (2001) J Comput Syst Sci 63(3):355
6. Altaisky M (2001). arXiv preprint quant-ph/0107012 (2001)
7. Fei L, Baoyu Z (2003) Neural networks and signal processing. In: 2003 Proceedings of the 2003 international conference on networks and signal processing, vol 1. IEEE, pp 539–542
8. Sagheer A, Zidan M (2013). arXiv preprint arXiv:1312.4149v1
9. Siomau M (2012). arXiv preprint arXiv:1210.6626

Memorization Approach to Quantum Associative Memory Inspired by the Natural Phenomenon of Brain

B. Venkat Raman, K. Chandra Shekar, Ranjith Gandhasiri and Sudarshan Gurram

Abstract An attempt to present the functionality of quantum associative memory, mainly memorization operation to mimic the natural phenomenon of the brain (more remembering of frequently observed patterns and gradual forgetting the non-recalled patterns) is done. Quantum neural network has features such as understanding, awareness, and consciousness. Quantum Neural Networks use content-addressable quantum associative memory for its "memorize" and "recall" operations. We attempt to mimic the memorization process to that of the natural phenomenon of the brain, wherein the memorization of patterns and their storage is based on the time and the frequency of its recall.

Keywords Quantum Neural Networks · Associative memory · Memorization · Recall · Patterns · Hopfield neural network · Hadamard gate

1 Introduction

We understand that the brain's memory is content addressable, since we recall any fact or memory by remembering a small part of it. Brain stores the more observed or recalled information more than the less frequently observed information. It is our

B. Venkat Raman (✉)
Osmania University, Hyderabad, India
e-mail: venkat521@yahoo.co.in

K. C. Shekar
JNTUH, Hyderabad, Telangana, India
e-mail: chandhra2k17@gmail.com

B. Venkat Raman · R. Gandhasiri · S. Gurram
Department of Computer Science & Engineering, RGUKT, Basar, Hyderabad, India
e-mail: ranji.gandhasiri123@gmail.com

S. Gurram
e-mail: aaryaas311@gmail.com

© Springer Nature Singapore Pte Ltd. 2019
H. S. Saini et al. (eds.), *Innovations in Computer Science
and Engineering*, Lecture Notes in Networks and Systems 74,
https://doi.org/10.1007/978-981-13-7082-3_40

natural tendency that as we if do not recall information or patterns for a long time, we forget it.

Many approaches for building a Quantum Neural Networks (QNNs) have been proposed, out of which Quantum Associative Memory is one. An associative memory is a content-addressable memory, which can search for patterns based on its content or portion of its content.

The main operations on a content-addressable memory are namely, "Memorization" and "Recall". Memorization is the process of storing a pattern in the memory and recall resembles the act of retrieving a particular pattern(s). In this paper, we concentrate on memorization operation to work as the biological brain does in abstract.

2 Literature Review

There are many QNN models, which are proposed by different authors. Some models are mentioned in [1].

Kak [2] proposed the QNN model by introducing the idea of "quantum neural computation". After Subhash K. Kak's proposal, Men-neer and Narayanan proposed Quantum-Inspired Neural Network [3].

Elizabeth Behrman and co-workers proposed a QNN model (using the concept of quantum dots), which uses only one quron to create the network [4].

Martinez and Ventura [5] proposed a model ("quantum associative memory") that is important in quantum associative memory models. Trugenberger [6, 7] uses a positive approach to render Ventura and Martinez' pattern completion algorithm into associative memory.

Among all the perceptron models, Altaisky's [8] introduction of a quantum perceptron is the first one. Recently, a different quantum perceptron model is proposed by Siomau [9].

Among the stated approaches, Quantum Associative Memory has a distinct way of representing a quantum neural network to mimic a natural brain. It is one of the most propitious approaches to quantum neurocomputing [10]. In general, computer memory consists of key and value pair. Values can be retrieved by using keys. But human brain works using content-addressable memory, if we recall some part of information about a pattern which is already stored in our brain we can get the whole pattern. Our brain searches for pattern which is closest to the information that we recall and retrieve that pattern. Normally, we forgot the patterns which we do not recall for a long period of time. Quantum associative memory follows the same scenario.

It stores the patterns in a quantum system by superposition of all patterns and retrieves the pattern using content-addressable memory. Classically, computer memory is located with address but in Quantum computer that can be happened with content and it is also known as content-addressable memory (quantum associative memory).

Quantum associative memory is divided into two parts which are memorization and recall. The memorization consists of storing patterns in the memory. Recall does the retrieval of patterns that are stored in the memory according to the partial or noisy input.

Memorization process may be understood as the process of learning patterns. The process begins by relating the number of qubits capable of representing a number of patterns at different states. The number of qubits N to represent 2^N patterns that can be represented with M. Let $X^P = \left\{ \left| x_1^{(1)}, \ldots, x_N^1 \right\rangle, \left| x_1^{(P)}, \ldots, X_N^{(P)} \right\rangle \right\}$ here p is no. of patterns. The memory superposition reads

$$|M\rangle = 1/\sqrt{P} \sum_{p=1}^{P} \left| x_1^{(P)}, \ldots, X_N^{(P)} \right\rangle. \tag{1}$$

The above formula is used to distribute the equal probabilities to each pattern to make one as total probability. Dan Ventura and Tony Martinez proposed an algorithm for memorization which uses basic operations namely "FLIP", "SAVE" and registers labeled as x, g, c [5].

Recall is another operation done by the quantum associative memory. It recollects or fetches the patterns that have been already learned. It is used to retrieve the patterns among all patterns, which are in superposition. One of the best and mostly used algorithms for recall is Grover's search algorithm, which was initially proposed by Martinez and Ventura [5]. It works on unsorted items N and it can complete this search in $O(\sqrt{N})$ time.

The main advantage of a Quantum Associative Memory compared to classical Hopfield associative memory is it is able to store 2^N patterns into an N-qubit system. Grover's algorithm [11] is the fastest searching algorithm in quantum computing.

3 Proposed Method

We propose the memorization algorithm assuming associative memory, which can store patterns of N qubits, initially a parallel application of the Hadamard gate on a system of "n" qubits, $H^n|0\rangle^N$ is applied [12].

$$H^n|0\rangle^N = 1/2^{N/2} \sum x_{x=0}^{2N-1} |x\rangle \tag{2}$$

This can create 2^n patterns that can be memorized for n-bit of memory. All the patterns are created with equal weights representing equal synaptic weights of memorization. By that, the total amplitude of all patterns sums up to 1.

We consider the amplitudes of the different states of the qubit as the synaptic weight, which resembles the probability of its remembering.

$$\sum_{i=0}^{2^{N-1}} \mathbf{W_i} = 1 \tag{3}$$

Whenever a pattern "p" is to be memorized, all existing patterns except "p" would have their weights deducted by "μ" units. This resembles that, when we learn new things, we gradually forget the old patterns which are not recollected.

$$\mathbf{W_i} - \mathbf{j} * \mu \tag{4}$$

where $i = 0 \ldots 2^{N-1}, j = 1 \ldots 2^{N-1}; i! = p$

"i" represents all patterns excluding the currently memorizing pattern and j represents the order of distinct weights of "i" in descending order. Sum the subtracted weights of the corresponding patterns

$$\mathbf{V} = \sum_{i=1}^{2^{N-1}} \mathbf{j} * \mu \tag{5}$$

where $j = 1 \ldots 2^{N} - 1; j! = k$; V is memorizing weight.

As a pattern needs to be memorized the corresponding weight is increased, p_k, where k is the pattern number. V is added to the current pattern (p_k) weight.

$$\mathbf{W_{pk}} + \mathbf{V} \tag{6}$$

Thus, the newly learned pattern gets its weight increased. This resembles that the knowledge we acquire freshly can be recollected early.

4 Results

We have simulated a 2-qubit storage QAS and memorized the following patterns in sequential order.

Initially, the Hadamard gate was applied to $|00\rangle$. This yields four patterns in the form of states $|00\rangle, |01\rangle, |10\rangle, |11\rangle$. All the states would be of equal amplitude.

$$H|00\rangle = 1/4|00\rangle + 1/4|01\rangle + 1/4|10\rangle + 1/4|11\rangle$$

We consider a sequence of patterns to be memorized are 10,11,00,10,01 and also considered μ value to be 1/10.

To demonstrate the algorithm numerically, let us take an instance of patterns with the following amplitudes. After memorizing pattern $|10\rangle$ state, the result will be

$$3/20 |00\rangle + 3/20 |01\rangle + 11/20 |10\rangle + 3/20 |11\rangle$$

Table 1 Demonstrates change of weights of all patterns in the entire memorization process

Entered pattern	Pattern weights			
	00 1/4	01 1/4	10 1/4	11 1/4
10	3/20	3/20	11/20	3/20
11	−1/20	−1/20	9/20	13/20
00	11/20	−7/20	5/20	11/20
10	9/20	−11/20	13/20	9/20
01	5/20	−1/20	11/20	5/20

Assume the pattern to be memorized is $|11\rangle$ state, then there are 2 distinct weights excluding $|11\rangle$ state which are 3/20 and 11/20. Therefore, j would increase from 1 to 2.

Weight of the pattern $|10\rangle = 11/20 - 1 * (1/10)$
Weight of the pattern $|00\rangle$ and $|01\rangle = 3/20 - 2 * (1/10)$

The subtracted weights would be added to memorization pattern $|11\rangle$ state.

$$3/20 + (1 * (1/10) + 2 * (1/10) + 2 * (1/10))$$

Resulting in the state (Table 1),

$$-1/20|00\rangle - 1/20|01\rangle + 9/20|10\rangle + 13/20|11\rangle$$

5 Conclusion

By using this approach, we can achieve or represent the functionality of the quantum associative memory, namely memorization to mimic the natural phenomenon of the brain.

Due to increased amplitude of the pattern observed frequently, the recall operation of such patterns becomes more efficient using Grover's quantum search algorithm.

References

1. Petruccione F, Schuld M, Sinayskiy I (2014) The quest for a quantum neural network
2. Kak SC (1995) Adv Imaging Electron Phys 94:259
3. Narayanan A, Menneer T (1995) Tech Rep 329:1995
4. Nash L, Behrman EC, Steck JE, Skinner SR, Chandrashekar V (2000) Inf Sci 128(3):257
5. Martinez T, Ventura D (2000) Inf Sci 124(1):273
6. Trugenberger CA (2001) Phys Rev Lett 87:067901
7. Trugenberger CA (2002) Quantum Inf Process 1(6):471–493

8. Altaisky M (2001). arXiv preprint quant-ph/0107012
9. Siomau M (2012). arXiv preprint arXiv:1210.6626
10. Ventura D (1998) Artificial associative memory using quantum processes. In: Proceedings of the international conference on computational intelligence and neuroscience, vol 2, pp 218–221
11. Grover LK (1996) A fast quantum mechanical algorithm for database search. In: Proceedings of the 28th annual ACM symposium on the theory of computation, pp 212–219
12. Castro LN (2006) Fundamentals of natural computing

Supervised Machine Learning Classifier for Email Spam Filtering

Deepika Mallampati, K. Chandra Shekar and K. Ravikanth

Abstract Email is the omnipresent and persistent consequence applied to a regular base by a huge number of individuals worldwide. However, as a result of social support systems and advertisers, the majority of the emails hold unnecessary information called spam. This concern not just affects typical users of the net, but additionally causes an enormous setback for companies and organizations as it costs a massive amount of money in mislaid productivity, wastage of user's time, and network bandwidth. In recent times, various parallel researchers have presented several email spam classification techniques, but it is extremely tough to eradicate the spam emails completely, while the spammers transform their techniques frequently. The proposed method is an efficient technique to classify the email spam messages using Support Vector Machines (SVM). Here, we present an SVM handling separation of nonlinear data using a Kernel function, which is an advanced machine learning technique in R to improve the accuracy of the model. Finally, we present a generic template for a working of kernel function in SVM that can be built in R.

Keywords Email spam · Support vector machines · Classification · Machine learning

D. Mallampati (✉)
Department of CSE, Sreyas IET, Hyderabad, India
e-mail: mokshhyd@gmail.com

K. Chandra Shekar
Department of CSE, JNTUH, Hyderabad, India
e-mail: chandhra2k7@gmail.com

K. Ravikanth
Department of CSE, RGUKT, Basar, India
e-mail: ravikanth27787@gmail.com

© Springer Nature Singapore Pte Ltd. 2019
H. S. Saini et al. (eds.), *Innovations in Computer Science and Engineering*, Lecture Notes in Networks and Systems 74,
https://doi.org/10.1007/978-981-13-7082-3_41

1 Introduction

The first email message recognized as "spam" was sent in 1978 to the users of ARPANET, and represented a little greater than a minor aggravation. Electronic mail is one of the universal ways of communication among Internet users. Spam can be defined as an unsolicited commercial email (UCE) or unsolicited bulk email (UBE) [1, 2]. The first mathematical apparatus applied to the filtering systems is the Bayes algorithm, which was used first by Sahami et al. in 1996, and then by the other researchers [3].

Major contributions of our proposed system are to find an efficient spam filter that classifies the mail into spam and ham. The rest of the paper has been organized as follows. Section 2 of the paper concentrates on a review of related work. The proposed email spam filtering technique using SVM is described in Sect. 3. The predicted results in R are discussed in Sect. 4 of the paper and finally the conclusion of the paper is presented in Sect. 5.

2 Related Work

A memory-based (or 'instance-based') machine learning method is proposed in [4], which uses the k-Nearest Neighbor algorithm. Paper [5] proposes an artificial neural network (ANN), a synthetic version of the biological mind. In paper [6], Support vector machines (SVMs) map training times right into a better dimensional feature area by means of a few nonlinear feature, and then calculated the standard hyperplane, which maximizes the margin between the information points inside the positive class and the records points at the negative class. The hyperplane can then be used to classify new instances (i.e., it's miles the selection boundary) [7–9]. In practice, it includes zero computations within the function area; kernel functions permit the development of the hyperplane without mapping the training statistics into the better dimensional function area [5, 6, 10–12].

3 Proposed Approach

This section presents the proposed approach and introduction of kernel function of SVM in R. The purpose is to build a spam detection. If the support vector machine analyzes a single mail then it returns a 0 else it returns a 1. We considered the dataset from the UC Irvine Machine Learning Repository for spam emails.

3.1 Support Vector Machine (SVM)

Support vectors are the observations that supports hyperplane on either side. As a point in an n-dimensional space (where n is the number of features) with the value of feature being the value of a particular coordinate. We perform the classification by finalizing the right hyperplane that differs the two classes very well. It is just a line in 2D/3D. SVM helps to discover a hyperplane (or separating boundary), which supports two classes of outputs. SVM handles separation of nonlinear data by using a kernel function. The kernel function transforms the data into a higher dimensional feature space to make it possible to perform the linear separation.

3.2 Building SVM in R

Step 1: Installing Packages

There are certain functions that enable R to support SVM. They are library (caret), library (kernlab), library (doMC), library (caTools). The kernlab is the acronym for kernel-based machine learning, which implements classification and regression (Fig. 1).

Step 2: Reading the Data file and Data Names from datasets

The datasets downloaded from Machine repository has to renamed and saved it in CSV files. The following is the command to read CSV files in R (Fig. 2).

Data_Email<-read.csv("spambase.csv", stringsAsFactors = FALSE, header = FALSE)
Data_Name<-read.csv("names.csv", stringsAsFactors = FALSE, header = FALSE)

Fig. 1 Installing packages in R for SVM

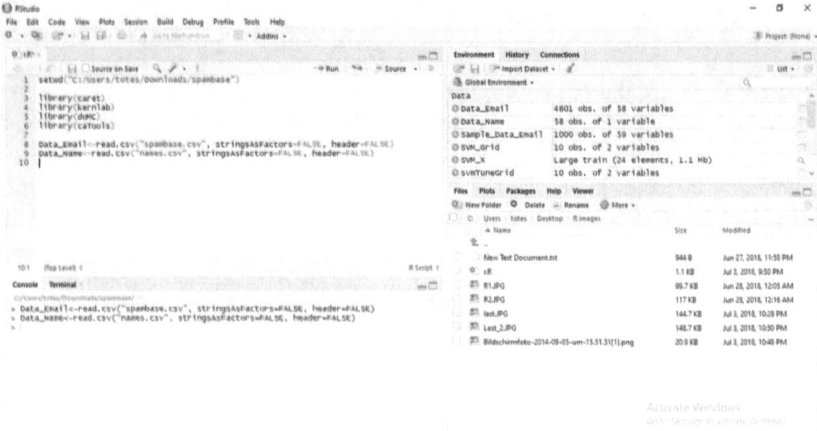

Fig. 2 Reading CSV files from datasets

Step 3: Transforming the Dependent (Y) from numeric to factor variable, to make SVM predict a classification output

Transforming the Y column which is a numeric is used to factor values, making the SVM provide a classification output. The command used is *Data_Email$Y<-as.factor(Data_Email$y)* (Fig. 3).

Step 4: Split the data into a training set and test set

An important step is to train the SVM using the SVM train() function of caret package. We can define the data which we want to use and the method to create the model (Fig. 4).

Fig. 3 Transforming column values to factors

Fig. 4 Training the SVM model

Fig. 5 Evaluating and predicting the resullts

Step 5: Evaluation

The model is created and used to classify emails as spam or bacon for performing a binary classification. The following is the command used to predict the output. To evaluate the model, we use datasets and predict the output of the caret package (Fig. 5).

pred<-predict(SVM-X, Test_Data[])

4 Result Evaluation

The performance of the proposed SVM algorithm is evaluated by implementing a filtering technique using kernel function in R. The figure below shows the confusion matrix and various statistics of the dataset using the proposed approach (Fig. 6).

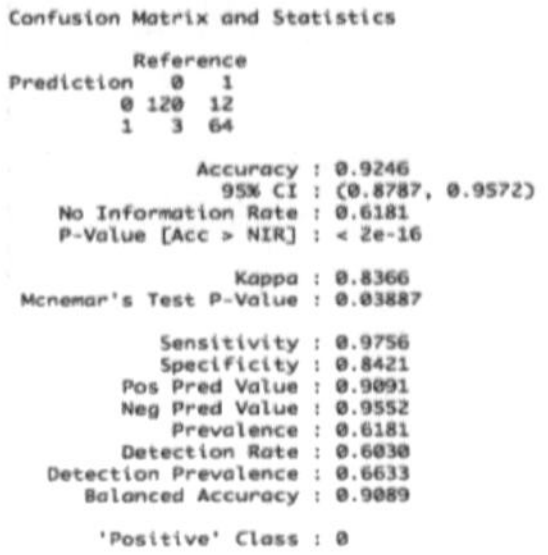

Fig. 6 Predicted results of
the proposed approach

```
Confusion Matrix and Statistics

            Reference
Prediction   0    1
         0  120   12
         1   3   64

         Accuracy : 0.9246
           95% CI : (0.8787, 0.9572)
No Information Rate : 0.6181
P-Value [Acc > NIR] : < 2e-16

            Kappa : 0.8366
Mcnemar's Test P-Value : 0.03887

      Sensitivity : 0.9756
      Specificity : 0.8421
   Pos Pred Value : 0.9091
   Neg Pred Value : 0.9552
       Prevalence : 0.6181
   Detection Rate : 0.6030
Detection Prevalence : 0.6633
 Balanced Accuracy : 0.9089

    'Positive' Class : 0
```

$$\text{Accuracy} = \frac{\text{Number of predictions made}}{\text{Total number of predictions}}$$

5 Conclusion

A supervised machine learning classifier for email spam filtering is proposed, which is based on SVM classifier where a kernel function is used for handling nonlinear data. It is implemented using R to improve the performance of the classification and prediction. SVM is an effective technique that is frequently used by data scientists for beating the accuracy benchmark of even the best of individual algorithms. SVM is a kind of advanced supervised machine learning method that looks at data and sorts it into one of the two categories. Our future direction is to plan for the ensemble classifiers using the concepts of bagging and boosting so that the accuracy of the model can be further improved.

References

1. Taufiq Nuruzzaman M, Lee C (2011) Independent and personal SMS SPAM filtering. In: International conference IEEE, Oct, 2011
2. Youn S, Mcland D (2007) A comparative study for email classification. In: Advances and innovations in systems, computing sciences and software engineering
3. Liu C, Wang G (2016) Analysis and detection of spam accounts in social networks. In: 2nd IEEE international conference on computer and communications
4. Jain K, Agarwal S (2014) A hybrid approach for spam filtering using local concentration based K means clustering. In: Confluence the next generation information technology summit, IEEE 5th national conference
5. du Toit, T., Kruger, H. (2012) Filtering spam e-mail with generalized additive neural networks, ©2012 IEEE
6. Feng W, Sun J, Zhang L, Cao C (2016) A support vector machine based naive Bayes algorithm for spam filtering. IEEE
7. Caruana R, Niculescu-Mizil A, Crew G, Ksikes A (2004) Selection from libraries of models. In: Proceedings of the twenty-first international conference on machine earning, pp 137–144

8. Renuka DK, Rajamohana PVS (2017) An Ensemble classifier for email spam classification in hadoop environment. Appl Math Inf Sci Int J
9. El-syed M., El-Ally, Fares S (2008) Fuzzy Similarity approach for spam filtering. IEEE
10. Sharma A, Rastogi V (2014) Spam filtering using K mean clustering with local feature selection classifier. Int J Comput Appl 108(10)
11. Awad WA, ELseuofi SM (2011) Machine learning methods for spam e-mail classification. Int J Comput Sci Inf Technol (IJCSIT) 3
12. Blanzieri E, Bryl A (2008) A survey of learning-based techniques of email spam filtering. Technical Report #DIT-06-056

A Hybrid Feature Selection Approach for Handling a High-Dimensional Data

B. Venkatesh and J. Anuradha

Abstract We proposed a Hybrid Feature selection method, the combination of Mutual Information (MI) a filter method, and Recursive Feature Elimination (RFE) a wrapper method. The methodology combines the strengths of both filter and Wrapper method. Performance of the proposed method is measured on three benchmark datasets (Ionosphere, Libras Movement, and Clean) from the UCI Repository. We compared the classification accuracy of the proposed Hybrid method with MI, RFE, Original Features by using random forest classifier. The performance are compared using four classification measures i.e. 1. F1-Score 2. Recall 3. Precession 4. Accuracy. It is evidence from the result analysis that the proposed hybrid method has out performed other methods.

Keywords Classification · Mutual information · Feature selection · RFE · Random forest classifier

1 Introduction

In today's world, the datasets that are used for machine learning applications are rapidly increasing in its dimensions both row and column wise. The size of the data has high impact on the computation power, processing time, and also reduces the accuracy of classification algorithms. This is due to the presence of noisy, irrelevant, and redundant features. The solution for these problems simply discards those features by using feature selection methods during the data cleansing process. This phenomenon is termed as dimensionality reduction. In these reduction processes, the best optimal feature subsets are selected from the original feature set. In general, feature selection techniques are divided into two categories (Filter and Wrapper

B. Venkatesh · J. Anuradha (✉)
SCOPE Vellore Institute of Technology, Vellore 632014, Tamil Nadu, India
e-mail: januradha@vit.ac.in

B. Venkatesh
e-mail: venkatesh.cse88@gmail.com

© Springer Nature Singapore Pte Ltd. 2019
H. S. Saini et al. (eds.), *Innovations in Computer Science
and Engineering*, Lecture Notes in Networks and Systems 74,
https://doi.org/10.1007/978-981-13-7082-3_42

methods) based on the interaction with learning algorithm. Feature selection in filter methods uses statistical measures like Pearson's Correlation, Linear discriminant analysis, ANOVA, and Chi-Square but no interaction with the learning algorithm, so it requires low computational time. Some of the filter methods [1] are Correlation-based Feature Selection, ReliefF [2], Fast Correlated Based Filter method, and Mutual Information [3].

In Wrapper method [4], a subset of features are selected using optimal search algorithms and a classification algorithm is used as evaluation criteria for subset selection. some of the Wrapper feature selection methods are Recursive Feature Elimination [5], Forward Feature Selection, Backward Feature Elimination.

There are some disadvantages with both filter and wrapper method. In filter method, the correlation between features and classifiers are not considered and its classification accuracy is less when compared with the Wrapper method. Wrapper method has more computational complexity, searching overhead and overfitting. But apart from these disadvantages, there are some advantages with these methods, filter method is computationally faster when compared with wrapper method and wrapper method is more accurate when compared with filter method. So, by combing the filter and Wrapper method, we can achieve both classification accuracy and reduce the computational cost.

Hybrid method uses the strength of both methods. First the dimensions of the data are reduced by using filter method followed by wrapper method for the selection of best feature subset. Usually, these methods has more classification accuracy and high computational speed when compared with filter and wrapper methods. So, in this work, we combine mutual information a filter method, and RFE a wrapper method, to form a hybrid method called MI-RFE. Random Forest classification model had been used for comparing the classification accuracy across three benchmark datasets from the UCI repository [6] using proposed MI-RFE Hybrid Feature selection method.

Further, this paper was organized as follows: the related work is provided in Sects. 2, 3 provides the methodology used, the implementation and experimental results are discussed in Sect. 4 and finally we conclude the work in Sect. 5.

2 Related Work

Hsu et al. [7] proposed a hybrid feature selection by combining wrapper method and filter method. This method removes the irrelevant and redundant features by using two filter methods namely, F-score and information gain. The feature sets resulted by these methods are combined to form a candidate features. These candidate feature set are passed through a wrapper method for selecting the final feature subset by using sequential floating search method (SFSM).

Lin et al. [8] proposed a hybrid method MI-SVM-RFE by combining mutual information (MI) and Support vector machine-recursive feature elimination (SVM-RFE). Artificial variables and MI have been used for filtering the non-informative and noise variable, for finding the best feature subset using SVM-RFE method.

Solorio-Fernández et al. [9] proposed a hybrid feature selection method based on ranking methodology for unsupervised learning. In this approach, during filter phase the features are ranked based on Laplacian score and the top rank features are selected based on some threshold value. Then, in wrapper stage, the selected features are indexed by using Calinski-Harabasz index.

Lu et al. [10] proposed a novel hybrid feature selection algorithm by combining Mutual Information Maximization (MIM) filter method and the 'Adaptive Genetic Algorithm' (AGA) wrapper method, for reducing the dimensions and redundant samples from the gene expression data.

Rouhi and Nezamabadi-pour [11] proposed R-m-GA method by the combination of Relief, mRMR (Minimum Redundancy and Maximum Relevance) and a genetic algorithm for reducing the dimensions of microarray data.

Vijayanand et al. [12] proposed a hybrid feature selection method GA + MI for intrusion detection system (IDS) for providing security in wireless mesh network. In GA + MI method, mutual information based filter method has been used and genetic algorithm has been used as wrapper method.

3 Methods

Figure 1 represents the proposed architecture of the Hybrid feature selection. It is a two-stage process. The proposed method is obtained by combining Filter and Wrapper feature selection methods.

In the first stage, k-best features are selected out of the all features by using mutual information between the actual variable and class variable. The best k value is selected based on the best threshold value. At the end of the first stage, we get k-best features which are used as input to further process. To select the optimal number of features, these K features are passed through RFE (Recursive Feature Elimination) Wrapper method. In Fig. 2 shows the cross-validation score, i.e., no. of

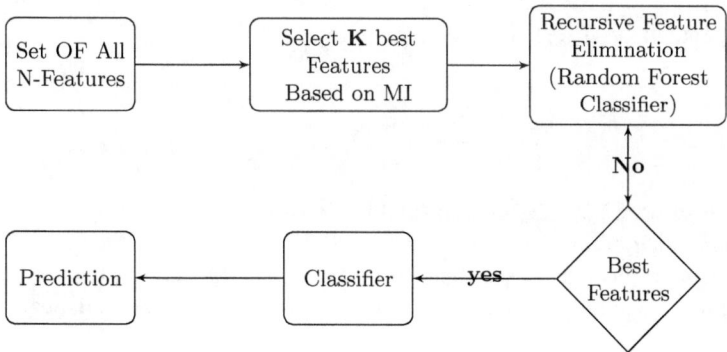

Fig. 1 Architecture

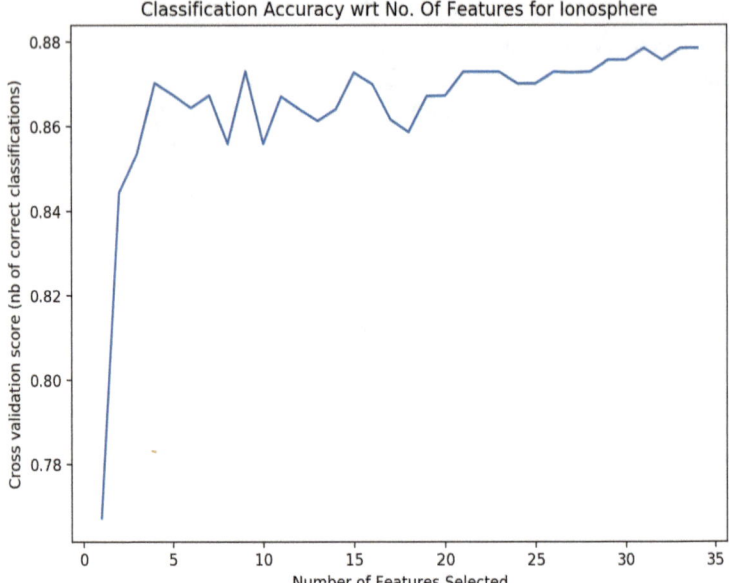

Fig. 2 Cross-validation score of ionosphere dataset

correct classification with respect to the no. of features selected, The optimal number of features are selected based on these cross-validation scores.

3.1 Mutual Information

Mutual Information is used to calculate how much information about a variable contains about another variable. Let us consider a dataset having (f, l) where f represents the features and l represents the class-label and the MI between f and l can be denoted by $I(f, l)$ and is calculated by using Eq. 1 [13].

$$I(f; l) = \sum_{n=1}^{j} \sum_{m=1}^{i} p(f_m, l_n) \log \left(\frac{p(f_m, l_n)}{p(f_m) \cdot p(l_n)} \right) \tag{1}$$

where $p(f_m, l_n)$ is *joint probability* function and $p(f_m)$ and $p(l_n)$ are *marginal probability* distribution function.

In this work, we have used the univariate mutual_info_classif feature selection method for selecting k-highest score features by using univariate statistical tests.

3.2 Recursive Feature Elimination

RFE is a recursive process in which the features are ranked based on some underlying Random Forest classification model using feature importance [14]. Algorithm 1 is the pseudocode for RFE.

Algorithm 1: Pseudocode for the Recursive Feature Elimination (RFE) Algorithm

---

I/P : Training set X

     Set of n features $Y = \{f_1, f_2, \ldots, f_n\}$

     A(X,Y) # Ranking method

O/P: Final Ranks R

for j *in* $1 : n$ **do**

    | Rank set Y using A(X,Y)

    | $f^* \leftarrow$ last ranked feature in Y

    | $R(n - j + 1) \leftarrow f^*$

    | $Y \leftarrow Y - f^*$

end

RFE is a recursive iteration process where features are ranked based on their feature importance. In Algorithm 1, X represents the training set and n no. of features, Y represents the original set of features, Method A(X,Y) is used for ranking features based on feature importance. R represents the final ranking of the features. Feature importance is measured at each iteration and the features with less relevant are discarded.

The optimal number of feature value is selected by using the cross-validation score as shown in Fig. 2. After finding the best number of features, the optimal subset is selected by using Algorithm 1.

4 Experimental Results and Implementation

Here, in this work, we embedded the filter and wrapper method to increase the accuracy of the machine learning model. We have used the standard benchmark datasets from UCI Repository [6] for implementation purpose. Table 1 represents the Dataset description that are used.

4.1 Implementation

Here, the implementation has been proposed in two ways; in the first phase, the candidate features are selected by using a filter method, and in the second phase,

Table 1 Description of datasets

Dataset	No. of attributes	No. of instances
Ionosphere	33	351
Libras movement	90	360
Clean	167	476

the candidate features further goes to select the optimal subset of features by using wrapper method.

In MI filter method, all features are ranked based on the mutual information between the features and the class labels. Then the features are sorted in descending order based on their ranks. The best k features are selected based on threshold value of the mutual information scores (p). For $p \geq 0.05$, we obtain the best k features. These features are used as input for the second phase.

In second phase, the k-candidate features undergo with RFE for selecting the optimal feature subsets. The optimal number of feature subsets are selecting based on the cross-validation score of Ionosphere dataset over their no. of features is shown in Fig. 2. After finding the optimal no. of features, the best feature subset has been selected using RFE wrapper feature method. The accuracy for the optimal features are measured by using Random Forest Classifier.

Four Classification measures such as 1. F1-Score, 2. Recall, 3. Precession, and 4. Accuracy are used to compare the classification accuracy for the proposed Hybrid Method (both MI + RFE Feature selection) and other methods 1. Original feature set 2. Mutual (By applying only Mutual Information feature selection) 3. RFE (Only RFE feature selection).

Table 2 Classification measures with respective to different feature selection methods

Dataset	Method	No. of features selected	F1. Score	Recall	Precession	Accuracy
Inosphere	Orginal	33	93.13	92.71	93.71	93.40
	Mutual	25	94.08	93.51	94.96	94.36
	RFE	20	93.13	92.71	93.71	93.40
	Hybrid	**15**	**95.09**	**94.65**	**95.70**	**95.28**
Libros Moment	Orginal	90	80.73	81.94	81.30	82.22
	Mutual	65	80.96	81.82	82.29	82.22
	RFE	50	84.40	85.99	85.71	85.56
	Hybrid	**40**	**86.60**	**87.50**	**87.02**	**87.78**
Clean	Orginal	167	89.28	88.71	90.09	89.92
	Mutual	120	90.13	89.14	90.23	90.23
	RFE	100	86.66	86.26	87.18	87.39
	Hybrid	**75**	**90.21**	**89.78**	**90.79**	**90.79**

4.2 Results

Table 2 represents the classification measures of the proposed hybrid method and other methods for the datasets. From the Table 2, it is observed that the classification accuracy is more in case of hybrid method (**95.28%**) for Ionosphere dataset when compared with original features (93.40%), MI (94.36%) and RFE (93.40%) methods.

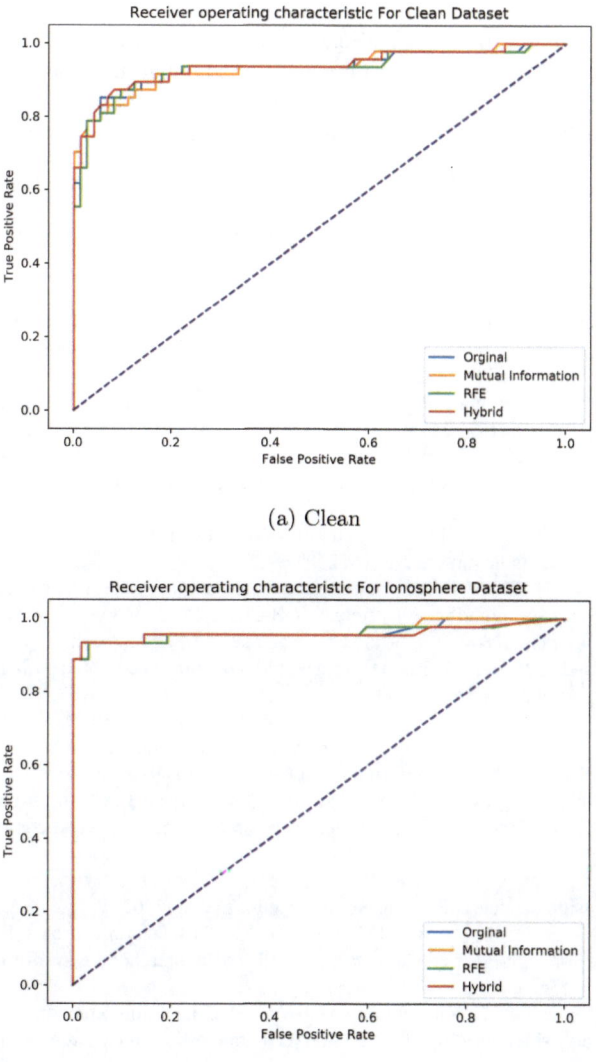

(a) Clean

(b) Ionosphere

Fig. 3 ROC curve of clean and ionosphere dataset

Similarly for Libras movement and clean datasets, the classification accuracy is more in case of hybrid method when compared with other methods. Figure 3 represents the Receiver Operating Characteristic (ROC) Curve for Clean and Ionosphere datasets with respect to different methods used.

5 Conclusion

In this paper, we proposed a hybrid feature selection method by combining the MI and RFE methods. From the experiments, we conclude that the proposed hybrid method has more classification accuracy (**95.28%**) when compared with MI filter method (94.36%) and RFE wrapper method (93.40%).

References

1. Sánchez-Maroño N, Alonso-Betanzos A, Tombilla-Sanromán M (2007) Filter methods for feature selection—a comparative study. In: International conference on intelligent data engineering and automated learning. Springer, pp 178–187
2. Robnik-Šikonja M, Kononenko I (2003) Theoretical and empirical analysis of relieff and rrelieff. Mach Learn 53(1–2):23–69
3. Peng H, Long F, Ding C (2005) Feature selection based on mutual information criteria of max-dependency, max-relevance, and min-redundancy. IEEE Trans Pattern Anal Mach Intell 27(8):1226–1238
4. Chuang LY, Ke CH, Yang CH (2016) A hybrid both filter and wrapper feature selection method for microarray classification. arXiv:1612.08669
5. Yan K, Zhang D (2015) Feature selection and analysis on correlated gas sensor data with recursive feature elimination. Sens Actuators B Chem 212:353–363
6. Dheeru D, Karra Taniskidou E (2017) UCI machine learning repository
7. Hsu HH, Hsieh CW, Lu MD (2011) Hybrid feature selection by combining filters and wrappers. Expert Syst Appl 38(7):8144–8150
8. Lin X, Yang F, Zhou L, Yin P, Kong H, Xing W, Lu X, Jia L, Wang Q, Xu G (2012) A support vector machine-recursive feature elimination feature selection method based on artificial contrast variables and mutual information. J Chromatogr B 910:149–155
9. Solorio-Fernández S, Carrasco-Ochoa JA, Martínez-Trinidad JF (2016) A new hybrid filter-wrapper feature selection method for clustering based on ranking. Neurocomputing 214:866–880
10. Lu H, Chen J, Yan K, Jin Q, Xue Y, Gao Z (2017) A hybrid feature selection algorithm for gene expression data classification. Neurocomputing 256:56–62
11. Rouhi A, Nezamabadi-pour H (2017) A hybrid feature selection approach based on ensemble method for high-dimensional data. In: 2017 2nd conference on swarm intelligence and evolutionary computation (CSIEC). IEEE, pp 16–20
12. Vijayanand R, Devaraj D, Kannapiran B (2018) A novel intrusion detection system for wireless mesh network with hybrid feature selection technique based on ga and mi. J Intell Fuzzy Syst 34(3):1243–1250
13. Liu H, Sun J, Liu L, Zhang H (2009) Feature selection with dynamic mutual information. Pattern Recogn 42(7):1330–1339

14. Granitto PM, Furlanello C, Biasioli F, Gasperi F (2006) Recursive feature elimination with random forest for ptr-ms analysis of agroindustrial products. Chemom Intell Lab Syst 83(2):83–90

Predicting the Entire Static Load Test by Using Generalized Regression Neural Network in the United Arab Emirates

A. K. Alzo'ubi and Farid Ibrahim

Abstract In the UAE, continuous flight auger piles (CFA) are the most commonly used type of foundations. To minimize the risk of failure, of these CFA piles, mandatory expensive field tests need to be performed and the most important one is the Static Pile Load Test (SPLT). This paper proposes using a General Regression Neural Network (GRNN) to predict the pile performance ahead of any test. Thousands of loading points in over one hundred projects from Dubai, Abu Dhabi, and Al Ain cities are used to develop a GRNN capable of predicting SPLT curves with reasonable accuracy. The friction angle, unconfined compressive strength, depth, soil type, groundwater table, pile's diameter, and pile's length are the parameters that are input to predict the load–displacement curves of the SPLT. This approach can complement conventional SPLT and provide engineers with sufficient insight into the pile performance ahead of the actual test.

KEYWORDS Static load test · Piles · GRNN · United Arab Emirates · CFA piles

1 Introduction

The upper geological formation in the United Arab Emirates (UAE) is a soft ground consisting of sand, silty sand, or soft rocks with a high groundwater table. This geotechnical environment is ideal for using CFA piles to safely transmit the load into the ground. These piles are easy and fast to construct, and in order to predict the pile capacity, two approaches are utilized; the empirical method and/or field tests. The first approach is conservative due to many variables controlling the pile capacity while the latter is so expensive and it adds to the total cost of a project.

A. K. Alzo'ubi (✉)
Civil Engineering Department, Abu Dhabi University, Al Ain, UAE
e-mail: abdel.alzoubi@adu.ac.ae

F. Ibrahim
Information Technology Department, Abu Dhabi University, Al Ain, UAE
e-mail: farid.ibrahim@adu.ac.ae

© Springer Nature Singapore Pte Ltd. 2019
H. S. Saini et al. (eds.), *Innovations in Computer Science and Engineering*, Lecture Notes in Networks and Systems 74,
https://doi.org/10.1007/978-981-13-7082-3_43

Many factors contribute to the most effective pile design; the size of the project, risk, and the significance of the project, to name a few [1]. In the UAE, conducting the static load tests is an important part of designing these CFA piles and inspecting their performance to avoid the risk of foundation failure. According to the code of building in UAE, a minimum of 5% of the piles in a project are selected and tested under the static load test after finishing the construction of all piles. The static pile test load is applied in increments up to 1.5 times the design load for piles, or more in rare cases, as required by the engineer. To accelerate the construction process in UAE, engineers started using CFA piles (Fig. 1). However, to minimize the risk of partial or complete failure of the foundation system, over designing was the solution.

This paper proposes to use the General Regression Neural Network (GRNN) to predict the load–displacement curves of the static load test prior to conducting the field test. This approach would help engineers to minimize the risk of under testing piles failure and consequently reducing the overall cost. The tests in this paper are based in three major cities, the two coastal cities of Abu Dhabi and Dubai, and the city of Al Ain. The geological stratification in the three cities are very similar. The diameter of the piles, however, was limited to half meters, while the pile's lengths range between 6 and 16 m. In the building code, a single pile that exceeds the limit of the acceptable downward movement of 1% of its diameter at 1.5 times the design load, is deemed as a failure, hence it was excluded from our study. This GRNN would allow engineers to predict the load–displacement curves prior to the actual test. This approach can reduce the need for multiple static load tests in the field, and therefore reducing the risk of failure while using a reasonable factor of safety. Coduto [1] stated that the analytic methods are not as expensive as the static load test, and thus the former is preferred by the designers.

Fig. 1 CFA piles underneath the current building in UAE

2 Artificial Neural Networks

The artificial intelligence approaches such as the artificial neural network (ANN) is commonly used in many geotechnical engineering applications such as foundations, soil type, and liquefaction. One kind of such networks is the Backpropagation Neural Network (BPNN) [2]. Another kind of neural network models is the GRNN. This prediction model is based on function approximation. It is a feed forward network where regression is performed using a probability density function applied to the input data. It has four layers; input, pattern, summation, and output layers. The input layer facilitates inputting all the attributes as a vector. These inputs are passed to the pattern layer, which consists of units that store cluster centers that are determined using a particular clustering technique such as the k-means. The objective of these centers is to group the sample inputs so that each group can be represented in one unit of the pattern layer that measures the distance of the input vector from the cluster center [3]. These units are used to subtract each new vector X from the stored cluster centers. The output of the pattern layer is stored as two parts in the summation layer, which consists of two units. The first unit stores the value that is calculated by (1):

$$\sum_{j=1}^{n} yj \, exp\left(-\sum \frac{(xi - xij)^2}{2\sigma^2}\right) \tag{1}$$

The second unit, however, stores the value that is calculated by (2):

$$\sum_{j=1}^{n} exp\left(-\sum \frac{(xi - xij)^2}{2\sigma^2}\right) \tag{2}$$

where $(xi-xij)$ represents the distance between a new point xi and a training sample xij. Finally, the output unit will compute the division of the values stored in the two summation units to produce the predicted value.

3 Related Works

In this research, we are proposing to use GRNN to predict the static load test of CFA piles, while most of the work that was carried out was based on using BPNN to predict the friction and/or end bearing capacity of driven piles. Goh [4] used an ANN model to estimate the friction capacity of driven piles in cohesive soils. He compared his results to the actual ones and achieved good agreement, he also compared the results with other empirical approaches. Lee and Lee [5] attempted to forecast driven pile capacity by using ANN approaches and they calculated an error of 20% between the prediction and actual results. Their focus was on the pile capacity and not the entire static load test. Abu-Kiefa [6] also focused on predicting the pile capacity by using three artificial neural network models. Additionally, Goh [7] developed a new neural network model to forecast pile capacity in cohesionless soils. Shahin et al. [8] discussed many applications of the artificial neural network in Geotechnical Engineering and listed many applications including the ones for predicting the driven pile ultimate capacity.

Benali and Nechnech [9] and Maizir and Kassim [10] utilized artificial neural networks for forecasting the driven pile capacity. In another attempt [11] artificial neural network model was used to predict the pile setup in the field. In a recent paper, Alzo'ubi et al. [12] suggested a framework for a new model that associates the artificial neural network model to forecast the pile performance with the geographic information system (GIS) data [12].

In all of the above studies, the focus was on the pile capacity but not to predict the entire load–displacement curves of the static load test, moreover, no artificial neural network experiments had been performed on CFA piles. This research demonstrates that the entire load–displacement curves with two cycles of the static load test can be predicted by GRNN if the training data is available and comprehensive.

4   General Geology

Fookes and Knill [13] conducted a geological survey along the coastline of UAE and classified the region as "the base plain". The upper layer in the UAE coastline consists of sand dunes, silt, and loess. Wind-blown material predominates the movement and transportation of great quantities of fine materials such as silt and sand if the high wind period occurs. Water also contributes to this movement and deposition of Silty Sand material all over the area. This soil layer depth varies from 3 to 10 m. This Silty Sand layer has a high friction angle and the piles capacity depends on two components; friction at the sides, and end bearing at the bottom of a pile. In some cases, up to 20 cm thick salt crust can be built, these salt layers may affect the pile's capacity through friction reduction. Underneath the sand and silt layers' rock materials can also be found, in other cases, the sand and silt layers cover interbedded sandstones, conglomerates, limestone, and siltstones. In this paper, the pile's length ranges from 6 to 16 m, so all the layers from 0 to 20 m will be classified as shown in the next section.

5   Classification of Materials

To classify the soil layers and identify their mechanical properties, field and labora-tory test results are conducted and summarized for each borehole in each project. The Unified Soil Classification System was utilized to classify the ground materials by conducting sieve analysis and Atterberg Limits tests. This classification was used as an attribute in the Neural Network to increase model accuracy and precision. To do

Table 1 Coding of some soil layers as will be used in the GRNN

Soil type	Soil profile number	ANN code
Silty SAND (SM)	1	10000000
Poorly graded SAND with silt (SP-SM)	2	01000000
MUDSTONE: Very weak to weak	3	00100000

so, we coded the soil layer classification as a binary number. Table 1 shows samples of this coding which will be included in the input data of GRNN as a separate unit for each soil type at each respective depth.

6 Static Load Test

In the UAE, the most commonly recommended type of foundation is CFA piles. Pile load tests are considered the most reliable method to evaluate the capacity of a pile under axial load. In these tests, two cycles of loading and unloading are usually used to produce complete load–displacement curves for each static load test.

The British Standard Code of Practice 8004, on specially constructed piles, usually used to test CFA piles that are installed before the start of the construction of a structure. In the UAE, the test of the piles usually conducted after the construction of all piles for quality assurance and not for design purposes. As a result of using the test for quality check and not design, the CFA piles are usually overdesigned by the engineers to avoid failure in the testing stage.

In this paper, the BS8004 [14] was used for all the test as the testing procedure. The pile is loaded up to 1.5 times the design load for working piles while monitoring the displacement. The load was increased in stages and maintained until a settlement has stopped. In this model, the loading/unloading cycles were coded in binary numbers to reflect the status of the cycle. For example, the first loading cycle was coded as 1000, while the first unloading was coded as 0100. These codes were considered as 4 input units in the GRNN.

This testing procedure is used in the UAE following the 2013 Abu Dhabi International Building Code. According to this code, if one pile during testing failed at a particular site, the CFA piles must be removed from the site. These code requirements pushed engineers to overdesign the CFA piles to guarantee a successful test. The proposed approach in this paper hopes to reduce the cost of designing CFA piles through predicting the load–displacement curves, of a static load test, based on geotechnical data available at a site, the pile dimensions, and the groundwater table at that specific site.

7 The Network Model

This research argued that it is likely to predict the displacement at a load using GRNN model that is trained over data identified by 28 input attributes constituting the soil profile (17 units), pile configuration (2 units), groundwater (1 unit), load (1 unit), loading status (4 units), strength properties (2 units) and depth (1 unit), while the displacement (1 unit) is to be the output predicted attribute. Coduto [1] pointed out that the soil profile and the pile configuration are extremely important factors in controlling the pile capacity and performance under loading and unloading cycles. In order to produce an acceptable model, a huge amount of data is needed for training to guarantee the model's performance and its ability to forecast the displacement of the new pile at a new site with different or similar soil profile. Actual pile test data were collected and verified from 100 sites in the field and were implemented in the network. Moreover, we verified the data for completeness and filtered it to remove any outliers. As a result of this process, 6437 load–displacement points were used to train the model. The model must generalize the mapping between input and output and should be able to forecast the output for cases that are not included in the training data.

For this experiment, MATLAB Simulink 2016 software [15] was utilized to construct the GRNN. The sample data was migrated to MATLAB worksheet at which they were transposed into a matrix P in the MATLAB command shown in (3):

$$net = newgrnn\,(P, T) \tag{3}$$

This command generates the network: net, and it requires two parameters: the input transposed matrix P and the target predicted vector values T corresponding to the average displacement Two layers are created by newgrnn as shown in Fig. 2. The first layer has as many units as there are inputs in P. The weight W in this layer is set to P'. The unit's weighted input is computed as the distance between the input vector and its weight vector. The bias b is set to a column vector of 0.8326. The second layer has as many units as the target vector but with different weight setting W [15].

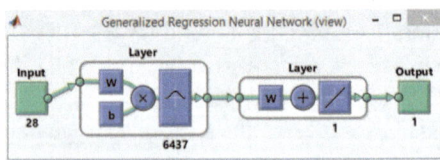

Fig. 2 The GRNN as produced by MATLAB

8   Results

Many factors may control the pile performance, so predicting the pile's entire load--displacement curve has been a difficult task for engineers. Hence, the static load test is conducted commonly in the UAE in every new project constructed by CFA piles. This research presents a GRNN model to predict the loading–unloading curves of a static load test for CFA piles under axial loading.

Though neural network models have been utilized to forecast the pile ultimate capacity, as discussed earlier in Sect. 3, no previous attempt has yet been made to predict the entire static load test. To test the proposed model, a site was arbitrarily picked with an entire pile test that was not included in the training set to guarantee the independence of the test and the ability of the model to predict the two cycles of loading and unloading. The geotechnical data, groundwater elevation, pile length and diameter, loading cycle (loading versus unloading), and the value of the load were input in the model, while the displacement field was left out to be predicted using the MATLAB command shown in (4):

$$V = sim\ (net,\ P) \tag{4}$$

where P is the input transposed matrix of the tested pile, and V is the resulting output vector.

Once the testing stage is finished, different statistical methods are to be used to compare the predicted displacement values against the actual field ones. The coefficient of determination and root mean square error, for example, will be used to show the relationship between the predicted values and the actual ones. Figure 3 shows the preliminary results of the test that was conducted using the setup discussed above. The results show good agreement between the actual and predicted displacement at each load point and in both cases of loading and unloading.

Fig. 3 Static load test results with two cycles of loading and unloading

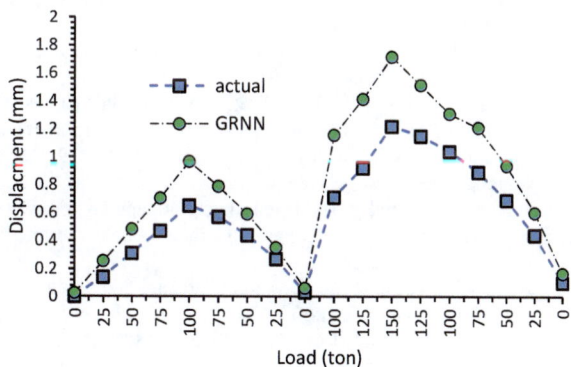

9 Conclusion

This paper introduced a general regression neural network model to forecast the displacement at each load of a static load test curves that consist of two loading–unloading cycles. The artificial network is based on input parameters including geomaterial properties (such as friction angle), soil classification and type, groundwater table elevation, pile configuration, pile load, and displacement. The attributes are used to build and train the network, and later test it to predict the displacement. This research introduces, for the first time, a GRNN approach that is capable of simulating the static load test. The GRNN results were compared with existing field tests that were arbitrarily excluded from the training set of data. This approach is a tool for engineers to forecast static load test results with acceptable accuracy from a practical point of view to interfere at the right time and alter the design of a pile or increase/decrease the number of static pile load tests required at the field.

Acknowledgements The authors would like to express their gratitude for the office of scholarship and sponsorship programs at Abu Dhabi University, grant number 19300093.

References

1. Coduto (2001) Foundation design: principles and practices. Prentice Hall, New Jersey
2. Han J, Kamber M (2012) Data mining, concepts and techniques, 3rd edn. Morgan Kaufman Publishers, MA, USA
3. Specht DF (1991) A general regression neural network. IEEE Trans Neural Netw 2(6):568–576
4. Goh ATC (1995) Empirical design in geotechnics using neural networks. Geotechnique 45:709–714
5. Lee IM, Lee JH (1996) Prediction of pile bearing capacity using artificial neural networks. Comput Geotech 18(3):189 200
6. Abu Kiefa MA (1998) General regression neural networks for driven piles in cohesionless soils. J Geotech Geoenviron Eng ASCE 124(12):1177–1185
7. Goh ATC (1996) Pile driving records reanalyzed using neural networks. J Geotech Geoenviron Eng ASCE 122(6):492–495
8. Shahin MA, Jaska MB, Maier HR (2001) Artificial neural network applications in geotechnical engineering. Aust Geomech 49–62
9. Benali A, Nechnech A (2011) Prediction of the pile capacity in purely coherent soils using the approach of the artificial neural networks. In: International seminar, innovation & valorisation in civil engineering & construction materials, N°: 5O-239, University of sciences and technology, Algiers, Algeria
10. Maizir H, Kassim K (2013) Neural network application in prediction of axial bearing capacity of driven piles. In: Proceedings of the international multiconference of engineers and computer scientists, Hong Kong
11. Tarawneh B (2013) Pipe pile setup: database and prediction model using artificial neural network. Soils Found 53(4):607–615
12. Alzo'ubi AK, Ati M, Ibrahim F (2015) Smart framework for predicting drilled shaft capacity based on data mining techniques and GIS data. In: Manzanal D, Sfriso AO (Eds), From fundamentals to applications in geotechnics, the pan american conference on soil mechanics and geotechnical engineering, 15th PCSMGE/8th SCRM/ 6th IS-BA 2015, 15–18, Nov, pp 1909–1915

13. Fookes PG, Knill JL (1969) The application of engineering geology in the regional development of northern and central Iran. Eng Geol 3:81–120
14. British standard (2015). BS 8004, Code of practice for foundations, BSI
15. MathWorks, MATLAB & SIMULINK (2016) Generalized regression networks. https://www.mathworks.com/help/nnet/ug/generalized-regression-neural-networks.html

Predicting the Risk of Readmission of Diabetic Patients Using Deep Neural Networks

G. Siva Shankar and K. Manikandan

Abstract One of the major concerns these days are hospital readmissions, as they are expensive and mark the shortfalls in health care. The United States alone has spent 41.3 billion dollars between January and November 2011 to treat patients readmitted within 30 days of discharge, according to the Agency for Healthcare Research and Quality (AHRQ). There are already several attempts made in the same field using Machine learning methods to leverage public health data to build a system for identifying diabetic patients facing a high risk of future readmission. This paper predicts whether a patient discharged from the hospital will return within 30 days or not. The best possible feature engineering pipeline is chosen to process the data so that it can be learnt by the model in the best manner in determining the most important evaluation metric. A classifier is built using the traditional machine learning algorithms such as Linear SVM, Random Forest. We have also built a deep neural network based on a specific and optimized sequential architecture.

Keywords Binary classification · Deep neural nets · Feature engineering · Machine learning · Perceptron · Scikit-Learn

1 Introduction

The dataset represents a decade worth of clinical care data from 130 hospitals in the United States, as well as integrated delivery networks. In order for an entry to qualify for the dataset, it must be a hospital admission and the admission must be diabetic related with the length of the stay to be more than 1 day but less than 2 weeks. The dataset's features comprise of a mix of nominal and numeric data types. The features span a multitude of patient information categories such as race, gender,

G. Siva Shankar (✉) · K. Manikandan
Vellore Institute of Technology, Vellore, Tamil Nadu, India
e-mail: shivashankar89@outlook.com

K. Manikandan
e-mail: kmanikandan@vit.ac.in

© Springer Nature Singapore Pte Ltd. 2019
H. S. Saini et al. (eds.), *Innovations in Computer Science
and Engineering*, Lecture Notes in Networks and Systems 74,
https://doi.org/10.1007/978-981-13-7082-3_44

age, weight, and time spent in a hospital. Some of the more specific features include information such as medications administered (miglitol, metformin, tolazamide, etc) and a number of lab procedures. First and foremost, a classification model that can accurately predict whether a diabetic inpatient will return to the hospital has several implications in itself. A similar concept applies to the patients themselves; patients who can know they may be at risk for rehospitalization can take extra precautions once they are discharged in an attempt to prevent further emergencies. And finally, if hospital administrators know that a certain number of patients will be readmitted, they could use this information in order to better plan room reservations, medical supply projections, and further logistical planning [1–3].

2   Related Works

American hospitals spend over 41 billion dollars on diabetic patients in 2011, who got readmitted within 30 days of discharge, according to the Agency for Healthcare Research and Quality (AHRQ). Researchers have attempted to find predictors of readmission rate and among other factors; medication change upon admission has also been shown to be associated with lower readmission rates. The predictors of medication change could point towards interesting insights on what patient characteristics and admission conditions might influence whether physicians change their medication or not. According to the Hospital Readmissions Reduction Program which aims to improve the quality of care for patients and lower healthcare spending. A patients visit to a hospital may be constituted as a readmission if that patient is admitted to a hospital within 30 days after being discharged from an earlier hospital stay. Using a medical claim dataset, we planned to answer the questions: what are predictors of medication change in diabetics who gets admitted to a hospital and how can we build a predictive model to lower the readmission rates? A large number of previous researches have presented the risk factors that can help to identify and predict hospital readmissions of diabetic patients [3, 6, 9, 13, 17] out of which significant ones are chosen and discussed. The acute and chronic glycaemic control influenced the risk of readmission in a dataset of more than 29,000 patients over the age of 65 [17]. Other factors such as demographic and socioeconomic factors also have an influence on readmission rates. The study, made on the impact of HbA1c, have done analysis on hospital readmissions without targeting any specific disease. The dataset chosen in this case, has attributes of demographic, clinical procedure-related, and diagnostic-related features. It also has drug information of patients over 65 of age. The prediction is done for readmission within 30 days(short-term readmission prediction) [16, 18].

3 Methodology

From the traditional machine learning algorithms, we used to build the baselines models that were Support Vector Machine and Random Forest algorithm.

One of the supervised machine learning algorithms is Support Vector Machine. The plot of each data point is an item and a point in n-dimensional space, where n represents the number of features along with its value and each feature represents the value of particular coordinate. On the other hand, Random forest is an ensemble technique which uses many decision tree models at the same time. There are a lot of unique features of the Random Forest Algorithm. Random Forest is one of the most widely used supervised machine learning algorithm.

First, in Random Forest a random seed is chosen which selects a collection of observations from training data by taking care of class distribution. When this dataset is selected, a set of the random variables are chosen randomly based on the user-defined values and not all the variable are chosen in the random forest because of enormous computation and also the over fitting chances are high. Based on the attribute selection, the best possible split is given by Gini index and the decision tree is devolved based on this attribute selection.

A sequential deep neural network model is used along with Keras wrapper and classes with Tensor Flow at the backend. The sequential model is a linear stack of layers connected to each other by calling the sequential function then a series of layers can be added till the output layer. The model needs to know what input shape it should expect. For this reason, the first layer in a sequential model (and only the first, because the following layers can do automatic shape inference). Layers of different type are a few properties in common, specifically their method of weight initialization and activation functions. The type of initialization used for a layer is specified in the initialization argument which can take uniform, normal, and zero as initializations. Another thing that Keras provides is a range of standard neuron activation function, such as SoftMax, reLu, tanh, and sigmoid. The main objective would be to build a deep optimized neural network model and compare it with the Scikit-Learn multi-perceptron Layer.

4 Implementation and Results

Dataset is taken from https://archive.ics.uci.edu/ml/datasets.html [4]. The dataset is downloaded from UCI Machine Learning Repository. It is a publicly available dataset. The original dataset creates three distinct class boundaries for hospital readmittance: 1. *NO*: No record of readmission. 2. <30: Patient was readmitted in less than 30 days. 3. >30: Patient was readmitted in more than 30 days.

It would be far more feasible to narrow the scope to a binary classification. Instead of the three distinct classes seen above, a simplification could prove more practical

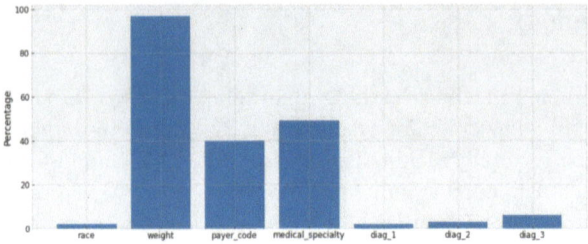

Fig. 1 Percentage of missing values in dataset

and informative as it will help to balance the dataset as the dataset is highly imbalanced.

1. NO: No record of readmission 2. >30: If the patient was readmitted in less than 30 days.

Some features like encounter ID and patient ID are removed because they do not contribute any relevant information to our classification dataset and machine learning model would not understand the ID which a unique feature for each patient is, we also decided to remove a few other features that actually have conceptual significance to the dataset. For example, the weight attribute has many potential ties to qualifying overall health. However, 97% of the weight data was missing, and imputing 97% of the data for a category is not possible.

4.1 Data Preprocessing

The preprocessing of a dataset is done by converting the label to required form, find out missing values in the dataset, and drop irrelevant columns based on human expertise of the data set and impute missing values. The next objective will be involving the feature selection, model building, model optimizations, and evaluation. Figure 1 describes the percentage of missing values in the variables in the dataset.

The numerical missing values are replaced by their mean and median and categorical missing values by mode. The exploratory data analysis is done for various variables, for e.g., number of medications with and without diabetic medications as in Fig. 2.

The diabetic medication is also analyzed with the race, gender, and age of the patients as shown in Figs. 3, 4 and 5. It can be inferred that two classes which are Caucasian and American-African are in majority. On the basis of gender, we can conclude that females count is more than male patient counts. On the other hand, from the histogram we can infer that diabetic and nondiabetic patients are spread equally.

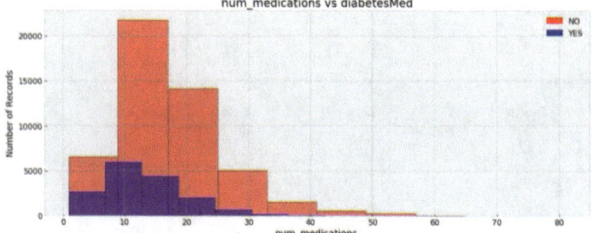

Fig. 2 Number of medications with and without diabetic

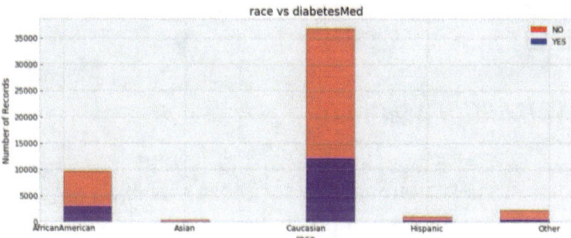

Fig. 3 Race versus diabetic medication

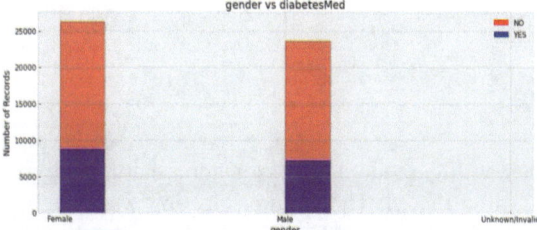

Fig. 4 Diabetic medication with gender

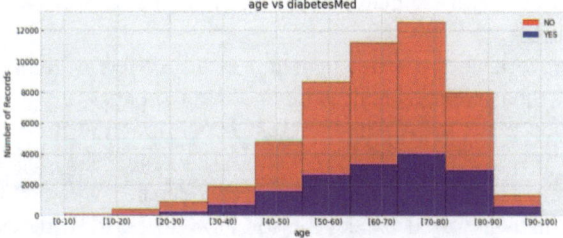

Fig. 5 Age versus diabetic medication

Table 1 Random forest

Test-A: 0.99 (Random forest)				
Classification report:				
	Precision	Recall	f1-score	Support
0.0	1.00	0.99	0.99	27,085
1.0	0.99	1.00	0.99	30,702
Avg/total	0.99	0.99	0.99	7787
Confusion matrix				
26,706		379		
0		30,702		

4.2 Cross-Validation Methods

By looking at our class distribution, it is seen that we have a large class imbalance. 82.8% of our entries consist of patients who were not readmitted, while 17.2% of our dataset was readmitted. A k = 10-fold cross-validation will utilize 90% of the data in each iteration. A 10-fold stratified cross-validation will maximize the generalization performance of our trained model [12, 14, 15].

4.3 Modeling

Next step was the baseline model for classification that is built for the data to train and create a prediction model. The feature selection was done using F regression along with the normalization of the data using Z-score method also converting the string representation values into categorical buckets. To make sure that there is no overfitting because of the imbalance dataset; Oversampling was implemented to balance the data before implementing the SVM classifier as well as voting classifier (Tables 1, 2).

For deep network, the experiment is done with different numbers of network layers. We will then integrate both wide and deep network architectures into one network and then experiment with different network architectures [7, 8, 10, 11] (Table 3).

Next, we compare the performance analysis of deep nets with Scikit-Learn Multi-Layer Perceptron. We would expect Tensor Flow to perform better because it is optimized to perform better with GPU's. It does not prioritize training speed and instead prioritizes performance. While using Sklearns MLP, due to timing constraints, we must limit the number of cross-validation folds. With more time, we would be able to provide a higher confidence in our generalization performance when compared to Scikit-Learn. We were able to get only 53–56% cross-validation score. So, from the

Table 2 Ensemble

Test-A: 0.96 (Ensemble)				
Classification report:				
	Precision	Recall	f1-score	Support
0.0	1.00	0.91	0.95	27,085
1.0	0.92	1.00	0.96	30,702
Avg/total	0.96	0.96	0.96	57,787
Confusion matrix				
24,589		2496		
0		30,702		

Table 3 Linear SVM

Test-A: 0.60 (Linear SVM)				
Classification report:				
	Precision	Recall	f1-score	Support
0.0	0.58	0.57	0.57	27,085
1.0	0.62	0.63	0.63	30,702
Avg/total	0.60	0.60	0.60	57,787
Confusion matrix				
15,456		11,629		
11,383		19,319		

Table 4 Performance analysis

Model	Accuracy
Scikit-learn MLP	55.97
Deep neural network	83.99

performance of Scikit-Learn MLP classifier, we can conclude that the Tensor Flow implementation is superior in performance [3, 5] (Table 4).

5 Conclusion

Irrespective of the advances in health care, the quality control offices in hospitals and health organizations still use traditional techniques and predefined variables to check the probability patient readmission. Machine Learning and Predictive Modeling using big data could provide clues to improve the quality of healthcare delivery. The main objective of the research was to explore the different machine learning models and data mining techniques to propose a solution to identify and predict the hospital readmissions. Combining Machine Learning and Predictive Modeling with preventive measures would also engage patients, physicians, and payers to participate proactively in improving the health facilities and hence the level of wellness. A data-driven deep learning based predictive model is developed to predict readmission rates in diabetic patients. Cases and controls were compiled based on 30-day readmission evidence of readmission. Compared to the existing repertoire of predictive models to assess readmission, our model shows better accuracy using optimized architecture of the neural network. Feature selection provides insights into several novel factors that could help to delineate readmission rates associated.

References

1. Agrawal R, Srikant R et al (1994) Fast algorithms for mining association rules. In: Proceedings of 20th international conference very large data bases, VLDB, vol 1215, pp 487–499
2. Breiman L (2001) Random forests. Mach Learn 45(1):5–32
3. Cooper GF, Herskovits E (1991) A Bayesian method for constructing Bayesian belief networks from databases. In: Uncertainty proceedings, pp 86–94. Elsevier
4. Dheeru D, Karra Taniskidou E (2017) UCI machine learning repository. http://archive.ics.uci.edu/ml
5. Dungan KM (2012) The effect of diabetes on hospital readmissions
6. Eby E, Hardwick C, Yu M, Gelwicks S, Deschamps K, Xie J, George T (2015) Predictors of 30 day hospital readmission in patients with type 2 diabetes: a retrospective, case-control, database study. Curr Med Res Opin 31(1):107–114
7. Freund Y, Schapire RE (1997) A decision-theoretic generalization of on-line learning and an application to boosting. J Comput Syst Sci 55(1):119–139
8. Fukushima K (1988) Neocognitron: a hierarchical neural network capable of visual pattern recognition. Neural Netw 1(2):119–130
9. George M, Bencic S, Bleiberg S, Alawa N, Sanghavi D (2014) Case study: delivery and payment reform in congestive heart failure at two large academic centers. In: Healthcare, vol 2, pp 107–112. Elsevier (2014)
10. Hall M, Frank E, Holmes G, Pfahringer B, Reutemann P, Witten IH (2009) The weka data mining software: an update. ACM SIGKDD Explor Newslett 11(1):10–18
11. Hasan M (2001) Readmission of patients to hospital: still ill defined and poorly understood. Int J Qual Health Care 13(3):177–179
12. Hosseinzadeh A, Izadi MT, Verma A, Precup D, Buckeridge DL (2013) Assessing the predictability of hospital readmission using machine learning. In: IAAI
13. Howell S, Coory M, Martin J, Duckett S (2009) Using routine inpatient data to identify patients at risk of hospital readmission. BMC Health Serv Res 9(1):96
14. Kim H, Ross JS, Melkus GD, Zhao Z, Boockvar K (2010) Scheduled and unscheduled hospital readmissions among diabetes patients. Am J Manage Care 16(10):760
15. Krizhevsky A, Sutskever I, Hinton GE (2012) Imagenet classification with deep convolutional neural networks. Adv Neural Inf Process Syst 1097–1105
16. Lichman M et al (2013) Uci machine learning repository
17. Osmanbegović E, Suljić M (2012) Data mining approach for predicting student performance. Econ Rev 10(1):3–12
18. Silverstein MD, Qin H, Mercer SQ, Fong J, Haydar Z (2008) Risk factors for 30-day hospital readmission in patients 65 years of age. In: Baylor University Medical Center Proceedings, vol 21, pp 363–372. Taylor & Francis

Reliable Healthcare Monitoring System Using SPOC Framework

P. Ramya, P. Naga Sravanthi and Morampudi Mahesh Kumar

Abstract The m-healthcare system can benefit medical users by providing high-quality pervasive healthcare monitoring, the growing of m-healthcare system is still strange on how we fully understand and manage the challenges facing in this m-healthcare system, especially during a medical emergency. In this paper, we propose a new secure and privacy-preserving opportunistic computing framework, called SPOC, to address this challenge. With the help of our proposed SPOC framework, each medical user who is in an emergency can achieve the user-centric privacy access control to allow only those qualified helpers to participate in the opportunistic computing to balance the high reliability of PHI process and minimizing PHI privacy disclosure in m-Health care emergency. We introduce an efficient user-centric privacy access control in SPOC framework, which is based on an attribute-based access control and a new privacy-preserving scalar product computation (PPSPC) technique, and allows a medical user to decide who can participate in the opportunistic computing to assist in processing his great PHI data.

Keywords Healthcare system · Privacy-preserving · Framework · User-centric · Attribute-based

P. Ramya (✉)
Department of CSE, Gudlavalleru Engineering College, Gudlavalleru, India
e-mail: mothy274@gmail.com

P. Naga Sravanthi
Department of IT, Gudlavalleru Engineering College, Gudlavalleru, India
e-mail: puppalasravanthi822@gmail.com

M. M. Kumar
Department of CSE, NIT Warangal & IDRBT, Warangal, Telangana, India
e-mail: morampudimahesh@gmail.com

© Springer Nature Singapore Pte Ltd. 2019
H. S. Saini et al. (eds.), *Innovations in Computer Science and Engineering*, Lecture Notes in Networks and Systems 74,
https://doi.org/10.1007/978-981-13-7082-3_45

1   Introduction

The patient monitoring system (PMS) is a very critical process that is mandatory these days. It is used for monitoring the physiological parameters including electro-cardiograph (ECG), respiration, invasive and noninvasive blood pressure, and oxygen saturation in human blood (Spo2), body temperature, etc. In PMS, the multiple sensor and electrodes are used for receiving physiological signals like ECG electrodes, Spo2 Finger Sensor, Blood Pressure Cuff, and Temperature Probe to measure the physiological signals. During treatment, it is highly important to continuously monitor the vital physiological signs of the patient. Therefore, patient monitoring systems have always been occupying a very important position in the field of medical devices. The continuous improvement of technologies not only helps us to transmit the vital physiological signs to the medical personnel but also simplifies the measurement and as a result raises the monitoring efficiency of patients.

1.1   Classes of Patient Monitoring System

This can be classified into two types:

- Single-Parameter Monitoring Systems
- Multiple Parameter Monitoring Systems.

1.1.1   Single-Parameter Monitoring Systems

The single-parameter monitoring systems are available for measuring blood pressure of a human body, ECG monitor, Spo2 monitor, etc.

1.1.2   Multiple Parameters Monitoring Systems

A multi-parameter monitoring system is used for multiple critical physiological signs of the patient to transmit vital information.

In the traditional system, with the pervasiveness of smartphones and the advancement of wireless body sensor networks (BSNs), mobile health care(m-healthcare), which extends the operation of healthcare provider into a pervasive environment for better health monitoring, has attracted considerable interest recently. However, the flourish of m-healthcare still faces many challenges including information security and privacy preservation.

Limitations of Traditional Approach

- The flourish of m-healthcare still faces many challenges including information security and privacy preservation.
- The smartphone's energy could be insufficient when an emergency takes place.

2 Literature Survey

Lin et al. [1] proposed a method "GSIS, A secure and privacy-preserving proto-col for vehicular communications", where the authors identify some unique design requirements in the aspects of security and privacy preservation for communications between different communication devices in vehicular ad hoc networks. The authors then proposed a secure and privacy-preserving protocol based on group signature and identity (ID)-based signature techniques and demonstrated that the proposed protocol can not only guarantee the requirements of security and privacy, but can also provide the desired traceability of each vehicle in the case where the ID of the message sender has to be revealed by the authority for any dispute event.

Avvenuti et al. [2] proposed a method "Opportunistic Computing for Wireless Sensor Networks" based on the idea of partitioning the application code into a number of opportunistically cooperating modules. Each node contributes to the execution of the original application by running a subset of the application tasks and providing service to the neighboring nodes. But the authors did not address the problem of storing and executing an application that exceeds the memory resources available on a single node.

Ren et al. [3] proposed a method "Monitoring Patients Via A Secure And Mobile Health Care System" in which the authors present several techniques that can be used to monitor patients effectively and enhance the functionality of telemedicine systems, and discuss how current security strategies can impede the attacks faced by wireless communications in healthcare systems and improve the security of mobile healthcare.

Conti et al. [4] proposed a method "From Opportunistic Networks to Oppor-tunistic Computing" in which the authors discussed the evolution from opportunistic networking to opportunistic computing; the authors survey key recent achievements in opportunistic networking, and describe the main concepts and challenges of oppor-tunistic computing. Finally, envision further possible scenarios and functionalities to make opportunistic computing a key player in the next-generation Internet.

Li et al. [5] proposed a method "Scalable and Secure Sharing of Personal Health Records in Cloud Computing Using Attribute-Based Encryption" in which the authors proposed a novel patient-centric framework and a suite of mechanisms for data access control to PHRs stored in semi-trusted servers. To achieve fine-grained and scalable data access control for PHRs, we leverage attribute-based encryption (ABE) techniques to encrypt each patient's PHR file. Different from previous works in secure data outsourcing, the authors focused on the multiple data owner scenarios, and divide the users in the PHR system into multiple security domains that greatly reduces the key management complexity for owners and users.

Passarella et al. [6] proposed a method "Performance Evaluation of Service Execution in Opportunistic Computing" in which the authors presented an analytical model that depicts the service invocation process between seekers and providers. Specifically, the authors derived the optimal number of replicas to be spawned on encountered nodes, in order to minimize the execution time and optimize the computational and bandwidth resources used. Performance results show that a policy operating in the optimal configuration largely outperforms policies that do not consider resource constraints.

3 Proposed Method

3.1 Architecture for Monitoring System

In this paper, we propose a new secure and privacy-preserving opportunistic computing framework, called SPOC, to address this challenge. With the proposed SPOC framework, each medical user in an emergency can achieve the user-centric privacy access control to allow only those qualified helpers to participate in the opportunistic computing to balance the high reliability of PHI process and minimizing PHI privacy disclosure in m-Health care emergency. We introduce an efficient user-centric privacy access control in SPOC framework, which is based on an attribute-based access control and a new privacy preserving scalar product computation (PPSPC) technique, and allows a medical user to decide who can participate in the opportunistic computing to assist in processing his overwhelming PHI data.

Advantages of the New Approach

- SPOC framework allows a medical user to decide who can participate in the opportunistic computing to assist in processing his overwhelming PHI data.
- The user-centric privacy access control to allow only those qualified helpers to participate in the opportunistic computing to balance the high reliability of PHI.
- The attributed-based access control can help a medical user in an emergency to identify other medical users.

Figure 1 shows the proposed method and it consists of the following steps:

1. Medical users
2. Body Sensor Network (BSN)
3. Smartphone communication
4. Healthcare Center.

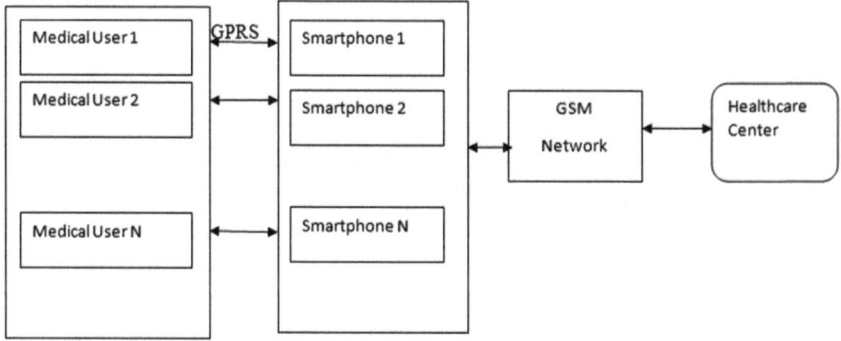

Fig. 1 Architecture for monitoring system

3.1.1 Medical Users

Normally, the medical user personal healthcare information (PHI) is mainly invented for monitoring the patients without direct interaction with doctors. In an m-healthcare system, medical users are no longer needed to be monitored within the home or hospital environments. Instead, after being equipped with smartphone and wireless body sensor network (BSN) formed by body sensor nodes, medical users can walk outside and receive high-quality healthcare monitoring from medical professionals anytime and anywhere.

3.1.2 Body Sensor Network

This sensor will be equipped directly in the medical user. This BSN will transmit the user details for every time period that we have indicated. For example, each mobile medical user's personal health information (PHI) such as heartbeat, blood sugar level, blood pressure, and temperature and other details will be captured by the medical users Smartphone (Fig. 2).

3.1.3 Smartphone Communication

For each data transmitted from BSN will be aggregated by the smartphone that the medical users having with them using Bluetooth communication. This received

Fig. 2 Body sensor network

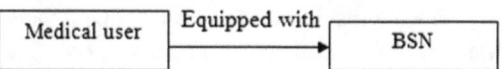

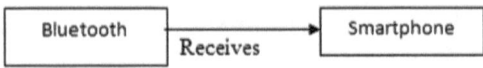

Fig. 3 Smartphone communication using Bluetooth

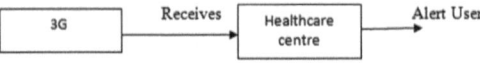

Fig. 4 Healthcare center

medical information or symptom will be transmitted to healthcare center periodically with the help of 3G network (Fig. 3).

3.1.4 Healthcare Center

We propose SPOC, a secure and privacy-preserving opportunistic computing framework for m-healthcare emergency. With SPOC, the resources available on other opportunistically contacted medical users' smartphones can be gathered together to deal with the computing-intensive PHI process in an emergency situation. Since the PHI will be disclosed during the process in opportunistic computing, to minimize the PHI privacy disclosure, SPOC introduces a user-centric two-phase privacy access control to only allow those medical users, who have similar symptoms to participate in opportunistic computing (Fig. 4).

4 Result Analysis

Figures 5, 6 and 7.

5 Conclusion

A prototype of a home health monitoring system has been achieved, offering a system implemented in a commercial-off-the-shelf (COTS) router that is able to collect data from any wireless sensors, is able to send vital signs data to any remote caretaker and does not require any technical intervention because it offers the auto setup through a remote auto-configuration server. Also, this is a robust system because any of its update features do not interfere with the vital sign measurements in progress. Modular software architecture allows easy organization when creating applications, modification of system segments, and adding new extensions to the system. The system supports any type of measurement of vital signs as well as through this mod-

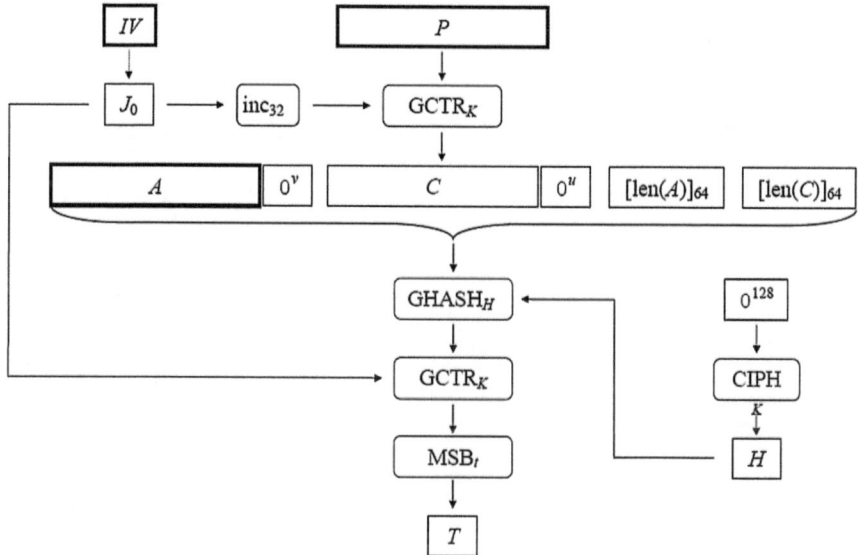

Fig. 5 GCM-AEK $(IV, P, A) = (C, T)$

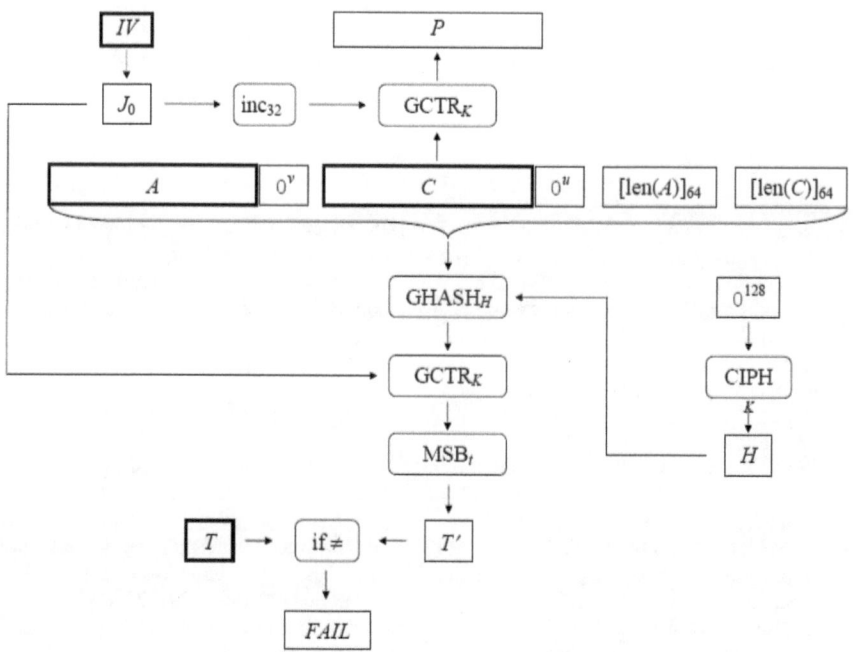

Fig. 6 GCM-ADK $(IV, C, A, T) = P$ or *FAIL*

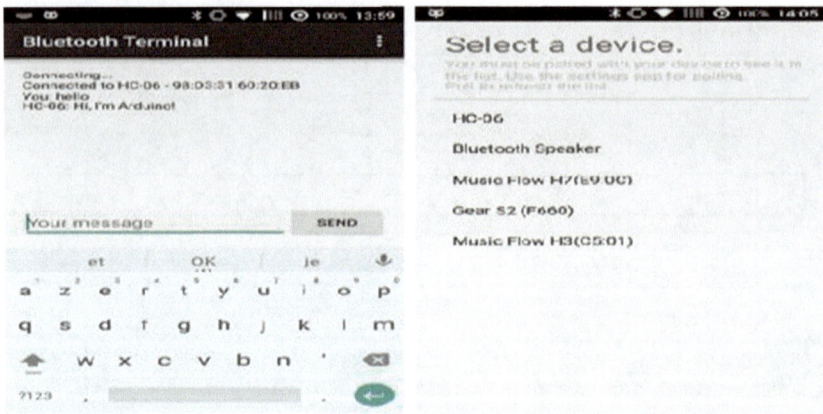

Fig. 7 Selecting a Bluetooth device

ularity, it is easy to develop sensor bundles using other communication protocols, as long the gateway has the necessary hardware. One of the gains during this project research was the chance to create bundles and learn how to handle the OSGi framework, taking advantage of the resources it offers, and strengthen the functionality of this framework.

References

1. Xiaodong L, Sun X, Ho P-H, Shen X (2007) GSIS: a secure and privacy-preserving protocol for vehicular communications. IEEE Trans Veh Technol 56(6):3442–3456
2. Avvenuti M, Corsini P, Masci P, Vecchio A (2007) Opportunistic computing for wireless sensor networks. In: IEEE international conference on mobile adhoc and sensor systems, MASS, pp 1–6. IEEE
3. Ren Y, Werner R, Pazzi N, Boukerche A (2010) Monitoring patients via a secure and mobile healthcare system. IEEE Wirel Commun 17(1)
4. Conti M, Giordano S, May M, Passarella A (2010) From opportunistic networks to opportunistic computing. IEEE Commun Mag 48(9)
5. Li M, Shucheng Yu, Zheng Y, Ren K, Lou W (2013) Scalable and secure sharing of personal health records in cloud computing using attribute-based encryption. IEEE Trans Parallel Distrib Syst 24(1):131–143
6. Passarella A, Conti M, Borgia E, Kumar M (2010) Performance evaluation of service execution in opportunistic computing. In: Proceedings of the 13th ACM international conference on modeling, analysis, and simulation of wireless and mobile systems, pp 291–298. ACM
7. Lu R, Lin X, Liang X, Shen X (2011) A secure handshake scheme with symptoms-matching for mhealthcare social network. Mob Netw Appl (special issue on wireless and personal communication) 16(6):683–694
8. ManogaranG, Varatharajan R, Lopez D, Kumar PM, Sundarasekar R, Thota C (2018) A new architecture of Internet of Things and big data ecosystem for secured smart healthcare monitoring and alerting system. Future Gener Comput Syst 82:375–387

9. Shi M, Hanxiang W, Zhang J, Han M, Meng B, Zhang H (2017) Self-powered wireless smart patch for healthcare monitoring. Nano Energy 32:479–487
10. Baig MM, GholamHosseini H, Moqeem AA, Mirza F, Lindén M (2017) A systematic review of wearable patient monitoring systems—current challenges and opportunities for clinical adoption. J Med Syst 41(7):115

Computer-Aided Lung Parenchyma Segmentation Using Supervised Learning

G. N. Balaji and P. Subramanian

Abstract Advances in the fields of image processing and information technology have led to the use of computers for the diagnosis of diseases. This has led to the emergence of Computer-Aided Diagnosis (CAD) systems for disease diagnosis. This research work focuses on improving the performance of CAD systems that use Computed Tomography (CT) of the chest for diagnosis of lung disorders. This improvement has been achieved by developing techniques for determining the significance of image features for discrimination among different diseases of the lung and a technique for segmentation of lung parenchyma in chest CT irrespective of the presence or absence of peripherally placed Pathology Bearing Regions (PBRs). Another major challenge in CAD of lung disorders based on analysis of chest CTs is accurate segmentation of lungs especially in the presence of peripherally placed PBRs. In this research work, a segmentation algorithm has been developed to extract the complete lung parenchyma even in the presence of severe peripherally placed PBR in chest CT. The proposed system has been found to improve the diagnostic performance of CAD systems for diagnosis of lung disorders based on analysis of chest CT slices. This would aid the physicians to perform better diagnosis, which would result in choosing the appropriate treatment strategy, thereby reducing the mortality rate.

Keywords Computer-aided diagnosis (CAD) · Computed tomography (CT) · Pathology bearing regions (PBRs)

G. N. Balaji
CVR College of Engineering, Hyderabad, India
e-mail: balaji.gnb@gmail.com

P. Subramanian (✉)
Guru Nanak Institute of Technology, Chennai, India
e-mail: 69subbu@gmail.com

© Springer Nature Singapore Pte Ltd. 2019
H. S. Saini et al. (eds.), *Innovations in Computer Science and Engineering*, Lecture Notes in Networks and Systems 74,
https://doi.org/10.1007/978-981-13-7082-3_46

403

1 Introduction

Lung is the most important organ in the respiratory system of human beings. It takes in oxygen from the air, gives it to the bloodstream and releases carbon dioxide. The lungs are covered by a thin tissue layer called the pleura. Human beings have two lungs, a right lung and a left lung. The right lung consists of three lobes. The left lung is smaller, providing accommodation for the heart and consists of two lobes. Although the term, lung parenchyma, refers solely to alveolar tissue, it has now been used normally to refer to any form of lung tissue including bronchioles, bronchi, blood vessels, interstitium, and alveoli [1]. In order to deliver oxygen to the body, air is breathed in through the nose, mouth, or both. After entering the nose or mouth, air travels down the trachea or "windpipe". The trachea divides into a left and right breathing tube, and these are termed bronchi. The left bronchus enters the left lung and the right bronchus enters the right lung. The bronchi divide into smaller tubes called bronchioles. The bronchioles end in tiny air sacs called alveoli. All alveoli are not in use at the same time, so that the lung has many to spare in the event of damage from disease, infection, or surgery.

1.1 Lung Diseases

Lung disease refers to any disease or disorder of the lung that affects the lung. Smoking [2], infections [3] and genetics [4] are the major factors responsible for most of the lung diseases. Lung diseases are classified into three broad categories, namely, airway diseases, lung tissue diseases, and lung circulation diseases. Many lung diseases involve an overlap of these three categories. Airway diseases [5] affect the airways that carry oxygen and other gases into and out of the lungs. They are characterized by the presence of persistent inflammation. These diseases usually cause a narrowing or blockage of the airways. They are also associated with alterations in tissue architecture, leading to impaired lung function and reduced quality of life. Examples include asthma, bronchiectasis, emphysema, and chronic bronchitis.

As per WHO statistics 2018, four respiratory disease categories, namely, lower respiratory infections, Chronic Obstructive Pulmonary Disease (COPD), trachea/bronchus/lung cancer, and tuberculosis, appear in the top ten causes of mortality all over the world. They account for one in six deaths and one in ten Disability-Adjusted Life Years (DALY) lost. Among the leading causes of death worldwide, lower respiratory infections take the third position, COPD takes the fourth position, trachea/bronchus/lung cancer take the seventh position, and tuberculosis takes the eighth position. All these lung diseases put together, they form the leading cause of death. The most common causes of DALYs lost worldwide is lower respiratory infections. The report also states that "These figures confirm that lung diseases have remained globally important causes of death and disability during the past two decades." According to WHO, COPD is underdiagnosed and will become the third leading cause of death in 2030.

1.2   Diagnostic Techniques

Chest radiography is the most commonly used imaging technique as it can be obtained at a minimum cost and the amount of exposure to radiation is acceptable by most experts. It remains invaluable [6] for the initial investigations of diseases of the lung, pleura, mediastinum, and chest wall. CT has been used as a standard procedure for the diagnosis of lung diseases. However, as different diseases exhibit different radiological signs and characteristics, exactly which CT technique has to be used depends greatly on the disease. The main strength [7] of CT is that it is capable of representing interstitial structures that are needed for the differential diagnosis of connective tissue diseases of the lung. CT is used to evaluate the shapes, borders, and densities of nodules [8]. When temporal analysis has to be done on images taken regularly over an interval of time, Magnetic Resonance Imaging (MRI) is preferred due to the absence of ionizing radiation. The role of MRI is significant in the diagnostic investigation of vascular diseases such as pulmonary hypertension, or complex diseases such as cystic fibrosis [7].

The authors in [9] have stated that clinical and radiological criteria would be sufficient for diagnosing about two third of the cases affected by Idiopathic Pulmonary Fibrosis (IPF) and that surgical lung biopsy is needed in the remaining one-third of cases to achieve the ultimate diagnosis. This requires multidisciplinary cooperation and biopsy has to be taken from at least two lobes due to the histological variability. Kaarteenaho [9] has reported that lung biopsy was performed in about 30–60% of the cases in large clinical trials conducted during the past decade and that the most serious complication of lung biopsy is mortality within 30 days after the procedure, with a frequency of about 3–4% reported in most studies.

Sputum smear microscopy [10, 11] is used for the diagnosis of pulmonary tuberculosis (TB). It is a simple, rapid, and inexpensive technique which is widely used in various populations with different socioeconomic levels. However, the sensitivity is grossly compromised when the bacterial load is less than 10,000 organisms/ml sputum sample and some patients may not come the following day. Hence, patient drop-outs are high. Shafiyabi et al. [11] have demonstrated that two spot sputum samples collected on a single day at an hour interval were equally effective as the 2-day method in detection of sputum smear-positive TB cases.

1.3   Computer-Aided Diagnosis

CAD is a software system for interpreting digital images or laboratory tests to provide a diagnosis. CAD system aids the physicians in performing diagnosis by giving a "second opinion" based on known similar cases by applying rule-based inference techniques, Artificial Intelligence (AI)techniques or Content-Based Image Retrieval (CBIR). The use of CAD combined with the physician's diagnosis has been shown to improve the diagnostic accuracy of radiologists, lighten the burden of increasing workload, reduce cancer missed due to fatigue, overlooked or data overloaded and improve inter- and intra-reader variability [12].

1.4   Role of Image Processing and Machine Learning in CAD

In CAD of lung disorders based on analysis of chest CT various image processing techniques are essential for dataset preparation. The chest CTs must first be made suitable for further processing by removing the noise and improving the quality of the image. This is achieved by using image enhancement techniques. Once the image is enhanced, the Regions of Interest (ROIs) have to be extracted and this involves the use of image segmentation techniques. The ROIs should then be described quantitatively and this involves the use of feature extraction techniques.

Machine learning methods have become one of the dominant approaches in CAD for medical imaging [13, 14]. Classifiers are developed and trained to label new cases according to a set of features derived from already labeled data and this approach comes under the field of supervised learning. The features may include the raw image data, higher level descriptive features derived from the images, or additional clinical data. However, using too many features in the classification algorithm can lead to problems, particularly if there are irrelevant features. This can lead to overfitting, in which noise or irrelevant features may influence the classification decisions because of the modest size of the training data. This requires the use of feature selection or dimensionality reduction techniques.

As per WHO statistics 2011, lung diseases form the leading cause of death. The mortality rate can be reduced if the disease is diagnosed at an early stage. CT images of the chest are the commonly used standard procedure for the diagnosis of lung diseases. There are chances that even an experienced radiologist misses small PBRs that appear at an early stage. They may also find discrimination among diseases difficult in certain cases.

With the advances in image processing and information technology, computer software that could aid the radiologist in diagnosis by providing quantitative analysis and detecting even small abnormal structures are being developed. Two major issues have been identified with such computer-aided diagnosis systems and these issues motivated this research work. First, the diagnostic accuracy of CAD systems depend on the image features chosen for analysis; but the mapping of radiological features into image features could be imprecise. Second, the diagnostic accuracy of CAD systems depend on the accuracy of segmentation result, which in turn depends on the location and size of the PBRs, for instance, the segmented lung may not include large PBRs that are attached to the chest wall. These two issues in CAD of lung disorders have been handled in this research work with an aim to aid the medical community in improving the diagnostic accuracy of the radiologists in diagnosis of lung disorders based on analysis of chest CT and thereby help the society to get treated at the early stages for the relevant lung disorder, if any. This could improve the survival rate of patients affected by lung diseases [15]. This paper aims at the development of approaches to improve the diagnostic accuracy of computer-aided diagnosis of lung disorders based on analysis of chest CT by focusing on both image processing and data analysis.

When a CAD system is used for analysis of CT images, image processing techniques, namely, image enhancement and image segmentation can be applied to extract both small and large ROIs at the same time. This will aid the radiologist in detecting the PBRs at an early stage. In addition, the ROIs can be analyzed quantitatively by computing quantitative measures of their features and applying machine learning techniques to derive the diagnosis. This diagnostic result can be considered as a second opinion by the radiologist thereby improving the diagnostic accuracy of the radiologist.

2 Literature Review

The authors of [16] in their work have proposed a fully automatic method for identifying the lungs in three dimensional (3D) CT images. They have achieved it in three steps. First, gray-level thresholding has been applied to extract the lungs. Then, the lung border has been identified using dynamic programming and finally, a set of morphological operations have been applied to smoothen the lung border. They have tested their proposed method by processing 3D CT datasets of eight normal subjects. Elizabeth et al. [17] in their work have developed a CAD system for the detection of bronchiectasis from CT images of the chest. They have used iterative thresholding followed by morphological operations for segmentation [18] and region growing for ROI extraction. They have used Gray-Level Co-occurrence Matrix (GLCM) texture features for describing the characteristics of the ROIs. They have performed diagnosis using two techniques, namely, Probabilistic Neural Network (PNN) and Mahalanobis distance. They have concluded that the system shows higher efficiency with a probabilistic neural network to classify the images as diseased or not. Sluimer et al. [19] in their work have proposed a segmentation by-registration scheme for segmentation of lung from chest CT in which an atlas-based segmentation of the pathological lungs is refined by applying voxel classification to the border volume of the transformed probabilistic atlas. Lai and Ye [20] in their work have proposed an active contour based lung parenchyma segmentation approach with prior knowledge about shape to fit the lung boundary. Hu et al. [16] in their work have proposed an automated method for segmentation of lung parenchyma from chest CT. Darmanayagam et al. [21] in their work have proposed a supervised approach for segmentation of lung from chest CT. This involves training a Back Propagation Neural Network (BPNN) with shape features to distinguish between the complete and incomplete lung and using it to determine if the lung extracted by thresholding and morphological operations is complete.

From the literature discussed in this section on segmentation approaches that can be used in CAD systems for diagnosis of lung disorders, it can be inferred that existing segmentation approaches are not efficient in segmenting the lung in the presence of peripherally placed PBR.

3 Methodology

Segmentation of lung parenchyma is one of the challenging tasks in the CAD of lung disorders based on analysis of chest CT. In this paper, a system is proposed using which the initial lung region is obtained by applying a combination of iterative thresholding and morphological operations. The shape features of the resulting lung region are applied to a backpropagation neural network (BPNN) that is trained using a dataset to determine whether the segmented lung forms a complete lung. Figure 1 shows the steps involved in the extraction of lung parenchyma.

A set of 100 chest CT slices of different patients were taken for the experiment. The images were segmented using iterative thresholding and morphological operations. The relationship between the shape features of the left lung and that of the right lung has been used to determine the completeness of the segmented lung. In order to determine this relationship, three possible sets of shape features that could discriminate among the complete and incomplete lungs were identified.

The dataset consisting of 100 chest CT slices of different patients were collected from Rajah Muthaiah Medical Hospital, Annamalai University for this study. The images were segmented using iterative thresholding and morphological operations. The relationship between the shape features of the left lung and that of the right lung has been used to determine the completeness of the segmented lung. In order to determine this relationship, three possible sets of shape features that could discriminate among the complete and incomplete lungs were identified.

If the initial lung is not complete, the following operations are performed: First, the longer of the two connected components in the initial lung region is determined. The longest connected component is then folded and translated horizontally. The two lung regions are then converted to a single connected component and the convex hull is obtained. The convex hull is interpolated to obtain the outer convex edge. The outer convex edge thus obtained is superimposed on the binary image obtained by folding and translation and used as the initial contour for the ACM. It is also ensured that the number of components does not exceed two. This method is adaptive, in that the number of iterations of ACM is not fixed and is based on the image for which it is applied. This method of lung segmentation has been compared with the conventional iterative thresholding method, convex hull based algorithm and supervised algorithm for segmentation.

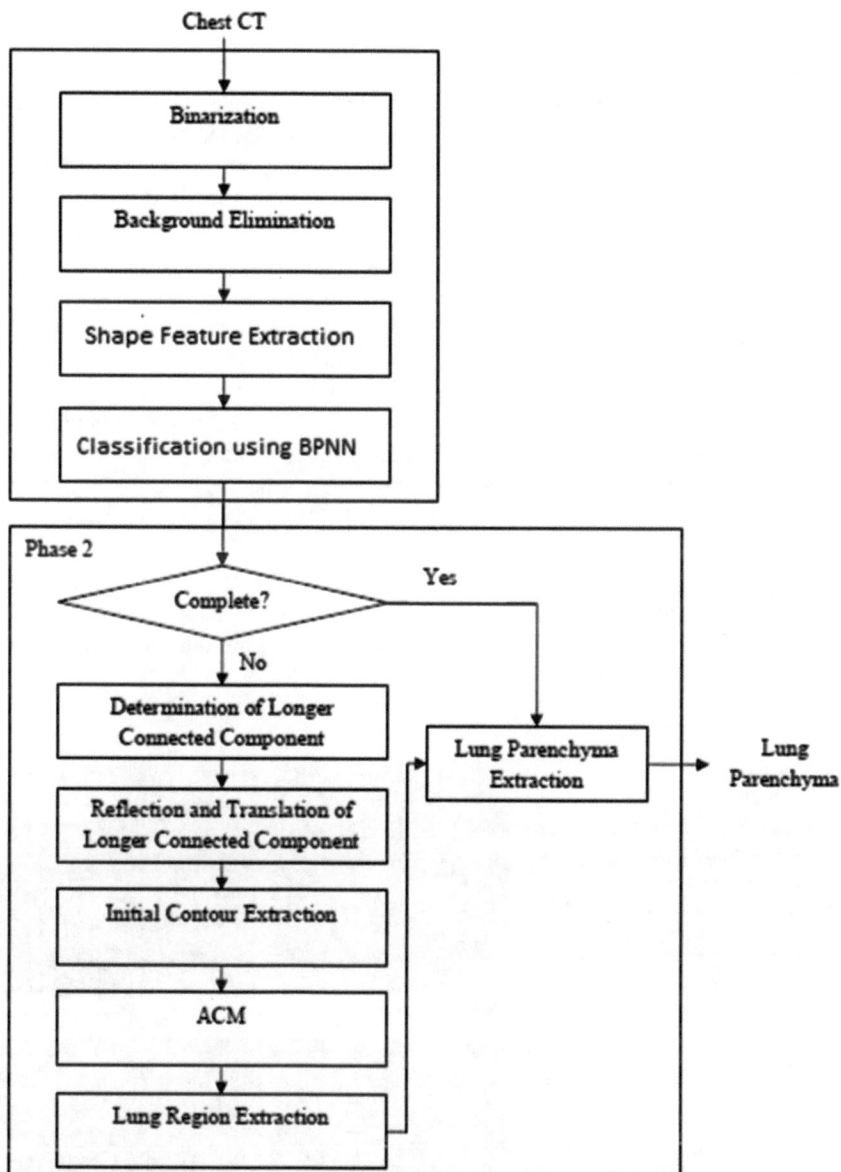

Fig. 1 Proposed lung parenchyma segmentation

4   Results and Discussion

The proposed two-phase segmentation algorithm has been tested with chest CT images of subjects affected by various types of lung disorders, namely, bronchiectasis, tuberculosis, and pneumonia. It has been compared with the conventional iterative thresholding method, convex hull based algorithm, and supervised algorithm for segmentation.

The test dataset consisted of 20 chest CT slices of bronchiectasis affected patients, 20 chest CT slices of tuberculosis affected patients, 10 chest CT slices of pneumonia affected patients, and 20 chest CT slices of normal patients. Figure 2 shows the results obtained for chest CTs of patients affected by bronchiectasis.

Quantitative evaluation of the accuracy of the lung segmentation algorithm has been done by computing the overlap percentage. The best case is 100, which means in cases where there are no juxta-pleural nodules, the discontinuity between the lung and the outer chest region is clear, and hence the lung segmented by all the algorithms considered in this work perform effectively with 100% overlap. The worst case is an overlap of 55.3% with the proposed two-phase supervised lung segmentation algorithm, 37.83% with iterative thresholding based segmentation, 25.82% with convex hull based segmentation, and 54.25% with supervised segmentation. Thus, the proposed algorithm is found to perform better compared to the other two algorithms with which it has been compared.

Fig. 2 **a** Input CT image, **b** threshold-based segmentation, **c** convex hull based segmentation, and **d** the proposed segmentation method

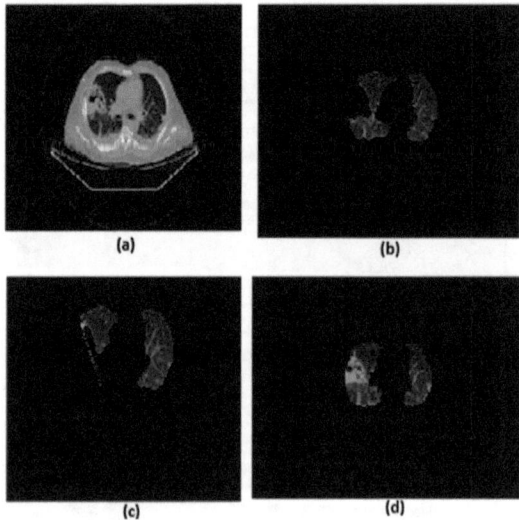

5 Conclusion

In this paper, a supervised segmentation algorithm has been proposed for segmentation of lungs from chest CT images. This has been found to be a challenging task especially in cases of subjects, whose lung consists of peripherally placed PBRs as discussed in the literature. The proposed segmentation algorithm combines the relative advantages of iterative thresholding method, convex hull based algorithm, supervised algorithm, and ACM that have been used in the literature for segmentation of lungs and hence found to produce better segmentation results compared with each of the individual algorithms. Segmentation of lungs is a crucial and challenging step in CAD of lung disorders based on analysis of chest CT. Hence, segmentation techniques that result in accurate segmentation results are preferred, thereby providing further scope for research in this area.

References

1. https://www.emeraldinsight.com/doi/abs/10.1108/09504120710719671
2. Murin S, Bilello KS (2005) Respiratory tract infections. Clevel Clin J Med 72(10):916–920
3. Panda BN (2004) Fungal infections of lungs: the emerging scenario. Indian J Tuberculosis 51:63–69
4. Machado RD, Eickelberg O, Elliott CG, Geraci MW, Hanaoka M, Loyd JE, Newman JH, Phillips III JA, Soubrier F, Trembath RC, Chung WK (2009) Genetics and genomics of pulmonary arterial hypertension. J Am College Cardiol 54(1), Supplement:S32–S42
5. Kraft M (2011) Approach to the patient with respiratory disease. In: Goldman L, Schafer AI (eds) Cecil medicine, 24th edn, Chap 83. Saunders Elsevier, Philadelphia, PA
6. Stark P (2011) Imaging of the lungs, mediastinum, and chest wall. In: Goldman L, Schafer AI (eds) Cecil medicine, 24th edn, Chap 84. Saunders Elsevier, Philadelphia, PA
7. Welte T (2014) Imaging in the diagnosis of lung disease: more sophisticate methods require greater interdisciplinary collaboration. Deutsches Arzteblatt Int 111(11):170–180
8. Bunyaviroch T, Coleman RE (2006) PET evaluation of lung cancer. J Nuclear Med 47(3):451–469
9. Kaarteenaho R (2013) The current position of surgical lung biopsy in the diagnosis of idiopathic pulmonary fibrosis. Respir Res 14:43
10. Desikan P (2013) Sputum smear microscopy in tuberculosis: is it still relevant? Indian J Med Res 137(3):442–444
11. Shafiyabi S, Ravikumar R, Krishna Ramaprasad S (2013) Study of same day sputum smear examination in diagnosis of pulmonary tuberculosis under RNTCP. Sch Acad J Biosci 1(7):352–356
12. Fujita H, Uchiyama Y, Nakagawa T, Fukuoka D, Hatanaka Y, Hara T, Lee GN, Hayashi Y, Ikedo Y, Gao X, Zhou X (2008) Computer-aided diagnosis: the emerging of three CAD systems induced by Japanese health care needs. Comput Methods Programs Biomed 92:238–248
13. Way TW, Hadjiiski LM, Sahiner B, Chan HP, Cascade PN, Kazerooni EA, Bogot N, Zhou C (2006) Computer-aided diagnosis of pulmonary nodules on CT scans: segmentation and classification using 3D active contours. Med Phys 33(7):2237–2323
14. Balaji GN, Subashini TS, Chidambaram N (2016) Detection and diagnosis of dilated cardiomyopathy and hypertrophic cardiomyopathy using image processing techniques. Eng Sci Technol Int J 19(4):1871–1880

15. Depeursinge A, Iavindrasana J, Cohen G (2008) Lung tissue classification in HRCT data integrating the clinical context. In: Proceedings of the twenty-first IEEE international symposium on computer-based medical systems, pp 542–547
16. Hu S, Hoffman EA, Reinhardt JMM (2001) Automatic lung segmentation for accurate quantitation of volumetric x-ray CT images. IEEE Trans Med Imaging 20(6):490–498
17. Elizabeth DS, Kannan A, Nehemiah HK (2009) Computer aided diagnosis system for the detection of bronchiectasis in chest computed tomography images. Int J Imaging Syst Technol 19(4):290–298
18. Balaji GN, Subashini TS (2013) Detection of cardiac abnormality from measures calculated from segmented left ventricle in ultrasound videos. In: Mining intelligence and knowledge exploration. Springer, Cham, pp 251–259
19. Sluimer I, Prokop M, Ginneken BV (2005) Toward automated segmentation of the pathological lung in CT. IEEE Trans Med Imaging 24(8):1025–1038
20. Lai J, Ye M (2009) Active contour based lung field segmentation. In: Proceedings of the international conference on intelligent human machine systems and cybernetics, pp 288–291
21. Darmanayagam SE, Harichandran KN, Cyril SRR, Arputharaj K (2013) A novel supervised approach for segmentation of lung parenchyma from chest CT for computer-aided diagnosis. J Digit Imaging 26(3):496–509

Comparative Study of Machine Learning Approaches for Heart Transplantation

Shruti Kant and Vandana Jagtap

Abstract Heart failure is a severe medical case, where the heart is not able to function properly to maintain blood flow. The surgical heart transplant procedure accomplished with the last stage of failure of the heart. Machine learning approaches account an ability to handle large datasets systematically and are extensively used in a biomedical research field. With the help of machine learning algorithms, tools are developed that helps specialist as a successful mechanism. The objective of this study is to learn different machine learning approaches for analyzing the heart transplantation dataset by using suitable classification algorithm. Also, the theoretical and the experimental comparative study of different machine learning techniques, using heart transplantation data. This study provides basic guidelines on machine learning technique. The results provide an overview of machine learning technique. We have used a WEKA machine learning software for evaluation and analysis to get an easy way to understand the result.

Keywords Artificial neural network · Classification · Data analysis · Heart transplantation · Machine learning

1 Introduction

Heart failure is a severe medical case, happens when the heart cannot direct sufficiently to keep up blood flow to address the body's issues. A coronary infraction is an ordinary, expensive, and potentially critical and very important condition. Organ shortage and clinical complications after cardiac therapy make the appropriate donor and recipient choice fundamental for a successful and fruitful outcome. The surgical

S. Kant (✉) · V. Jagtap
Department of Computer Engineering, MAEER'S Maharashtra Institute of Technology, Pune, Maharashtra, India
e-mail: shruti.kant02@gmail.com

V. Jagtap
e-mail: vandana.jagtap@mitpune.edu.in

© Springer Nature Singapore Pte Ltd. 2019
H. S. Saini et al. (eds.), *Innovations in Computer Science and Engineering*, Lecture Notes in Networks and Systems 74, https://doi.org/10.1007/978-981-13-7082-3_47

heart transplant procedure accomplished with the last stage of failure of the heart. The interest for heart transplantations has been terribly expanding for a few reasons that include: increasing population of middle age people, increasing the ratio of obesity, etc. Anticipating the endurance of the person being treated for a heart transplant is an imperative, still a difficult dilemma because of the critical process of figuring out the corresponding identical donor and a recipient. Organ transplantation ensures the two important things, i.e., long-life survival of the organ and patient and the quality of life.

Medical field uses Model for the End-stage Liver Disease (MELD) score for a precise and superior assignment of the organ. It helps in the organ assignment using the sickest first policy. Utilizing the sickest first policy the first waiting list patient receive the organ first without thinking about the features of the donor patient and receiver patient. However, after using MELD score for a patient undergoing transplantation it is hard to estimate and predict the correct outcome for transplantation.

The machine learning approaches like Artificial Neural Networks (ANN) may accurately anticipate graft outcomes by combining donor and recipient data and accordingly utilize to optimize donor-recipient selection [1]. The benefit of ANNs is that they are able to manage a huge amount of unpredictable, fragmented, incomplete, and often uncertain information and are able to recognize better solutions considering all pertinent information [2]. ANNs have been used in several clinical operations such as diagnosis of disease, classification and staging, analysis of images, and outcome prediction of disease.

2 Machine Learning Approaches

The basic detail of different machine learning approaches used in this study for analyzing heart transplantation dataset is given below.

2.1 Multilayer Perceptron Neural Network

The basic and important property of Artificial Neural Network (ANN) is prediction. The structure of the ANN classification model is similar to biological neural systems, the neurological components are interconnected by neurotransmitters. ANN anticipates new information by learning from history or existing information [3]. The predictive models are build by learning from historical data. The commonly used type of neural network is Multilayer Perceptron (MLP) neural network, which put the nodes in layers. The input layer is the first layer. The output layer is the outermost layer and the middle from two layers, come one or more layers these are known as the hidden layer [4–6].

2.2 K-Nearest Neighbor (K-NN)

K-Nearest Neighbors (K-NN) is a supervised classification algorithm in which the k-closest neighbors of a point are picked, by limiting a closeness measure. The classification is done in view of the closest neighbors. To decide the class of unlabeled case tuples, K-NN figures its closeness to the remaining (labeled) tuples, and decides its k-closest neighbors and the respective labels. The unlabeled tuples are then classified either by popularity voting the predominant class in the neighborhood or by a weighted popularity, where a greater weight is given to the points closer to the unlabeled tuples. The disadvantage of K-NN is that it is a lazy learning algorithm, i.e., if there is no "model": the training data is not utilized to play out any classification. This requires a precise investigation of the dataset and the improvement of different K-NN models, with a specific end goal to accomplish the best outcomes [7].

2.3 Naive Bayesian

This model is a statistical pattern recognition method which has an exact presumption about how the information is created. Naive Bayes (NB) classifier considers the probability distribution of the patterns in each class to make a choice, accepting that there is a probabilistic relationship between features and the output class. With exact preprocessing, it can compete to ML approaches such as the support vector machine and other similar methods. Another application of naive Bayes algorithm is in the automatic medical diagnosis. The benefit of this method is enabling the specialist to incorporate his domain experience in the demonstrating procedure of NB classifiers. The error rate of Bayesian classifier is also minimum in contrast to all other classifiers [8].

2.4 Random Forest

Random Forest is a supervised learning algorithm. Random forest constructs different decision trees and combines them together to get a more precise and exact prediction. It is additionally standout amongst the most used algorithms, because of its simplicity and also the way that it can be used for both regression tasks and classification. The benefit of a random forest approach is it runs effectively on huge databases. Also, another benefit is it can deal with thousands of input features without feature deletion. The random forest machine learning algorithm is excellent or magnificent in accuracy among current algorithms [9, 10].

3 Materials and Methods

3.1 Dataset Description

The dataset used in this study was supplied by the Stanford Heart transplant program. This dataset comprises the data about the endurance of the waiting list patients for the heart transplantation scheme. Dataset comprises of 172 number of instances and 8 attributes.

3.2 Performance Measures

For computing the efficiency of the various classification algorithms the following performance measures are available. These all are as follows:

$$Accuracy = TP + TN/(TP + TN + FP + FN) \tag{1}$$

$$Sensitivity = TP/(TP + FN) \tag{2}$$

$$Specificity = TN/(TN + FP) \tag{3}$$

where TP, TN, FP, FN indicates the true positive, true negative, false positive, and false negative, respectively [11, 12].

3.3 K-Fold Cross-Validation

K-fold cross-validation is also known as rotation estimation. This is for analyzing the predictive accuracy of two or more techniques [13, 14]. In k-fold cross-validation, the entire database is arbitrarily partitioned into k fundamentally unrelated subsets of the almost same amount. The training and testing of the classification model is done k times. Every time, it is trained on all except one-partition and tested on the remaining single partition [11, 15].

4 Result and Analysis

The evaluation and analysis are done using WEKA software. It is open source data analysis software supplied by the Waikato University. We have captured or record

Table 1 The comparative study table of classification models

Parameters	Multilayer perceptron	K-nearest neighbor	Naive Bayesian	Random Forest
Time taken to build model (s)	1.18	0.01	0.03	0.67
Accuracy (%)	80.23	72.93	81.39	80.23
Precision	0.803	0.720	0.823	0.803
Recall	0.802	0.721	0.814	0.802
F1-score	0.802	0.720	0.811	0.802

performance of the multilayer perceptron neural network, k-nearest neighbor, naive Bayesian and random forest classifiers along with the tenfold cross-validation.

Four different classifiers are applied or used on heart transplant dataset with tenfold cross-validation. These classification algorithms are compared based on the time required to build the model and performance measures. The comparative results of four classifiers are shown in Table 1.

We have compared the four classification model performances based on performance measures. From this result, we observed that the accuracy of naive byes, multilayer perceptron and random forest algorithm is significant than K-nearest neighbor model. The other performance measures such as precision, recall, and F1-score for these algorithms are somewhat same to some extent. The simulation of multilayer perceptron and the other three classification algorithms are carried out with WEKA machine learning software and simulation results are provided in Table 1. The training of all four classification algorithms is carried out with tenfold cross-validation. It is observed from the results that the K-nearest neighbor algorithm accuracy is very less as compared to other three classification algorithms even though it requires minimum time to build.

Whereas it is observed that the accuracy result derived for MLP, random forest and naive Bayes are similar to some extent, MLP algorithm can imply higher efficiency compared to that of random forest and naive Bayes, since MLP handles complex clinical problems efficiently and gives the outstanding result by creating complex patterns. Also, the neural network works best or better when having a larger database.

5 Conclusion

Heart transplantation is imperative medical care for the serious and incurable condition of the heart. The survival analysis is a suitable procedure to discover the impact of such a task. The objective of this study is to learn different machine learning approaches for analyzing the heart transplantation dataset by using suitable classification algorithm. The dataset used for the study is supplied by the Stanford Heart Transplant Program. The analysis of heart transplantation data is achieved using four machine learning algorithm multilayer perceptron neural network, k-nearest neighbor, naive Bayesian, and random forest. These techniques are evaluated based on the criteria of precision, accuracy, recall, and F1-measure. Also, time appropriated for the model construction is also considered for evaluation. The result of comparison shows that accuracy result derived for MLP, random forest, and naive Bayes are similar, MLP algorithm implies higher efficiency compared to that of random forest and naive Bayes, since MLP handles complex clinical problems efficiently and gives outstanding result by creating complex patterns.

References

1. Ramírez MC, Martínez CH, Fernández JC, Briceño J, la Mata M (2013) Predicting patient survival after liver transplantation using evolutionary multi-objective artificial neural networks. Elsevier J Artif Intel Med 58(1):37–49
2. Petrovsky N, Tam SK, Brusic V, Russ G, Socha L, Bajic VB Use of artificial neural networks in improving renal transplantation outcomes, vol 5, Issue 1, Feb 2002. SAGE Publications
3. Dag A, Oztekin A, Yucel A, Bulur S, Megahed FM (2017) Predicting heart transplantation outcomes through data analytics. Elsevier J Dec Support Syst 19:42–52
4. Gardner MW, Dorling SR (1998) Artificial neural networks (the multilayer perceptron)—a review of applications in the atmospheric sciences. In: Elsevier science, atmospheric environment, vol 32, no 14/15, pp 2627–2636
5. Zhang M et al (2012) Pretransplant prediction of posttransplant survival for liver recipients with benign end-stage liver diseases: a nonlinear model. PLoS One 7(3), Art. no. e31256
6. Caocci G, Baccoli R, Vacca A, Mastronuzzi A et al (2010) Comparison between an artificial neural network and logistic regression in predicting acute graft -vs-host disease after unrelated donor hematopoietic stem cell transplantation in thalassemia patients. Elsevier J Exp Hematol 38(5):426–433
7. Rao V, Behara RS, Agarwal A Predictive modeling for organ transplantation outcomes. In: IEEE international conference on bioinformatics and bioengineering (BIBE). Boca Raton, USA, Nov 2014
8. Lin RS, Horn SD, Hurdle JF, Goldfarb-Rumyantzev AS (2008) Single and multiple time-point prediction models in kidney transplant outcomes. Elsevier J Biomed Inf 41(6):944–952
9. Kaur H, Wasan SK (2006) Empirical study on applications of data mining techniques in healthcare. Citeseerx J Comput Sci (2):194–200. ISSN 1549-3636
10. Lawrence L, Yamuna K, Benjamin R, Jones et al (2017) Machine-learning algorithms predict graft failure after liver transplantation. J Transp Soc Int Liver Transp Soc 101(4):e125–e132
11. Oztekin A, Delen D, Kong Z(James) (2009) Predicting the graft survival for heart–lung transplantation patients: an integrated data mining methodology. Elsevier Int J Med Inf 78(12):e84–e96
12. Raji CG, Vinod Chandra SS (2017) Long-term forecasting the survival in liver transplantation using multilayer perceptron networks. IEEE Trans Syst Man Cybern Syst 47(8)
13. He H, Garcia EA (2009) Learning from imbalanced data. IEEE Trans Knowl Data Eng 21(9):1263–1284
14. Kotsiantis S, Kanellopoulos D, Pintelas P (2006) Handling imbalanced datasets: a review. GESTS Int Trans Comput Sci Eng 30
15. Raji CG, Vinod Chandra SS (2016) Predicting the survival of graft following liver transplantation using a nonlinear model. Springer J Publ Health 24(5):443–452

Hybrid Method for Speech Enhancement Using α-Divergence

V. Sunnydayal, J. Sirisha Devi and Siva Prasad Nandyala

Abstract A hybrid method for speech enhancement based on Non-Negative Matrix Factorization (NMF) and statistical modeling is presented for using speech and noise bases with online updating is proposed. In the presence of nonstationary noises, template-based approaches have shown better performance when compared to statistical modeling but these approaches depend on a priori information. To overcome the drawbacks of these approaches, a hybrid method is developed. The performance of the proposed method is further improved by considering speech bases as well as noise bases. In terms of Source-to-Distortion ratio (SDR) and Perceptual Evaluation of Speech Quality (PESQ) the proposed method have outperformed the traditional algorithms in nonstationary noise environment conditions.

Keywords Online-based update · Non-negative matrix factorization (NMF) · Source-to-distortion ratio (SDR) · Speech enhancement

1 Introduction

Speech enhancement is having many applications like hearing aids, mobile phones. Spectral subtraction is the noise reduction technique and is suffered from musical noise [1]. Input signal estimates the parameters in these assessments [2, 3]. Prior training is required in case of statistical method based methodologies. The Wiener

V. Sunnydayal (✉)
Vellore Institute of Technology AP, Amaravathi, Andhra Pradesh, India
e-mail: sunny.dayal@vitap.ac.in

J. Sirisha Devi
Institute of Aeronautical Engineering, Hyderabad, India
e-mail: siri.cse21@gmail.com

S. P. Nandyala
Model Based Design, ABU, Tate Elxsi Limited, Bangalore, India
e-mail: sivaprasad.n@tataelxsi.co.in

© Springer Nature Singapore Pte Ltd. 2019
H. S. Saini et al. (eds.), *Innovations in Computer Science and Engineering*, Lecture Notes in Networks and Systems 74,
https://doi.org/10.1007/978-981-13-7082-3_48

channel performs well for stationary commotion, however, it flops in nonstationary natural conditions [2].

Information like statistics or patterns, regarding speech and noise is exercised in advance in template-based approaches [4, 5]. NMF falls under this category. In template-based method, Power Spectral Densities (PSD) of speech and noise are obtained from their estimates of magnitude spectrum [6]. The PSDs are used for computing gain functions in the style of the Wiener filter. In [7], Wiener filter gives an output based on template-based algorithm. A combination of template-based and statistical-based approach is the advantage of this method. Since there is no update of the method, the output of the Wiener filter might be distorted. The Voice active detector (VAD) and NMF-based speech enhancement were proposed in [8]. In this methodology, the execution debases if the prepared commotion method is not the same as the genuine noise environment. The main disadvantage of this method depicts in the cases where each set communicates with a separate set; simultaneous updating of multiple sets of databases is not possible. In this paper [9] Amari's α-divergence is considered as an inconsistency assessment and thoroughly infers a multiplicative redesigning calculation, which minimizes the divergence.

In [10], template-based and statistical method based approaches are combined in parallel with updating the bases are proposed. The proposed algorithm does not use speech and noise prototypes which are present in the training dataset. It uses a combination of [9] and [10] which will decrease the reliance on from the earlier data and it can deal with nonstationary noises.

The following paper is organized as, in the Sect. 2, spectral gain used in NMF-Based speech enhancement is explained. In Sect. 3, statistical and α-NMF-based approach's combination is used for online update is discussed. Section 4 explains the experimental results. Section 5 concludes with a conclusion.

2   NMF-Based Speech Enhancement

The summation of speech signal S[n] and noise signal N[n] is a noisy speech signal Y[n] and given in Eq. (1).

$$|Y(f, t)| = |S(f, t)| + |N(f, t)| \tag{1}$$

where f denotes the frequency and t denotes the time frame.

The NMF factorizes $\mathbf{X}$ as two non-negative lattices, $\mathbf{W} = [\mathbf{w}_1, \mathbf{w}_2, \ldots, \mathbf{w}_R] \in R^{f \times R}$ and $\mathbf{H} = \left[h_1^T, h_2^T, \ldots, h_R^T \right]^T \in R^{R \times t}$ and R represents latent components. Weight of X is given by Eq. (2)

$$X = WH = \sum_{z=1}^{R} W_z H_z \tag{2}$$

Assume

$$W = \left[|W_s(f, t)| \; |W_n(f, t)| \right] \in R^{M \times (r_s + r_n)},$$

where $W_s \in R^{M \times r_s}$, $W_n \in R^{M \times r_n}$, r_s and r_n represent speech and noise bases matrices and their corresponding base vectors. Similarly, $H = \left[|H_s^T(f, t)| \; |H_n^T(f, t)| \right] \in R^{(r_s + r_n) \times N}$.

The estimate of speech $\left| \hat{S}(f, t) \right|$ and the estimate of noise $\left| \hat{N}(f, t) \right|$ spectra magnitude are given in [10].

In the proposed approach, $|G(f, t)|$ is derived from $\left| \hat{S}(f, t) \right|$ and $\left| \hat{N}(f, t) \right|$.

Minimum Mean Square Error-Log Spectral Amplitude (MMSE-LSA) method [2] is incorporated and the gain is obtained using [10].

The a priori $\xi(f, t)$ signal to noise ratio (SNR) and a posterior SNR $\gamma(f, t)$ are estimated by using [10].

Finalized enhanced speech is acquired by Eq. (3)

$$\hat{S}^{Final}(f, t) = G(f, t)Y(f, t) \tag{3}$$

3 Proposed Method with Alpha-NMF

The proposed technique comprises of two stages. In stage 1, Statistical Model-based Enhancement (SE) and in stage 2, online NMF bases update are proposed as appeared in Fig. 1. The yield of SE gives pre-upgraded motion with an enhanced yield SNR The came about the pre-improved flag is utilized for stage 2. Alpha-NMF analysis includes KL-divergence ($\alpha = 1$), Hellinger divergence ($\alpha = 0.5$), and χ^2-divergence (Pearson's distance, $\alpha = 2$) as a distance metric. $X(f, t) = \left| Y'(f, t) \right|$ is given as input to NMF. In the NMF investigation, the redesigned bases at tth casing are utilized for the $(t + 1)$th outline.

3.1 Speech and Noise Online Update Bases

The bases for Alpha-NMF is updated, after obtaining $\left| \hat{S}(f, t) \right|$ and $\left| \hat{N}(f, t) \right|$. The variations in speech and noise can be addressed by using online updated bases. Since $Y'(f, t)$ gives distorted output components, it may mislead the speech enhancement

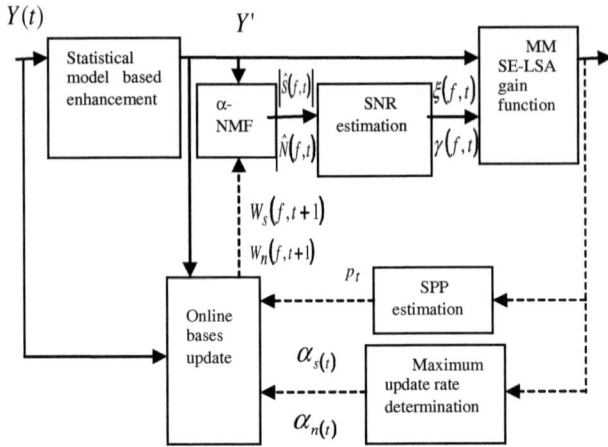

Fig. 1 Proposed speech enhancement approach

procedure if we feed this output to the stage. Hence bases must be updated. The bases online can be updated using $\tilde{X}(f, t)$ and is the concatenation of $Y(f, t)$ and $Y'(f, t)$.

The α-divergence includes KL-divergence when $\alpha = 1$, Hellinger divergence when $\alpha = 0.5$, and χ^2-divergence (Pearson's distance when $\alpha = 2$). Multiplicative updates for α-NMF are given as [9]

$$W_{ij} \leftarrow W_{ij} \left[\frac{\sum_k H_{jk}(Y_{ik}/[WH]_{ik})^\alpha}{\sum_l H_{jl}} \right]^{\frac{1}{\alpha}} \tag{4}$$

$$H_{ij} \leftarrow H_{ij} \left[\frac{\sum_k W_{ki}(Y_{ik}/[WH]_{kj})^\alpha}{\sum_l W_{li}} \right]^{\frac{1}{\alpha}} \tag{5}$$

where $H \in R^{(r_s+r_n)\times 2}$ and $\tilde{W} = \left[\tilde{W}_s \; \tilde{W}_n \right] \in R^{M\times(r_s+r_n)}$.

When the speech signal is sparse in nature, $\tilde{W}_s$ may not give accurate speech bases. This can be overcome by using SPP $p_t \in R^{M\times 1}$. The update of bases can be controlled by p_t. p_t is estimated using as given in [10].

3.2 Determination of Maximum Update Rates

Maximum update rates $\alpha_{s(t)}$ and $\alpha_{n(t)}$ are defined by [10]. From frame to frame, to avoid variations in $e(t)$, $H(f, t)$ is acquired from NMF stage and smoothed as given in Eq. (6)

$$\tilde{e}(t) = \tau_e \tilde{e}(t-1) + (1 - \tau_e)e(t) \qquad (6)$$

where τ_e is smoothing constant $\tilde{e}(t)$ represent smoothed reconstruction error $\alpha_{s(t)}$ and $\alpha_{n(t)}$ are obtained as in [10].

4 Experimental Results

TIMIT speech data is used for experimentation [11]. The noise samples (street and babble noise) were taken from Noisex database [12] and embedded into the speech signals. The input SNRs of 0, 5, 10, 15 dB is considered. The sampling rate is $f_s = 16$ KHz. Hanning window of length 1024 samples is used in the experiment. Speech and noise have 100 each ($r_s = r_n = 100$) number of bases vectors.

NMF algorithm is applied with different values of α ($\alpha = 0.5, 2$). The parameter values were $\alpha_s^{max} = 0.5$, $\alpha_n^{max} = 0.6$, $\tau_s = 0.4$, $\tau_e = 0.98, \tau_n = 0.8$.

The traditional methods used to evaluate proposed method are Alpha-NMF (α-NMF) [9], Itakura-Saito NMF (IS-NMF) [13], Wiener filtering [14], constrained version of NMF (CNMF) [15], SE + (α-NMF) without the online bases update, Speech Enhancement (SE) [3] and SE + (α-NMF) + OU: SE and NMF with online update. PESQ [16] and SDR [17] are considered as the parameters to evaluate the proposed method.

In this experiment, NMF with $\alpha = 0.5$ (Hellinger Divergence) yields better performance. Since most of the discriminative characteristics are revealed when $\alpha = 0.5$. NMF with $\alpha = 2$ (χ^2-Divergence or Pearson's Distance) shows higher PESQ and SDR values compared to other existing NMF methods as shown in Tables 1, 2. The proposed method gives PESQ values of 2.32, 2.62 and 2.95 SDR values of 9.39, 16.23 and 20.97, respectively, for 0 dB, 5 dB, and 10 dB.

Tables 1 and 2 show that Wiener filter performance is poor for nonstationary noise. IS-NMF, CNMF and α-NMF approaches also give poor results since their bases are not updated online and hence these approaches cannot address the variations in noise and speech. The proposed method overcomes the disadvantages of traditional methods by updating the bases online.

Table 1 PESQ and SDR for χ^2-divergence (Pearson's distance, when $\alpha = 2$) in babble noise

Method	Evaluation	Input SNR (in dB)			
		0	5	10	15
Wiener [14]	PESQ	2.61	3.12	3.42	3.63
	SDR	2.14	6.00	9.53	10.76
SE [3]	PESQ	2.67	3.24	3.44	3.70
	SDR	2.31	6.06	9.68	10.82
IS-NMF [13]	PESQ	2.82	3.29	3.48	3.61
	SDR	2.54	6.37	10.06	11.06
CNMF [15]	PESQ	2.51	3.08	3.50	3.74
	SDR	2.64	7.84	10.24	10.85
α-NMF [9]	PESQ	2.90	3.32	3.56	3.79
	SDR	4.84	9.39	10.68	11.54
SE + (α-NMF)	PESQ	2.83	3.40	3.76	3.94
	SDR	10.19	14.25	17.32	19.69
SE + NMF [10]	PESQ	3.02	3.47	3.73	4.17
	SDR	10.51	14.64	17.73	20.38
Proposed SE + (α-NMF) + OU	PESQ	**3.08**	**3.52**	**3.79**	**4.24**
	SDR	**10.56**	**14.70**	**17.81**	**20.43**

Table 2 PESQ and SDR for Hellinger divergence (when $\alpha = 0.5$) in babble noise

Method	Evaluation	Input SNR (in dB)			
		0	5	10	15
Wiener [14]	PESQ	2.61	3.12	3.42	3.63
	SDR	2.14	6.00	9.53	10.76
SE [3]	PESQ	2.66	3.24	3.44	3.70
	SDR	2.31	6.05	9.68	10.82
IS-NMF [13]	PESQ	2.82	3.29	3.48	3.61
	SDR	2.54	6.37	10.07	11.05
CNMF [15]	PESQ	2.51	3.06	3.50	3.74
	SDR	2.64	7.84	10.24	10.85
α-NMF [9]	PESQ	2.97	3.40	3.63	3.86
	SDR	4.90	9.48	10.76	11.65
SE + (α-NMF)	PESQ	2.96	3.49	3.89	4.05
	SDR	10.28	14.31	17.39	19.76
SE + NMF [10]	PESQ	3.19	3.61	3.83	4.32
	SDR	10.61	14.73	17.84	20.52
Proposed SE + (α-NMF) + OU	PESQ	**3.24**	**3.67**	**3.91**	**4.38**
	SDR	**10.70**	**14.82**	**17.92**	**20.58**

5 Conclusion

In this paper, a hybrid of Non-Negative Matrix Factorization (NMF) and statistical modeling with online bases updates is proposed. It is known that the performance of Wiener filter degrades when a speech signal consists of nonstationary noise. NMF approaches like IS-NMF, CNMF, and α-NMF are more robust to nonstationary noise but their bases cannot be updated online.

References

1. Miyazaki R, Inoue T, Takahashi K, Kondo K, Saruwatari, H, Shikano Y (2012) Musical-noise-free speech enhancement based on optimized iterative spectral subtraction. IEEE Trans Audio Speech Lang Process 20(7):2080–2094
2. Ephraim Y, Malah D (1985) Speech enhancement using a minimum mean-square error log-spectral amplitude estimator. IEEE Trans Acoust Speech Signal Process ASSP-33:443–445
3. Loizou PC, Rangachari S (2006) A noise-estimation algorithm for highly non-stationary environments. Speech Commun 48:220–231
4. Wilson KW, Smaragdis P, Raj B (2008) Regularized non-negative matrix factorization with temporal dependencies for speech denoising. Interspeech, pp 411–414
5. Smaragdis P, Mohammadiha N, Leijon A (2013) Supervised and unsupervised speech enhancement using nonnegative matrix factorization. IEEE Trans Audio Speech Lang Process 21(10):2140–2151
6. Mohammadiha N, Leijon A, Gerkmann T (2011) A new linear MMSE filter for single channel speech enhancement based on nonnegative matrix factorization. In: 2011 IEEE workshop on applications of signal processing to audio and acoustics (WASPAA), pp 45–48
7. Lee SJ, Park JH, Kim HK, Kim SM, Lee YK (2012) Non-negative matrix factorization based noise reduction for noise robust automatic speech recognition. Lect Notes Comput Sci 7191:338–346
8. Rinaldo R, Canazza S, Montessoro PL, Cabras G (2010) Restoration of audio documents with low SNR: a NMF parameter estimation and perceptually motivated bayesian suppression rule. In: Proceedings of sound and music computing conference, pp 314–321
9. Hyekyoung Lee N, Eungjin Choi AC, Kim Y-D (2008) Nonnegative matrix factorization with α–divergence. Pattern Recognit Lett 29(9):1433–1440
10. Kwon K, Kim NS, Shin JW (2014) Speech enhancement combining statistical models and NMF with update of speech and noise bases. In: IEEE international conference on acoustics, speech and signal processing, 4–9 May. Florence, Italy, pp 7053–7057
11. Garofolo JS (1988) Getting started with the DARPA TIMIT CD-ROM: an acoustic phonetic continuous speech database. National Institute of Standards and Technology (NIST), Gaithersburg, MD, USA
12. Steeneken H, Varga A (1993) Assessment for automatic speech recognition: II. NOISEX-92: a database and an experiment to study the effect of additive noise on speech recognition systems. Speech Commun 12:247–251
13. Durrieu J-L, Fevotte C, Bertin N (2009) Nonnegative matrix factorization with the Itakura-Saito divergence: with application to music analysis. Neural Comput 21(3):793–830
14. Ephraim Y, Malah D (1984) Speech enhancement using a minimum mean square error short-time spectral amplitude estimator. IEEE Trans Acoust Speech Signal Process 32(6):1109–1121
15. Browne M, Berry MW, Langville AN, Plemmons RJ, Pauca VP (2007) Algorithms and applications for approximate nonnegative matrix factorization. Comput Stat Data Anal 52(1):155–173

16. Loizou P, Hu Y (2008) Evaluation of objective quality measures for speech enhancement. IEEE Trans. Speech Audio Process 16(1):229–238
17. Vincent E, Fevotte C, Gribonval R (2006) Performance measurement in blind audio source separation. IEEE Trans Audio Speech Lang Process 14(4):1462–1469

Early Detection of Brain Tumor and Classification of MRI Images Using Convolution Neural Networks

Kumbham Bhargavi and Jangam J. S. Mani

Abstract The early detection of brain tumor can drastically improve the survival rate of patients. The MRI images of brain tumor Meningioma and Glioma are used for classification. With the help of Gray-Level Co-occurrence Matrix (GLCM) and Discrete Wavelet Transform (DWT), different features are extracted from the image. Then, the tumor segmentation is done to discover which part of the tumor area is affected after the detection of tumor. Tumor classification is done using CNN (Convolution Neural Networks). The effects could be pretty helpful for the specialists and radiologists for early detection and if the classifier does not identify any tumor, then it concludes that there is no tumor, if it locates any type of tumor, then we are able to find out the location affected also. Accuracy, sensitivity, and specificity were used to evaluate the proposed approach. A GUI (Graphical User Interface) has been created for the usage of the MATLAB 2013a.

Keywords Gray-level co-occurrence matrix (GLCM) · Discrete wavelet Transform (DWT) · Segmentation · CNN (convolution neural networks)

1 Introduction

A brain tumor is a collection of odd cells that grow in or across the brain. Tumors can without delay spoil wholesome brain cells. They can also in a roundabout way harm healthful cells by the way of crowding different components of the brain and inflicting inflammation, brain swelling, and stress inside the cranium. Brain tumor segmentation techniques have already shown splendid ability in detecting and study-

K. Bhargavi
Department of Computer Science, Keshav Memorial Institute of Technology, Hyderabad, India
e-mail: bhargavikumbham@kmit.in

J. J. S. Mani (✉)
Department of Computer Applications, KTS Govt Degree College, Rayadurg, Anantapuramu, Andhra Pradesh, India
e-mail: jsmani.jangam@gmail.com

© Springer Nature Singapore Pte Ltd. 2019
H. S. Saini et al. (eds.), *Innovations in Computer Science and Engineering*, Lecture Notes in Networks and Systems 74,
https://doi.org/10.1007/978-981-13-7082-3_49

427

ing tumors in the clinical snapshots, and this trend will certainly be followed in the future [1].

The major trouble in tumor detection arises due to every tumor being of different form, size, location, and depth. Guide detection of brain tumor calls for human interplay and is time ingesting. Additionally,, it depends on the potential of the observer to discover the location, form, and length of the tumor [2]. Biomedical image segmentation is a complicated and very specific task. Picture segmentation does a chief role in biomedical imaging packages inclusive of the enumeration of tissue volumes analysis, confinement of pathology evaluation of anatomical shape, remedy planning, partial quantity improvement of realistic imaging information, and PC-integrated surgery [3]. A principal purpose of picture segmentation is to acknowledge systems inside the photo that place unit predicted to correspond to scene gadgets. The goal of the segmentation is to separate a photograph into nonintersecting areas supported by intensity or textural records [4]. Single-layered CNN used for the classification is less when compared to deep networks [5].

This research work mainly concentrates at the early detection of tumor for the MRI images efficiently. So that, the additional treatment can be involved in order to save the life of the human. The presentation of the paper is specified as follows: Sect. 1 explains the problem and brain tumor detection. Section 2 explains the proposed methodology. Section 3 discusses about the proposed experimental results and the performance evaluation of our proposed methodology. Finally, in Sect. 4, conclusion and future scope is discussed.

2 Proposed Approach

The proposed approach to identify the tumor in early stages. To discover the brain tumor of MRI image, we use the system based totally on laptop imaginative and prescient. The contribution to do this is preprocessing in which image enhancement and elimination of the noise. The detection and segmentation of brain tumor system architecture are shown in Fig. 1. The proposed system is divided into the following steps: Image Acquisition, Image Preprocessing, Feature Extraction with GLCM and DWT using the multi-class convolution neural network, and histogram-based segmentation. The images of a patient obtained by MRI scan is displayed as an array of pixels (a two-dimensional unit based on the matrix size and the field of view) and stored in the database.

The detection and segmentation of brain tumor system structure are shown in Fig. 1. The whole proposed technique division is as follows: Image Acquisition, Brain Image preprocessing, feature extraction using GLCM and DWT with the help of convolution neural network, and histogram-based segmentation. The dataset is split into trained data and testing data (75 and 25%).

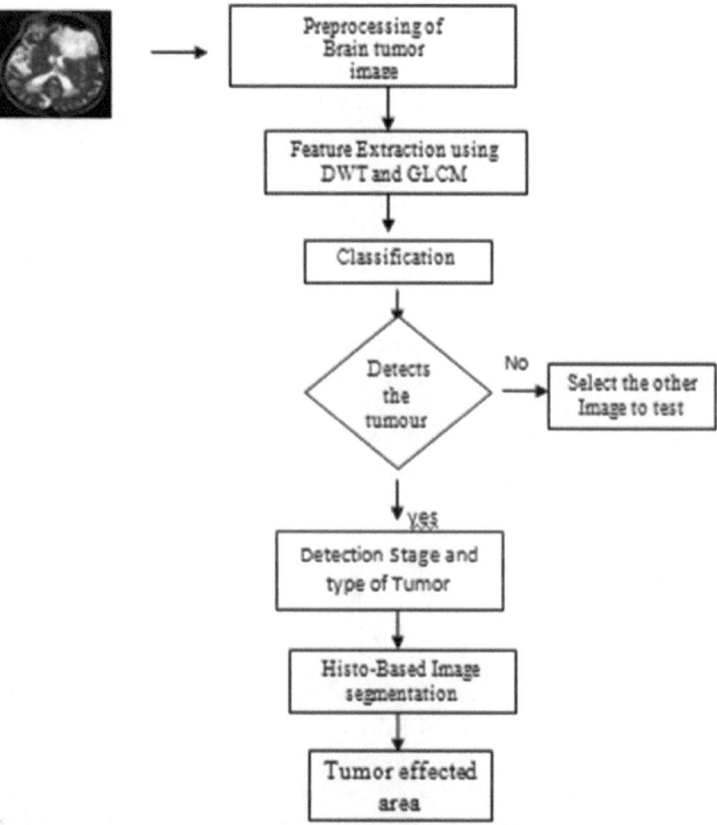

Fig. 1 Brain tumor early detection system architecture

2.1 Preprocessing of Brain Tumor Image

In preprocessing, the tumor image is selected from the database to do the preprocessing. First, we convert the MRI image into gray level. After converting the tumor image into gray coloration image, it is divided into two identical parts alongside its principal axis. Grayscale tumor image is combined solely of shades of gray, varying from black at the weakest intensity to white at the strongest. Each grayscale picture is resized to the size of 256 × 256. Inside the preprocessing brain tumor image, the nonlocal median filter is used to lessen the noise inside the above-resulted noise picture that is used to show the correct captured brain tumor image and noiseless images from every channel and restore it [6]. The noise might be eliminated by way of using the non-neighborhood median filter out which does now not replace a pixel's value with an average of the pixels around it, as an alternative updates it the usage of a weighted common of the pixels judged to be maximum kindred. The weight of every pixel depends on the gap among its depth gray-level vector and that of the target pixel. Denoised image of each pixel of the nonlocal manner is computed with Eq. (1). Where N(i, j) is the denoised image and i, j are the two elements, W(i, j) the filter function and $B_D(i, j)$ is the filtered image.

$$N(i, j) = \sum_{j \in B_D} w(i, j) \, B_D(i, j) \tag{1}$$

A. Feature Extraction Using DWT

The proposed system makes use of the discrete wavelet remodel (DWT) coefficients as a feature vector. The wavelet is an effective mathematical device for function extraction, and has been used to extract the wavelet coefficient from MR pics. Wavelets are localized foundation functions, which might be scaled and shifted versions of a few constant mother wavelets. The major benefit of wavelets is that they provide localized frequency facts about a characteristic of a sign that is specifically useful for classification.

To extract the characteristics of the brain tumor image, here, we are using the 2D discrete wavelet transform (DWT) technique. Discrete wavelet transform (DWT) is used to extract the wavelets and coefficients from brain tumor image. The size of the tested image is 256 × 256 has been filtered and decomposed into four subbands. 2D discrete wavelet transform was implemented and that led to four (4) separate subbands Low-Low, High-Low, Low-High, and High-High with the two-level wavelet decomposition of region of interest (ROI). Once ROI is found, then it divides into next low-level and high-level frequency of the image. So, the LH1 represents the first level and LH2 represents the second level. Here, these two specify the low-frequency part of the images. The other parts LH1, HL1, HH1, LH2, HL2, and HH2 specifies the high-level frequency and provides the guidelines for first and second level also.

B. Feature Extraction Using GLCM

GLCM is a properly mounted statistical approach for extracting the features from the gray-level brain tumor. Haralick introduced the co-occurrence matrix and texture features, which can be the most famous second-order statistical features today. The calculations of texture use the information of the GLCM to give a degree of the change in depth on the pixel of interest. GLCM helps us to do the texture analysis reorganization about ordinary and deviating from the original tissues without difficulty for human visual perception and machine learning [7]. It moreover gives difference among malignant tissue and ordinary tissues, which won't be identified by way of using the human. It improves the accuracy by a manner of the usage of powerful measurable characteristics for early analysis.

With the help of two techniques, here, we are extracting features of the brain tumor image. The first step is computing the co-occurrence matrix and the second step is to assesses the texture characteristics with the help of the co-occurrence values. Usually, the co-occurrence matrix is calculated primarily depending on two parameters, which were the absolute distance from one from the other pixel pair d measured in pixel variety and is compared to the notion of the inner orientation. Commonly, it is measured in four directions (e.g., 0°, 45°, 90°, and 135°), despite the fact that diverse different combinations might be feasible.

C. Classification Using Convolution Neural Networks (CNNs)

Convolution neural networks (CNNs) encompass more than one layer of other related specifications. CNN are size in less than normal neuron collections which technique portions of the given picture. The returns of those gathering are then tiled in order that their input areas overlap, to attain a higher representation of the original photo, which is repeated for each such layer. Tiling allows CNNs to tolerate translation of the input photographss. Convolution networks may encompass nearby or global pooling layers, which combine the outputs of neuron clusters [8]. One foremost advantage of the convolution networks is using shared weight in convolution layers, this means that the same clear out (weights monetary group) is used for each pixel in the layer, and this reduces each memory footprint and improves the overall performance [9]. CNN exploits distinct features found in an image as well as more international contextual capabilities simultaneously [10].

Step one of a CNN, that we are working on the brain images here, which an essentially two-dimensional arrays, we're the usage of convolution 2-D, we will use convolution 3-D at the same time as coping with videos, where the other dimension size may be time. For the classification purpose, here, we consider the two layers of feedforward neural network, wherein learning assumes the availability of a categorized (i.e., ground truthed) set of training records.

D. Histogram-Based Image Segmentation

The histogram of the brain image is a representation of the frequency of incidence of every gray stage in the image. The purpose of segmentation is to pick out and adjust the illustration of a picture into something that is greater significant and simpler to evaluate. Here, we have got the two entities, namely one is background and the alternative is MRI brain tumor picture. The background occupies maximum of the photograph and because of this, gray degree is the massive height inside the certain histogram picture and the alternative one is smaller peaks that specifies the picture within the histogram [11].

Image histogram characterizes the frequency of prevalence of gray values. Especially, the histogram h[i], (i = 0, ..., 255) is the opportunity of an arbitrary pixel taking the gray levels i and represented as follows: h[i] = wide variety of pixels of gray degree, i is general number of pixels and cumulative feature described as j

$$H[J] = \sum_{i=0} h[i], (j = 0, 1, \ldots, 255) \tag{2}$$

So H[255] = 1;

ROI is a preferred separation of pattern within a statistics set identified for a meticulous function. The MRI image has the simplest two values, namely binary zero and binary 1(255).

3 Experimental Setup, Results, and Discussion

The proposed method samples are collected from the Whole Brain Atlas (WBA). The performance parameter of the proposed approach is measured in this section using the MRI dataset and the results are presented. The datasets have been collected from the digital imaging and communication from the medicine dataset. Here, the new approach applied the histogram-based segmentation with the help of DWT and GLCM to extract the features of brain tumor image. We have taken 100 brain MRI images here and on which, it included different kinds also. We divided our input data into two parts. 80% of training data and 20% of validation data is used for training and validation. From the given dataset, few images were considered for training to obtain the performance. Figure 2 specifies the dataset of MRI images.

Once we completed the training, neural networks start working on those trained data. The purpose of this is to analyze the values of the neurons and for classification also.

It helps us to realize the unambiguous on the case or MRI image, i.e., the input selected. Here, we are selecting the image from the dataset, and then it converts into gray-level image and remove the noise using the median filter. Then, we are applying the DWT and GLCM method to get the features of the image. The results of that are as specified in Fig. 3.

Then, we are using the convolution neural network to train the brain tumor images. If the tumor exists, then we are identifying the stage of the brain tumor and type here

Fig. 2 The different test images

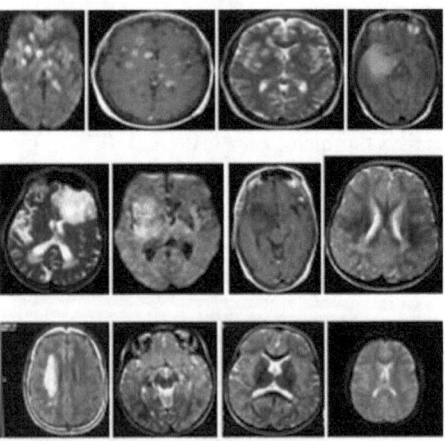

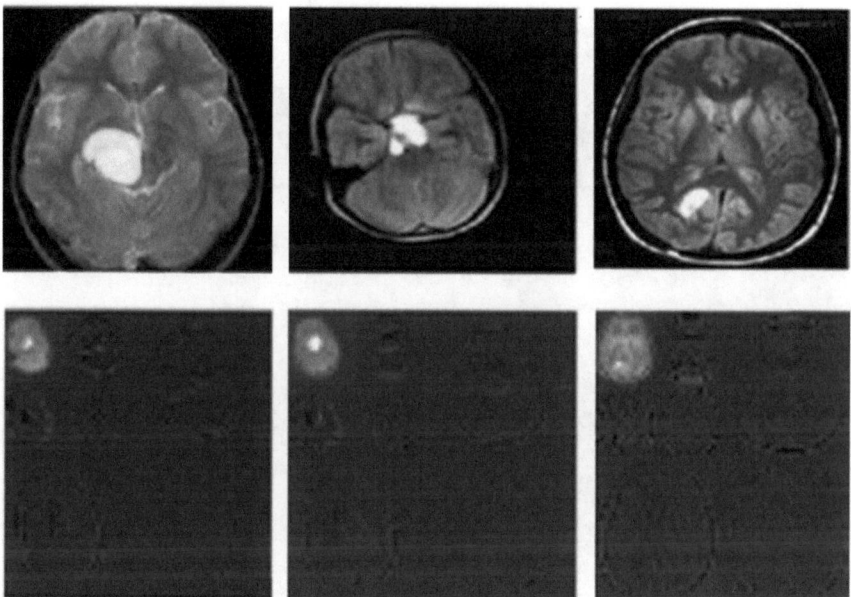

Fig. 3 Feature extraction of brain tumor image

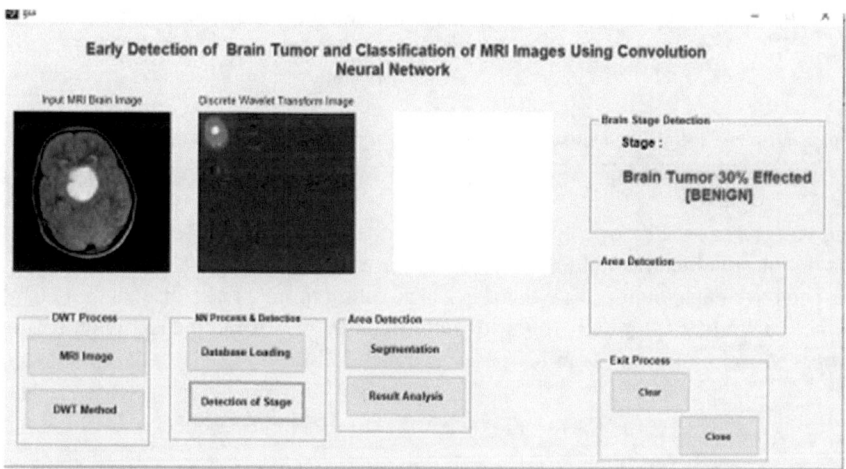

Fig. 4 Stage of the brain tumor

itself as shown in Fig. 4. Brain Tumor 30% affected and Benign. Then, we are using a segmentation technique to identify the part of the tumor.

Then decomposing the Brain tumor image as follows shown in Fig. 5.

The user has to select the segmented images which are displayed in Fig. 5. When you chose an image from the segmented images, it is displayed on the third handler,

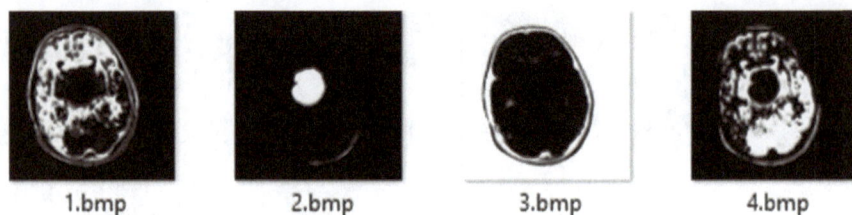

1.bmp   2.bmp   3.bmp   4.bmp

Fig. 5 Brain tumor sub-images

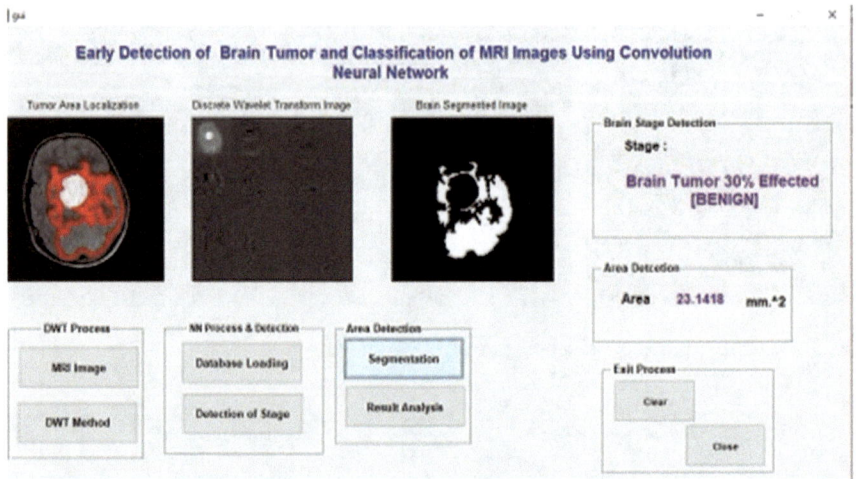

Fig. 6 Displays segmented image, tumor area localization, and square area detection

and you can see the area detected in square mm under the area detection tag and also the tumor area localization on the first handler as shown in Fig. 6.

Thus, we can conclude that there exists a benign tumor. The subsequent metrics, namely accuracy as in and similarities are undertaken for the measurement of the proposed technique and its miles is described as follows:

$$Accuracy(A) = (TP + TN)/(TP + TN + FP + FN) \tag{3}$$

where TP—True Positives, TN—True negatives

$$Sensitivity(SE) = True\,Positive(TP)/(True\,Positive + False\,Negative) \tag{4}$$

$$Specificity(SP) = True\,Negative/(True\,Negative + False\,Positive) \tag{5}$$

The results of brain tumor image performance are shown in Fig. 7.

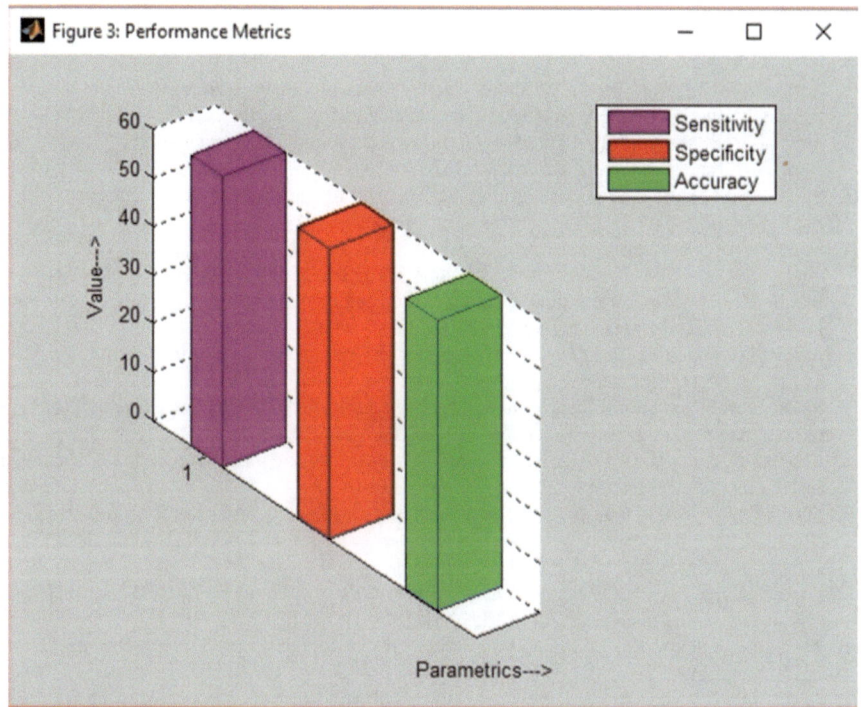

Fig. 7 Performance of brain tumor images

4 Conclusion

Detecting the existence of brain tumors from MRI in a fast, accurate, and repro-
ducible way is a challenging problem. Early detection of brain tumor classification
system is implemented using convolution neural networks. In this paper, we have
used brain MRI images to eliminate the noise and smoothen the photograph, and
preprocessing is used which additionally has a outcome inside the development of
signal-to-noise ratio. Next, we have used discrete wavelet rework that decomposes
the pixel and textural features were extracted from gray-level co-occurrence matrix
(GLCM) accompanied by using morphological operation. Convolution neural community classifier is used for the type of tumors. From the observation effects, it could
be absolutely expressed that the detection of brain tumor is speedy and accurate. The
overall performance elements is evaluated and it also suggests that it gives better
final results.

References

1. Verma K, Mehrotra A, Pandey V, Singh S (2013) Image processing techniques for the enhancement of brain tumor patterns. Int J Adv Res Electr Electron Instrum Eng 2:1611–1614
2. Shivani P, Deshmukh, Rahul D (2014) Ghongade. Detection and segmentation of brain tumor from MRI images. Int J Eng Sci Adv Technol 4:426–430
3. Sharma N, Aggarwal LM (2010) Automated medical image segmentation techniques. J Med Phys 35:3–14
4. Zijdenbos A, Forghani R, Evans A (2002) Automatic pipeline analysis of 3D MRI data for clinical trials: application to multiple sclerosis. IEEE Trans Med Imaging 21:1280–1291
5. Pereira S et al (2016) Brain tumor segmentation using convolutional neural networks in MRI images. IEEE Trans Med Imaging 35:1240–1251
6. Jafari M, Shafaghi R (2012) A hybrid approach for automatic tumor detection of brain MRI using support vector machine and genetic algorithm. Global J Sci Eng Technol 3:1–8
7. Sivasankari S, Sindhu M, Sangeetha R, Shenbaga Rajan A (2014) Feature extraction of brain tumor using MRI. Int J Innovative Res Sci Eng Tech 3:10281–10286
8. Kharat KD, Kulkarni PP (2012) Brain tumor classification using neural network based methods. Int J Comput Sci Inform 1(4):2231–5292. ISSN (PRINT)
9. Sheejakumari V, Gomathi S (2015) MRI brain images healthy and pathological tissue classification with aid of improved particle swarm optimization and neural network (IPSONN). Comput Math Methods Med 2015:1–12
10. Havaei M, Davy A, Warde-Farley D, Biard A, Courville A, Bengio Y et al (2017) Brain tumor segmentation with deep neural networks. Med Image Anal 35:18–31
11. Tobias OJ, Seara R (2002) Image segmentation by histogram thresholding using fuzzy sets. IEEE Trans Image Process 11:1457–1465
12. Arizmendi C, Vellido A, Romero E (2011) Binary classification of brain tumors using a discrete wavelet transform and energy criteria. IEEE 2011, pp 1–4

A Hybrid Machine Learning and Dynamic Nonlinear Framework for Determination of Optimum Portfolio Structure

Sayan Gupta, Gautam Bandyopadhyay, Sanjib Biswas and Arun Upadhyay

Abstract Capital market investment is a growing stream of the economic literature. It has been a prime concern of a large number of investors belonging to different clusters or income groups for two reasons mainly. First, the construction of a portfolio, which deals with the selection of the stocks. Second, the formulation of an appropriate investment strategy, which calls for minimizing the risk while maximization of the return, i.e., optimization of the constructed portfolio. Following the broad framework as suggested in the seminal work of Markowitz [1], this research attempts to address the issue of portfolio optimization based on risk and return parameters while dynamically allocating the weights to the constituent stocks. In the first part of this study, k-means clustering is applied to a heterogeneous sample of 53 number of stocks enlisted with the NSE during the year 2012–2017. The purpose is to classify the stocks in three categories (such as low stock price, medium stock price, and high stock price) based on their monthly closing return. In the second phase, this study focuses on finding out the distribution of weights among the stocks belonging to the portfolio by using the generalized *reduced gradient* (GRG) method under the dynamic environment. Finally, this study attempts to validate the results by applying perception mapping. We have found eight stocks in the cluster of low stock price which is the sample studied in this research. We have observed that dynamic allocation of weights led to minimization of risk and the finding is validated through a perceptual map.

Keywords Portfolio · Nonlinear optimization · K-means clustering · Generalized reduced gradient method

S. Gupta (✉) · G. Bandyopadhyay
Department of Management Studies, NIT Durgapur, Durgapur, India
e-mail: sg.17ms1101@phd.nitdgp.ac.in

G. Bandyopadhyay
e-mail: gautam.bandyopadhyay@dms.nitdgp.ac.in

S. Biswas
Calcutta Business School, Calcutta, India

A. Upadhyay
NSHM Business School, Durgapur, India

© Springer Nature Singapore Pte Ltd. 2019
H. S. Saini et al. (eds.), *Innovations in Computer Science
and Engineering*, Lecture Notes in Networks and Systems 74,
https://doi.org/10.1007/978-981-13-7082-3_50

1 Introduction

The construction of the optimal portfolio plays a critical role behind any investment decision as it is premised on the concept of maximization of the return on investment. It is imperative to contemplate on the selection of appropriate stocks in the right proportion for constructing a portfolio with an objective to increase return while minimizing risk. The essence of effective portfolio management lies in optimum utilization of the capital in terms of allocation of the stocks constituting the portfolio and balancing the same taking risk-return trade-off into account, since risk tolerance level and expectation of return vary from investor to investor. Stated in this field, Markowitz's mean–variance approach [1] is considered to be the first systematic attempt to formulate a diversified investment decision with an objective of ensuring high return at low-risk level [2, 3]. In tune with this work, Tobin [4] postulated the concept of *efficient frontier* and *capital market line*, wherein the author argued for the persistence of the portfolio structure irrespective of risk tolerance level at a consistent expectation level; only relative proportions of stocks change otherwise. Extending the work, Markowitz [5] introduced expected return–semi-variance based analysis which considers both the extremes while determining the efficiency of the portfolio. Sharpe [6, 7] further worked on Markowitz's analysis and noted that optimal portfolio (among all efficient portfolios) depends on expected return and risk preference of the investors. These classical approaches paved the way to the *modern portfolio theory* (MPT). MPT aims to maximize the expected return of the portfolio at a given risk level or minimizing portfolio risk at a given return by optimally allocating the total available fund to different assets [8]. However, while forming the portfolio, it is essential to consider the influence of the assets on each other, i.e., only on the basis of individual stock performance one cannot form a portfolio [9]. In effect, the distribution of the returns decides effective optimization of the portfolio. In the context of portfolio management, a risk is perceived as the total risk of a portfolio which has two components; *systematic risk* or *market risk* and *unsystematic risk* or *diversifiable risk*. By prudent stock selection and distribution of appropriate weightage for them within a portfolio, unsystematic risk can be reduced to a considerable extent [10].

In line with the seminal work of Markowitz [1], over the years, several studies have been made on portfolio selection and optimization. There has been a growing stream of alternative methods suggested by several researchers and practitioners in the stated field. Examples include artificial neural network [11], genetic algorithm [12], particle swarm optimization [13], simulated annealing [14], ensemble [15], decision tree [16], clustering [17], multi-criteria decision-making approach [18] to name a few. It is evident from these researches that in order to optimize the portfolio and correctly predict its return, selection of the right number of right stocks in the right proportion is of paramount importance. There has been a plethora of research conducted on portfolio selection using unsupervised learning methods like clustering. Clustering finds its importance in pattern identification, classification, and detection of an anomaly while selecting stocks to form a portfolio. Further, it is useful in finding interrelationship or co-movements of the stocks [19]. Although hierarchical clustering is a dominant method in this regard, k-means clustering and C-means clustering have also drawn significant attraction of the researchers [20].

In this study, the authors have used k-means clustering for segregating 53 heterogeneous companies with different sectors enlisted with the NSE during the year 2012–2017 on the basis of monthly stock return (60 months) into three distinct clusters; *low stock price* (LSP), *mid stock price* (MSP), and *high stock price* (HSP) in order to construct a portfolio. Further, GRG method has been used to optimize the constructed portfolio of the stocks belonging to the MSP cluster through dynamic allocation of the weights to different stocks with an objective to minimize risk while maximizing return. The rest of this paper is organized as follows. In Sect. 2, data and methodology are discussed. Section 3 presents the findings. Finally, Sect. 4 concludes the paper while highlighting some of the implications and future scope.

2 Data and Methodology

The broad objective of this study is to find out the distribution of the stocks under the portfolio ensuring maximum return at minimum risk. It requires appropriate weight assignment to individual stocks forming the portfolio.

2.1 Sample

The sample for this study consists of 53 numbers of heterogeneous companies with different sectors enlisted with the National Stock Exchange (NSE), India through convenience sampling (refer Table 1). The study period is the year 2012–2017. The data were collected from the published secondary database. Monthly returns of those stocks for 60 months have been considered for analysis. In order to calculate monthly returns in case of the Index or Stock (since the data collected is month wise), we have used the formula $Ln(P_1/P_0)$, $Ln(P_2/P_1)$ and so on since the behavior of data is continuous. After clustering, we have worked on MSP cluster.

2.2 Methods

In this study, a three-stage approach has been followed. First, k-means clustering, an unsupervised learning technique has been applied in order to classify the stocks based on the monthly returns for constructing the portfolio; second, GRG, a nonlinear optimization technique has been selected for deciding weights for the stocks with an objective to minimize risk while maximizing return. Finally, we have applied perception mapping using mean and standard deviation of the monthly returns of the stocks belong to MSP.

Table 1 Stocks under the study (primary level)

Sector(s)	Companies	Total
Automobile	AMARA RAJA BATT; MRF; TVS MOTOR; TATA MOTORS; HERO MOTOCORP; BAJAJ AUTO; MARUTI SUZUKI; BHARAT FORGE; M&M; ASHOK LEYLAND	10
Banking	CANARA BANK; SBI; BANK OF BARODA; HDFC BANK; YES BANK; INDUSIND BANK; PNB; FEDERAL BANK	8
Oil and natural gas	IOCL; GAIL; RIL; NTPC; ONGC; BPCL	6
FMCG sector	ITC; HUL; GODREJ CP; MARICO; COLPAL; DABUR; P&G; GSK CP; EMAMI LTD; GODREJ IND	10
Food and beverages	MCDOWELL; TATA GLOBAL; UBL; UB (H) L; BRITANNIA; JUBILANT FOODWORKS	6
IT/ITes	KPIT TECH; HCL TECH; INFOSYS; WIPRO; TCS; TECH MAHINDRA; MINDTREE; OFSS	8
Others	TATA ELXSI; RIL INFRA; HIND PETRO (HPCL); POWER GRID CORP; TATA POWER	5
Grand total		53

The expected return on a portfolio is computed as follows:

$$E\left(R_p\right) = \sum_{i=1}^{N} w_i E\left(R_i\right) \tag{1}$$

where

$E(R_p)$ the expected return on the portfolio
N the number of Index or stocks in the portfolio
w_i the proportion of the portfolio invested in Index or ith Stock
$E(R_i)$ the expected return on ith Stock.

The risk is calculated as follows:

$$\sigma_p^2 = (w_A)^2 \sigma_A^2 + (w_B)^2 \sigma_B^2 + 2 w_A w_B \, \sigma_{A,B} \tag{2}$$

where σ_p: Standard deviation of the portfolio; σ_A and σ_B: Standard deviation of the stocks A and B; w_A and w_B: Weights assigned to the stocks A and B.

K-means Clustering. It is a type of unsupervised learning, which distribute the unlabeled data into a specific number of groups or clusters (represented by the variable K) on the basis of feature similarity. Each such cluster is represented by its centroid which is a collection of feature values pertaining to that cluster. This algorithm takes a set of "m" number of data points in "n" dimensions into "K" number of clusters through an iterative process. The objective is to minimize within-cluster sum of squares [21]. It starts with an initial estimate of clusters wherein a particular data point x_j(where, $j = 1, 2, \ldots$ m) is allocated to a particular cluster c_i(where, $i = 1, 2, \ldots$ k) depending on the squared Euclidean distance between the data point and the centroid of the respective cluster. The points which find their distances with respect to the centroid of a particular cluster minimum are assigned to that particular cluster. Then, the process recomputes all the centroids and again finds the distances of all the data points with respect to each newly constructed centroids for assigning the data points to newly formed clusters. This process repeats itself until limiting condition is reached, i.e., either there is no change of clusters by the data points (i.e., within a particular cluster all data points are homogeneous) or within group sum of squares are minimized or the prefixed maximum number of iterations is conducted.

Generalized Reduced Gradient Method (GRG). This is a general version of the reduced gradient method, which solves optimization problems with nonlinear constraints and arbitrary bounds. The method is described as follows as explained by Lasdon et al. [22].

The structure of general nonlinear problem is given by

$$\text{Minimize } f_0(X) \tag{3}$$

$$\text{subject to the constraints } g_i(X) = 0; \text{ where, } i = 1, 2, \ldots m \tag{4}$$

$$\text{Where, and } l_j \leq X_j \leq u_j; \text{ where, } j = 1, 2, \ldots n \tag{5}$$

Here, u_j and l_j indicate the upper boundary and lower boundary, respectively, where, $u_j > l_j$ assuming m < n in order to avoid infeasibility of the solution or unique solution. The above forms are general since inequality constraints get transformed to equalities by adding slack variables. Following this, for solving the problem, basic variables (m) get expressed in terms of remaining nonbasic (n-m) variables. If $\bar{X}$ denotes a feasible solution point and Y be the vector representing basic variables and Z be the vector of nonbasic variables on X, then after partitioning, Eqs. (4) and (5) can be written as

$$X = (Y, Z), \bar{X} = (\bar{Y}, \bar{Z}) \tag{6}$$

$$g_i(Y, Z) = 0 \tag{7}$$

Here, it is assumed that both the objective and constraint functions are differentiable. The transformed objective function is given as

$$F(Z) = f(Y(Z), Z) \tag{8}$$

Accordingly, the nonlinear problem is transformed at least for Z close to $\bar{Z}$, to a reduced problem given by

$$\text{Minimize } F(Z) \tag{9}$$

$$\text{Subject to } l' \le Z \le u' \tag{10}$$

The GRG method actually solves the original problem as stated above by solving Eqs. (9) and (10). This study addresses a nonlinear problem since in our study, it has been observed that the monthly rate of return is having quadratic and cubic nature. Hence, the data suits the applicability of GRG method in our case.

3 Results and Discussions

Table 2 describes the distribution of the stocks (Table 1) into three defined clusters such as LSP, MSP, and HSP.

In our study, we have selected MSP for further analysis since in LSP cluster we have 44 companies which make the portfolio too stretched for a common investor. Also, HSP cluster shows opposite nature and therefore, it has not been considered. Also, LSP signifies comparatively prematured or poor performance at the market and HSP indicates a bit saturated performance and comparatively less growth. We have considered to form a portfolio based on the stocks belonging to the MSP cluster. Table 3 lists out the stocks under study along with their monthly expected rate of return and standard deviations.

Table 2 Number of companies in each cluster

Cluster	No. of companies
LSP	44
MSP	8
HSP	1
Total	53

Table 3 Stocks under MSP cluster

	Hero Moto-corp	Bajaj Auto	Maruti Suzuki	Britannia	P&G	GSK CP	TCS	OFSS
Monthly expected rate of return (AVROR)	0.01148	0.00746	0.03127	0.03742	0.02045	0.00902	0.01276	0.00364
Standard devia-tion (SD)	0.06651	0.06602	0.08861	0.07129	0.05079	0.07688	0.06413	0.05535

Table 4 Normality test

	Hero Moto-corp	Bajaj Auto	Maruti Suzuki	Britannia	P&G	GSK CP	TCS	OFSS
Kolmogorov-Smirnov Z	0.760	0.427	0.762	0.461	0.510	0.986	0.740	0.430
Asymp. Sig. (two-tailed)	0.610	0.993	0.607	0.984	0.957	0.285	0.644	0.993

Table 5 Correlation matrix

	Hero Moto-corp	Bajaj Auto	Maruti Suzuki	Britannia	P&G	GSK CP	TCS	OFSS
Hero Moto-corp	1							
Bajaj Auto	0.541032	1						
Maruti Suzuki	0.371683	0.420821	1					
Britannia	0.251049	0.325695	0.203601	1				
P&G	0.016686	0.066044	0.287807	0.271632	1			
GSK CP	0.076709	0.056452	0.244738	0.50983	0.374539	1		
TCS	0.207421	−0.00987	−0.16676	0.082849	0.04115	0.185365	1	
OFSS	0.210708	0.342858	0.312976	0.184763	0.171558	0.129388	0.257402	1

Further, we have performed a normality test (refer Table 4) in order to comply with the conditions of Markowitz [1]. It is seen from the result that the rate of returns of the stocks satisfies normality condition.

In order to understand whether diversification is possible among the stocks, we have conducted a correlation analysis (refer Table 5). The determinant value (0.16548) obtained from the correlation matrix suggests that the stocks under the portfolio can be diversified. Hence, it is a problem of dynamic allocation of weights among the stocks.

Table 6 shows the variance (diagonal values) and covariance (off-diagonal values) analysis for determining risk.

Table 7 shows the weights of the stocks calculated dynamically using GRG method. The optimum risk of the portfolio under study is 0.11% as calculated through the GRG method.

For validation purpose, we have performed perception mapping (refer Fig. 1). Table 8 shows the combined values of **AVROR** and SD. These two values are calculated as follows:

Table 6 Variance and covariance table

	Hero Motocorp	Bajaj Auto	Maruti Suzuki	Britannia	P&G	GSK CP	TCS	OFSS
Hero Motocorp	0.004350	0.002336	0.002154	0.001171	0.000055	0.000386	0.000870	0.000763
Bajaj Auto	0.002336	0.004286	0.002421	0.001507	0.000218	0.000282	−0.000041	0.001232
Maruti Suzuki	0.002154	0.002421	0.007721	0.001265	0.001274	0.001640	−0.000932	0.001509
Britannia	0.001171	0.001507	0.001265	0.004997	0.000967	0.002748	0.000372	0.000717
P&G	0.000055	0.000218	0.001274	0.000967	0.002537	0.001438	0.000132	0.000474
GSK CP	0.000386	0.000282	0.001640	0.002748	0.001438	0.005813	0.000899	0.000541
TCS	0.000870	−0.000041	−0.000932	0.000372	0.000132	0.000899	0.004044	0.000898
OFSS	0.000763	0.001232	0.001509	0.000717	0.000474	0.000541	0.000898	0.003013

Table 7 Distribution of the weights to the stocks under the portfolio

Companies	Hero Motocorp	Bajaj Auto	Maruti Suzuki	Britannia	P&G	GSK CP	TCS	OFSS	Total
Weights (%)	10.3	12.3	0.0	3.9	35.2	2.6	19.5	16.2	100.0

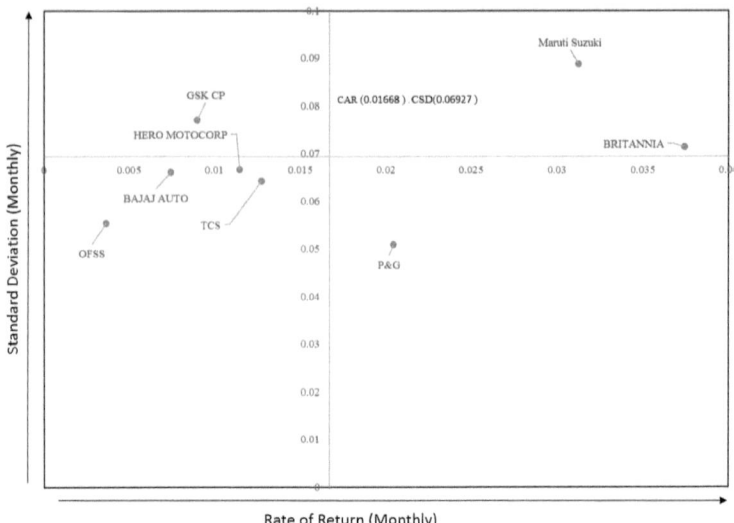

Fig. 1 Perceptual map

Table 8 Combined AVROR and SD	CAR	0.0166888
	CSD	0.0692789

$$\text{Combined AVROR(CAR)} = \text{Average(AVROR)} \qquad (11)$$

$$\text{Combined SD(CSD)} = \frac{\sqrt{\sum(d_i^2 + SD_i^2)}}{n}; \qquad (12)$$

where

d_i Average return for ith Stock-CAR; i = 1, 2, ...n.

Further, we have used the values of CAR and CSD to construct a perceptual map, wherein the axes are shifted from the origin to the point having CAR and CSD values. This results into generation of four new quadrants representing (High Return, High SD), (High SD, Low Return), (Low SD, High Return), and (Low SD, Low Return), respectively.

It is seen from the above figure that only P&G falls in the third quadrant, which characterizes high growth and low risk. It suggests that out of the stocks considered for the study, P&G stands alone, matched with the stated requirements. Therefore, it is logical to state that the earlier findings are being validated by the perceptual map.

4 Conclusion

In this study, we have attempted to assign optimal weights to the individual stocks constituting a portfolio in a dynamic environment. In order to select a portfolio, we have applied k-means clustering on a heterogeneous set of stocks listed in the NSE which are selected based on convenient sampling. Next, we have used a nonlinear optimization technique such as GRG for allocating weights with an objective to minimize the risk while maximizing the return. Allocation of the weights to the individual stocks forming the portfolio is having a significant impact on return on investment from the same. Further, in order to validate the results, a perception mapping of stocks under study has been performed which commensurate the earlier findings. This study is unique in the sense that initially, it started with unsupervised learning, but in the process, it came up with the bridging effect of unsupervised learning to a supervised learning validating all the way. This study may further be extended to analyze the clusters in the same way other than MSP for investigating any relation with the present one. Further, sector-wise portfolio performance may also be assessed for a comparative analysis in a dynamic and uncertain environment.

References

1. Markowitz H (1952) Portfolio selection. J Fin 7(1):77–91
2. Steinbach MC (2001) Markowitz revisited: mean-variance models in financial portfolio analysis. SIAM Rev 43(1):31–85
3. Rubinstein M (2002) Markowitz's "portfolio selection": a fifty-year retrospective. J Fin 57(3):1041–1045
4. Tobin J (1958) Liquidity preference as behavior towards risk. Rev Econ Stud 25(2):65–86
5. Markowitz H (1959) Portfolio selection. Yale University Press, New Haven, Connecticut
6. Sharpe WF (1963) A simplified model for portfolio analysis. Manage Sci 9(2):277–293
7. Sharpe WF (1966) Mutual fund performance. J Bus 39(1):119–138
8. Fabozzi FJ, Gupta F, Markowitz HM (2002) The legacy of modern portfolio theory. J Invest 11(3):7–22
9. Elton EJ, Gruber MJ, Brown SJ, Goetzmann WN (2009) Modern portfolio theory and investment analysis, 6th edn. Wiley
10. McClure B (2017) Modern portfolio theory: why it's still hip. Investopedia. http://www.investopedia.com/articles/06/MPT.asp#axzz1g3JQY7nY. Accessed 21 June 2018
11. Nazemi A, Abbasi B, Omidi F (2015) Solving portfolio selection models with uncertain returns using an artificial neural network scheme. Appl Intel 42:609–621. https://doi.org/10.1007/s10489-014-0616-z
12. Chang TJ, Yang SC, Chang KJ (2009) Portfolio optimization problems in different risk measures using genetic algorithm. Expert Syst Appl 36(7):10529–10537
13. Zaheer KB, Aziz MIBA, Kashif AN, Raza SMM (2018) Two stage portfolio selection and optimization model with the hybrid particle swarm optimization. Matematika 34(1):125–141
14. Crama Y, Schyns M (2003) Simulated annealing for complex portfolio selection problems. Eur J Oper Res 150(3):546–571
15. Nagy G, Barta G, Henk T (2015) Portfolio optimization using local linear regression ensembles in RapidMiner. arXiv:1506.08690
16. Tucker CS, Kim HM (2009) Data-driven decision tree classification for product portfolio design optimization. J Comput Inf Sci Eng 9(4):041004

17. Nanda SR, Mahanty B, Tiwari MK (2010) Clustering Indian stock market data for portfolio management. Expert Syst Appl 37(12):8793–8798
18. Ehrgott M, Klamroth K, Schwehm C (2004) An MCDM approach to portfolio optimization. Eur J Oper Res 155(3):752–770
19. Liao SH, Chou SY (2013) Data mining investigation of co-movements on the Taiwan and China stock markets for future investment portfolio. Expert Syst Appl 40:1542–1554. https://doi.org/10.1016/j.eswa.2012.08.075
20. Pai GV, Michel T (2009) Evolutionary optimization of constrained k-means clustered assets for diversification in small portfolios. IEEE Trans Evol Comput 13(5):1030–1053
21. Hartigan JA, Wong MA (1979) Algorithm AS 136: a k-means clustering algorithm. J R Stat Soc Series C (Applied Statistics) 28(1):100–108
22. Lasdon LS, Fox RL, Ratner MW (1974) Nonlinear optimization using the generalized reduced gradient method. Revue française d'automatique, informatique, recherche opérationnelle. Recherche opérationnelle 8(V3):73–103

Machine Learning Technique for Automatic Intruder Identification and Alerting

B. K. Uday, Anirudh Vattikuti, Kailash Gogineni and P. Natarajan

Abstract Security has become an important factor. Intruders have become prominent factors for all the data/property theft. The basic idea in this paper is to identify the intruder and alert owner/administrator in different possible ways. This paper discusses different ways such as "a message (SMS)", "WhatsApp message", "location of intruder", "an immediate call", and "intruder's image to owner's/administrator's WhatsApp" to alert owner/administrator. For identifying the intruder, machine learning algorithm is used. A camera placed at the locality is trained such that it can identify the familiar people and it is "on" all the time. Whenever an unknown/unidentified person comes to the vicinity of the camera, all the above-said features get activated and the owner gets alerted. The idea can be applied in many real-life situations, like thief identification near the house.

Keywords Intruder detection · Alerting owner · Machine learning · Image processing · Thief identification

1 Introduction

Identifying and catching intruders has become a difficult task as intruders/thieves found intelligent ways to tackle techniques used to catch them. It is similar to the story of a mosquito. Mosquitos are getting habituated for every new method to kill

B. K. Uday (✉) · A. Vattikuti · K. Gogineni · P. Natarajan
School of Computer Science and Engineering, Vellore Institute of Technology, Vellore 632014, Tamil Nadu, India
e-mail: uday.nbausj@gmail.com

A. Vattikuti
e-mail: vattikuti.anirudh97@gmail.com

K. Gogineni
e-mail: gkailashnath1998@gmail.com

P. Natarajan
e-mail: pnatarajan@vit.ac.in

© Springer Nature Singapore Pte Ltd. 2019
H. S. Saini et al. (eds.), *Innovations in Computer Science and Engineering*, Lecture Notes in Networks and Systems 74,
https://doi.org/10.1007/978-981-13-7082-3_51

449

them. On this basis, the solution to the problem is found. In this paper, new ways are introduced to identify the intruder and alert owner/administrator. This paper discusses five different ways to catch intruder/thief viz "a message (SMS)", "WhatsApp message", "location of intruder", "an immediate call", "and Intruder's image to owner's/administrator's WhatsApp". This paper discusses how face recognition is used to detect intruders/thieves. All the familiar and known faces are trained first. If an unknown face is detected, then the owner/administrator is alerted. Machine learning is used for facial identification. Facial images of familiar persons are captured in every angle to recognize familiar persons in any angle. There are many algorithms for facial recognition. Some of them are Eigen faces [1], Fisher faces [2], and Local binary pattern histogram [3, 4].

The main purpose of the paper is to discuss and how to identify the intruder and alert owner/administrator. Intruders are mostly the ones, who does not belong to the same locality/does not belong to the same institution. Among all the methods of identifying the intruder using facial recognition, this paper discusses "Local binary pattern histogram". The system is trained in such a way that as soon as it identifies the intruder, "a message (SMS)", "WhatsApp message", "location of intruder", "an immediate call", and "intruder's image to WhatsApp" is sent. The implementation of the project is simple and more effective.

The distribution of the paper in different sections is as follows. Section 2 deals with the literature survey. In Sect. 3, description of the algorithm is provided. Section 4 deals with the technology to be used. Section 5 deals with methodology. Section 6 deals with process design. Section 7 deals with snapshots. Section 8 deals with the scope of the paper. Section 9 deals with conclusions. Section 10 deals with references.

2 Literature Survey

Florian Schroff et al. proposed that all the images are directly mapped to Euclidean distances [5]. Here, distances directly correspond to the similarity of the images. Once all the calculations are done, it is very easy to calculate the detection and recognition. Here, they mainly applied the clustering algorithm. They mainly used deep convolutional network to optimize the embedded faces and distances. The optimization has reached to that level that each face takes 128-bytes. Iti s one of the best algorithms used in "Face Recognition" history. The accuracy has reached to 99.68%, almost-human level performance.

Changxing et al. proposed a comprehensive deep learning method using neural networks [6]. The set of neural networks extracts the face features from multimodal data. Then, the extracted features are concatenated to form a high-dimensional feature vector. 9000 different subjects regarding face are considered and trained. The accuracy is about 98.4%. Almost human-level performance. These systems achieve 99.0% when we train the naturally available training dataset.

Kshirsagar proposed a methodology for recognizing the faces using principal component analysis and feature extraction [1]. The main goal is that it recognize the face from large dataset with some real-time variations as well. Eigenfaces use PCA for face recognition.

John Wright et al. proposed that human should express all the feelings in front of a camera [7]. The algorithm trains itself. The cast recognition problem is one of the classifiers among multiple linear regression models and argues that new theory from sparse signal representation offers the key to addressing this problem. Based on a sparse representation computed by 1 minimization, we propose a general classification algorithm for (image-based) object recognition. This new framework provides new insights into two crucial issues in face recognition: feature extraction and robustness to occlusion. For feature extraction, this shows that if sparsity in the recognition problem is properly harnessed, the choice of features is no longer critical.

Yaniv Taigman et al. proposed 3D modeling transformation from nine-layer neural network by revisiting the important steps in face recognition [6]. This deep face recognition needs 120 million parameters. So, Yaniv et al. trained the algorithm over the biggest dataset to date. Over 4000 identities were covered. This method reached an accuracy of 97.35% on Labeled Faces in the Wild (LFW) dataset, almost approaching human-level performance.

3 LBPH Algorithm

LBPH stands for local binary pattern histogram. It is a descriptor used for facial recognition. The main algorithm behind this is LBP (local binary pattern). It is one of the best algorithms for texture or feature extraction. LBPH algorithm is a combination of LBP and HOG (histogram of orientated gradients) [8].

3.1 Concept

The process of LBPH algorithm is as follows:

The given image is converted into a matrix. Consider a pixel "a". Find the neighbors of the pixel (8-way connected).

- The formula for calculating the LBPH is

$$\text{LBP}_{p,r}(N_c) = \sum_{n=1}^{\infty} (N_p - \bar{N}_c)2^p$$

- Binary threshold function g(x) is

$$g(x) = \begin{cases} 0, & x < 0 \\ 1, & x \geq 0 \end{cases}$$

where N_p is neighbor pixel, N_c is pixel "a".
P = 0, 1, 2, ..., 7 for 3 * 3 matrix, r = 1 for 3 * 3 matrix (radius).
- Now, LBP is combined with HOG.

4 Technology Stack

The paper deals with identifying the intruder and alerting owner/administrator. It implies that it recognizes known faces. It also implies that it detects the unknown faces. Technology stack includes machine learning and image processing.

- **Machine Learning**: It is the advanced technology in computer science, which helps the system to behave as a human. The base of the Machine learning is pure mathematics, which helps to perform wonderful tasks. "Haarcascades" are used to detect the structure of the face. Later, "LBPH face recognizer" is used to train the images for recognition. Now, the machine is trained to recognize some faces. Recognized data is loaded into the program and tested against the known and unknown faces.
- **Image Processing**: Image processing helps to do some processing techniques which help to improve the accuracy of face recognition.

5 Methodology

The methodology contains step by step to identify the intruder.

The first step in the process is to scan the faces of all known members of the vicinity. Each person will have a name and a unique ID number. He/She shall enter her details. After registering each person, training him/her is done. Every time this process is continued. The next step is the recognition. This whole process of the scanning, training, and recognition are made possible through GUI [9].

Later, the implementation code is dumped into Raspberry Pi, so that it becomes a real-time system. For scanning, training and recognizing machine learning is used. An algorithm named as Local Binary Pattern Histogram (LBPH) [*for recognition*] and haarcascades [*for detection*] are used. If an intruder is detected, automatically "a message (SMS)", "WhatsApp message", "location of intruder", "an immediate call", and "intruder's image to owner's/administrator's WhatsApp"is sent to the owner/administrator. The messaging/calling service is done by importing the TWILIO [10] library. TWILIO is a cloud platform for sending trial messages and calling services. It provides five trial messages and calls per day.

6 Designing the Process

The design includes the following steps.

The first step is the implementation of detection, training, and recognition. Later, the GUI part is integrated into the code. Then, the code is dumped into the Raspberry pi. The process of detection is started. The first step is to scan the faces of all the members of the vicinity. For each person, 25 images are captured and stored as a dataset. This set of images is trained. Each person has an ID value. Then, the next step is the recognition. When the trained members are recognized, the ID of the person will be displayed. Whenever the intruders are detected, then all the above-said ways to catch intruder are activated.

7 Output Screen Shots

7.1 Message Service and WhatsApp Service Snapshots

Figure 1 explains the core architecture and workflow. Figure 2 explains the message service. The message is triggered when an intruder is detected. Twilio [10] is used for this free message service. Figure 3 explains the WhatsApp service which triggers when an intruder is detected. Customized message can be sent to the owner/administrator. Selenium [11] module is used for this automation. The message in Fig. 2, "Sent from your Twilio trial account—Intruder Detected. Please call 100" is sent to the owner with the help of Twilio, when an intruder is detected. The message as shown in Fig. 3, "Intruder Detected Please Find Help" is sent from an automated system to the owner (*Samba*), when an intruder is detected.

8 Scope

The work might be extended as—if an intruder is identified, location and call can be sent to the nearby police station. Image of the intruder can also be sent to police mobile, this will help police to catch thief/intruder easily. By this, the probability of catching an intruder/thief will be much easier. By installing this kind of systems, the security will enhance exponentially.

9 Conclusion

Security is utmost important and face recognition plays a very important role. Using technology to solve real-world problems is the actual use of technology. This paper discussed how machine learning and image processing can be used to find intruders.

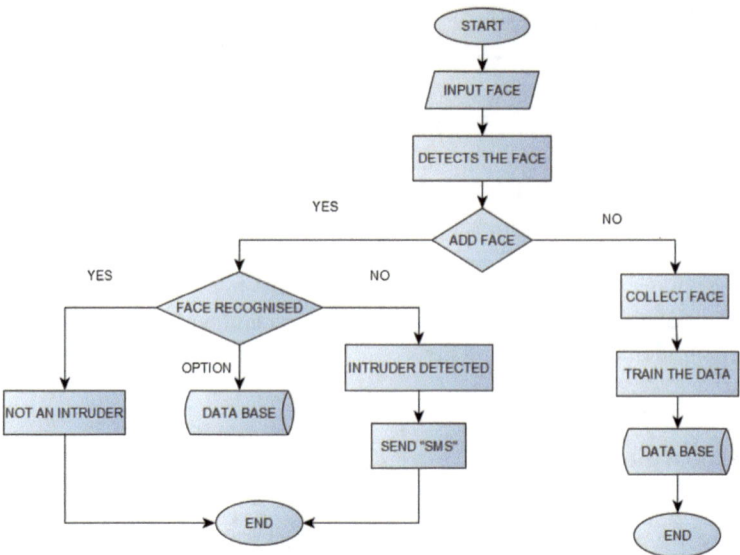

Fig. 1 Represents the architecture of intruder detection system

Fig. 2 Screenshot of
message service

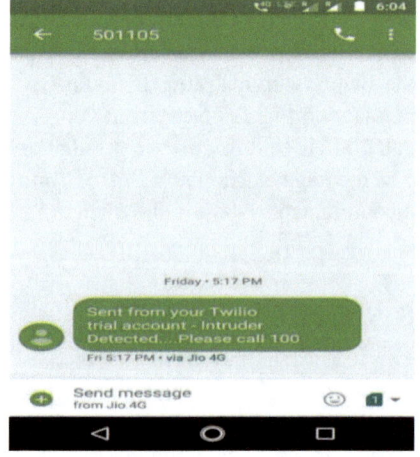

Machine learning algorithm local binary pattern histogram (LBPH) is used. Usually, it is very hard to identify intruders/thieves without any proofs, but with features such as "a call", "a message", "whatsapp messages", "location of intruder", and "Image of intruder" becomes very convenient and easy for police department to catch intruders. "Call" feature will serve as an immediate alert, which helps the owner/administrator to act quickly. If the image of an intruder is not known, it becomes very hard to find thieves. "Image of Intruder" will help to identify intruder easily. It also works as a proof to find the intruder. "Location" will help us to know where exactly the theft has

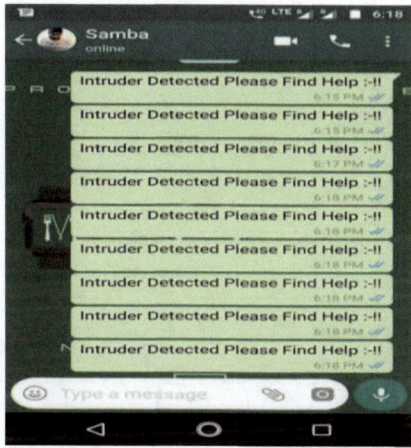

Fig. 3 Screenshot of WhatsApp service

taken place. With these features, it is very much possible to find the intruder and get back all the property/information. This paper concludes how intruder is efficiently identified with mentioned features.

References

1. Kshirsagar VP Prof, Baviskar MR, Gaikwad ME Face recognition using Eigenfaces
2. Lee H, Lee W-S, Chung J-H Face recognition using fisherface algorithm and elastic graph matching
3. Ahonen T, Hadid A, Pietikäinen M Face description with local binary patterns: application to face recognition
4. Ahonen T, Hadid A, Pietikäinen M Face recognition with local binary patterns
5. Schroff F et al FaceNet: a unified embedding for face recognition and clustering
6. Ding C et al Robust face recognition via multimodal deep face representation
7. Wright J et al Robest face recognition via sparse representation
8. Pang Y, Yuan Y, Li X et al Efficient HOG human detection
9. https://wiki.python.org/moin/TkInter
10. https://www.twilio.com/
11. https://www.seleniumhq.org

Unstructured Data Analysis with Passphrase-Based REST API NoSQL for Big Data in Cloud

Sangeeta Gupta

Abstract The synthetic word big data is emerging from almost every source in the modern world pertaining to any and every field, either technical or nontechnical. This big data generating at a rapid pace is difficult to analyze and yield productive decisions using the traditional tools like relational stores. NoSQL data stores stand as an alternative to deal with streaming big data. Though a wide set of NoSQL databases exist, column-oriented store Cassandra and document-based MongoDB are chosen to carry out the current work due to their simplicity and easy to set up environment. Both the chosen NoSQL databases are highly scalable with minimal security added at a single node level. However, the scalability exists at either load or retrieval levels but not both as desired by an enterprise to satisfy their end users and attain huge profits out of minimal investments. Moving toward this end, there is a need to design a system in which scalable load and retrieval operations with strengthened security is to be developed. Also, it is essential to enlighten the big data users to choose an appropriate NoSQL store among abundant available ones. Hence, in this work, Passphrase-Based REST API model is developed to enhance the security for huge scalable data. Among a huge set of NoSQL data stores, Cassandra and MongoDB are chosen to carry out the research which proved worth efficient in achieving better results for drastically increasing the number of records. Also, the proposed security implemented is to further strengthen the entire system from being tampered.

Keywords NoSQL · Big data · Cloud · Cassandra · MongoDB · Encryption · Passphrase · REST API

Sangeeta Gupta (✉)
Vardhaman College of Engineering, Hyderabad, India
e-mail: ss4gupta13@gmail.com

© Springer Nature Singapore Pte Ltd. 2019
H. S. Saini et al. (eds.), *Innovations in Computer Science and Engineering*, Lecture Notes in Networks and Systems 74,
https://doi.org/10.1007/978-981-13-7082-3_52

457

1   Introduction

Any computer with an Internet connection makes life simple for the cloud users to take the advantage of its services hosted across the web. Cloud computing deals with the ability to provide anytime, anywhere access to the users working on a single project with resources geographically scattered [1].

Big data at the other end is enormous data emerging from a wide set of sources, difficult to fit into a traditional system. To deal with such data, traditional data stores almost failed when scalability, security are the prime factors to be focussed on. Hence, emerging NoSQL stores [2] are found to be more comfortable to deal with any nature of the emerging big data (structured/unstructured), emerging from any source (across the web) and at any rate (fast/medium/slow) [3].

Among a huge set of cloud service providers like Amazon, Microsoft Azure, Google Cloud Platform, VMware, IBM Cloud, Rackspace, Dropbox, etc., which offer a mix of open source and commercial solutions based on the criticality of the problem, Amazon Web Services (AWS) and Bitnami cloud solutions are used to develop the work and establish connectivity with the chosen databases [4].

The paper is structured as follows: the initial focus is onto the introduction section that presents the modern terms like NoSQL, big data, and their purposes in recent times. Later on, a discussion is made on various existing works in Sect. 2. Next, in the third section, the need to move toward the proposed model for big data using NoSQL is identified and designed with a prime focus on strengthening the security through Passphrase-based Representational State Transfer (REST) API. Experimentation is highlighted in Sect. 4 with effective results achieved in proving the worth of Cassandra over MongoDB. Finally, the work is concluded in the fifth section with an ideology to further extend the proposed work in future.

2   Related Work

A compact set of existing works on NoSQL databases like casandra, MongoDB, and relational data store MySQL is taken to carry out background study where the parameters taken to prove the security aspect may change from one case to the other based on the length of the data set and the development environment used.

In [5], the authors expounded comparative analysis of a NoSQL store MongoDB with relational store MySQL. The methods discussed outlined the significance of selecting MongoDB over MySQL without any proof of implementation is done.

The authors in [6] described modern security threats and solutions in the field of NoSQL data stores at a theoretical pace without any evaluation represented for the same. At the other end, the authors in [7] used Bayesian algorithm specific to filter spam in social networking sites. The focus is in a single direction with a compromise on either authentication or security.

The authors in [8] expound shared NoSQL security aspects with a prime focus on encryption schemes without throwing any light on the implementation of the same. Also, the authors in [9] show the flexibility to add new set of classes to the existing functionality to enhance security of NoSQL store MongoDB suitable for shorter key lengths, without any inputs provided to deal with semi structured and unstructured big data.

The authors in [10] use index and replication-based techniques applied to MongoDB and expounded the efficiency of MongoDB as compared to MySQL for Climatic dataset. However, the access is restricted to load operations without throwing any ight on other set of operations like select, scan, delete, etc. At the other end, in [11], the authors expound combination of the existing techniques which work in isolation in order to extract better results without providing support to varying natured big data. Also, the authors in [12] carry out an analysis to deal with a few security features like confidentiality, integrity, etc., at a theoretical level without any experimentation carried out to analyze the same.

By going through the above data, it can be concluded that majority of the works developed in the stream of NoSQL with big data in cloud environment are of extreme theoretical inputs rather than practical orientation. Hence, in this work, an attempt is made to propose a new model to strengthen the security of Cassandra for big data in cloud.

3 Proposed Model

The proposed security is implemented using Passphrase-based REST API to strengthen the entire system from being tampered. The data arrives in the form of tweets collected through a basic web crawler to speed up the load process. The nature of data is analyzed before loading into appropriate virtual machines pertaining to Cassandra or MongoDB. If data is highly unstructured, then it is initially converted to structured format, and then the Passphrase-Based Rest API (PBRA) technique is applied to load and retrieve huge records set. Figure 1 represents the overall architecture with the flow of data from one module to the other [13].

REST APIs with get, put, post, and delete methods are integrated with the data before being stored in the appropriate data stores [14]. The performance changes are recorded in Sect. 4. An improvement in performance is achieved using Passphrase-based security is depicted in Sect. 4.

4 Experimental Evaluation

Experiments are conducted to build a cluster with five nodes where a pair of nodes deals with each data store. One node is dedicated to test scalability aspect and the other to deal with security aspects without any chaos in connection establishment.

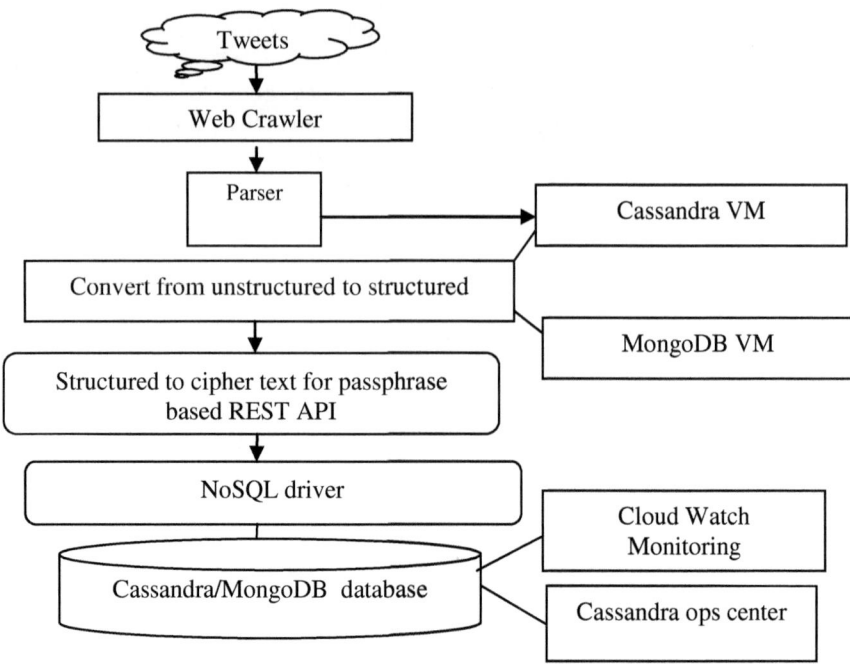

Fig. 1 Proposed architectural diagram

The fifth node is dedicated for commit log data maintenance for easy retrievals in the event of failures [4].

4.1 Encryption and Decryption Results

To represent the results, a tweet dataset with 100,000 records is taken. A few entries in the dataset are represented as in Table 1 and Table 2, respectively, where Table 1 presents a sample unstructured data as generated from tweets and Table 2 represents the structured data format for Table 1 [15].

The time taken to load and retrieve data from Cassandra and MongoDB data store is expounded in Tables 3 and 4.

Table 1 A sample Twitter streaming unstructured data

Anderson Andy Italy 12 123 234 Middle East 17-4-2014 10-12-2015
Mariadizou Mary China 98 678 84 East 14-6-2017 20-2-2018
Parinda Pari India 45 56 43 North 11-1-2016 25-5-2017

Table 2 Unstructured data converted into structured data

Tweet name	Real name	Location	Followers	Friends	No of tweets	Time zone	Created on	Last tweeted on
Andy	Anderson	Italy	12	123	234	Middle East	17-4-2014	10-12-2015
Mary	Mariadizou	China	98	678	84	East	14-6-2017	20-2-2018
Pari	Parinda	India	45	56	43	North	11-1-2016	25-5-2017

Table 3 Passphrase-based REST API (PBRA) Evaluation for Cassandra

No. of data items	Weight on nodes time (ms)	PBRA weight time (ms)	Select time (ms)	PBRA select time (ms)
10	1	1	1	10
100	1	4	2	20
1000	3	10	3	30
10,000	19	20	10	100
100,000	85	100	150	300

Table 4 Passphrase-based REST API (PBRA) evaluation for MongoDB

No. of data items	Weight on nodes time (ms)	PBRA weight time (ms)	Select time (ms)	PBRA select time (ms)
10	1	10	10	10
100	2	20	20	20
1000	5	40	300	300
10,000	15	80	450	500
100,000	200	500	700	800

The comparative analysis between Cassandra and MongoDB is presented in Table 5, respectively. It is observed from the results presented in Table 5 that Cassandra performs better over MongoDB for streaming big data set.

Table 5 Comparative analysis of Cassandra PBRA with MongoDB PBRA

No. of data items	PBRA weight time (ms) for Cassandra	PBRA weight time (ms) for MongoDB	PBRA select time (ms) for Cassandra	PBRA select time (ms) for MongoDB
10	1	10	10	10
100	5	20	20	20

(continued)

Table 5 (continued)

No. of data items	PBRA weight time (ms) for Cassandra	PBRA weight time (ms) for MongoDB	PBRA select time (ms) for Cassandra	PBRA select time (ms) for MongoDB
1000	10	40	30	300
10,000	20	80	100	500
100,000	100	500	300	800

5   Conclusion and Future Work

In this work, Passphrase-Based REST API model is developed to enhance the security for huge scalable data. Among a huge set of NoSQL data stores, Cassandra, and MongoDB are chosen to carry out the research which proved worth efficient in achieving better results for drastically increasing number of records. With a set of 1,00,000 tweets, PBRA-based Cassandra outperformed over PBRA-based MongoDB as shown in Sect. 4. In future, security can be further strengthened to deal with critical applications to serve as the need of the hour for efficient big data analysis with millions and billions of parsed data items.

References

1. Sandholm T, Lee D (2014) Notes on cloud computing principles. In: Sandholm and Lee (eds) Journal of cloud computing: advances, systems and applications, vol 3, no 21. Springer
2. Gudivada VN, Rao D, Raghavan VV (2014) NoSQL systems for bigdata management. In: Proceedings of 10th world congress on services, vol 42. IEEE, pp 190–197. https://doi.org/10.1109/services
3. Agarwal D, Das S, Abbadi AE (2011) Bigdata and cloud computing: current state and future opportunities. In: EDBT 2011/ACM, March 22–24 2011, Uppsala, Sweden
4. Amazon EC2. https://console.aws.amazon.com/ec2/
5. Zhao G, Huang W, Liang S, Tang Y (2013) Modelling MongoDB with relational model. In: Proceedings of fourth international conference on emerging intelligent data and web technologies, vol 25. IEEE, pp 115–121. https://doi.org/10.1109/eidwt
6. Okman L, Gal-oz N et al (2011) Security issues in NoSQL databases. In: Proceedings of international joint conference of IEEE TrustCom, pp 541–547
7. Xu L, Zheng X, Rong C (2013) Trust evaluation based content filtering in social interactive data. In: The proceedings of 2013 international conference on cloud computing and BigData. IEEE, pp 538–542
8. Zahid A, Masood R et al (2014) Security of sharded NoSQL databases. In: Proceedings of conference on information assurance and cyber security. IEEE, pp 1–8
9. Bang Tian XG, Huang B, Wu M (2014) A transparent middleware for encrypting data in MongoDB. In: Proceedings of IEEE workshop on electronics, computer and applications, pp 906–909
10. Ameri P, Grabowski U, Meyer J, Streit A (2014) On the application and performance of MongoDB for climate satellite data. In: The proceedings of 13th international conference on trust, security and privacy in computing and communications. IEEE, pp 652–659

11. Murugesan P, Ray I (2014) Audit log management in MongoDB. In: The proceedings of 10th world congress on services. IEEE, pp 53–57
12. Srinivas S, Nair A (2015) Security maturity in NoSQL databases-are they secure enough to haul the modern IT applications? In: Proceedings of the 2015 international conference on advances in computing, communications and informatics. IEEE, pp 739–744
13. Yunhua G, Shu S, Guansheng Z (2011) Application of NoSQL database in web crawling. Int J Digital Content Technol Appl 5(6)
14. Chen X, Ji Z, Fan Y, Zhan Y (2017) Restful API architecture based on Laravel framework. IOP Conf Series J Phys Conf Series 910:012016. https://doi.org/10.1088/1742-6596/910/1/012016
15. Twitter API. https://dev.twitter.com/rest/public

Discovery of Web Services Using Mobile Agents in Cloud Environment

T. Aditya Sai Srinivas, Somula Ramasubbareddy and K. Govinda

Abstract Recently, there is an increase in the use of web services and also increase the interest in the use of cloud computing technologies. A huge set of services is published on the internet every day. To discover and utilize these services based on human approach is quite time consuming and it also requires continuous user interaction. In modern network, there is rapid growth in web service application. So, the main task is how to discover and use proper web services. In this paper, the web service discovery and selection model based on the mobile agent is studied. The mobile agents are self-adaptable, intelligent, and collaborative, and so on. The system is composed of two clouds. The first cloud deals with keyword-based research and second cloud deals with the selection and filtering of web services found by search mobile agents. The selection algorithm is implemented using OpenMP.

Keywords Cloud computing · Mobile agent · Web service discovery · OpenMP

1 Introduction

This cloud computing is an emerging technology among other technologies in IT. These years, cloud computing has a greater impact on the IT industry and it is very important for distributed computing. The discovery, as well as selection of web services, is the most essential application over the internet. Nowadays, applications based on web services is increased therefore there is need for better web service discovery mechanism for effective discovery of web services based on the user's requirements. So, the mobile agents can play an important role in such scenarios. The

T. Aditya Sai Srinivas · S. Ramasubbareddy (✉) · K. Govinda
Department of Computer Science and Engineering, VIT University, Vellore, Tamil Nadu, India
e-mail: svramasubbareddy1219@gmail.com

T. Aditya Sai Srinivas
e-mail: saircew@gmail.com

K. Govinda
e-mail: kgovinda@vit.ac.in

© Springer Nature Singapore Pte Ltd. 2019
H. S. Saini et al. (eds.), *Innovations in Computer Science and Engineering*, Lecture Notes in Networks and Systems 74,
https://doi.org/10.1007/978-981-13-7082-3_53

system based on mobile agents is very useful to discover the efficient web services over the internet. The mobile agent is a combination of software and data. The mobile agents can move from one host to another host and can perform their tasks. Mobile agents are self-adaptive and tolerant to network faults which makes them one of the important part in applications of the distributed system. Web services are used in web service applications, so the important thing is that the user's needs must be satisfied by knowing how to discover and use the proper web service. There are methods have been proposed for discovering the web services. But there are defects about these techniques and it also does not satisfy the user's needs. Automating the process of web service discovery is not as per the user's requirements and it also does not satisfy the user's requirements. These techniques are not responded in the real time. These techniques use more data to discover the particular web services which on the other hand increases the overhead on the network. Efficiency is very bad and also not flexible. So to solve all oboe problems, the method based on agent technology can be used to automate and efficiently discover the web services in minimum time and without increasing the overhead on the network. OpenMP (Open Multi-Processing) is used for parallelizing the program to work efficiently on multi-processor/core and shared memory machines [1].

In this paper, we are presenting a new approach for the discovery of web services using mobile agents in the cloud environment. Cloud computing is the new technology and with the help of it, the users can share resources. We used OpenMP for parallelizing the selection algorithm in order to work efficiently on multi-core processors. We chose cloud computing technology for creating and deploying our system which has a significant effect on clients and the system also. Our system makes the resources available to users in instant access.

2   Literature Survey

There are some existing works proposed on web services discovery mechanism in recent years. In this paper, Jingliang and Zhe [2] proposed an architecture which is based on mobile agents and a set of situated agents. The situated agents also called static agents do the tasks such as management of the system, analyzing the request, keep an update of each mobile agent, and analyzing the result. The syntactic search of web services is done by the mobile agents. There is a problem using this approach and it does not consider the different position of words in the description of web services. In this paper, Chen et al. [3] proposed an approach which uses WordNet for vector extracting. This approach allows to insert the semantic information to this representation, space, and size vectors also reduces to calculate the similarity based on kernel the authors proposed a set method, which are based on kernel and using these methods, the similarity between web services can be estimated. Aversa et al. [4] proposed a mobile agent-based architecture to discover and access the web services in an efficient way. Session Initiation Protocol (SIP) and UDDI technology permit mobile agents and users to get access to the application, as well as resources over the internet at heterogeneous locations, and it also permits to search those resources by dynamically configuring the sessions of interactions as well as functionalities of services. These can be performed by means of terminal's characteristics as well as quality of interconnection. He et al. [5] proposed a architecture for managing the mobile services in grid computing, which is based on mobile agents and service discovery mechanism in detail. This architecture uses a hierarchical model of similar agents. The task of each agent is to provide and request services in the grid. When these agents move from one location or host to another host, then by calling registration and deregistration processes which makes the discovery even more effective and faster. Wagener et al. [6] proposed an approach based on cloud computing on an open standard "Extensible Messaging and Presence Protocol (XMPP)" for bioinformatics, service discovery. There is no need of an external register for the discovery of XMPP cloud services. Rajendran and Balasubramanie [7] proposed a model for the discovery of web services based on QoS which integrates augmented UDDI registry for publishing the QoS information. By designing a new framework, we can improve service discover process that improves the retrieval algorithms. Jingliang and Zhe [2] proposed a architecture for the discovery of web services. The architecture comprises the use of mobile agents. The mobile agents are intelligent, self-adaptive, mobility as well as collaborative and many more. Mobile agents discover the effective web services according to the requirements of web service provided by the user and return the appropriate web service. Baousis et al. [8] make use of mobile agents used to invoke the web service which is semantically matched to the given web service. It does not need the online presence of requestor who need a particular web service. The registries contain the information of web services, and the authors used these registries for semantic matching. Ketel [9] proposed a Mobile Agent Framework based on mobile agent technology for the integration of web services. They made use of capabilities of mobile agents for invoking web services and there is no need for the presence of service requestor.

3 Parallel Selection Algorithm (PSA)

Input: Query String, Decomposed web service
Output: Degree of each matched web service

Algorithm

```
{ Query_keywords[]; //keywords in user's query
      No_of_query_keywords;
  Ws_keywords[]; //keyword presents in web service description
  Webservice[];
  No_of_webservices;
  Match_count[];
  Synonym[]; //synonym of each keyword
  For all web services in webservice[]
    {
For all keywords in query_keywords[]
{
Synonym = Aceess_dictionary(query_keyword);
Flag=0;
For all words in synonym[]
{
For keywords in ws_keywords[]
{
If(match_str(synonym[word],ws_service[keyword]))
{
Match_count[webservice]++;
Flag=1
Break; }
If(flag==1) Break;
}}}
Degree[webservice]=(match_count[webservice]/no_of_query_keywords) * 100;
 }
Sort_degree_of_ws(degree);
For all webservices in webservice[]
{
If(degree[webservice]>50)
Then show webservice in results.
}}
```

4   System Architecture

The system architecture shown Fig. 1 contains two clouds, mainly cloud A and cloud B.

The mobile agents are used to transfer the information from one host to another host in the network. In the architecture, we have used four types of mobile agents.

a. **Input Agent**: The main task of the input agent is to get the request from the client who wants to search the particular web service.

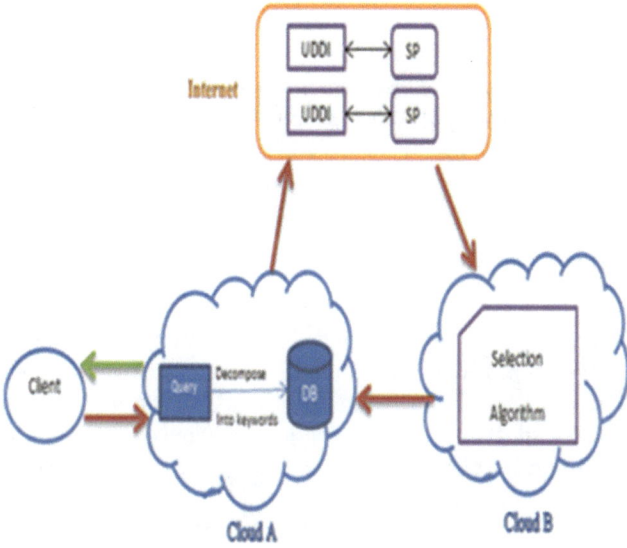

Fig. 1 System architecture

b. **Output Agent**: The main task of the output agent is to show the selected web service information to the client.
c. **Search Agent**: The main task of the search agent is to search the particular web service on the internet. The search agent is created when the user request came in the cloud A and these agents are sent to the internet to search web services UDDI.
d. **Dispatcher Agent**: The search agent creates dispatcher agent when it found web service matching with client's requirements. The main task of the dispatcher agent is to transfer the information of founded web service to cloud B.

The cloud A has one database which stores the keyword from the client's request. When the client request came, it is divided into keywords and these keywords are stored in a database with the unique identifier of each request. So that we can identify each request. The search agent is created as soon as the client's request came and with the help of keywords, it searches the web services in the UDDIs. The main advantage of search agent is that it can create a dispatcher agent when it found any relevant web service. When the search agent found any web service, it creates dispatcher agent to transfer information about web service to cloud B. The Cloud B region has WordNet [10] database contains a set of synonyms. The parallel selection algorithm as shown above is used to select the efficient web service based on the degree of web service. This algorithm takes inputs as a keyword in the user's query and the number of web services. Using the WordNet, it selects the synonym of each keyword present in the user's query. Then, the synonym of a keyword is matched with web service keywords of each individual web service. According to the match count, the matching degree of web services is calculated. The matching degrees of all web services are sorted in descending order. This algorithm returns all web services having a matching degree greater than 50. The selected web service information is then transferred to cloud A by dispatcher agent. And finally, cloud A shows this web service to the client using an output agent.

5 Experimental Analysis

In the experimental analysis, the selection algorithm is implemented using OpenMP (Open Multi-Processing) and its execution is tested on two core and four core machines (Table 1).

The above table shows the serial execution time and parallel execution time of selection algorithm according to the number of keywords presents in the web service description.

The above Fig. 2 clearly indicates that the execution time of the algorithm is significantly reduced using OpenMP. The execution time on four core machine is less than that of two core machine, it shows the parallel algorithm.

Table 1 Execution time

No of keywords	Serial execution time (MS)	Parallel execution time (2 core)	Parallel execution time (4 core)
5–7	180	123	89
7–10	187	125	91
11–14	193	128	93
14–17	198	131	94
18–20	201	134	97

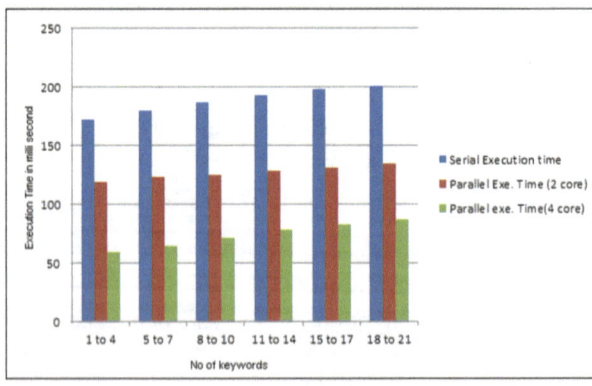

Fig. 2 Comparison graph

6 Conclusion

In this paper, we have proposed a parallel selection algorithm using OpenMP for selecting efficient web services using mobile agent technique. OpenMP helps the program to execute very fast on multi-core processors. The use of two clouds in system architecture helps to serve the different client's request within a short period of time. The use of mobile agents is the most important part of architecture because mobile agents are self-adaptive and tolerant to network faults that it can operate between client and server without an active connection and if some failure happens, then it can take its own decision not depending on any other entities.

References

1. OpenMP. https://computing.llnl.gov/tutorials/openMP/
2. Jingliang C, Zhe M (2010) Research on web service discovery based on mobile agents. In: International conference on computer design and applications
3. Chen L, Yang G, Wang D, Zhang Y (2010) WordNetpowered web services discovery using kernel based similarity matching mechanism. In: Fifth IEEE international symposium on service oriented system engineering
4. Aversa R, Di Martino B, Mazzocca N (2004) Terminal-aware grid resource and service discovery and access based on mobile agents technology. In: Proceedings of the 12th Euromicro conference on parallel, distributed and network-based processing. IEEE Computer Society
5. He YW, Wen HJ, Liu H (2005) Agent-based mobile service discovery in grid computing. In: Proceedings of the 2005 the fifth international conference on computer and information technology. IEEE Computer Society
6. Wagener J, Spjuth O, Willighagen LE, Wikberg JES (2009) XMPP for cloud computing in bioinformatics supporting discovery and invocation of asynchronous web services. BMC Bioinform BioMed Central
7. Rajendran T, Balasubramanie P (2009) An efficient framework for agent-based quality driven web services discovery. In: Proceedings of IAMA
8. Baousis V, Zavitsanos E, Spiliopoulos V, Hadjiefthymiades S, Merakos L, Veronis G. Wireless web services using mobile agents and antologies
9. Ketel M (2008) Mobile agents based infrastructure for web services composition. IEEE
10. WordNet: a lexical database for English. http://wordnet.princeton.edu/

Big Data: Scalability Storage

Aruna Mailavaram and B. Padmaja Rani

Abstract As the growth of enterprise data accelerates, the task of protecting data becomes more challenging. Data scalability, availability, and reliability are the major issues. In this paper, we have covered a few architectures such as Amazon EFS, GPFS IBM, Google GFS, HDFS, Blobseer, and AFS based on a distributed file system and their characteristics. A few important features of different file systems are compared. Here, we discuss a set of technologies, which are used in the market and are most relevant and represents the state of the art in the field of the distributed file system.

Keywords Amazon EFS · Data storage · Blobs · GFS · AFS · GPFS · Blobseer · Big data · HDFS

1 Introduction

Big data [1, 2] deal with very large data sets, which are available in a structured, unstructured, and semi-structured format meant for some useful purpose. The big data value chain is classified into Data Generation, Acquisition, Storage, and Analysis.

Data is generated from various sources of electronic devices like sensors, cell phones, laptops, cameras, etc., the data gathered has to be preprocessed before storage in order to avoid replication, which is known as Data Acquisition. Large-scale data sets are managed in data storage. The storage system is divided into hardware infrastructure and data management. The hardware infrastructure dynamically reconfigures and handle different types of data. Data management software is deployed on this hardware infrastructure for maintenance of large-scale data sets. Data Analytics is an analysis of past data to forecast the future.

A. Mailavaram
TKR College of Engineering and Technology(K9), Hyderabad, India
e-mail: aruna0949@gmail.com

B. Padmaja Rani (✉)
JNTUH College of Engineering, Hyderabad, India
e-mail: padmaja_jntuh@yahoo.co.in

© Springer Nature Singapore Pte Ltd. 2019
H. S. Saini et al. (eds.), *Innovations in Computer Science
and Engineering*, Lecture Notes in Networks and Systems 74,
https://doi.org/10.1007/978-981-13-7082-3_54

The big data structure is divided into three layers, namely Infrastructure layer, Computing layer, and Application layer. In Infrastructure layer, a pool of resources is organized using cloud computing infrastructure and these resources are exposed to upper layer systems. Here, the resources are allocated in such a way to meet the demand of big data. The Computing layer encapsulates different data tools for data integration, data management, and programming model. Integration of data is done from disparate sources and integrates into a unified form. Programming model abstracts the application logic and provides facilities for data analysis applications. Application layer is the interface between programming models and various data analysis functions, which includes querying, clustering, and classification, etc.

Few big data application domains are public sector administration, retail, global manufacturing, and personal location data. Companies like Facebook [3], Twitter, Google, AT&T, Amazon, etc., deal with big data. The technology used here is NoSQL [4]. NoSQL is "Not only SQL" which supports structured, unstructured, and semi-structured data. NoSQL databases are fast, highly scalable, and reliable for handling big data. A basic classification based on data model is key-value stores, document-based databases, column-oriented databases, graph databases, and multi-model databases.

According to the current development trends, data will reach from petabyte to exabyte. Handling these data can be done by increasing the capacity of the system power in terms of CPU speed and RAM, which is Vertical Scalability. And by adding more machines into the pool of resources is Horizontal scalability. The objective of this paper is to review, discuss, and compare the main characteristics of some major technological orientations existing in the market, such as Amazon Web Services (AWS), Google File System (GFS) and IBM General Parallel File System (GPFS) or Hadoop Distributed File System (HDFS), Blobseer, and Andrew File System (AFS).

2 Literature Survey

A DFS is a system which stores and retrieves data in the distributed systems. Data is divided into parts and named for consistency purpose. In order to maintain a reliable system, people has to concentrate on the design issues of DFS. The major design issues are scalability, heterogeneity, replication, transparency, migration, etc. These features make the system reliable while sharing the data in distributed system. DFS should be able to manage large data sets. Scalability is an important challenge in cloud computing. Fault tolerance can overcome by replication of data, because replication provides transparency to users and provides availability of the system. Security is also one of the major issues of DFS. Generally, DFS provides security with authentication, authorization, and privacy.

3 DFS Architectures

3.1 Andrew File System (AFS) [5]

AFS Architecture

- *Vice* – implements flat file system on server
- *Venus* – intercepts remote requests, pass to vice

Andrew file system is a distributed file system. It is developed at Carnegie Mellon University. This support reliable service for all network clients by transparent and homogenous namespace files locations. The authentication used in AFS is Kerberos. AFS uses a weak consistency model, for performing read and write operations updates are made only in the local cache. Once the modified file is closed, the updated portions are copied back to the file server by call-back mechanism. The important features of AFS are its volume, a tree of files, and subdirectories AFS mount points. AFS files are cloned to read-only copies, and the users will retrieve data from read-only copies and has no access to modify the file.

The file name is divided into two namespaces, one is sharing name space and the other one is local namespace, the local namespace is unique to the workstation. Sharing name space is common to all workstations and stores temporary files needed for workstation initialization.

3.2 Google File System (GFS) [6]

The GFS is a scalable distributed file system and handles large volumes of data. Fault tolerance is avoided to a great extent, and provides availability to a large number of users. The component failure is a norm rather than an exception. Because of use of inexpensive storage machines at the client side, they virtually guarantee that they will not recover from the current failure or some will not function at any given time. So, error detection, constant monitoring, fault tolerance, and automatic recovery are needed to the system. GFS is designed with a single master node and multiple chunk servers which are accessed by multiple clients. Files are divided into equal sized chunks and each chunk is identified by a global unique 64-bit given by master at the time of creation of chunk. The chunks are maintained by the chunk servers and for reliability purpose, replicas of chunks are kept on the chunk servers, and by default generally, three replicas are created.

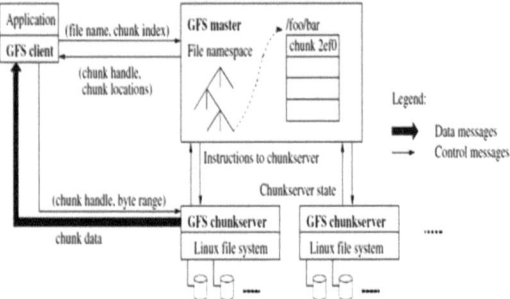

The chunk information is maintained by the master which maintains the metadata of the file system, including namespaces, access control information, and mapping from files to chunks and current location. The master periodically communicates with chunks in order to know chunk status.

3.3 Blobseer Architecture [7, 8]

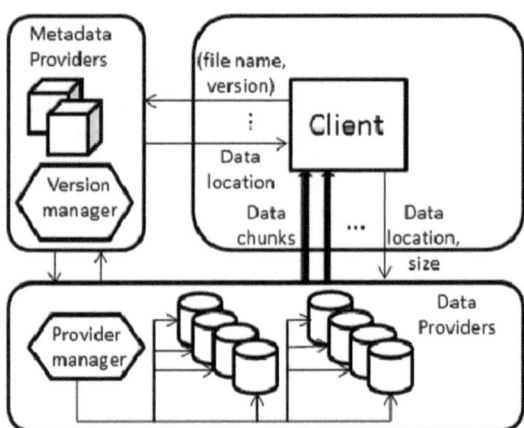

Blobseer is a large-scale-distributed storage service. It stores data in TB. It provides fine-grain access control in the order of MB. All snapshots are accessible by using versioning. Data and metadata both are decentralized. It achieves high throughput under heavy concurrent access concurrency in any combination of read and write. It is flexible for handling heterogeneous data quickly. To achieve scalability and improve the performance of the file system, data acquisition and computations are distributed over large-scale infrastructures which comprises of hundreds and thousands of machines. Blobseer is an efficient distributed data management service that addresses the issues like, scalable storage, fine grain access to data subsets, high throughput, fault tolerance, concurrency, versioning, and high quality of service.

3.4 Hadoop Distributed File System [9–11]

HDFS is developed to handle a very large amount of data using a distributed file system. It is designed using low-cost hardware and high fault tolerance. To maintain a huge amount of data, files are distributed over multiple machines. They are stored in a redundant fashion for data availability. HDFS runs on master–slave principle. A cluster in HDFS consists of a single name node, a master server that handles all the files accessed by clients and there are a number of data nodes. The name node manages operations like opening, closing, renaming files, and directories. The data node and their respective blocks are mapped. The data nodes are responsible for read and write operations. HDFS is a reliable storage of very large-scale files across multiple machines in a large cluster. It stores each file as a sequence of blocks of equal size, leaving the last block. The blocks are replicated in order to overcome fault tolerance.

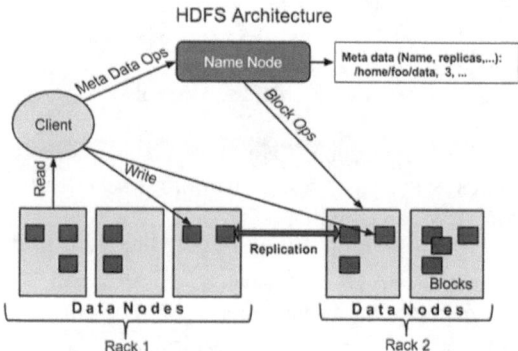

3.5 GPFS IBM [12]

GPFS achieve greater performance by providing access to multiple computers at once. GPFS provides good performance for reading and writing operations by data striping. Other features of GPFS are high availability, disaster recovery, security, and support to heterogeneous clusters. The files are divided into blocks of size less than 1 MB each, and is distributed over different nodes across disk array which results in high reading and writing speed because of combined bandwidth from different servers. To prevent loss of data, RAID controllers are used which makes multiple copies of data and stores at the physical disk on the individual nodes. We can even go for out of RAID-replicated blocks, instead of that, it makes two copies of data and stores at different file system nodes. Another important feature of GPFS is partition aware. On network failure, file system gets divided into two or more groups and this will be known only in the group which makes some nodes remain working.

3.6 Amazon Elastic File Service

The Amazon Elastic File System provides a reliable, scalable system by using Amazon EC2 instances in the AWS cloud. Amazon EFS is simple to integrate for easy access and one can create and configure file system quickly. Its elastic nature expands and shrinks according to the incoming data.

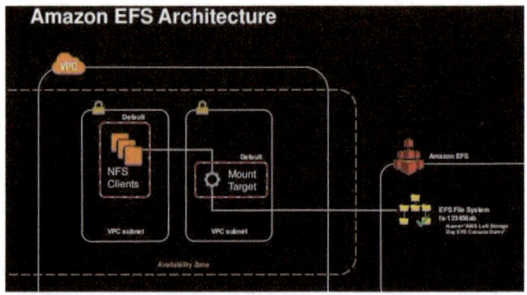

When we mount EFS on Amazon EC2 instances, it provides a standard file system interface and semantics for system access, which allows seamlessly integrate Amazon EFS with the existing tools and applications.

Amazon EFS can serve multiple Amazon EC2 instances at the same time. By using Amazon VPC with AWS Direct Connect, we can mount Amazon EFS on our data center servers, where we can backup our on-premises data to EFS.

Other features of Amazon EFS are high performance, durability, high availability, provides a broad spectrum of uses, including web and content serving, media processing workflows, container storage, and big data and analytic applications.

4 Comparative Analysis of Different File Systems

In this paper, we have discussed about various file systems and important characteristics of distributed file storage. Amazon S3 stores data as objects within resources called as buckets. LINUX is the operating system used by many file systems. AWS EC2 uses elastic cache for storing frequently used data. Data scalability is the main important characteristics of a distributed system which is achieved by almost all file systems. The concurrency of data access is high in AWS. As the data grows from petabyte to exabyte, it is important for us to scale the in vertically and horizontally. Load balancing and job scheduling according to priority wise are the important features of a distributed system [13] where we can optimize scalability, these things will be discussed in the next sequence of this paper.

5 Conclusion

In this paper, we have discussed various architectures and their features of the different distributed file system. One of the important features is scalability because based upon the incoming data the resources has to be allotted which improves system efficiency and maximum resource utilization. Apart from scalability, fault tolerance and data availability are the major concerns of the distributed file system. Data striping and RAID are used to optimize concurrent data. A DFS with minimum cost and effort and working with multiple operating systems is desired.

	Amazon EFS	GFS by Google	GPFS IBM	HDFS	Blobseer	AFS
Data scalability	Yes	Yes	Yes	Yes	Yes	Yes
Metadata scalability	Yes	No	Yes	No	Yes	No
Fault tolerance	Auto scaling	Chunk replication, fast recovery, master replication	Synchronous and asynchronous data replication Clustering features	Block replication Secondary name node	Chunk replication Metadata replication	No
Data access concurrency	High concurrency	Optimized for concurrent appends	Distributed byte-range locking	Files have strict to one writer at any time	Yes	Byte-range file locking
Snapshots	Yes	Yes	Yes	Yes	Yes	No
Versioning	Yes	Yes	Unknown	No	Yes	No
Data striping	Disk striping	64 MB chunks	Yes	Yes (Data blocks of 64 MB)	64 MB chunks	No
Storage as blobs	Buckets	No	No	No	Yes	No
Supported OS	Linux	Linux	Linux distributions, Windows Server 2008, AIX, Red Hat, SUSE, Debian	Linux and Windows supported, Mac OS/X, BSD, Open Solaris is known to work	Linux	Solaris, Linux, Windows, FreeBSD, NetBSD, OpenBSD, AIX, Mac OS X, Darwin, HP-UX, Irix
Dedicated cache	ElastiCache	No	Yes by AFM technology	Yes (Client)	No	Yes

References

1. Toward scalable systems for big data analytics: a technology tutorial (2017)
2. Data storage in big data context: a survey (2016)
3. Beaver D, Kumar S, Li HC, Sobel J, Vajgel P. Facebook Inc: finding a needle in Haystack: Facebook's photo storage

4. Catell R (2011) Scalable SQL and NoSQL data stores. SIGMOD Rec 39(4):12–27
5. OpenAfs. www.openafs.org/
6. Ghemawat S, Gobioff H, Leung S-T. Google*: the Google file system
7. Blobseer. blobseer.gforge.inria.fr/
8. Nicolae B, Antoniu G, Bouge L, Moise D, Carpen-Amarie A (2010) BlobSeer: next generation data management for large scale infrastructures
9. Mavani M (2013) Comparative analysis of Andrew files system and Hadoop distributed file system
10. Hadoop. hadoop.apache.org/
11. Shvachko K, Kuang H, Radia S, Chansler R (2010) Yahoo!: the Hadoop distributed file system
12. Fadden S (2012) An introduction to GPFS version 3.5. Technologies that enable the management of big data
13. Leue S (2001) Distributed systems—fall

Study Report on Using IoT Agriculture Farm Monitoring

G. Balakrishna and Moparthi Nageshwara Rao

Abstract The Internet of things (IoT) is rebuilding an agribusiness empowering the ranchers with the extensive variety of strategies, for example, accuracy and supportable farming to confront challenges in the farm. IOT innovation helps in gathering data regarding a situation like climate, dampness, temp, and richness of soil, monitoring crop through internet by farmer empowers discovery of weed, level of water, bug recognition, creature interruption into the field, trim development, and farming. IOT use agriculturists to get associated with his ranch from anyplace and whenever. Remote sensor systems are utilized for observing the ranch conditions and smaller scale reviewer are utilized to control and mechanize the homestead forms. To see remotely the conditions as picture and video, remote cameras have been utilized. An advanced mobile phone enables the rancher to keep refreshed with the continuous states of his rural land utilizing IOT whenever and any piece of the worldwide. IOT innovation can lessen the cost and upgrade the efficiency of conventional cultivating.

Keywords Component · Formatting · Style · Styling · Insert · Temp · IoT · Smart agriculture

1 Introduction

Internet of Things (IoT) has the ability to change the world we live in; increasingly proficient businesses, associated autos, and more astute urban communities are on the whole parts of the IoT condition. Be that as it may, the use of innovation like IoT in farming could have the best effect. The worldwide populace is set to touch 9.6 billion

G. Balakrishna (✉) · M. Nageshwara Rao
Department of Computer Science and Engineering, Koneru Lakshmaiah Education Foundation, Guntur, India
e-mail: me2balu@gmail.com

M. Nageshwara Rao (✉)
Velagapudi Ramakrishna Siddhartha Engineering College, Vijayawada, India
e-mail: Rao1974@gmail.com

© Springer Nature Singapore Pte Ltd. 2019
H. S. Saini et al. (eds.), *Innovations in Computer Science and Engineering*, Lecture Notes in Networks and Systems 74, https://doi.org/10.1007/978-981-13-7082-3_55

by 2050. Along these lines, to bolster this much populace, the cultivating business must grasp IoT. Against the difficulties, for example, outrageous climate conditions and rising environmental change, and ecological effect coming about because of serious cultivating hones, the interest for more sustenance must be met. Shrewd cultivating in light of IoT advances will empower cultivators and agriculturists to decrease waste and upgrade efficiency extending from the amount of manure used to the quantity of adventures the homestead vehicles have made.

In IoT-based keen cultivating, a framework is worked for observing the product field with the assistance of sensors [1] (light, stickiness, temperature, soil dampness, and so on.) and mechanizing the water system framework. The agriculturists can screen the field conditions from anyplace. IoT-based brilliant cultivating is exceedingly productive when contrasted and the customary approach. The uses of IoT-based shrewd cultivating target traditional, huge cultivating tasks, as well as be new levers to elevate other developing or basic patterns in agrarian like natural cultivating, family cultivating (perplexing or little spaces, specific dairy cattle as well as societies, safeguarding of specific or top-notch assortments and so forth.), and upgrade very straightforward farming. In terms of ecological issues, IoT-based savvy cultivating can give awesome advantages including more proficient water use, or improvement of sources of info and medicines. Presently, we should talk about the real utilization of IoT-based keen cultivating that are changing horticulture. In view of the suggestions by different specialists, it is the request of the examination to assemble a choice help structure on the cloud to dissect the IoT gathered information.

1.1 Stages of Agriculture and Technologies

A farmer will determine appropriate crops for his land supported integrated soil, water, air analysis. Sensors collect weather information, which may be analyzed and shared with farmers. Mechanized system may be used to chop–chop and accurately plant saplings in the field. Water and plant food distribution system may be automatic by sensors-equipped system Weed removal systems connected to sensors will mechanically spot weeds and acquire obviate the Crop care may be designed supported prognosticative trends any supported analyzed information from the past. Harvesting machinery may be created into a lot of economic and automatic system supported SmartTech. Crop grading and cleansing systems may be customized, improved and automatic. Warehousing and crop transport system's temperature, wetness and air internal control system.

1.2 Challenges in Smart Agriculture

For Farmers, it is difficult for them to understand technical terms and usage of technology, and also it is a cost effective affair. It is a challenge to balance both. A low budget to hold outcomes because of the dependence on the harvest. Cost optimization can be achieved by minimizing number of people involved in the hierarchy in between farmer and government. Global temperature change is important for management weather and ambient conditions.

2 Literature Survey

Xiao [2] and Magno [3] anticipated farming checking framework utilizing remote sensor arrange (WSN). The conditions that can be observed continuously are temperature, light force, and mugginess. The test includes the equipment and programming plan of the manufactured modules, arrange topology, and system correspondence convention with the difficulties. The configuration clarifies how the hub can accomplish agrarian condition data accumulation and communication. The framework is smaller in outline work, lightweight, great in execution, and activity. It enhances the farming generation proficiency naturally.

Haule [4], Ofrim et al. [5] have proposed a trial that clarifies the utilization of WSN utilized as a part of mechanizing water system. Water system control and rescheduling in light of WSN are capable answers for ideal water administration through programmed correspondence to know the dirt dampness states of water system plan. The procedure utilized here is to decide the correct recurrence and time of watering are essential to guarantee the productive utilization of water, high caliber of harvest recognition postpone throughput, and load. Reenactment is improved the situation horticulture by OPNET. Another plan of WSN is sent for water system framework utilizing Zig honey bee convention, which will affect battery life. There are a few downsides as WSN is still a work in progress arrange with temperamental correspondence times, delicate, control utilization, and correspondence can be lost in the horticultural field. Remote light sensors have been utilized by computerized water system framework. WSN utilizes low power and a low information rate and consequently vitality productive innovation. Every one of the gadgets and machines controlled with the assistance of information sources is got by means of sensors which are blended with soil. Agriculturists can investigate whether the framework performs in ordinarily or a few activities are should be performed.

Dan [6], Michael [4] and Shuntian [7] proposed nursery checking agribusiness framework in view of Zigbee innovation. The framework performs information, taking care of, transmission and social occasion capacities. Purpose of their test is to check whether the algorithm reduces the cultivating cost under various tropical conditions or not. IOT innovation here depends on the B-S structure and cc2530 utilized like handling chip to work for remote sensor hub and organizer. The passage has a Linux working framework and cortex A8 processor go about as center. By and large, the plan acknowledges remote astute observing and control of nursery and furthermore replaces the customary wired innovation to remote, likewise diminishes labor cost.

Li [8] and Tam [9, 10] have proposed a framework that utilizes Zigbee innovation. This examination manages equipment and the product of the system facilitator hub and the sensor hubs. The hypothetical and viable outcomes demonstrate that the framework can effectively catch nursery ecological parameters, including temperature, dampness, and carbon dioxide fixation and furthermore clears the typical correspondence amongst hubs and the system organizer, great system steadiness. The execution investigated values is utilized as a part of the intricate nursery ecological observing

2.1 Water Monitoring

Ji-hua [10, 11] led an examination on the development of oat edit seedlings, and in addition, the status and pattern of their development. This paper presented the outline, techniques utilized, and usage of a worldwide yield development observing framework, which fulfills the need of the worldwide harvest checking on the planet. The framework utilizes two techniques for checking, which are constant product development observing and edit developing procedure observing. Constant product development checking could get the yield developing status for a certain period by looking at the remote detected information of the period with the information of the period in the history. The differential outcome was ordered into a few classifications to mirror the situation at distinction level of yield developing. In this framework, both continuous product development checking and edit developing procedure observing are done at three scales, which are state (region) scale, nation scale, and landmass scale. Worldwide product development checking framework was found in this plan and assembled a framework that can screen the worldwide harvest development with remote detecting information. The framework demonstrated the qualities of quick, compelling, high validity, and operational in its run.

Kim [12], Balamurali [13] have proposed the outline for remote sensor arrange for a water system control and observing that is made out of various sensor hubs with a systems administration ability that is conveyed for an impromptu for the motivation behind continuous checking. Data collected by sensors will be sent to the base station. The future framework offers a smaller amount of power utilization with high unwavering quality in view of the outcome. The utilization of high-power wireless sensor networks is appropriate for assignments in businesses including gigantic region observing like assembling, mining developing, and so on. The framework talked about here is anything but difficult to introduce and the base station can be set at the nearby living arrangement near the region of checking where a man requires negligible preparing toward the start of the framework establishment.

2.2 Farm Monitoring

Tirelli, Borghese [14] found that checking nuisance bug populace is right now an issue in trim insurance. The framework here is as of now in view of a disseminated imaging gadget worked by means of a remote sensor arrange that can naturally catch and transmit pictures of caught zones to a remote host station. The station approves the thickness of bug advancement at various homestead areas and produces an alert when creepy crawly thickness goes over the limit. The customer hubs are spread in the fields, which go about as checking stations. The ace hub facilitates the system and recovers caught pictures from the customer hubs. Amid a checking time of a month, the system working frequently predicts a nuisance creepy crawlies' populace bend connected to day-by-day assessment got by visual perceptions of the trap and subsequently the practicality is resolved.

Suresh et al. [15] proposed framework will gauge the estimations of N, P, and K from the dirt and furthermore screen the level of soil supplements substance and in like manner apportion, the required amount of the manures through water system framework. Every one of the information will be refreshed to the client through email.

Suresh et al. [15] proposed a framework in which N, P, K, and PH estimations of a soil test are estimated continuously and contrasted, and the pre-put away qualities got from the rural division. The framework additionally gives the data about the yields that can be developed in particular soils.

Parameswaran and Sivaprasath [16] proposed a robotized water system frame-work in light of soil moistness. The water system status is refreshed to the server or neighborhood have utilizing PC. This availability is conveyed by IoT. Different restrictions like the water system and not the NPK esteems are refreshed to the client through IoT [16].

Londhe and Galande [17] proposed a mechanized water system framework utiliz-ing ARM processor. The framework screens and controls every one of the exercises of water system framework. The valves are turned on or off and naturally give the deliberate information in regards to the dirt pH and supplements like nitrogen along-side the best possible proposals and give the correspondence interface.

Moparthi et al. [18] proposed Water Quality location framework utilizing IoT for the most part centering to make a more perfect air contamination identification frame-work while wipes out some downside of the customary frameworks and arrangement are one of the basic systems for programming imperfection detection [18].

Joshi et al. [19] proposed a sensible little scale cultivating utilizing IoT. Utilizing the proposed display, an individual can keep up his own homestead in little gardens. A large portion of the equipment used is effortlessly obtained and taken as a toll productive. Fusing picture preparing to stay away from weed development apparently is solid and the outcomes appear above are predictable with this reality. The proposed basic strategy to perceive shadow or soaked soil as foundation supposedly is powerful and block the utilization of complex closer view extraction methods. As a major aspect without bounds work, it can be stretched out to vast scale cultivating (Table 1).

Table 1 Comparative table

Done by/Parameters	K. A. Patil	Chiyurl Yoon	S Rajeswari	Vaishali S	Ruby Roslin
Knowledge based	Data collection from sensors	Collection and transmission without any losses	Data collection, storing on cloud, data analytics is performed	Data collection motor on/off based on values	Sensor-based automated irrigation
Approach	Remote monitoring system	Nodes—LPWAN, LPBluetooth, RS485 communication to the server through MQ telemetry transport	Cloud-based data analytics	Remote irrigation using sensor values	Remote irrigation using sensor values
Objectives	Collect real time data from sensors	Minimum Loss by sending 10 identical data in wireless transmission	Increase crop production and control agricultural cost using analyzed data	Control the water supply and monitor the plants through a smartphone	Prevention of crops from spoilage during rains and recycling rainwater
Advantages	Decision support, alerts solar power	Wired and wireless transmissions, any number of devices can be added easily	Temperature and rainfall factors are predicted, Crop patterns	Water management	PIR sensor is used to detect motions, rainwater is recycled
Protocol	Zigbee	LPWAN, LPBluetooth, RS485	Zigbee, prediction using Naïve Bayesian classification, map reduce	Raspberry Pi	Arduino, GSM
Future	Pest detection	Wireless transmission in kms	Interfacing different soil nutrient sensors and collect the data	Outdoor utilization system	Detect plant disease, crop theft
People	Not required	Not required	Minimal requirement	Not required	Not required
Done by/Parameters	*T K Rana*	*Prathiba Sri*	*Shreyas B*	*Nikesh Gond-chawar*	*Gireesh Babu*

(continued)

Table 1 (continued)

Done by/Parameters	K. A. Patil	Chiyurl Yoon	S Rajeswari	Vaishali S	Ruby Roslin
Knowledge based	Remote control vehicle for monitoring	Data collection and capturing images	Data collection through sensors and transmission to web app	GPS based mobile robot, decision making, warehouse management	Sensor modules, mobile app module, farm cloud, govt. and agro module
Approach	Monitoring, data collection and irrigation based on values	Monitoring environmental factors using sensors	Remote monitoring system	Remote controlled robot	Mobile computing, big data analytics
Objectives	Automation of monitoring, watering, data collection	Wireless monitoring of field	Monitoring and controlling in real time	Automated irrigation, decision-making warehouse management	Give farmer required fertilizers based on soil sample
Advantages	Solar power for sensors	Human effort is reduced	FLASK eliminates the requirement of databases	Auto monitoring, warehouse management	Provides details about the latest agricultural schemes and products
Protocol	AVR micro controller, Zigbee	CC3200	Raspberry Pi, FLASK	AVR micro controller, Zigbee	Farm cloud
Future	GPS based vehicle	Irrigation method, solar power	Precision storage, capturing live images	Improve crop yield based on analysis	Using different sensors and data analytics for accurate results
People	Not required	Not required	Not required	Not required	Not required

3 Conclusions

IoT empowers the farmer to trim check the man power for cost optimization and to improve the productivity which leads to profit maximization. WSN and sensors of various sorts are utilized to accumulate the data of yield conditions and normal changes and these data are transmitted through the framework to the rancher/contraptions that starts remedial exercises. Ranchers are related and mindful of the states of the provincial field at whatever point and anyplace on the planet. A couple of preventions in correspondence must be overpowered by pushing the advancement to expend less essentialness and moreover by affecting UI to accommodation.

References

1. Zhang L, Yuan M, Tai D, Oweixu X, Zhan X, Zhang Y (2010) Plan and execution of storage facility checking framework in view of remote sensor arrange hub. In: 2010 International conference on measuring technology what's more, mechatronics automation
2. Xiao L, Guo L (2010) The realization of precision agriculture monitoring system based on wireless sensor network. In: 2010 international conference on computer and communication technologies in agriculture engineering
3. Murphy FE, Popovici E, Whelan P, Magno M (2015) Improvement of heterogeneous wireless sensor network for instrumentation and analysis of beehives. In: 2015 IEEE international conference instrumentation and measurement technology conference (I2MTC)
4. Haule J, Michael K (2014) Organization of remote sensor systems (WSN) in robotized water system administration and planning frameworks: an audit. In: Pan African conference on science, computing and telecommunications (PACT)
5. Ofrim DM, Ofrim BA, Săcăleanu DI (2010) Enhanced ecological screen and control utilizing a remote clever sensor arrange. In: 2010, third International Symposium on Electrical and Electronics Engineering (ISEEE)
6. Dan L, Xin C, Chongwei H, Liang J (2015) Savvy operator nursery condition observing framework based on IOT innovation. In: 2015 international conference on intelligent transportation, big data and smart city
7. Qiu W, Dong L, Yan H, Wang F (2014) Outline of intelligent greenhouse environment monitoring system based on ZigBee and implanted innovation. In: 2014 IEEE International gathering
8. Li L-L, Yang S-F, Wang L-Y, Gao X-M (2011) The greenhouse environment monitoring system based on wireless sensor network technology. In: Proceedings of the 2011 IEEE international conference on cyber technology in automation, control, and intelligent systems, March 20–23 2011, Kunming, China
9. Bhanu B, Rao R, Ramesh JVN, Hussain MA (2014) Horticulture field monitoring and analysis utilizing wireless sensor networks for enhancing crop production. In: 2014 Eleventh international conference on wireless and optical communications networks (WOCN)
10. Zhuang W-Y, Junior MC, Cheong P, Tam K-W Surge monitoring of distribution substation in low-lying areas using wireless sensor network. In: Proceedings of 2011 international conference on system science and engineering, Macau, China – June 2011 (NESEA), 2012 IEEE third International Conference
11. Fan C-L, Guo Y (2013) The application of a ZigBee based wireless sensor network in the LED street lamp control framework. In: 2014 IEEE international conference on college of automation and electronic engineering, Qingdao University of Scientific and Technology, Qingdao, China embedded technology, Consumer Electronics—China

12. Kim Y, Evans RG, Iversen WM Remote sensing and control of an irrigation framework using a distributed wireless sensor network. In: IEEE exchanges on instrumentation and estimation, vol 57, no-7, July 2008, Member, IEEE
13. Balamurali R, Kathiravan K (2015) An analysis of various routing protocols for precision agriculture utilizing wireless sensor network. In: IEEE international conference on technological innovations in ICT for agriculture and rural development (TIAR 2015)
14. Tirelli P, Borghese NA, Verma S, Chug N, Gadre DV (2010) Remote sensor network for crop field monitoring. In: 2010 international gathering on recent trends in information, telecommunication and computing
15. Suresh DS, Jyothi Prakash KV Rajendra CJ (2013) Robotized soil testing device. ITSI Trans Electr Electron Eng (ITSI-TEEE) 1(5). ISSN (PRINT): 2320–8945
16. Parameswaran G, Sivaprasath K (2016) Arduino based smart drip irrigation system utilizing internet of things. IJESC 6(10)
17. Londhe G, Galande SG (2014) Robotized irrigation system by using ARM processor. IJSRET 3(2). ISSN 2278– 0882
18. Moparthi NR, Mukesh Ch, VidaySagar P (2018) Water quality monitoring system using IoT. In: An International Conference by IEEE, pp 109–113
19. Joshi J, Kuma P, Polepally S Machine learning based cloud integrated farming. ACM. ISBN 978-1-4503-4828-7/17/01

Partitioning in Apache Spark

H. S. Sreeyuktha and J. Geetha Reddy

Abstract Apache Spark performs in-memory computation. The data structure used is Resilient Distributed Datasets (RDDs). These RDDs are partitioned using inbuilt Hash and Range Partitioning. We propose a partition scheme which uses modular division on keys of elements with numbers from 2 to 10. This scheme works on smaller datasets in order to enhance the execution time.

Keywords Partition · Hash partition · Range partition · Resilient Distributed Datasets (RDDs)

1 Introduction

Apache Spark [1] is a fast and general data-processing engine. It performs batch processing and allows us to leverage memory space by performing in-memory computation. This allows Real-Time Processing (RTP) up to 100 times faster than MapReduce in some cases using DStreams. Spark allows us to input files like variable which is not possible in Hadoop's MapReduce. Spark offers an abstraction called Resilient Distributed Datasets (RDDs) to support these applications efficiently. RDDs can be stored in memory between queries without requiring replication. They rebuild lost data on failure using lineage. Each RDD remembers how it was built from other datasets (by transformations like map, join, or groupBy) to rebuild itself. RDDs allow Spark to outperform existing models by up to 100x in multi-pass analytics.

H. S. Sreeyuktha (✉) · J. Geetha Reddy
Department of Computer Science and Engineering, Ramaiah Institute of Technology,
Bangalore 560054, India
e-mail: sreeyuhs@gmail.com

J. Geetha Reddy
e-mail: geetha@msrit.edu

© Springer Nature Singapore Pte Ltd. 2019
H. S. Saini et al. (eds.), *Innovations in Computer Science and Engineering*, Lecture Notes in Networks and Systems 74,
https://doi.org/10.1007/978-981-13-7082-3_56

2 Background and Related Works

2.1 Apache Spark Architecture

Apache Spark has a well defined and layered architecture [2]. It is loosely coupled and has two main abstractions: 1. Resilient Distributed Datasets (RDDs). 2. Directed Acyclic Graphs (DAGs)

Resilient Distributed Datasets (RDDs) represents an immutable partitioned collection of elements that can be operated in parallel. As the name suggests, Resilient implies the provision of fault tolerance via lineage graph. Distributed means data is present in multiple nodes in a cluster (slaves). Dataset means a collection of partitioned data with primitive values (Fig. 1).

Apache Spark does lazy evaluation, i.e., RDDs will not be transformed unless an action triggers the execution of the transformation. Spark also provides persistence which results in fast computation, i.e., users can specify which RDDs they want to reuse and select desired storage like in-memory, cache or a hard disk. Each RDDs holds a reference to partition objects. Each partition object reference is a subset of the data. Partitions are signed to nodes in clusters. Each partition will be in RAM by default. This increases the performance by up to 100x.

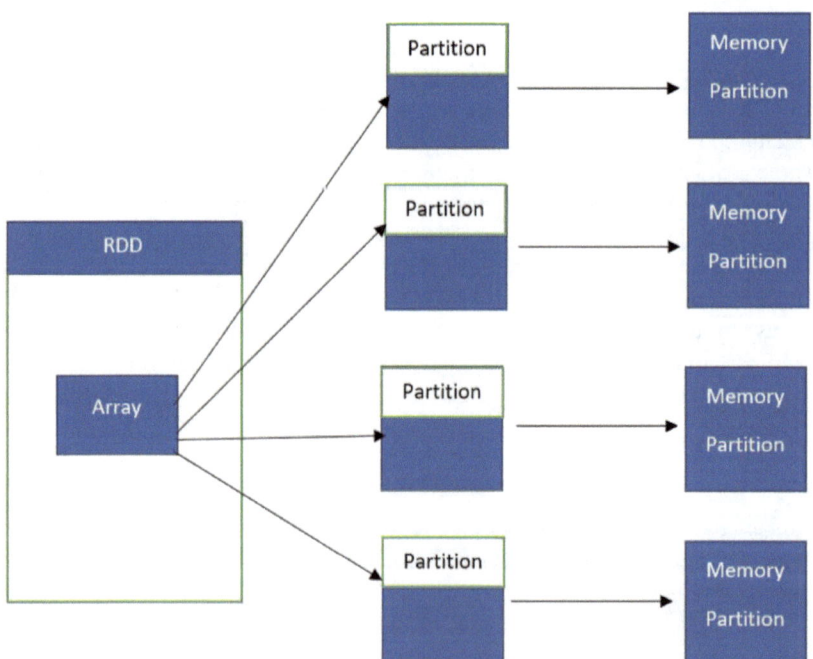

Fig. 1 Structure of resilient distributed dataset—RDD

Directed Acyclic Graphs RDDs are nothing but partitions of the given datasets. Apache Spark is better than Hadoop because it involves the use of DAGs. Here, Direct means transformation of partitioned dataset A to partitioned dataset B. Acyclic means transformation cannot return to older partition. This helps eliminate Hadoop's MapReduce multistage execution model and provides performance enhancements over Hadoop. $RDD_1 \rightarrow RDD_2 \rightarrow \ldots \ldots \rightarrow RDD_N$. This is a DAG.

2.2 Wide Versus Narrow Dependencies

Dependencies [3] are the relationship between RDDs in graph of computation, which is used a lot in shuffling. Narrow Dependency means each partition of the parent RDD is used by at most one partition of child RDD. This is fast because no shuffle is necessary and optimization like the pipeline is possible. Thus, the transformations which have narrow dependency are fast. Wide Dependency means each partition of parent RDD may be used by multiple child partitions. This is slow because shuffle is necessary for all or some data over the network. Thus, transformations involving wide dependencies are slower.

3 Problem Definition and Approach

3.1 Partitioning Techniques

The large datasets are partitioned (or divided) into multiple parts across a cluster. This works on the principle of locality. The worker nodes take data for processing that are nearer to them because network I/O will be reduced so processing is faster. To divide data into partitions, it must be stored first. Hence, Spark stores data in the form of RDDs. So say RDD_1 can have partitions like P_1, P_2, P_3; RDD_2 can have partitions like P_4, P_5, P_6...so on. There are two types of partitioning: 1. Hash Partitioning. 2. Range Partitioning

Hash Partitioning [4] uses Java's Object. HashCode() method. In this technique, the Hashcode of the keys are divided by the number of partitions required; i.e., (keys.hashcode() % numofpartitions). It is the default Spark partitioner. Hash Partitioner takes a single argument which defines the number of partitions. The values are assigned to partitions using a hash of keys. If the distribution of keys is not uniform, we can get situations where part of the cluster is idle. Hash Partitioner is neither injective nor surjective. Multiple keys can be assigned to a single partition and some partitions can remain empty. Hash Partitioner shuffles the data. Unless partitioning is reused between multiple operations, it does not reduce the amount of data to be shuffled.

Range Partitioning [5] involves sorting the keys, and then dividing them into equal ranges depending on the number of partitions is required. Ranges are determined by sampling the content of RDDs passed in. Basically, we compute the range boundaries and construct a partitioner from these range boundaries which gives you a function from key K to partition index i.

4 Proposed Methodology

The proposed partition technique will do partitioning of small datasets say 10k keys by applying the modulus operation to the keys with the numbers ranging from 2 to 10. We start from the largest divisor, i.e., 10 and then go backwards; this is done so that a larger set of numbers are filtered when dividing with lower numbers and prevents data skew. Figure 2 shows that our algorithm is successful for elements between 5 and 30k. Table 1 contains the runtime values obtained by execution of the proposed algorithm and the Existing Hash Partitioner.

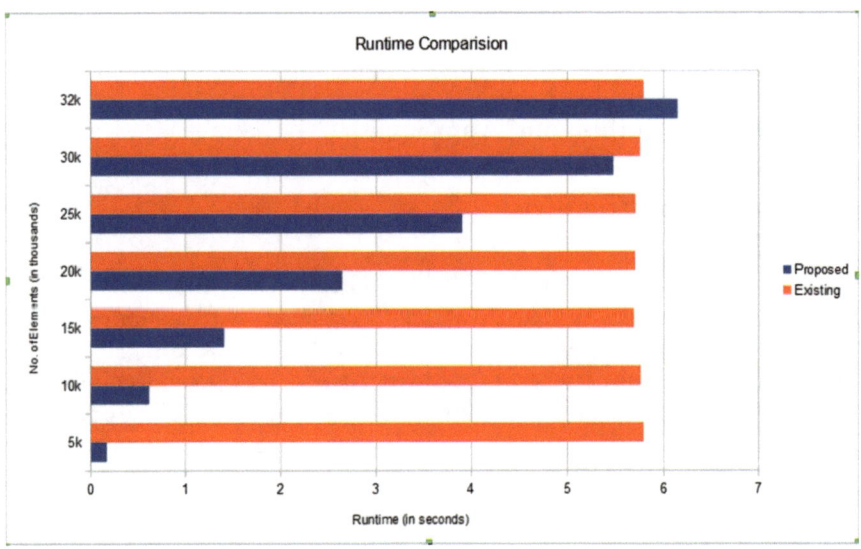

Fig. 2 Comparison of execution time of hash partitioning versus proposed partitioning technique

Table 1 Comparison of execution time of hash partitioning versus proposed partitioning technique	Number of elements	Proposed partitioning	Hash partitioning
	5000	0.172	5.7893
	10000	0.62	5.7593
	15000	1.412	5.6893
	20000	2.6452	5.7023
	25000	3.9022	5.7063
	30000	5.4773	5.7523
	32000	6.1464	5.7913

Algorithm 1: Proposed Algorithm

1: **procedure** PARTITIONING
2: d ← empty dictionary
3: **loop1**:
4: i ← 10 to 2
5: list l ← empty
6: **loop2**:
7: j ← list of keys
8: **if** j%i == 0 **then** append key j to list l. Remove that key from list of keys
9: goto loop2.
  insert l in dictionary d
10: i ← i- 1
11: goto loop1
  At the end append the remaining elements of keys list as prime in dictionary d
12: close

5 Evaluation and Result

System specification: 8 GB RAM, Processor: Intel(R) Core(TM) i3-2350 M CPU @ 2.30 GHz 2.30 GHz, 64-bit Windows 7 OS Application Specification: Java version: 8, PySpark. Our approach involves partitioning of small datasets; keys undergo modulo division with the numbers ranging from 2 to 10. We start from the largest divisor, i.e., 10 and then go backwards till 2; this is done so that a larger set of numbers are filtered when dividing with lower numbers. For example, if we do (keys % 10), then all numbers divisible by 10 and 5 will be grouped in the partition set of 10, then when divided by 5, we get lesser number of keys in the partition set of 5; else we can imagine that the partition set will become skewed say if it starts dividing[1] from the lower number first. This implementation we will be partitioning a dataset into 10 partitions only.

This partitioning technique holds good for smaller datasets. In our approach, the runtime for partitioning on PySpark is 0.62 s for dataset of size 10 k. The runtime for hash partitioning in PySpark, i.e., the default partitioner <pys-park.rdd.Partitioner> is 5.93 s for dataset of size 10 k, which is drastic. Figure 2 shows that our algorithm is successful for elements between 5 and 30 thousand. The two methods (Proposed and Hash) meet (which means runtime matches) in between 31 and 32 thousand elements.

6 Conclusion

The current work highlights the drastic reduction in runtime for partitioning of smaller datasets. We would like to incorporate the required algorithmic changes in our future work to reduce the execution time for partitioning of larger datasets. We still need to overcome the limitation that it works only on smaller datasets and not the bigger ones. Since the partitioning scheme needs to be more generalized, we are working toward a more optimal solution for partitioning for all sizes of datasets.

References

1. Zaharia M, Chowdhury M, Das T, Dave A, Ma J, McCauley M, Franklin MJ, Shenker S, Stoica I (2012) Resilient distributed datasets: a fault-tolerant abstraction for in-memory cluster computing. In: Proceedings of the 9th USENIX conference on networked systems design and implementation, pp 2–2. USENIX Association
2. Spark architecture. https://www.dezyre.com/article/apache-spark-architecture-explained-in-detail/338
3. Wide-vs-narrow-dependencies. https://github.com/rohgar/scala-spark-4/wiki/Wide-vs-Narrow-Dependencies
4. Hash partitioner API. https://spark.apache.org/docs/1.4.0/api/scala/index.html#org.apache.spark.HashPartitioner
5. Range partitioner API. https://spark.apache.org/docs/1.4.0/api/scala/index.html#org.apache.spark.RangePartitioner
6. Apache spark partitioning. https://github.com/apache/spark/blob/v1.4.1/core/src/main/scala/org/apache/spark/Partitioner.scala#L78
7. Partitioning in apache spark. http://parrotprediction.com/
8. Apache spark introduction—for installation. https://www.tutorialspoint.com/apache_spark/apache_spark_introduction.htm
9. Scott JA (2015) Getting started with apache spark-inception to production

---

[1] Partitioning in Apache Spark.

Workflow Scheduling Algorithms in Cloud Computing: An Analysis, Analogy, and Provocations

Shubham Jain and Jasraj Meena

Abstract Cloud computing is based on sharing of resources to make a service economical same as a public utility. Nowadays, the most prominent application of cloud computing is workflow scheduling. Workflow consists of a repeatable pattern of business activity that needs to be executed sequentially in a diagram or checklist, whether with the help of a colleague, some tool, or using another process. Scheduling means mapping of these business activities to available resources keeping in mind the QoS constraints to achieve. A various number of workflow scheduling algorithms have been proposed till now for scheduling workflows on cloud environment. This paper provides a comprehensive classification of some of the recently used algorithms in this sector. Besides the algorithm's illustration, the paper also states comparison in these algorithms for better clarification of their objectives and limitations as well as scope and research problems in this sector.

Keywords Cloud computing · QOS constraints · Scheduling algorithm ·
Workflow scheduling

1 Introduction

Cloud computing [1] is a technology that makes use of remote server and Internet to provide services to users. Cloud computing comes out as a new paradigm, where the storage [2] and business process which were once accessible to only large organizations can now be accessed by smaller companies too. Cloud computing comes with various features which attracts the users like on-demand self-service, broad network access, pooling, elasticity, etc. On-demand self-service states that a user will have to pay only for the resource been used and for the time period being used. Broad net-

S. Jain (✉) · J. Meena
Department of Information Technology, DTU Delhi, Rohini, India
e-mail: shubham81.sj@gmail.com

J. Meena
e-mail: jasrajmeena@dtu.ac.in

© Springer Nature Singapore Pte Ltd. 2019   499
H. S. Saini et al. (eds.), *Innovations in Computer Science
and Engineering*, Lecture Notes in Networks and Systems 74,
https://doi.org/10.1007/978-981-13-7082-3_57

work access states that cloud services can be accessed from anywhere using Internet. Resource pooling shows that there are unlimited resources for the customer to access. Different customers can make use of several similar storages, or other computing resources as per their need. Elasticity accounts for an added advantage that customers can make their scale of use of resources up or down as per their requirement. Virtualization is the main key of cloud computing which allows multiple VMs to reside on a single machine. Users make use of different instances of VMs to launch their applications and then the VMs execute the user tasks.

As stated earlier, workflow [3] consists of a repeatable pattern of business activities. It can be explained as a series of operations, work of a person or of a group, or work of a staff or of an organization. On a higher level, it can also be depicted as a view or representation of real work. Large workflows of organizations are represented in the form of directed acyclic graphs (DAG). Each node of the graph signifies a task or a process, and the edges between the nodes specify the dependency in tasks. Executing heavy workflows can always lead to errors in terms of uncertainty. Execution of workflows on cloud environment has always been a challenging task [4]. Sometimes tasks are in number of hundreds and one needs to perform their execution in such a way to manage the time and cost constraints. Executing a workflow comes in parts, submitting an application over VM, making the required input files accessible, transfer of data, and much more. Along with the ease of execution on cloud, there also comes risks like failure of VMs [5], data transfer failure, etc. So all such problems need to be rectified by taking proper actions.

Scheduling a workflow [6] as mentioned is one of the most important and demanding tasks in cloud computing. It is an effortful task and is the primary step of execution of an application. It is the deciding factor of performance of an application. Scheduling basically is a mapping of various workflow tasks on different VMs of various performance factors so as to achieve QoS of users. A workflow is a representation of various independent business tasks which are combined together and are contained in a flow using dependencies, which are a vital part of scheduling. However, workflow scheduling is an NP-hard problem in cloud computing which makes an optimal solution difficult to achieve. Although there can be numerous objectives of users, the most common of them is time and cost. Since there can be numerous requirements (task) of the user with numerous cloud resources, scheduling is done so as to achieve the users' objectives of cost and time by proper allocation of tasks to resources.

This paper discusses a complete survey of some of the workflow scheduling algorithms used in cloud computing. A survey of their different objectives, factors, and a simplified comparison among them is to get a proper clarification on their parameters, procedures, and limitations. Along with this, a simple classification is also presented in the paper for better understanding of algorithms. The rest of the paper is organized as follows: Sect. 2 includes problem denotation. Section 3 states analysis and discussion. Section 4 describes a comparative study of existing workflow scheduling algorithm with their scope and provocations [7]. Section 5 refers to the conclusion.

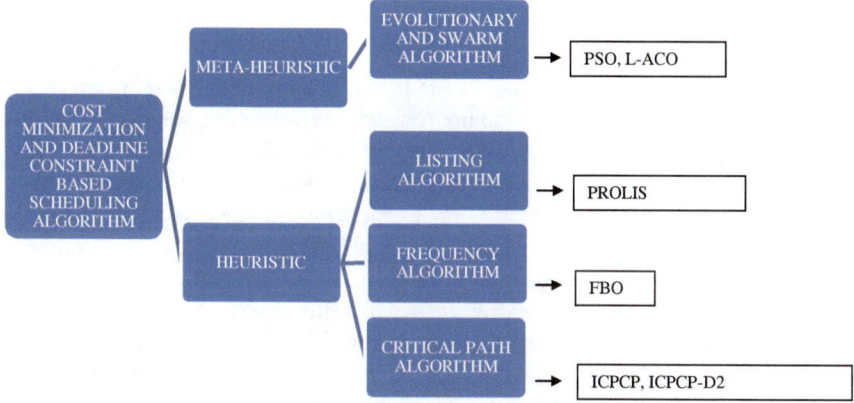

Fig. 1 Cost minimization and deadline constraint-based scheduling algorithm in cloud

2 Problem Denotation

Workflow scheduling [8] mainly aims at assigning resources to tasks and sequencing their order of execution while satisfying the QoS constraints. This paper presents an analysis of workflow scheduling algorithms to mainly address two problems, cost minimization and deadline constraint during workflow execution.

Workflow scheduling is broadly divided into two steps: As per requirement of tasks, first select a set of resources from the available pool and their distribution. Next step is to generate a schedule using desired strategy and mapping of tasks on these resources keeping in mind the QoS constraints.

The problem which is discussed through these algorithms is to perform the complete execution of workflow application (scheduling) within user given deadline (time) and also in the minimized budget (cost).

2.1 Classification

For a better and easy understanding of the user, the algorithms are classified on the basis of scheduling category under cost and deadline optimization constraints as shown in Fig. 1.

3 Analysis and Discussion

The study dignifies different types of algorithms used for scheduling workflows in cloud environment.

3.1 Particle Swarm Optimization (PSO)

Proposed by Eberhart and Kennedy, PSO [9] is an evolutionary algorithm. It is a computational method which tries to improve the efficiency of a solution iteratively. Here, each particle represents a solution and is depicted in the form of vectors of position and velocity. Position represents the solution to the problem and velocity specifies movement in search space, both of which are computed using mathematical formulas. Each particle has a local best position called pbest and a best position of population called gbest. Apart from finding pbest, the algorithm also guides the particle toward gbest solution. After each generation, these positions (pbest and gbest) are updated. This is done to guide the algorithm toward best solution. Once solution is achieved, the total execution cost [10, 11] and time of solution [12, 13] are calculated to satisfy the objectives. However, PSO is not so efficient with tighter deadlines and also has weaker performance in terms of cost and time when compared to other algorithms.

3.2 Frequency-Based Optimization (FBO)

Algorithm aims at choosing appropriate frequency so as to minimize the cost while meeting the deadline constraint [14, 15]. Cost of each resource is based on the frequency [16] allocated to a task. In this algorithm, three types of pricing model have been used to meet the goals: linear, sublinear, and superlinear, and the total cost of executing the workflow is calculated by subtracting the pricing model cost value from the execution time value of running tasks on maximum frequency. Two variants of the algorithm are proposed. In the first variant, a makespan-based scheduling [17] algorithm such as HEFT is used to generate the initial schedule. HEFT assigns tasks of workflow to maximum frequencies so as to ensure that deadline will be attained not considering the cost parameter. In the second variant of algorithm, the task, resource, and frequency allocation keep on changing iteratively so as to minimize the cost while ensuring about deadline too.

For the second variant, a weight table is prepared which contains values of all combinations of task, resource, and CPU frequency allocations. This reallocation continues until execution cost is reduced without violating deadline. A point to remember is that algorithm can try for reassignment of task to a resource and frequency pair only once. So when no untried combination will be left, the algorithm will terminate.

3.3 Probabilistic Listing (ProLiS)

ProLiS [18] also aims to provide a cost-effective schedule under deadline constraint. It distributes the deadline of workflow to individual tasks. Then it performs tasks prioritization followed by task selection. ProLiS performs deadline distribution on the basis of probabilistic upward rank. Upward rank of a task is the longest path from it to the end task. Sub-deadline is assigned to a task in proportion to the longest path from entry node to current task. Differentiation factor in probabilistic upward rank and upward rank is the inclusion of a boolean variable to ensure that tasks having larger transmission time are assigned on the same resources. Task-ordering algorithm uses the probabilistic upward rank methodology since data transmission in this may become zero. In the service selection step, initially, such service is searched that can minimize the cost under deadline constraint. If none of such service exists, then objective is narrowed down to minimization of execution time.

3.4 Laco

Laco makes use of ant colony optimization (ACO) to redefine task-ordering step of ProLiS. Laco also largely depends on heuristic and pheromone trail. Pheromone can be understood as an aspire to select a task just after another task. Same as ProLiS, Laco is divided into three parts. Deadline distribution, task ordering, and task selection are organized task lists depending on pheromone (unlike ProLiS which depends on probabilistic upward rank) and heuristic information. In order to ensure that the proposed solution of algorithm does not contain violation of dependencies between tasks, it adapts Kahn's algorithm for topological sorted ordered schedule. In order to meet the problem of deadline violation algorithm, make use of improved independent optimization method, where the deadline is relaxed in proportion to the iteration number of Laco and then comparison between solutions meeting the relaxed deadline is done on the basis of some parameter. With the increase in the iteration number, the relaxation goes on decreasing until the number reaches to terminating iteration number. In each iteration, after the local best solution is figured, pheromone trail is updated. At last, global best solution is returned.

3.5 IaaS Cloud Partial Critical Paths (ICPCP)

Most important concepts of ICPCP [19] are critical path, assigned node, and critical parents. Assigned node speaks for a node whose service has been selected. Critical parent of a task is an unassigned parent of the task whose data reaches the current task most lately. Partial critical path of workflow will be empty if task does not have any unassigned parents. Otherwise, the path contains the critical parent of the task as well as partial critical path of the parent if it exists. In algorithm, initially, some parameters are calculated, and entry and exit tasks are assigned followed by calling another algorithm for assigning parents. This called algorithm takes an assigned node as input and allocates all the unassigned parents of input node to some service before the start time of input node. Through this scheduling [20] strategy, the algorithm finds all critical paths of the workflow. Once done another procedure is approached for path assignment to services. This new procedure schedules received critical paths on applicable instances that can complete each task's execution before it is latest finish time with minimum price.

3.6 IaaS Cloud Partial Critical Path with Deadline Distribution (ICPCPD2)

ICPCPD2 works in two phases and has three main concepts to depend on which are partial critical path, critical parent, and assigned node. While the definition of first two concepts remains the same as in ICPCP, assigned node here stands for a node to which sub-deadline has already been assigned. As mentioned earlier the two phases are deadline distribution and planning. In deadline distribution phase, complete workflow deadline is distributed over individual tasks followed by calling separate procedure for parents assignment. This procedure is completely same as the one discussed in ICPCP. Once assignment is done, the execution shifts to another procedure for assigning path and this is where difference exists in ICPCP and ICPCPD2 algorithms. Unlike ICPCP, in ICPCPD2, the path assigning procedure performs assignment of sub-deadlines to all the unallocated parents of input node. For this, it distributes the deadline of path among tasks which are a part of path in proportion to their minimum execution time. In the planning phase, there exists another procedure which plans and assigns each task to cheapest applicable instance which can make task to finish its execution before the sub-deadline.

4 Comparative Study of Existing Workflow Scheduling Algorithms

4.1 Scope and Provocations

Due to the previous execution environments like grid and clusters where availability of resources was limited, world adopted a new technology for workflow scheduling called cloud computing as it provides a pool of unlimited resource which helps in executing multiple workflows simultaneously. The main reason for acceptance of this technology is its cheaper and scalable services. In cloud world, no advance booking of resource is required. As workflow applications require compatible and efficient environments for execution, cloud technology provides it such environments. As large workflow applications also need larger and complex resources, in brief larger demands, cloud has to deal with some provocations in workflow execution [25] (Table 1).

Table 1 Comparison of deadline-constrained and cost-optimized algorithms

Method	Laco	ICPCP	ICPCPD2	PSO	FBO	ProLiS
Objective	Cost & deadline optimization [21]	Cost & deadline optimization	Cost & deadline optimization	Cost & deadline optimization	Cost & deadline optimization	Cost & deadline optimization
Scheduling category	Meta-heuristic [22]	Heuristic [23]	Heuristic	Meta-heuristic	Heuristic	Heuristic
On-demand [24]	Yes	Yes	Yes	Yes	Yes	Yes
Resource type	Heterogeneous	Homogeneous	Homogeneous	Heterogeneous	Heterogeneous	Heterogeneous
Pay-As-You-Go	Yes	Yes (hourly & min)	Yes (hourly & min)	Yes	Yes	Yes
Performance variation of VMs	No	No	No	Yes	No	No
Acquisition delay	No	No	No	Yes	No	No
Fault tolerance	No	No	No	No	No	No
Termination delay	No	No	No	No	No	No
Stochastic	Yes	No	No	Yes	Yes	Yes
Applicable (Best-Fit) instance	No	Yes	Yes	No	No	No
Environment	Cloud	Cloud	Cloud	Cloud	Cloud	Cloud
Simulation tool	Java	Cloud environment with 10 services	Cloud environment with 10 services	CloudSim	Java	Java
Data transfer time	Yes	No	No	Yes	Yes	Yes

4.1.1 Scope

It states the benefits of some key features of cloud computing:

- Due to the presence of unlimited resources in cloud computing, the requirement and execution of workflow applications [26] get completed timely.
- Since cloud supports pay-per-use and on-demand mechanisms, dynamic allocation and resource utilization can be effectively carried out in cloud computing.
- One of many attracting features of cloud computing is heterogeneity in resources which means end user can get different resources for different applications.

4.1.2 Provocations

With the increase in workflow applications, the demands and complexity also increase. In the management of these demands and applications, there arise some provocations to be dealt with which are as follows:

- Need for a user-friendly workflow management system, where user can deploy their requirements easily and resource management, resource computation, task definition, and efficiency can be monitored.
- In today's world, applications are largely scaled which makes the cloud environment heavily loaded with data. As a result, movement of data from storages to computing devices is a big issue. An efficient solution needs to be figured out for this problem.

5 Conclusion

Cloud provides limitless resources for workflow systems which helps in applications timely execution and performing scientific experiments in a better way. In this paper, we have covered a survey of some of the workflow scheduling algorithms with the objective of deadline constraint and cost optimization. The paper contains a detailed discussion of each algorithm procedure, scheduling factors, pros, and cons. Apart from this, we present a summarized comparison of these algorithms based on their defined objectives, scheduling category, performance variation, acquisition and termination delay, tool, and environment. Other than this, a classification of these scheduling algorithms is also done for their better and simpler understanding. Finally, we proposed some of the problems which arise in a cloud environment while managing large-scale workflow applications and their complex and large demands and the scope of cloud technology in the present world. From the comparison and detailed discussion, it was clear that there is certain limitation in most of these algorithms like fault tolerance, termination and acquisition delay, and applicable instance utilization, which if improved can make these algorithms well more efficient.

References

1. Mell P, Grance T (2009, June) Draft NIST working definition of cloud computing. http://csrc.nist.gov/groups/SNS/cloud-computing/index.html
2. Sahi SK, Dhaka VS (2016) A survey paper on workload prediction requirements of cloud computing. In: 2016 3rd international conference on computing for sustainable global development (INDIACom), New Delhi, pp 254–258
3. Choudhary A, Govil MC, Singh G, Awasthi LK (2016) Workflow scheduling algorithms in cloud environment: a review, taxonomy, and challenges. In: 2016 fourth international conference on parallel, distributed and grid computing (PDGC), Waknaghat, pp 617–624
4. Harbajanka S, Saxena P (2016) Survey paper on trust management and security issues in cloud computing. In: 2016 symposium on colossal data analysis and networking (CDAN), Indore, pp 1–3
5. Bindu GBH, Janet J (2017) A statistical survey on vm scheduling in cloud workstation for reducing energy consumption by balancing load in cloud. In: 2017 international conference on networks & advances in computational technologies (NetACT), Thiruvananthapuram, pp 34–43
6. Rani BK, Babu AV (2016) Survey on workflow scheduling of scientific processes using cloud technology. In: 2016 2nd international conference on contemporary computing and informatics (IC3I), Noida, pp 375–380
7. Liu L, Zhang M, Lin Y, Qin L (2014) A survey on workflow management and scheduling in cloud computing. In: 2014 14th IEEE/ACM international symposium on cluster, cloud and grid computing, Chicago, IL, pp 837–846
8. Deelman E, Gannon D, Shields M, Taylor I (2009) Workflows and e-Science: an overview of workflow system features and capabilities. Future Gener Comput Syst 25:528–540
9. Rodriguez MA, Buyya R (2014) Deadline based resource provisioning and scheduling algorithm for scientific workflows on clouds. IEEE Trans Cloud Comput 2(2):222–235
10. Chen ZG, Du KJ, Zhan ZH, Zhang J (2015) Deadline constrained cloud computing resources scheduling for cost optimization based on dynamic objective genetic algorithm. In: 2015 IEEE congress on evolutionary computation (CEC), Sendai, pp 708–714
11. Lin B, Guo W, Chen G, Xiong N, Li R (2015) Cost-driven scheduling for deadline-constrained workflow on multi-clouds. In: 2015 IEEE international parallel and distributed processing symposium workshop, Hyderabad, pp 1191–1198
12. Arabnejad V, Bubendorfer K (2015) Cost effective and deadline constrained scientific workflow scheduling for commercial clouds. In: 2015 IEEE 14th international symposium on network computing and applications, Cambridge, MA, pp 106–113
13. Chen ZG et al (2015) Deadline Constrained cloud computing resources scheduling through an ant colony system approach. In: 2015 international conference on cloud computing research and innovation (ICCCRI), Singapore, pp 112–119
14. Arabnejad V, Bubendorfer K, Ng B, Chard K (2015) A deadline constrained critical path heuristic for cost-effectively scheduling workflows. In: 2015 IEEE/ACM 8th international conference on utility and cloud computing (UCC), Limassol, pp 242–250
15. Kaur G, Kalra M (2017) Deadline constrained scheduling of scientific workflows on cloud using hybrid genetic algorithm. In: 2017 7th international conference on cloud computing, data science & engineering - confluence, Noida, pp 276–280
16. Zheng W, Emmanuel B, Wang C, Qin Y, Zhang D (2017) Cost optimization for scheduling scientific workflows on clouds under deadline constraints. In: 2017 fifth international conference on advanced cloud and big data (CBD), Shanghai, pp 51–56
17. Juve G, Chervenak A, Deelman E, Bharathi S, Mehta G, Vahi K (2013) Characterizing and profiling scientific workflows. Future Gener Comput Syst 29:682–692
18. Wu Q, Ishikawa F, Zhu Q, Xia Y, Wen J (2017) Deadline-constrained cost optimization approaches for workflow scheduling in clouds. IEEE Trans Parallel Distrib Syst 28(12):3401–3412

19. Abrishami S, Naghibzadeh M, Epema DHJ (2012) Cost-driven scheduling of grid workflows using partial critical paths. IEEE Trans Parallel Distrib Syst 23(8):1400–1414
20. Cooper K, Dasgupta A, Kennedy K, Koelbel C, Mandal A, Marin G, Mazina M, Mellor-Crummey J, Berman F, Casanova H, Chien A, Dail H, Liu X, Olugbile A, Sievert O, Xia H, Johnsson L, Liu B, Patel M, Reed D, Deng W, Mendes C, Shi Z, YarKhan A, Dongarra J (2004) New grid scheduling and rescheduling methods in the GrADS project
21. Poola D, Garg SK, Buyya R, Yang Y, Ramamohanarao K (2014) Robust scheduling of scientific workflows with deadline and budget constraints in clouds. In: 2014 IEEE 28th international conference on advanced information networking and applications, Victoria, BC, pp 858–865
22. Nandhakumar C, Ranjithprabhu K (2015) Heuristic and meta-heuristic workflow scheduling algorithms in multi-cloud environments—A survey. In: 2015 international conference on advanced computing and communication systems, Coimbatore, pp 1–5
23. Raj RJS, Prasad SVM (2016) Survey on variants of heuristic algorithms for scheduling workflow of tasks. In: 2016 international conference on circuit, power and computing technologies (ICCPCT), Nagercoil, pp 1–4
24. Ye X, Liang J, Liu S, Li J (2015) A survey on scheduling workflows in cloud environment. In: 2015 International Conference on Network and Information Systems for Computers, Wuhan, pp 344–348
25. Wieczorek M, Prodan R, Fahringer T (2005) Scheduling of scientific workflows in the ASKALON grid environment. ACM SIGMOD Rec 34:56–62
26. Deelman E, Singh G, Su M, Blythe J, Gil Y, Kesselman C, Berriman GB, Good J, Laity A, Jacob JC, Katz DS, Mehta G, Vahi, K (2005) Pegasus: a framework for mapping complex scientific workflows onto distributed systems. Sci Program J 13:219–237

Big Data Clustering: Applying Conventional Data Mining Techniques in Big Data Environment

P. Praveen and Ch. Jayanth Babu

Abstract In the current data world, the term big data became popular term to define these massive data, increasing enormously. Efficient knowledge extraction techniques take major role in processing this big data. There are many techniques used to process big data; most importantly, the data mining technique–cluster analysis has many applications such as in information retrieval, image processing, machine learning, etc. and used widely. The clustering is the process of grouping similar data items into one class, dissimilar into another. The clustering of big data includes many overheads, particularly the designing new algorithms or conversion of efficient data mining algorithms for distributed environment. In this paper, we covered overview of traditional clustering techniques and trends in clustering models to process the big data, so that the current voluminous data can be processed and analyzed efficiently. Clustering of big data is the tremendous field of research in current trend that has huge scope for improvement of clustering algorithms.

Keywords Data mining · Big data clustering · HDFS · MapReduce

1 Introduction

In the current data world, the data are produced from various sources in various formats like structured, unstructured, and semi-structured. The growth of web usage through a smartphone, Internet of things, and social media is the major cause of massive data production. In day-to-day life, these data play an important role to

P. Praveen (✉)
Department of CSE, SR Engineering College, Warangal, India
e-mail: prawin1731@gmail.com

Department of CSE, Kakatiya University Warangal, 506371 Warangal, Telangana, India

Ch. Jayanth Babu
Department of CSE, Kakatiya Institute of Technology & Science, Warangal 506015, Telangana, India
e-mail: cjb.cse@kitsw.org

© Springer Nature Singapore Pte Ltd. 2019　　　　　　　　　　　　　　　509
H. S. Saini et al. (eds.), *Innovations in Computer Science and Engineering*, Lecture Notes in Networks and Systems 74,
https://doi.org/10.1007/978-981-13-7082-3_58

understand the things around the people and interactions. Every year the data generation is increasing enormously; it is doubling every year. By the year 2020, it may reach the size of 440 petabytes. The biggest challenge is understanding these data and reusing it appropriately.

There are some techniques which are failing to process data when data size is increased, such as operational methods like retrieval operations, process operations, analytical operations, etc. [1]. The era of data mining solved these problems by introducing efficient mining algorithms. The data mining is the process of discovering knowledge from data. It is used to analyze the data by searching the exact pattern from large data. The result generated by processing various heterogeneous data is stored in separate file. This analytical result set reviewed and displayed as the output for the better understanding purpose [2].

The term big data became very popular in the current digital data world. It means the data which is sound in 3Vs: volume, variety, and velocity. There are many Vs that are added by researchers like value [3] and veracity [4]. The term volume represents vast amount of generated minute, velocity represents the speed of the data generation, variety represents the various types of data (structured, unstructured, and semi-structured) from various sources of data, veracity represents the accuracy and quality of data which are less controllable, and value represents the data that to be turned into value. These 5Vs characteristics are increasing the complexity of data processing which can be called as big data processing [5].

The main goal of clustering the big data is to increase the processing speed of conventional clustering methods by proposing the efficient models and reducing the computational time, even though the data is complex, voluminous, and heterogeneous [6]. In the list of clustering techniques, some will run on a single node with those node resources and some techniques will run on multiple nodes, which have more scalability and quick processing capacity by using multiple node resources [7].

The rest of them is organized in the following sections: In Sect. 2, the concepts of cluster analysis and various categories of clustering techniques in data mining are explained including the possible ways of improving the performance of big data clustering. Section 3 deals with the techniques of big data clustering and generalized architecture for implementation and also various categories of clustering based on MapReduce technique. Applications of big data clustering are discussed in Sect. 4. Finally, the paper is concluded in Sect. 5.

2 Clustering Techniques

2.1 Cluster Analysis

The concept of cluster analysis is the study of similar data elements and their properties. The group of similar data elements is called clusters, and dissimilar data elements are called outliers that are added to another group used for outlier analysis. The goal of the clustering is "minimizing interclass similarity and maximizing interclass similarity." The cluster analysis plays an important role in many fields such as in data mining, machine learning, pattern recognition, biochemistry, and bioinformatics.

In general, many of the clustering techniques use the iterative process to find the best-fit set of data points (e.g., k-means) [8]. The most of the real-time data are complex and huge in size, so we cannot find the unique optimal solution for this type of problems [9]. The huge dataset contains huge number of records with multiple dimensions, considering all these factors as tedious task. It requires series of experiments with different processing models and with various algorithms. The conventional algorithms of clustering are not scaled with voluminous datasets and computationally costly in memory and time. To overcome these problems, a practical approach is parallelization of conventional algorithms [10].

2.2 Clustering Algorithms

There are many clustering algorithms available in the literature. But, based on the clustering model and properties of data, these are broadly categorized into the following types [11].

(a) *Partitioning*: This is the simple clustering method, in which the number of clusters is fixed before forming. It uses iterative process to find cluster centers and final clusters.
(b) *Hierarchical*: In this clustering method, we do not specify the number of clusters. This is a static approach; once data objects are assigned to the cluster, they cannot be reassigned to another cluster. Data objects are formed into tree of clusters by the top-down or bottom-up approach.
(c) *Density-based*: In this clustering method, data objects are clusters based on the regions of density. This clustering method is useful to identify the clusters of arbitrary shapes and outlier can be easily separated.
(d) *Grid-based*: In this clustering method, the data objects are divided into grids. Clustering algorithms are applied to these grids instead of database directly. The performance of these methods depends on the size of the grid.

2.3 Big Data Clustering

In general, most of the conventional clustering algorithms are designed to handle the structured and small data sizes. But, when dealing with data big data clustering, including with the size of the data, it is also important to consider the various data types like text, video, audio, sensor data, images, mobile data, etc. Most recently, processing of the online data becames more popular where clustering algorithms play an important role to handle data with high velocity. In the above categories of algorithms, most of the clustering algorithms will use same procedure at high level, i.e., every procedure initiated with random function, then it follows the iterative process until a specific condition fails. In case of big data clustering, this approach may take much time to complete the clustering process. So, the research issue is how to increase the performance of clustering algorithms to get qualitative clusters by reducing the processing time [12].

There are three possible ways to improve the performance of algorithms for big data clustering.

1. **Sampling**: The sampling is the well-known approach, in which sample dataset is processed instead of total dataset. In big data clustering, the usage of sampling-based algorithms will reduce the number of iterations to form a cluster as small dataset is taken into consideration instead of whole data. There are some problems answered by sampling-based algorithms, for example, the k-medoids algorithm needs exponential search space to form a cluster. To overcome this problem, sampling-based algorithms such as Clustering for Large Applications (CLARA), Clustering Algorithm based on Randomized Search (CLARANS), and Partitioning Around Medoids (PAM) are proposed [9, 11–13].

2. **Dimensionality Reduction**: In big data clustering, the dimension of the dataset is the important aspect. If dimension of the data set is high, the complexity of the process increases and the process time increased as well. The reduction of the dimensions will improve the performance of the clustering algorithms. Generally, the dimension reduction is done using the randomized approach. CMD, CX/CUR, and Colibri are randomized dimension reduction techniques proposed to reduce the execution time [7, 9, 11–13].

3. **Parallel Processing**: In parallel processing environment, multiple systems are connected in the network. The task given to these systems is divided into subtasks and processed. In conventional parallel processing applications, we think that the storage capacity of distributed environment is enough to store input data. But, in big data era, most of the applications are data intensive. The MapReduce is one of the popular parallel processing models for big data (Fig. 1).

Fig. 1 MapRedce processing model

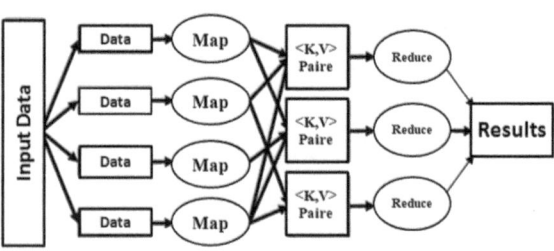

3 Clustering Techniques for Big Data

3.1 Clustering Based on Map Reduce Paradigm

The MapReduce is a parallel processing mechanism in Hadoop framework which works in cluster of commodity hardware. This processing mechanism is well suited for scalable applications with reliability. It becomes popular in many fields like education and industry because of its advantages [14]. The main advantage of this processing mechanism is providing the automated parallel processing environment, so that the user only concentrates on techniques of data processing.

There are two functions in MapReduce model called map() and reduce(). The map() function is applied to individual data nodes in cluster and produces (key, value) pairs [1]. The reduce() function shuffles and consolidates intermediate (key, value) pairs. The MapReduce process consists of set of sequences of steps. In the first step, the input data is loaded into Hadoop Distributed File System (HDFS). This file system supports structured and unstructured data storage [10]. Whenever the data is loaded, it is divided into blocks (block size 64/128 MB default) and each block is replicated into three copies to provide fault tolerance. These blocks of data are distributed to every node of the cluster. The second step is to apply desired MapReduce job on the blocks of data. In this step, map function performs generation of key, value pairs, shuffling, and sorting. The sorted records are sent to reduce () function. In this step, all sorted records are grouped based on the key. It is executed once for every key and it generates final output in sorted order (Fig. 2).

(a) *Partitioning-based Clustering using MapReduce*

The most popular clustering k-means is parallelized in Hadoop environment to cluster big data [3]. It used a generalized method of distribution into various nodes in Hadoop cluster environment. In this, the input data is loaded into HDFS. The data in the file

Fig. 2 General framework of big data clustering applications

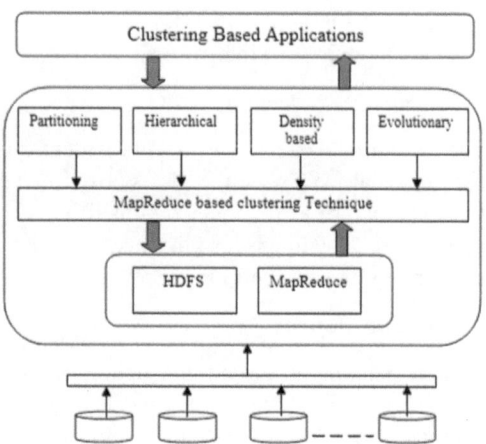

system is applied by MapReduce mechanism. The clustering at node level is made by using k-means algorithm; then, the resultant records are sent to reduce phase. The reduce phase produce centroids as output which can be used as input for the next iteration [8]. This example of clustering algorithm increases the scope of research on data mining algorithms to redesign for big data, K-medoids,

(b) *Hierarchical Clustering using MapReduce*

In hierarchical clustering, there are two strategies for clustering: Agglomerative: it is bottom-up approach to form a cluster and, Divisive: it is a top-down approach. These strategies can be combined for better results. Recently, co-occurrence techniques are proposed to implement these hierarchical clusterings using MapReduce. BIRCH, Diana, and Agnes are examples of hierarchical clustering algorithms.

(c) *Density-based Clustering using MapReduce*

One of the density-based algorithms—DBSCAN—is parallelized in Hadoop environment called MR-DBSCAN [15]. It is introduced to resolve the problems in traditional DBSCAN algorithm. The major drawbacks are load balancing, scalability, and portability, which are very much important in designing any algorithm. These drawbacks are resolved in MapReduce-based DBSCAN algorithm.

(d) *Evolutionary Clustering using MapReduce*

Particle Swarm Optimization (PSO) is an evolutionary clustering technique developed by Dr. Eberhart and Dr. Kennedy in 1995. It is having more similarities with genetic algorithm. Applying this technique to big data produces useful results. But it not scalable for huge datasets. To answer this drawback, the MR-CPSCO [12] is proposed in which MapReduce framework is used to parallelize the processing.

4 Applications of Big Data Clustering

There are many applications of big data clustering: (i) community detection, (ii) genetic mapping, (iii) image segmentation, and (iv) load balancing.

i. *Community Detection*

The community detection is useful for identifying similar group of people. For example, in social media such as Facebook, LinkedIn is used to find the people belong to same town or same organizations. It is very useful for mapping the people. The community detection is also like clustering, mapping real-world thing such as social networks. It is also useful to detect frauds in social networks.

ii. *Genetic Mapping*

Genetic mapping is a trend in genetic science. It has many practical uses in modern research of plants. The main concept in genetic mapping is linkage group, which combines genetic markers on chromosome. In these concepts, the clustering plays an important role to compute group-wise similarities. So, there is much scope of research in this field of research.

iii. *Image Segmentation*

The concept of image segmentation is used in classification of pixels of an image. It helps to identify image characteristics and its regions of similarity. But it is not an easy task due to the variations in color coding and image complexities. Many researchers proposed techniques on image segmentation, which is still a challenging task. Partitioning-based clustering algorithms are popular and suitable to solve these types of problems.

iv. *Load Balancing*

Nowadays, many applications run on cloud. Much data, processors, operating systems, software, and other components exist on virtual machines in this cloud environment. Load balancing is a curtail task in this network of heterogeneous virtual systems. There are many load balancing algorithms available, but not scaling with big data applications. The focus on load balancing algorithms for big data will produce useful outcomes.

5 Conclusion

In the study of literature, we came to know that in current trend of data world, conventional single-node algorithms are not suitable to handle the huge volume of data. Parallelizing the clustering algorithms is important for multi-node techniques. Parallel clustering techniques are potential for big data clustering. But implementation of such algorithms is a challenging for researchers. The MapReduce paradigm becomes popular to answer these complexities. The clustering as an essential task in data mining applied for big data analysis has many applications like community detection, load balancing, genetic mapping, and image segmentation.

References

1. Praveen P, Babu CJ, Rama B (2016) Big data environment for geospatial data analysis. In: 2016 international conference on communication and electronics systems (ICCES), Coimbatore, pp 1–6
2. Xu Z, Shi Y (2015) Exploring big data analysis: fundamental scientific problems. Ann Data Sci 2(4):363–372
3. Laney, D (2001) 3D data management: controlling data volume, velocity and variety. Technical Report 949, META Group. Accessed https://blogs.gartner.com/doug-laney/files/2012/01/ad949-3D-Data-Management-Controlling-Data-Volume-Velocity-and-Variety.pdf
4. Demchenko Y, Grosso P, De Laat C, Membrey P (2013, May) Addressing big data issues in scientific data infrastructure. In: Proceedings of the 2013 international conference on collaboration technologies and systems (CTS), pp 48–55. IEEE, Chicago

5. Fahad A, Alshatri N, Tari Z, Alamri A, Khalil I, Zomaya AY, Bouras A (2014) A survey of clustering algorithms for big data: Taxonomy and empirical analysis. IEEE Trans Emerg Top Comput 2(3):267–279

6. Kumar RR, Reddy MB, Praveen P (2017) A review of feature subset selection on unsupervised learning. In: 2017 third international conference on advances in electrical, electronics, information, communication and bio-informatics (AEEICB), Chennai, pp 163–167. https://doi.org/10.1109/aeeicb.2017.7972404

7. Shirkhorshidi AS, Aghabozorgi S, Wah TY, Herawan T (2014, June) Big data clustering: a review. In: International conference on computational science and its applications, pp 707–720. Q4 Springer International Publishing, Portugal

8. Praveen P, Rama B (2016) An empirical comparison of clustering using hierarchical methods and K-means. In: 2016 2nd international conference on advances in electrical, electronics, information, communication and bio-informatics (AEEICB), Chennai, pp 445–449. https://doi.org/10.1109/aeeicb.2016.7538328

9. Kaufman L, Rousseeuw PJ (2009) Finding groups in data: an introduction to cluster analysis, vol 344. Wiley, United States

10. Praveen P, Rama B (2018) A novel approach to improve the performance of divisive clustering-BST. In: Satapathy S, Bhateja V, Raju K, Janakiramaiah B (eds) Data engineering and intelligent computing, vol 542. Advances in intelligent systems and computing. Springer, Singapore

11. Kim W (2009) Parallel clustering algorithms: survey. CSC 8530 parallel algorithms, spring 2009. Q5

12. Aggarwal CC, Reddy CK (eds) (2013) Data clustering: algorithms and applications. CRC Press, United States

13. Ng RT, Han J (2002) CLARANS: A method for clustering objects for spatial data mining. IEEE Trans Knowl Data Eng 14(5):1003–1016

14. Zhang J (2013) A parallel clustering algorithm with MPI–Kmeans. J Comput 8(1):10–17

15. Chen CP, Zhang CY (2014) Data-intensive applications, challenges, techniques and technologies: a survey on big data. Inf Sci 275:314–347

Enhancement of Solar Cell Efficiency and Transmission Capability Using Nanotechnology with IoT

Mohd Niyaz Ali Khan, Mohammed Ghouse Ul Islam and Fariha Khatoon

Abstract The solar cell is unable to convert all the light energy into electricity as the efficiency of solar cell is in between 12 and 16%. The solar panel needs to improve by using the silicon surface of solar panel. Silicon is less efficient due to band gap. By improving the capture range, nanometals are induced on the silicon surface and then this nanomaterial with silicon increment of the productivity. The transmission of electricity from load to grid is done by nanomaterial quantum wire (nanowire) with coating of silicone rubber and the outdoor insulator is coated with nanofilter coating. The IoT system is implemented to gather the data and to operate the power plant which improves and lead to a modern technology. Nowadays, there is a rise in demand for the solar device because it is one of the affordable renewable energies.

Keywords Solar PV · Nanotechnology · IoT · Solar nanotechnology · Solar IoT

1 Introduction

In order to increase the efficiencies of scattering and absorption, the deposits of metal nanoparticles are used in a photovoltaic cell; it is found that semiconductor like silicon does not support efficient absorption [1, 2]. To use more light to be scattered across the surface, the nanoparticles are used which will help in the effective scattering of the incident light all over the surface. In the PV cell design, the material nanoparticles are introduced on the top layer of solar cell; when light hits the nanoparticles, the light scattered all over the silicon nanosurface. Thus, the light is made to be in contact with the whole substrate so that number of atoms can absorb more light. The electrons and the holes are separated as the photon in the solar cell is excited by absorbing light (Fig. 1).

M. N. A. Khan (✉) · M. G. U. Islam · F. Khatoon
Department of Electrical and Electronics, Hyderabad, India
e-mail: mohdniyazalikhan@gmail.com

© Springer Nature Singapore Pte Ltd. 2019
H. S. Saini et al. (eds.), *Innovations in Computer Science
and Engineering*, Lecture Notes in Networks and Systems 74,
https://doi.org/10.1007/978-981-13-7082-3_59

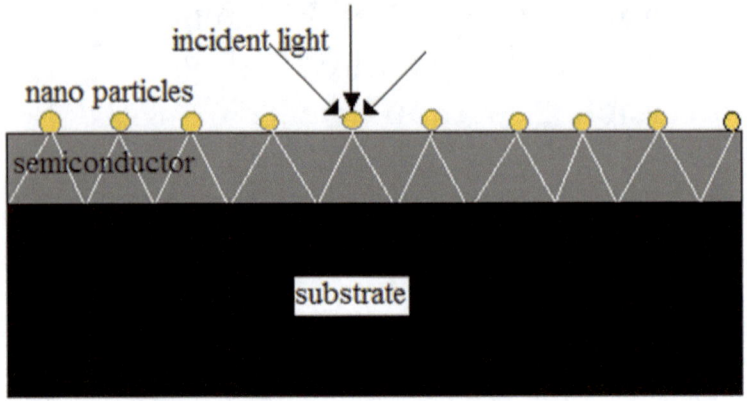

Fig. 1 Incidence and absorption of light

1.1 Nanomaterial Selection

The proper selection of metal nanoparticle is important for a large amount of light to be absorbed in the active layer [3]. The nanoparticles of Ag and Au are the most frequently used materials due to their tendency to support surface plasmon resonance efficiently but noble metals are rare and their implementation becomes impractical by there large scale so by using aluminum nanoparticles it supports surface plasmon resonance more efficient with low cost and earth abundant, and it helps solar cell in conversion. The Al nanoparticles are corrosion and wear resistant. The coating forms thermal barrier (Fig. 2).

Solar cell absorbs only the UV region of the sunlight but the visible and the IR region is left unabsorbed. The absorption region corresponding to different nanometals is given below.

S.No	Nanometal	Region absorbed
1	Aluminum	Ultraviolet
2	Silver	Ultraviolet
3	Gold	Visible
4	Copper	Visible

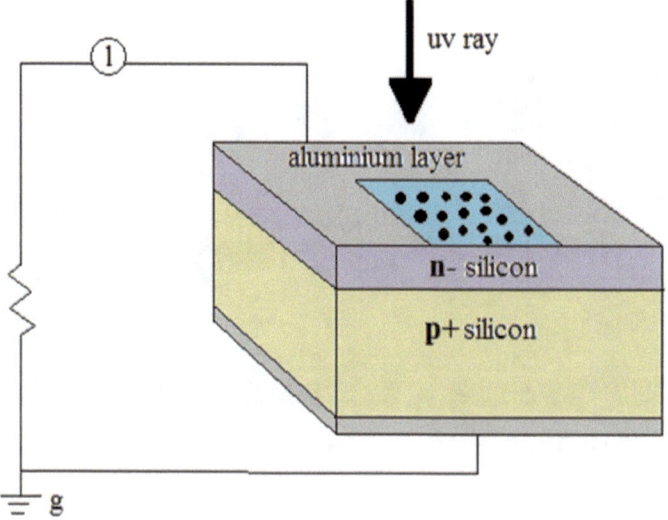

Fig. 2 Silicon wafer with Al nanoparticles

1.2 Implementing and Working Nanometals on Silicon Wafers

The process of deposition of nanometals on the silicon wafer is a tedious one; there are various innovative methods which have not been practically implemented. One of the methods is mechanism of spraying where nanometals are mixed with methanol and ethanol and compressed air. That mixture is taken in an atomizer and thick coating of nanometals is sprayed above the silicon wafer. The process contains spray, spray time, time taken to drying, and also the distance between the and atomizer should be maintained the other methods when coated adhere to the silicon surface. The choice of adhesive is crucial due to the chemical react and produce chemical changes [4]. The absorption and scattering of light across the silicon surface is more important for solar cell which is enhanced by the metal nanoparticles closer to the layer of nanometal; it absorbs more light by solar cell so the distance between the layer of nanoparticles and silicon substrate should be less for more light to be absorbed (Fig. 3).

The absorption of light is less due to the effect of reflection in the solar cell to eliminate the reflection; the antireflection layer is used in the solar cell which is identical; as antireflection layers used in optical lenses it is nothing but the layer of dielectric and also plays a role in trapping of light. The layer thickness of the reflecting layer should be one-third of the wave (Fig. 4).

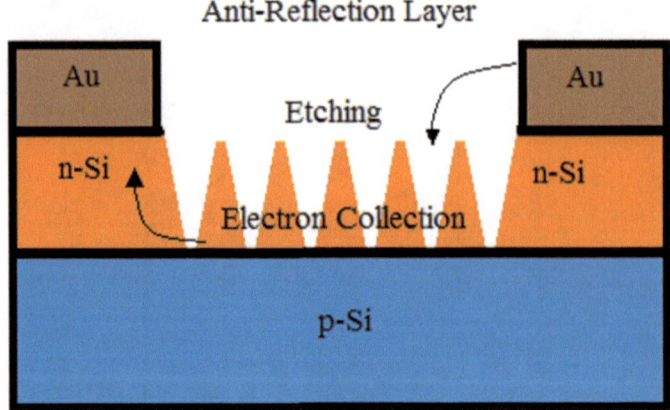

Fig. 3 Anti-reflection layer

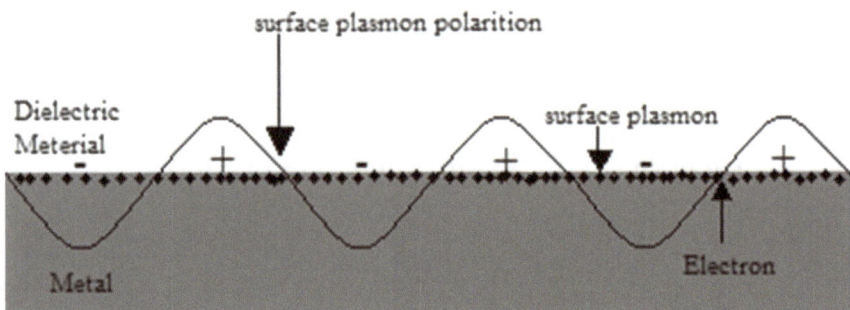

Fig. 4 Propagation electron density wave in the interference between metal dielectric

2 Transmission Using Nanomaterial

The transmission system supplies the power generated by solar from power station to the consumer; it consists of long transmission wires and electric insulators using nanomaterials. The transmission wires changed from a normal wire to quantum wires and electric insulator is coated with nanoparticle combining the transmission line and electric insulator. The maintainers and the losses are reduced and the transmission is enhanced.

2.1 Transmission Line and Quantum Wires

Transmission wires used in transmission types of wires commonly used are copper, aluminum, cadmium–copper alloys, phosphor bronze, etc. [4]. By changing the wire from normal to quantum wires, transfer electricity with significantly reduced losses at the point when contrasted with regular conductors, which control transmission at large voltage densities. Supplanting existing directing wire with quantum wire could alter the electrical matrix, as the electrical conductivity of quantum wire is higher than that of copper wire. Quantum wire is additionally one-sixth of the weight of copper and is twice as solid as steel. Aluminum leading wire with steel strengthening has dependably been the standard overhead channel for the transmission of electrical vitality.

$$R = \rho \frac{l}{A} \tag{1}$$

On the off chance that the breadth of a wire is adequately little, electrons will encounter quantum control the transverse way. Therefore, their transverse vitality will be constrained to a progression of discrete qualities. For wires, ρ is assumed as the capacity of materials, l is the length, and A is the wire's area of cross section.

2.2 Electric Insulator Coating with Nanoparticle

Electrical insulators, or dielectrics, are nonconducting materials that can withstand electrical current passage. Nanomaterial is applied to prevent from condition conductors so that current easily flows. These nanomaterials help to prevent short circuit and damages. Dielectric materials comprise substances with energy particles or electrons; they are compressed by chemical process. The voltage that transmits through the materials is impossible.

2.3 Hydrophobic Properties and Contact Angle of Insulator Surfaces

In nanotechnology items, there was a material to improve the hydrophobic conduct of surfaces. Hydrophobicity of a surface shows the capacity of the surface in repulsing water and is demonstrated by its contact edge [4]. Hydrophobic surface has contact point more than 90° while hydrophilic under 90°. Other than the benefits of lighter weight, higher vandalism opposition, and simplicity of taking care of when contrasted with clay encasings, non-earthenware separators have an extra legitimacy which is enhanced flashover execution under defiled conditions. The two classes of materials utilized broadly for non-earthenware protectors, in particular, silicone elastic and ethylene propylene elastic (EPR, other material families are additionally being used for HV outside protection), and the capacity to oppose water recording, ordinarily alluded to as hydrophobicity (Fig. 5).

Fig. 5 Pictures of non-coated [left] and RTV-coated [right] samples

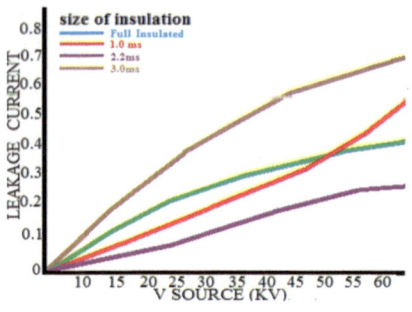 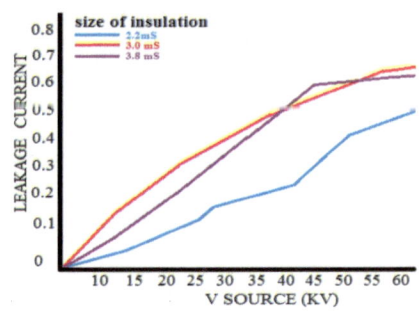

Fig. 6 Leakage current magnitude and leakage current (LC) magnitude

The aggregation of airborne defilement, maturing, crown action, and surface releases, and the materials can be expected to lose their initial hydrophobicity to varying degrees (Fig. 6) [5].

3   Internet of Things with Solar

Internet of things technology for overseeing sun-oriented power age can enormously upgrade the execution, checking, and support of the plant. With headway of advancements, the cost of sustainable power source hardware is going down all inclusive empowering substantial scale sun-oriented plant establishments [6]. This enormous size of nearby planetary group organization requires advanced frameworks for computerization of the plant observing remotely utilizing electronic interfaces as greater part of them are introduced in blocked-off areas and accordingly unfit to be checked from a committed area [7]. So here we propose a mechanized IOT-based sun-powered power-checking framework that takes into account computerized sunlight-based power observing from anyplace finished the web. We use ATmega controller-based framework to screen sunlight-based board parameters (Fig. 7).

3.1   ATmega, Current Voltage, and Sensor

The purpose of using ATmega is that it has high response, specification, and operation [7]. Its bridges in PV panel and IoT and supplies with light DC supply for operation. The voltage and current output all powers used by the load and give reading of used electricity in digital form to ATmega.

3.2   LCD and Wi-Fi Module

LCD is utilized for showing the item name and adds up to cost. At the point the item is put into truck in the wake of checking, it will demonstrate the cost and name and if second item is filtered, at that point second item cost will get included and it will be shown on LCD [7]. All the simulated data by ATmega is transferred by Wi-Fi module that store on IoT server or in cloud. In order to analyze the data on daily, weekly, and monthly basis, we are using popular IoT platform.

Fig. 7 IoT using solar power displaying system block diagram

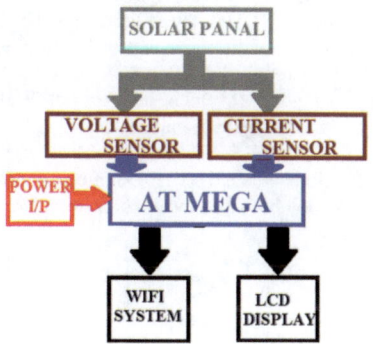

4 Results

Before the nanotechnology, the material used in generation and transmission was not coated with insulation or it was partially coated due to which the losses occur frequently. The proposed work illustrates results of the solar energy system by using aluminum layer of insulation that can increase the efficiency, and using quantum wire and silicon insulation on electric insulators, the losses are reduced. The solar energy monitoring system using IoT to gather the data is analyzed and operated the system in smooth manner. The application and the relation of privy method are upgraded by nanotechnology and IoT.

5 Conclusion

The cheep solar cell could revolutionize the electronics industry. Solar cell with nanotechnology would help preserve the atmosphere. The program of nanostructured solar cell is to reduce our reliance on fossil fuels and increase the efficiency of nano-based solar cell more than the convection solar cell the concept of nanotechnology with quantum wire, silicon, nanomaterial insulator and IoT the efficiency can be increased up to 45% the demand and reliability of whole structure is the concept to ensure the favorable outcome.

References

1. Khan MNA, Reddy DBG (2018) Increasing power transfer capacity using statcom. Int J Adv Res Innov Ideas Educ (Paper Id: 7376) 4(1):642–651. IJARIIE-ISSN(O)-2395-4396
2. Khan MNA, Khatoon F Self feeding system and smart pill box using internet of thing IEEE (Institute of Electrical and Electronics Engineers) Paper Id: 167. In: International conference at Osmania University
3. Chen C, Liu L, Lu Y, Kong ES-W, Zhang Y, Sheng X, Ding H (2006) A method for creating reliable and low-resistance contacts between carbon nanotubes and microelectrodes1:27
4. (2014) Lossless and efficient transmission of electrical energy using nanotechnology. Int J Eng Res Technol (IJERT) IJERTIJERT 3(11). ISSN: 2278-0181
5. Ganga S, Viswanath GR, Aradhya RS (2012, July) Improved performance of silicone rubber insulation with nano fillers. In: IEEE 10th international conference on the properties and applications of dielectric materials, Bangalore
6. Katyarmal M, Walkunde S, Sakhare A, Rawandale US (2018) Solar power monitoring system using IoT. Int Res J Eng Technol (IRJET) 5(3). e-ISSN: 2395-0056
7. Patil S, Vijayalashmi M, Tapaskar R (2017) Solar energy monitoring system using IoT. Indian J Sci Res 15(2):149–155. ISSN: 2250-0138
8. Hjortstam O, Isberg P, Oderholm SS, Dai H (2004) Can we achieve ultralow resistivity in carbon nanotube-based metal composites? Appl Phys A 78:1175–1179

An Inception Toward Better Classification Technique for Big Data

S. Md. Mujeeb, R. Praveen Sam and K. Madhavi

Abstract The hasty emerging technology in the field of information technology (IT) during past few years is "Big Data". One of the valuable tasks in a wide range of domains handling cumbersome database is classification. Correspondingly, this article presents the review of 10 research papers suggesting various techniques adopted for the productive big data classification, like Support Vector Machine (SVM) classifier, k-Nearest Neighbor (KNN), decision tree, association rule-based classifier, and fuzzy classifier. Subsequently, an effective technique must be designed to outperform present techniques for better management of big data. Additionally, an elaborative analysis is made by concerning the implementation tool and the adopted framework for classification of big data. Hereafter, the research issues and research gaps of abovementioned big data classification techniques are presented for aggrandizing the researchers for the better improvement of big data management.

Keywords Big data · MapReduce · Classification · Data mining · SVM · KNN · Decision tree

S. Md. Mujeeb (✉)
Jawaharlal Nehru Technological University Anantapur, Anantapur 515002,
Andhra Pradesh, India
e-mail: mujeeb.smd@gmail.com

R. Praveen Sam
Department of CSE, G. Pulla Reddy Engineering College, Kurnool 518007,
Andhra Pradesh, India
e-mail: praveen_sam75@yahoo.com

K. Madhavi
Department of CSE, JNTUA College of Engineering Ananthapuramu,
Anantapur 515002, Andhra Pradesh, India
e-mail: kasamadhavi@yahoo.com

© Springer Nature Singapore Pte Ltd. 2019
H. S. Saini et al. (eds.), *Innovations in Computer Science
and Engineering*, Lecture Notes in Networks and Systems 74,
https://doi.org/10.1007/978-981-13-7082-3_60

1   Introduction

During the past few years, the "big data" has grown as one of the attractive industries of the IT sector. The big data is a terminology generally used for exemplifying the challenges and benefits encountered during the accumulation and processing of cumbersome data [1]. The accurate definition of "big data" is the quantity of data which surpass the processing capabilities of a specific system in terms of memory and consumption of time. The big data has attracted the attention of wide range of fields like finance, medicine, retail, and industries which are handling massive amount of data. However, the process of extraction and analysis of knowledge is becoming difficult in most of the basic as well as advanced data mining tools [2]. One of the widely used tasks in the application of social media, marketing, and biomedicine is big data classification [7]. The commonly used model for solving the classification challenges of big data is traditional classification models [3]. This article mostly targets on the survey of various classification methodologies used for the classification of big data with the intention of improving the accuracy of classification. The survey is also made considering implementation tools and adopted frameworks for big data classification methods. The additional survey is performed to exploit the research issues and research gaps of abovementioned classification method. Hence, an inception is to bring out the better big data classification technique.

This article is systematized as follows: Sect. 1 provides the introduction about this article; Sect. 2 gives the literature survey of present big data classification techniques; Sect. 3 briefs about the research issues and research gaps identified; in Sect. 4, analysis of various tools and frameworks used is mentioned and Sect. 5 is the conclusion of this article.

2   Literature Survey on Various Big Data Classification Techniques

In this section, we review the various big data classification techniques for the keen big data management. The categorization of various big data classification methods is shown in Fig. 1. They are SVM, KNN, decision tree, association rule-based classification, and fuzzy classifier.

The improvement in the data accumulation has lead to enhance the volume of the data availability. This further becomes difficult in knowledge extraction and analysis. So to develop an efficient classification mechanism for big data is necessity.

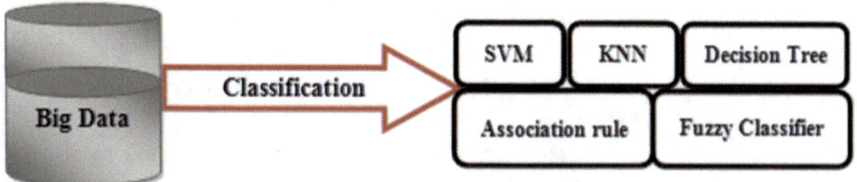

Fig. 1 Categorization of various big data classification techniques

2.1 SVM-Based Classifier

The research papers utilizing the SVM classifier for the big data classification are briefed below.

Sonali Agarwal et al. [4] proposed MapReduce (MR)-based SVM for the classification of large-sized dataset. The MapReduce is a distributed framework for dividing the large dataset into small chunks. The research was done to examine the influence of kernel parameters on the parallel SVM's performance. The results disclosed that the time utilized by the single-node cluster SVM is more compared to the proposed multi-node cluster.

Mohan et al. [5] proposed a DiP-SVM: Distribution Preserving kernel SVM approach which retains first- and second-order statistic in each of the data partitions helped in achievement of local decision vectors that were in accordance with the boundary of the global decision for reducing the liability of missing important global supported vectors.

2.2 KNN-Based Classifier

The research papers utilizing the KNN classifier for the big data classification are briefed below.

Triguero et al. [6] proposed an MR-based framework along with the KNN classifier for the prototype reduction (PR) during classification of data. The main aim of this research was to achieve an efficient classification of big data without much accuracy loss. This MRPR-based framework distributes the algorithm functions all over the clusters in the computing environment.

Maillo et al. [7] developed the MR model with the KNN classifier for performing classification of the big data. First, the map stage will figure out the KNNs in various splits of data. Second, definitive neighbors are computed in the reduce phase by utilizing the list of neighbors from the map phase. The implementation outcome of the adapted framework and the standard KNN classifier has the similar rate of classification.

2.3   Decision Tree-Based Classifier

The research papers employing the decision tree-based classifier for the big data classification are briefed below.

Bikku et al. [8] proposed a parallel-distributed model for decision tree by utilizing Hadoop framework with the idea to deal with the massive data classification. In comparison with other classification methods, this classification scheme does not require any prior knowledge of the accuracy of classification for various data characteristic types. The main advantage of this method of classification is to concentrate on both the data processing and efficiency capability.

Triguero et al. [9] proposed a decision tree model using Apache Spark for big data classification for handling the imbalance affair in the big data. The main aim of this approach is to resolve density problems by using Spark flexible framework. This scheme for classification adopted parallel in-memory operations to minimize the small-sized sample effect, and the results showed that it achieved an effective big data classification.

2.4   Association Rule-Based Classifier

The research papers employing the association rule-based classifiers for the big data classification are briefed below.

Bechini et al. [10] presented an MR framework for association-based classification known as MapReduce Associative Classifier (MRAC), which consist of three steps: Discretization, Classification Association Rules (CARs), and mining and pruning. The mining of the CARs was accomplished by making use of the properly distributed idea of Frequent Pattern (FP) growth algorithm. After finishing the CARs mining, the pruning is finished using distributed rule. This methodology acquired good scalability and serve as the suitable scheme for the handling huge datasets with balanced hardware.

Forkan et al. [11] developed an effective model for big data classification known as Big Data for Context-aware Monitoring (BDCaM). Through this BDCaM model, they analyze massive datasets generated by the Ambient-Assisted Living (AAL) models so as to make context-aware model to acclimate its behavior dynamically. This model provided the useful classification of big data with respect to accuracy and efficiency.

2.5   Fuzzy-Based Classifier

The research papers utilizing the fuzzy classifier for the big data classification are briefed as follows.

El Bakry et al. [12] developed a MapReduce paradigm with the help of the crisp and fuzzy methodologies. This method of classification comprised of two modules: Mapper and reducer. The dataset was divided into chunks by the mapper function for generating intermediate record in the scheme of [key, data] pair. The generated results were given as input to the reducer function for joining the results. The better classification precision was acquired by the fuzzy schemes compared to the crisp schemes.

Elkano et al. [13] proposed chi-fuzzy rule-based classification system for big data (Chi-FRBCS-BigDataCS). The basic feature of chi generating new rule for every input allowed achieving full potential of MR. Within the learning process, there were two stages for distributing both rule generation and rule weight computation processes for obtaining similar framework represented by classic chi algorithm.

3 Research Issues and Gaps Identified

In this section, we elaborated the various research issues and gaps faced by the above big data classification techniques.

The proposed MapReduce-based SVM was not able to process bulk-sized datasets [4]. The DiP-SVM experienced rundown of classification performance and high-rise communication overhead on classifying huge datasets [5]. The limitation of the MRPR-based framework for the KNN classifier was unable to handle the large size datasets with more dimensions [6]. The modeled MapReduce framework for the KNN classifier evidenced slow computation process [7]. The developed parallel-distributed model for decision tree classification could not classify the huge datasets having mixed attributes [8]. The developed decision tree model using Apache Spark was not able to deploy preprocessing mechanism like hybrid sampling and could not deal with the excessively imbalance problem during classification [9]. Using MRAC [10], they could not investigate speedup, load balance, and scalability for large datasets. The BDCaM model was not appropriate for the domain which requires greater context insight [11]. The fuzzy classifier first issue is to reduce execution time and second to improve accuracy and efficiency of computing resources [12]. The size up of issues was not addressed in the devised MapReduce-based Chi-FRBCS-BigDataCS framework [13].

4 Analysis

In this section, the analysis of various research works for the big data classification with respect to implementation tool and framework is discussed.

Table 1 Analysis with respect to the implementation tool used in classification schemes	Implementation tool	Research papers
	Java	[4, 8]
	Cloudera	[6, 7, 9]
	Matlab	[11]
	Other	[5] [12]

Table 2 Analysis with respect to the framework used for classification schemes	Framework	Research papers
	MapReduce framework	[4–7, 11, 12]
	Apache Hadoop framework	[8, 10, 13]
	Apache Spark framework	[9]

4.1 Analysis Based on Implementation Tool

This subsection tells about the various simulation tools employed in the above research papers. Table 1 displays the various implementation tools employed for the effective performance of the big data classification. The commonly used tools for implementation are Java, Cloudera, and Matlab. From Table 1, it is clear that the Cloudera is the frequently embraced implementation tool for effective classification of big data.

4.2 Analysis Based on the Framework Employed

This subsection tells about various adopted frameworks. Table 2 displays the various frameworks adopted for big data classification. From this analysis, it is clear that most commonly used framework is MapReduce framework for handling big data.

5 Conclusion

A survey on the various classification schemes used for the effective classification of the big data is discussed in this paper. The intention of this article is to review, learn, and categorize the different classification techniques used for big data by the analysis of 10 research papers from IEEE, Google Scholar, and various International Journals. The analysis was made with respect to implementation tool and employed framework. Finally, this survey also briefed the major gaps and issues in the present big data classification methods, and these gaps can be considered as the future scope for the inception of effective big data classification. With respect to discussion and analysis, it is concluded that the Cloudera is the frequently embraced implementation tool for effective classification of big data and the vastly adopted and used framework is MapReduce framework for handling big data.

References

1. Raghupathi W, Raghupathi V (2014) Big data analytics in healthcare: promise and potential. Health Inf Sci Syst 2(1):3
2. Wozniak M, Grana M, Corchado E (2014) A survey of multiple classifier systems as hybrid systems. Inf Fusion 16:3–17
3. Gandhi BS, Deshpande LA (2016) The survey on approaches to efficient clustering and classification analysis of big data. Int J Eng Trends Technol (IJETT) 36(1):33–39
4. Priyadarshini A (2015) A map reduce based support vector machine for big data classification. Int J Database Theory Appl 8(5):77–98
5. Singh D, Roy D, Mohan CK (2017) DiP-SVM: distribution preserving kernel support vector machine for big data. IEEE Trans Big Data 3(1):79–90
6. Triguero I, Peralta D, Bacardit J, García S, Herrera F (2015) MRPR: a MapReduce solution for prototype reduction in big data classification. Neurocomputing 150:331–345
7. Maillo J, Triguero I, Herrera F (2015) A MapReduce-based k-nearest neighbor approach for big data classification. In: Proceedings of 2015 IEEE conference on Trustcom/BigDataSE/ISPA, vol 2, pp 167–172
8. Bikku T, Rao NS, Akepogu AR (2016) Hadoop based feature selection and decision making models on big data. Indian J Sci Technol 9(10):1–6
9. Triguero I, Galar M, Merino D, Maillo J, Bustince H, Herrera F (2016) Evolutionary undersampling for extremely imbalanced big data classification under apache spark. In: Proceedings of 2016 IEEE congress on evolutionary computation (CEC), pp 640–647
10. Bechini A, Marcelloni F, Segatori A (2016) A MapReduce solution for associative classification of big data. Inf Sci 332:33–55
11. Forkan A, Khalil I, Ibaida A, Tari Z (2015) BDCaM: Big data for context-aware monitoring-a personalized knowledge discovery framework for assisted healthcare. IEEE Trans Cloud Comput 5(4):628–641
12. Hegazy O, Safwat S, El Bakry M (2016) A MapReduce fuzzy techniques of big data classification. In: Proceedings of IEEE SAI computing conference (SAI), pp 118–128
13. Elkano M, Galar M, Sanz J, Bustince H (2017) CHI-BD: a fuzzy rule-based classification system for big data classification problems. Fuzzy Sets Syst 2–37

A MapReduce-Based Association Rule Mining Using Hadoop Cluster—An Application of Disease Analysis

Namrata Bhattacharya, Sudip Mondal and Sunirmal Khatua

Abstract With the expansion and future potential in the field of healthcare industry, it is necessary to analyze a large amount of noisy data to obtain meaningful knowledge. Data mining techniques can be applied to remove inconsistent data and extract significant patterns. Association rule mining is a rule-based method which uncovers how items are associated with each other. Apriori algorithm is a broadly used algorithm for mining frequent itemsets for association rule mining. However, the performance of the apriori algorithm degrades with the large volume of data. So a parallel and distributed algorithm is required for efficient mining. In this paper, we provide an efficient implementation of the apriori algorithm in Hadoop MapReduce framework. We have considered medical data to produce rules which could be used to find the association between disease and their symptoms. These rules can be used for knowledge discovery to provide guidelines for the healthcare industry.

Keywords Association rule mining · Apriori algorithm · Hadoop MapReduce · Big data · Distributed computing

1 Introduction

Data mining [1], also known as Knowledge Discovery in Databases(KDD), is a process of extracting a meaningful pattern from massive transaction databases and other repositories. Popular data mining techniques are association rule mining [2], classification [3], sequential pattern analysis [4], data visualization [5], etc.

N. Bhattacharya (✉) · S. Mondal · S. Khatua
Department of Computer Science and Engineering, University of Calcutta, Kolkata, India
e-mail: namrata.bhattacharya03@gmail.com

S. Mondal
e-mail: sudip.wbsu@gmail.com

S. Khatua
e-mail: skhatuacomp@caluniv.ac.in

© Springer Nature Singapore Pte Ltd. 2019
H. S. Saini et al. (eds.), *Innovations in Computer Science and Engineering*, Lecture Notes in Networks and Systems 74,
https://doi.org/10.1007/978-981-13-7082-3_61

533

Agrawal and Srikant proposed apriori algorithm in the year 1994. Apriori algorithm is designed to operate on transactional databases (for example, list of symptoms caused by a particular disease among patients) for generating association rules. The rapid advancement of information technology has resulted in the accumulation of a large amount of data for organizations. Thus, data growing beyond the few hundreds of terabytes become unfeasible to manage, store, and analyze on a single sequential machine. The main limitation of apriori algorithm is the repeated scan of dataset and wastage of computing resources to store a large number of temporary candidate sets. Moreover, single processor system with limited computing resources is insufficient, which makes the algorithm performance inefficient. So improvements are needed in order to reduce the time complexity. The solution can be achieved by Hadoop MapReduce model for parallel and distributed computing that readily and efficiently process massive datasets [6] on large clusters of commodity hardware in a fault tolerance manner.

The medical data collected from healthcare unit are used to identify the nature of the disease from the list of possible diagnoses by examining the symptoms. This method is known as the differential diagnosis. The rules generated by association rule mining are helpful in identifying the symptoms corresponding to a disease [7]. Although they may not list all the symptoms, they can be used to narrow down some representative symptoms which can be used to diagnose the disease and select the initial treatment.

The paper is structured as follows. Section 2 briefly describes the association rule mining, apriori algorithm, and Hadoop MapReduce programming model. Section 3 briefly discusses the proposed MapReduce framework and Sect. 4 analyzes result. Finally, Sect. 5 presents conclusion and future scope.

2 Background

2.1 Association Rule Mining

In data mining, association rules are useful for analyzing frequent itemset data and using the criteria of support and confidence to find the recurring relationships among data called associations. Association rules are implication expressions of the form $\{A \rightarrow B\}$ where the intersection of set A and set B is a null set.

Support provides an estimate of the probability of occurrence of an event, i.e., $P(X \cup Y)$. The *support* of an itemset $\{X, Y\}$ is given by Eq. 1. If the *support* of an itemset is greater than or equal to the *minimum_support*, then that itemset is added to the set of frequent itemsets.

$$\textbf{Support(X, Y)} = \frac{No\ of\ transactions\ that\ contain\ X\ and\ Y}{No\ of\ transactions\ in\ the\ database} \qquad (1)$$

The *confidence* of rule $\{X \rightarrow Y\}$ indicates the probability of both antecedent and consequent appearing in the same transaction. *Confidence* is given by Eq. 2. If the *confidence* of the inference made by a rule is greater than or equal to the *minimum_confidence*, then that rule is added to the set of association rules [8, 9].

$$\textbf{Confidence}(\textbf{X} \longrightarrow \textbf{Y}) = \frac{Support(X \cup Y)}{Support(X)} \tag{2}$$

Among many applications of association rule mining, one of the major applications is disease analysis, which includes the mapping of illness to their symptoms [2]. For example, a patient who suffers from diabetes is somehow likely to suffer hypertension as well. Association rule mining is used to discover implications such as $\{Diabetes\} \rightarrow \{hypertension\}$. More formally, association rules are of the form given in Eq. 3.

$$\{Diabetes\} \longrightarrow \{Hypertension\} \quad [Support = 3\%, Confidence = 75\%] \tag{3}$$

In Eq. 3, *support* of 3% means that diabetes and hypertension occurred together in 3% of all the transactions contained in the database, while *confidence* of 75% means that the patients who suffer from diabetes, 75% of the times, suffer from hypertension as well.

2.2 Apriori Algorithm

The apriori algorithm proposed by Agrawal and Srikant in 1994 is one of the influential data mining algorithms which is used for mining the frequent itemsets from a given dataset containing a huge number of transactions.

Let $T = \{T_1, T_2, T_3 \ldots T_m\}$ be a set of transactions and $T_i = \{I_1, I_2, I_3 \ldots I_n\}$ be the set of items in transaction T_i for $1 \leq i \leq m$. Let $I = \{I_1, I_2, I_3 \ldots I_k\}, 0 \leq k \leq n$ be a subset of items formed by the items in a transaction that are obtained from the transactional database. If an itemset consists of k items, then this itemset is termed as k-itemset [1]. Apriori algorithm uses a level-wise frequent pattern mining where k-itemsets are used to explore (k + 1)-itemsets for mining association rules.

Apriori algorithm is based on apriori principle which states that "if an itemset is frequent, then all of its subsets must also be frequent." [1].

$$\forall X, Y : (X \subseteq Y) \implies Support(X) \geq Support(Y)$$

This principle holds true because of the anti-monotone property of *support*.

2.3 *Hadoop MapReduce Framework*

Hadoop is a Java-based open source and powerful tool that uses MapReduce architecture for store, design, and analysis of very large datasets like Google file system.
MapReduce programming model is applied for distributed storage and processing
of a large dataset. The distributed architecture consists of computer clusters built
from commodity hardware. The fundamental part of Apache Hadoop is Hadoop
Distributed File System (HDFS) and MapReduce programming model [10].

Google developed a MapReduce implementation that extends to large clusters of
machines comprising thousands of machines. It uses MapReduce for generating data
for many areas like web search service, sorting, data mining, machine learning, and
many other systems [11]. The evolution of huge data in health care has brought a lot
of challenges in terms of data transfer, storage, computation, and analysis. So big
data can be implemented in different dimensions of research as well as developments
in biomedical and health informatics, bioinformatics, sensing, and imaging will help
in future clinical research. Another important factor is study and analysis of health
data will contribute to the success of big data in medicine [12].

3 Distributed Apriori Algorithm for Association Rule Mining

3.1 *Problem Formulation*

Apriori algorithm suffers some major challenges despite being clear and simple. The
major limitation in mining frequent itemsets from large dataset is costly wasting of
time to hold the huge number of candidate itemsets and frequent itemsets satisfying
the *minimum_support* [1]. Long itemset will consist of a large number of combi-
natorics of smaller frequent itemset. For instance, a frequent itemset of length 100
will contain 100 frequent 1-itemset. Thus, the total number of frequent itemsets it
contains is given in Eq. 4.

$$^{100}C_1 + {}^{100}C_2 + \cdots + {}^{100}C_{100} \approx 1.27 \times 10^{30} \tag{4}$$

Thus, if a computer takes 1 millisecond to compute each itemset, then it will
take approximately 1.27×10^{30} millisecond which is huge. Also, if there are 10^4
frequent 1-itemsets, it may generate more than 10^7 candidate itemsets; thus, it will
scan the database many times repeatedly [1]. In Sect. 3.2, we have discussed a Hadoop
MapReduce-based solution for association rule mining.

3.2 Proposed MapReduce Framework for Association Rule Mining

This section will address the distributed association rule mining algorithm that has been used to generate the association rules. The paper performs an efficient data mining to group the frequently occurring symptoms for a specific disease as core symptoms and identify the significant association between them.

A MapReduce job splits the input transaction database into various blocks, and a mapper is invoked once for each transaction passed as arguments. The map task parses one transaction at a time and extracts each itemset included in the transaction it received as input. After processing, the mapper sends the itemset to the partitioner by emitting the itemset and frequency as <key, value> pair, where 'key 'is a candidate itemset and "value" is 1.

The partition task collects all the intermediate <key,value> pair emitted from the map task as its input and works like a hash function. Based on the key size of each key, i.e., the size of each itemset, the partitioner specifies that all the values for each itemset are grouped together and maps all the values of a single key go to the same reducer. The output pairs of all partitioner are shuffled and exchanged to make the list of values associated with the same key as <key,list(value)> pairs.

Reduce task collects each key passing all the values emitted against the same key as arguments, i.e., <key,list(value)> pairs emitted by partitioner task. Then, it sums up the values of respective keys. Candidate itemset whose sum of values is $supportcount \leq minimum\_support\_count$ is discarded. The result from all reducers is written to the output file. The whole process is shown in Fig. 1.

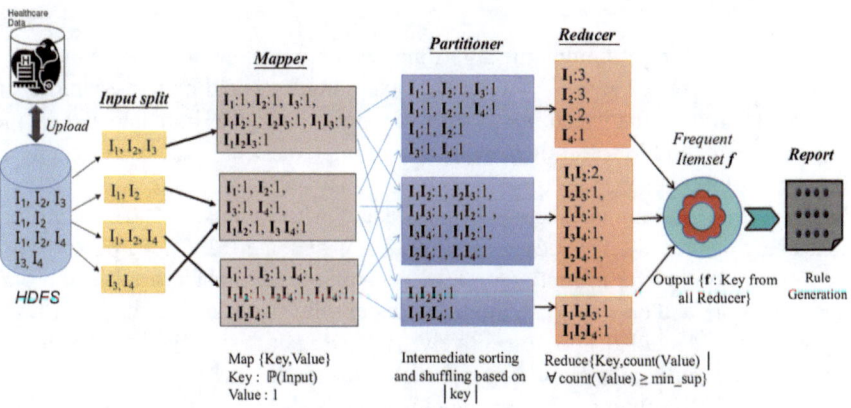

Fig. 1 Data flow in proposed MapReduce framework

4   Experiment and Results

In this section, we discuss the implementation specifications and experimental analysis performed.

4.1   Dataset and Experimental Setup

The dataset used for our experiment is a comma separated file that contains various types of disease and their symptoms. The file contains a collection of patient-wise disease-symptom transactions with disease representing a single transaction and each column in a row being the symptoms. We have generated five datasets with a varying number of transactions for the analysis. A total number of transaction in each dataset are 25,000, 50,000, 100,000, 150,000, and 200,000, and each transaction contains 10 items on average.

We have setup Hadoop single-node, 2-node, and 3-node cluster for our experiment, to compare the implementation of the distributed algorithm with centralized system. All experiments have been performed on machines that contain CentOS 7 64-bit, Eclipse Neon, and Hadoop version 2.7.1. Both the algorithms have been implemented in Java: JDK version is "1.8.0_131".

4.2   Performance Improvement of Association Rule Mining Algorithm Using Hadoop

We obtain Fig. 2 from running the algorithm, which is a graph that shows the comparison of the execution time of algorithm when implemented on non-MapReduce Java platform with the execution time of the algorithm when implemented on the Hadoop cluster and to test if there would be an improvement in the running time as the number of nodes increased.

When the algorithm is executed on the dataset which contains 200,000 transactions using the non-MapReduce code, it takes approximately 231 min. We are using the same algorithm for executing on Hadoop single-node cluster, it takes approximately 68 min, and on Hadoop multinode cluster with two and three nodes, it takes approximately 41 and 28 min, respectively. Thus, we can see in Fig. 2, 87.8% time is reduced by executing the algorithm in three-node cluster, 82.25% time in two-node cluster, and 70.5% time in single-node cluster with respect to the non-MapReduce algorithm. This is a significant improvement in the performance of the algorithm using distributed framework Hadoop. The same performance improvement using Hadoop can be witnessed for other datasets also. In our experiment, as the number of nodes increases, the execution time of the algorithm decreases drastically.

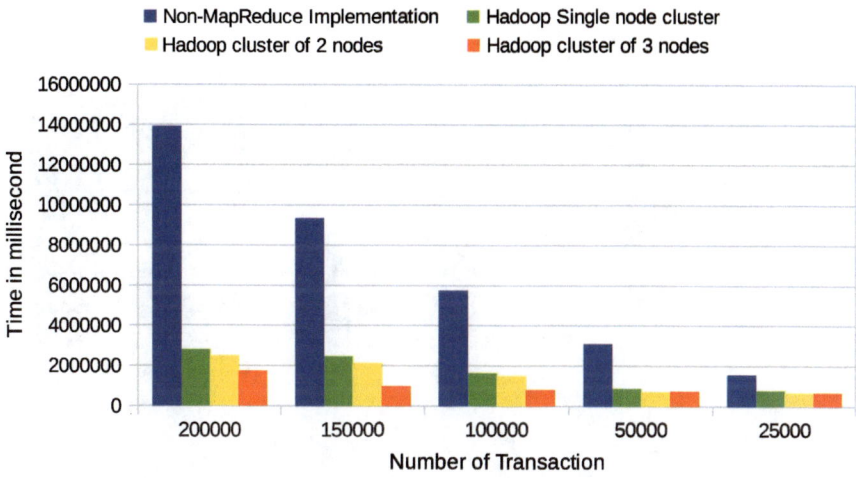

Fig. 2 Analysis of running time comparison based on number of transactions

We observe that the decrease in execution time is not always at the same rate. This depends upon the size of the data. For the dataset with 50,000 and 25,000 number of transaction, we can observe that the time required for executing the program in two-node cluster (11 and 10 min, respectively) is less than running the program in three-node cluster (12 and 11 min, respectively). This can be attributed to the fact that the input file is split into the block size of HDFS and the operations on the block are performed by the node it is assigned. If the number of nodes exceeds the number of blocks, it is a waste of resources. Thus, in some instances, increasing the number of nodes would not lead to the decrease in time. In almost every instance, a larger dataset in block size and executed on multiple nodes needs lower execution time rather than execute with a smaller dataset. It is because of Hadoop operates on larger HDFS block size to compute with its worker node.

We have obtained Fig. 3 by executing the apriori algorithm in the distributed platform, which is a graph for different itemsets that shows the *support* and *confidence* for patients. X-axis represents the diseases and Y-axis represents the value of support and confidence, respectively.

From Fig. 3, we can observe that risk of "influenza" is significant if the patient already carries "fever" since it gives association as 80% which is greater than the *minimum_confidence* value of 60%. But the converse is not always true, i.e., risk of "fever" is not significant if the patient already carries "influenza" since we can observe $\{influenza \rightarrow fever\}$ gives association as 44% which does not cross the *minimum_confidence* value.

The system can further be enhanced to find strong association rules. We can say the dataset {influenza,dry cough} contains strong association among its attributes as both the association rules, $\{influenza \rightarrow drycough\}$ and $\{drycough \rightarrow influenza\}$ range over the *minimum_confidence* value.

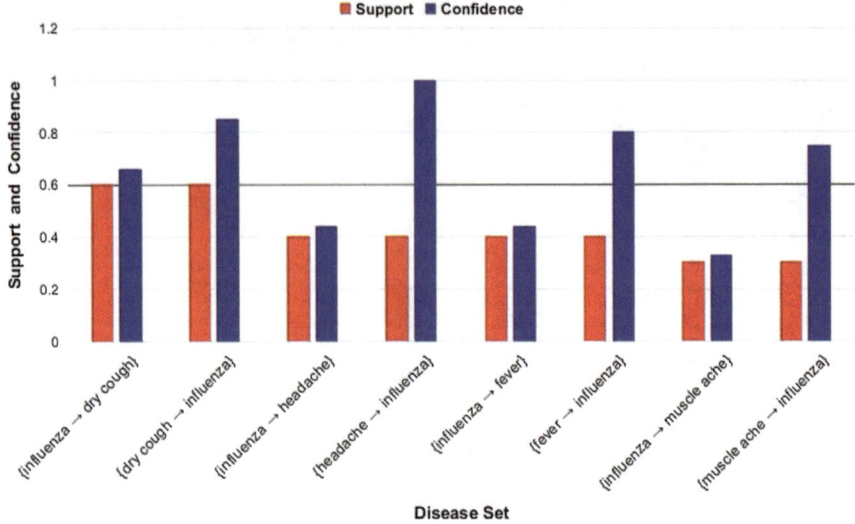

Fig. 3 Analysis of support and confidence for different itemsets

5   Conclusion and Future Scope

Efficient Apriori algorithm needs to be designed to mine large volume of data so that it can be implemented in parallel and distributed environment. The proposed model has applied data mining technique for production of association rule using Hadoop MapReduce in the distributed environment on a large clinical database. The result shows higher data processing time when the algorithm has been applied to the single centralized system over the distributed system. The proposed algorithm has developed on big data sets in the distributed system and it shows notable performance improvement. In future, we want to apply this algorithm to cloud computing system for better access and execute in real time.

References

1. Han J, Pei J, Kamber M (2011) Data mining: concepts and techniques. Elsevier, Amsterdam
2. Agrawal R, Imieliski T, Swami A (1993, June) Mining association rules between sets of items in large databases. In: ACM sigmod record, vol. 22, no. 2, pp 207–216. ACM
3. Chen G, Liu H, Yu L, Wei Q, Zhang X (2006) A new approach to classification based on association rule mining. Decis Support Syst 42(2):674–689
4. Koper A, Nguyen HS (2011, December) Sequential pattern mining from stream data. In: International conference on advanced data mining and applications, pp 278–291. Springer, Berlin
5. Keim DA (2002) Information visualization and visual data mining. IEEE Trans Vis Comput Graph 1:1–8

6. Yang HC, Dasdan A, Hsiao RL, Parker DS (2007, June). Map-reduce-merge: simplified relational data processing on large clusters. In: Proceedings of the 2007 ACM SIGMOD international conference on management of data, pp 1029–1040. ACM
7. Han J, Fu Y, Wang W, Koperski K, Zaiane O (1996, June) DMQL: a data mining query language for relational databases. In: Proceedings of 1996 SiGMOD, vol 96, pp 27–34
8. Agrawal R, Srikant R (1994, September) Fast algorithms for mining association rules. In: Proceedings of 20th international conference on very large data bases, VLDB, vol 1215, pp 487–499
9. Fan W, Bifet A (2013) Mining big data: current status, and forecast to the future. ACM sIGKDD Explor Newsl 14(2):1–5
10. Shvachko, K, Kuang H, Radia S, Chansler R (2010, May) The hadoop distributed file system. In: 2010 IEEE 26th symposium on mass storage systems and technologies (MSST), pp 1–10. IEEE
11. Han J, Fu Y (1999) Mining multiple-level association rules in large databases. IEEE Trans Knowl Data Eng 5:798–805
12. Dean J, Ghemawat S (2008) MapReduce: simplified data processing on large clusters. Commun ACM 51(1):107–113

Multi-keyword Ranked Fuzzy Keyword Search Over Encrypted Cloud Data

Saba and Shridevi Karande

Abstract The cloud servers help in sharing data among the people within an organization or outside it. The outsourcing of one's data over the Internet has become much easier with the help of the newer technology, the cloud servers. The cloud servers help in sharing data among the people within an organization or outside it. As security over the cloud has become a major challenge and also a very important aspect, providing security for the data that is being stored has become crucial. Hence, encryption of data that would be outsourced has become a very prominent talk recently. Similarly, fetching of the documents over the cloud should give faster and better results which would eventually give a better performance result. The number of keywords defined can be numerous, and hence the term multi-keyword.

Keywords Cloud · Encryption · Multi-keyword · Ranked search

1 Introduction

With the growing demand of cloud, the basic day-to-day software are making its way up. As its popularity is on a rise, almost every organization tries to store its data over the cloud. This has helped in the anywhere and anytime access concept of data retrieval. The same is done over Google, Amazon, and many other websites. The major difference between these websites is the deployment of such software on a public cloud. The use of this technology has spread to many major and minor companies and organizations [1]. As the data is saved over the Internet, it can be misused, stolen, or even used for receiving some amount of ransom. Many data over the Internet are too sensitive and cannot be compromised at any cost. Therefore, to overcome such a problem, encryption of the data can be done [2].

Saba (✉) · S. Karande
Computer Engineering, Maharashtra Institute of Technology, Pune, India
e-mail: saba.cs14@gmail.com

S. Karande
e-mail: shridevi.karande@mitpune.edu.in

© Springer Nature Singapore Pte Ltd. 2019
H. S. Saini et al. (eds.), *Innovations in Computer Science and Engineering*, Lecture Notes in Networks and Systems 74,
https://doi.org/10.1007/978-981-13-7082-3_62

543

2 Scope and Objective

- A new system was defined with multi-owner model for privacy-preserving keyword search over encrypted cloud data.
- An efficient data user authentication protocol, which prevents attackers from eavesdropping secret keys and pretending to be illegal data users performing searches, enables data user authentication and revocation.
- A secure search protocol enables the cloud server to perform secure ranked keyword search without knowing the actual data of both keywords and trapdoors [3].
- It increases efficacy and efficiency.

3 Mathematical Model

The system S consists of

$S = \{I, O, F, \emptyset\}$

I   Input given.
O   Output produced.
F   Functions performed over the input to get the output.
Ø   Case of success or failure.

- Data Owner

    I = Index of the file.
    {I1 = (k1, k2,..., kn) = Fi} I1 = Index of file Fi.
    Fi — File name.
    k1, k2,..., kn = Keyword list F = {F1, F2}.
    F1 = Hashing the index file using MD5. (I1, F1) = I2.
    F2 = Encryption of the index file using AES algorithm.
    (I2, F2) = I3.
    O = Secure Index I3.
    This is sent to the administrator server.

- Data User

    I = {k1, k2,..., kn} = K.
    {k1, k2,..., kn} = Keyword list F = {F1, F2}.
    F1 = Hashing the keyword list using MD5. (I, F1) = K1.
    F2 = Encryption of the keyword list using AES algorithm.
    (K1, F2) = K2.
    O = Secure Keywords K2.
    This is sent to the administrator server.

- Administrator Server I = {I3, K2}.

I3 = Encrypted index.
K2 = Encrypted keyword F = {F1, F2}.
F1 = Re-encryption of index I3.
F2 = Re-encryption of keyword K2.
O = Secure index and keyword.
This is sent to the cloud server.

- Cloud Server

    I = {I3, K2, N}.
    I3 = Re-encrypted index.
    K2 = Re-encrypted keyword.
    N = Number of files specified by the user. F = {F1, F2}.
    F1 = Compare I3 and K2.
    F2 = When a match is found, the counter is updated by one.

O = If a match is found, the file containing the keyword is sent back to the data user. The specified number of files is sent to the user.
Ø = If the file is found, it is considered as a case of success and if the file is not found, it is a failure case.

4   System Architecture

The system consists of four components: the data user, data owner, administrator, and the cloud server [4]. The data owners are the group of users who own the data and upload their data over the cloud server. The data users are the users who try to access the uploaded data. The administrator acts as a mediator between the data owner or data user and the cloud server. The cloud server is a huge collection of data, which is made available for all the data owners to upload files and documents and also for the data users for the successful retrieval of the data [5] (Fig. 1).

The data users and data owners have to authenticate beforehand. This will help in preventing any unauthorized trespassing [6, 7]. The administrator maintains a system for publishing the trapdoor whenever the data user asks for a specific file or folder [8]. The trapdoor is the encrypted keyword which is to be searched over the cloud server [9].

Once this is sent, it will help in safe retrieval of the documents without the cloud server getting to know about the plaintext. After the files are found, the cloud server sends the encrypted files to the respective data user [10]. The data user will have to decrypt the file with a key and access the contents present in it.

The hashing is done using the Message Digest 5 (MD5) algorithm. This algorithm gives an output of 128 bits and is unique to every single entry. The encryption algorithm which is used to encrypt the data is the Advanced Encryption Standard (AES) algorithm. There are various types of AES algorithms, but the 128-bit algorithm is faster computationally. When doing so, not only the security of the system gets

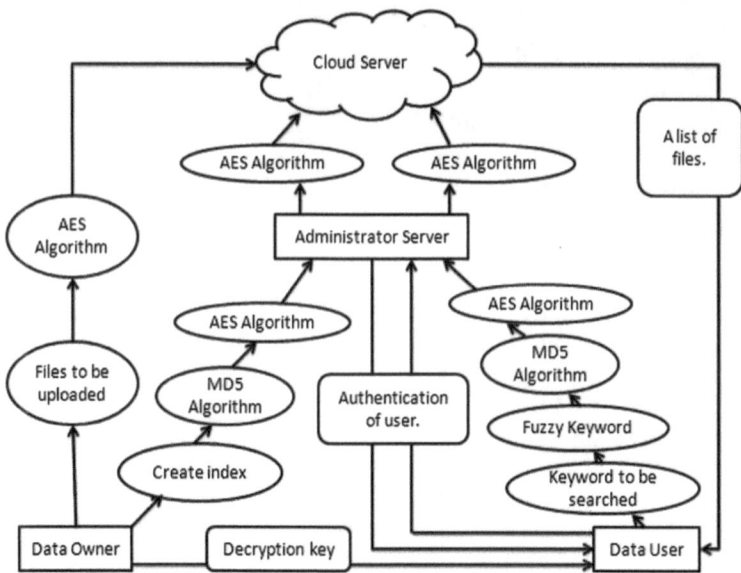

Fig. 1 System architecture

enhanced, the amount of time that is required to execute is lesser when compared to different algorithms that can be used in place of the AES algorithm.

The comparative analysis of RSA, 3DES, AES, and DES algorithm was discussed in [11]. This has highlighted a few of the features which make AES algorithm favorable in our case. AES algorithm is faster than the RSA, and hence this would help in faster recovery of the data that is stored over the cloud server. The next paper compares the MD5 algorithm and the SHA-1 [12]. Though SHA-1 is more secure for hashing, MD5 is efficient and faster than the SHA-1.

The fuzzy keywords can be defined as the keywords which are spelled wrong or have an error due to the manual input [13]. The keywords which are entered into the system can many a time be wrong. For example, if Castle is spelled wrong as Castel, the system will have to detect the error and return a file which would contain a keyword Castle.

5   Experiment and Analysis

The graphs were plotted after a thorough comparison of the various algorithms. The algorithms in place were AES-128, AES-192, and AES 256 [11]. The hashing algorithms included the MD5 and SHA-256 algorithms [12]. The times include the upload and download time of the file as well. This time is a very negligible and amounts to a variation of 3–6 ms (Fig. 2).

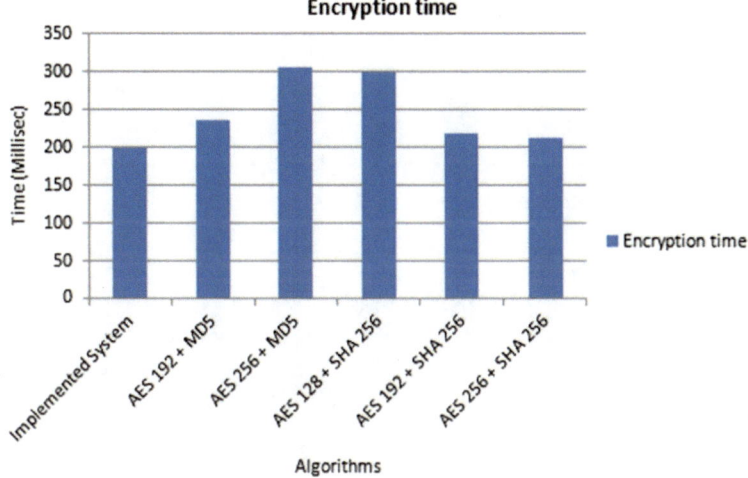

Fig. 2 Encryption time

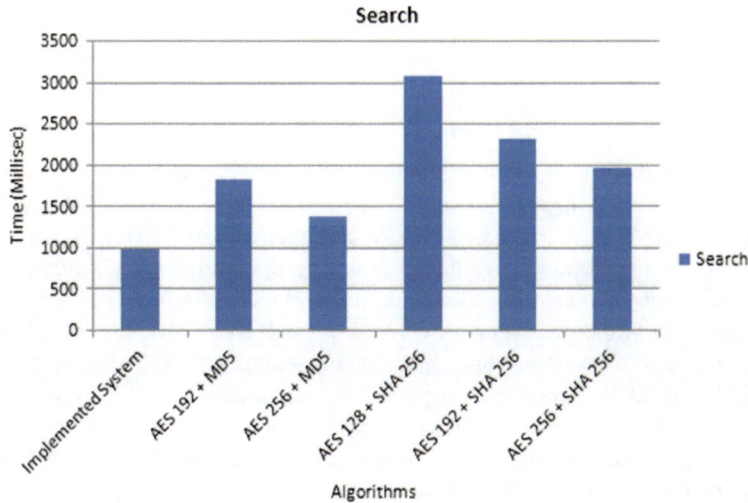

Fig. 3 Time taken to search a keyword

The implemented system, AES 128 with MD5, has a lower time for encrypting the data as well as the keys. The algorithms AES 256 with MD5 and AES 128 with SHA 256 take the longest time (Fig. 3).

The implemented system, AES 128 with MD5, has a lower time for searching the data. The algorithm AES 128 with SHA-256 take the longest time (Fig. 4).

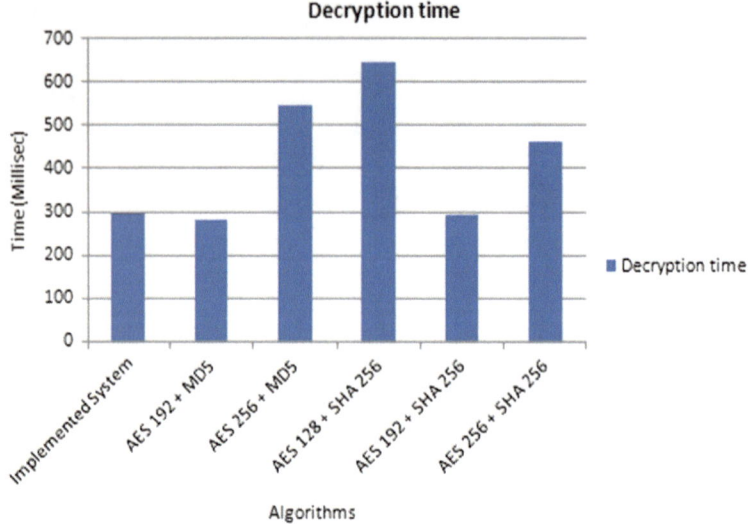

Fig. 4 Decryption time

6 Conclusion

As various encryption and hashing algorithms were compared, the best out of the compared algorithms turned out to be a combination of the AES 128 and MD5 algorithms. These algorithms are faster to compute and also give better results as experiment was performed over the system. The system helps in overcoming the maximum power given to the cloud service providers. As the data is encrypted before uploading, the data cannot be compromised. The possible downfall of the system would be the compromise of the administrator server. If this happens, the details of all the data users, data owners along with the keys would be compromised. The future work would include the conjugation of keywords. This would help in a better searching mechanism, and hence would give even a better search result. This model can also be extended to the public cloud in the future.

References

1. Zhang W, Lin Y, Xiao S, Wu J, Zhou S (2016) Privacy preserving ranked multi-keyword search for multiple data owners in cloud computing, IEEE Trans Comput 65(5)
2. Song D, Wagner D, Perrig A (2000) Practical techniques for searches on encrypted data. In: Proceeding 2000 IEEE Symposium on Security and Privacy, Nagoya, Japan, pp 44–55
3. Li H, Yang Y, Luan TH, Liang X, Zhou L, Shen X (2016) Enabling fine grained multi-keyword search supporting classified sub-dictionaries over encrypted cloud data. IEEE Trans Dependable Secur Comput 13(3):312–325

4. Wang C, Cao N, Li J, Ren K, Lou W (2010) Secure ranked keyword search over encrypted cloud data. In: Proceedings of IEEE Distributed Computing Systems, Genoa, Italy, pp 253–262

5. Armbrust M, Fox A, Griffith R, Joseph AD, Katz R, Konwinski A, Lee G, Patterson D, Rabkin A, Stoica I, Zaharia M (2010) A view of cloud computing. Commun ACM

6. Cao N, Wang C, Li M, Ren K, Lou W (2011) Privacy-preserving multi-keyword ranked search over encrypted cloud data. In: Proceedings of IEEE INFOCOM, Shanghai, China, pp 829–837

7. Cao N, Wang C, Li M, Ren K, Lou W (2014) Privacy-preserving multi-keyword ranked search over encrypted cloud data. IEEE Trans Parallel Distrib Syst 25(1):222–233

8. Zhang W, Xiao S, Lin Y, Zhou T, Zhou S (2014, June) Secure ranked multi-keyword search for multiple data owners in cloud computing. In: Proceedings of 44th annual IEEE/IFIP international conference on dependable systems and networks, pp 276–286

9. Zhang W, Lin Y, Xiao S, Liu Q, Zhou T (2014, May) Secure distributed keyword search in multiple clouds. In: Proceedings of IEEE/ACM 22nd international conference on quality service, Hong Kong, pp 370–379

10. Liu Q, Tan CC, Wu J, Wang G (2012) Efficient information retrieval for ranked queries in cost-effective cloud environments. In: Proceedings of IEEE INFOCOM, pp 2581–2585

11. Singh G, Supriya K (2013) A study of encryption algorithms (RSA, DES, 3DES and AES) for information security. Int J Comput Appl 67(19):33–38

12. Gupta P, Kumar S A comparative analysis of SHA and MD5 algorithm. Int J Comput Sci Inf Technol 5

13. Li J, Wang Q, Wang C, Cao N, Ren K, Lou W (2010) Fuzzy keyword search over encrypted data in cloud computing. In: Proceedings IEEE INFOCOM, San Diego, CA, USA, pp –15

A Review on Mobile App Ranking Review and Rating Fraud Detection in Big Data

L. Chandra Sekhar Reddy, D. Murali and J. Rajeshwar

Abstract In this smart world, mobile app fraud is growing rapidly. The fraud might be ranking, review, or rating. In play store, the purpose of fraud is promoting or bumping up the apps to top list. So the user might misunderstand while download-ing. Hence, a new mechanism is required in order to detect or prevent mobile fraud. Limited research work has been done on fraud detection. Mostly data mining tech-niques are applied for this work. But nowadays, large volume of data is getting generated which might be of different formats like structured or unstructured. In this paper, taking review from different researches and identifying the problems, machine learning algorithms are recommended as future scope for resolving these issues. Not only discovering the fraud within particular time but also the most frequent frauds which are committed on the mobile app should get detected.

Keywords Fraud app rank · Big data · Machine learning

1 Introduction

Today's applications are very modern, thanks to smartphones and Internet access for that. According to the latest apps survey in 2015, nearly 3.5 million apps are available in the App Store and play store. In order to attract users or market, applica-tion developers make false assessments, ratings, and classifications. To handle these

L. Chandra Sekhar Reddy (✉)
Department of CSE, Shri JJT University, Jhunjhunu, Rajasthan, India
e-mail: chandracse525@gmail.com

D. Murali
Department of CSE, Vemu Institute of Technology, Tirupathi, India
e-mail: dabbumurali@gmail.com

J. Rajeshwar
Department of CSE, Guru Nanak Institutions Technical Campus, Khanapur, Ibrahimpatnam, India
e-mail: prof.rajeshwar@gmail.com

© Springer Nature Singapore Pte Ltd. 2019
H. S. Saini et al. (eds.), *Innovations in Computer Science and Engineering*, Lecture Notes in Networks and Systems 74,
https://doi.org/10.1007/978-981-13-7082-3_63

types of frauds or counterfeit assessment, we provide a comprehensive view of fraud prevention system for mobile devices; we determine or classify assessments, ratings, and classifications of the fraud applications, instead of direct assessment of local classification and global classification. This allows the user to determine whether the application classifications are genuine or fraudulent, so that he can decide whether to recover or not to recover the app for mobile devices.

The conventional fraud rating detection technique for the mobile application creates large heuristic strategies. Based on these heuristics, a fraud classification decision will be made in two ways. In some cases, strategies will be made to determine if the case should be presented for evaluation. In other cases, a fraud checklist with marks will be provided for different fraud indicators. The combination of these scores at the cost of the declaration will determine if the file requests will be sent for analysis or not. The criteria for determining indicators and limits are statistical and are periodically re-calibrated.

But this classic heuristic approach may not be enough for the fraud detection and to achieve more accurate accuracy. So the machine learning techniques are recommended to improve the accuracy of fraud detection. Machine learning is an essential overview of artificial intelligence that facilitates computer learning without being explicitly programmed. This is particularly useful for data patterns and for the detection of anomalies in these models. This is a great strategy to prevent fraud.

Usage of ML algorithms can be defined as an approach, capable of improving the performance, through experiences. The experience is extracted from an environment in which knowledge is acquired by learning and adapted by evaluating its performance. In this specific context improvement of algorithm performance termed training and experiences are presented as data containing input–output pairs.

2 Literature Review

In this part, review of different author's researches toward fraud detection is discussed.

In Ref. [1], they found the fraud detection system for applications, but it is still being investigated. To deal with the following problem, we propose to develop the system of deception for applications. We also recognized several key challenges. The first real-life cycle of the application is not always fraud, so we should know when it is happening. Finally, because of the dynamic nature of the various shapes, it is difficult to find and confirm the evidence of it, which encourages us to find patterns of deceitful applications as evidence.

In Ref. [2], classification lowers the level of nuclear technology by implementing a standardized classification system that fits into contemporary standards and uses the latest approach to show and complete metal detectors with current generators. Ideas and concepts offer alternative ways of identifying things. The need for our approach is to reflect the central classification of a complete or partial explanation. Here, an algorithm is added to the completion of the management to access mediated information and use it to generate certificates for each item. The algorithm applies to the same review data. Because it is based on budget completion, it is strong in terms of noise. But has insufficient information.

In Ref. [3], fraudulent behavior in play store, the most popular Android app market, fuel search rank abuse, and malware proliferation are discussed. To identify malware, the previous work related to executing application and scan rights has been focused. In this article, we introduce FairPlay, a new system that detects malware and fraud cases with search results. It correlates assessment activities and combines unambiguous evaluation reports with linguistic and behavioral characteristics collected from play store application data (87,000 app, 2.9 million reviews, and 2.4 million reviews for more than a year). It achieves more than 95% accuracy when sorting standard golden datasets for malicious, fraudulent, and legitimate applications.

In Ref. [4], the author has proposed a new spammer detection system that attempts to influence audit evaluations on certain products or groups of target products. The main purpose of it is to detect users who generate spam notifications or to evaluate spammers. The author analyzes the behavior of spammers and uses them to abuse spammers. The author models some behaviors. First, they can target specific products or groups of products to maximize their impact. Second, they tend to deviate from other evaluators in their product evaluations. The author has proposed scoring methods to calculate the level of spam for each reviewer and to link them to a number of Amazon evaluation data. After evaluation, it shows the classification proposed. The controlled methods are effective to discover them.

In Ref. [5], to explore the conceptual documents of the Bayesian structure, the author demonstrates the new model "Model Spam City Model" (ASM). The proposed system is new because the existing methods are often dependent on the diagrams or add tags to detect spam. The Bayesian model allows the following characteristics of a variety of behavioral scenes using the hidden estimates. Statistical data results show that the proposed approach is effective and operates in competition.

In Ref. [6], in many educational areas, such as information return, collaborative analysis, and social selection, they encounter the problem of polling, which must include a series of preferences on the product in the list of opinions. The selection of units can be displayed in different formats, which makes the problem difficult. In this paper, we have explained a comparative method that fits all these types. This type of idea is very fast and is a solution for hundreds of problems.

In Ref. [7], the author investigates "spam on the web": the injection of artificial pages created to influence the search engine results to drive traffic to a particular page for fun or profit. This article discusses some techniques which are not specified previously for automatic spam detection controlling the effectiveness of these technologies individually by using a rating algorithm aggregated. Overall, our heuristics correctly identified 2,037 (86.2%) of the 2364 spam pages (13.8%) on our 17,616 verified pages, while 526 pp. spam and non-spam (3.1%) were identified incorrectly.

3 Problem Statement

Due to the popularity of the smartphones and availability of Internet, so many apps are developed and placed in play stores. Most of the smartphone users are depending on such apps. Generally, they will download apps from various play stores like play store, App Store, etc. Till now only 3.5 million apps are uploaded in play store. Due to the demand for apps, many app developers are committing fraud in order to promote their app. So users are unable to take decisions while downloading the apps.

According to the literature survey, all existing fraud detection methods are focused on only ranking fraud of app but it might be committed in review or rating also.

Webspam classification [7, 8]: Webspam scans spam to manipulate application classification to increase its position in the market.

Reviewing online spam [9, 10]: reviews give an opinion on the app, based on comments that are given by the user, if the application is good or bad. Referral spam web is used to change the content of reviews to establish the application that the user can provide a positive review to it. Online spam can be classified based on

Duplicate reviews and
Comments on spam.

Spam rating: in the spam index, the user will provide the highest rate in the app; in general, the app rating is average or individual for each rating.

4 Conclusion and Future Work

In this paper, we discussed various existing research methods, using historical data fraud discovered. Most of the existing work is focused only on fraud-ranking, but some users commit fraud using positive reviews and rating. So app fraud discovers should aggregate rank, rating, and review evidence, to overcome limitations proposing machine learning algorithms as it supports unstructured data and it detects fraud based on aggregate evidence.

References

1. Zhu H et al (2011) Discovery of ranking fraud for mobile apps 27(1):181–190. January 2015
2. Gleich DF, Lim L-H (2011) Rank aggregation via nuclear norm minimization, pp 60–68
3. Rahman M et al (2017) Search rank fraud and malware detection in google play 29(6):1329–1342
4. Lim E-P et.al (2010) Detecting product review spammers using rating behaviors, pp 939–948
5. Mukherjee A et al (2013) Spotting opinion spammers using behavioral footprints. In: KDD-13

6. Volkovs MN, Zemel RS (2012) A flexible generative model for preference aggregation, pp 479–488
7. Ntoulas A et al (2006) Detecting spam web pages through content analysis, pp 83–92
8. Spirin N, Han J (2012) Survey on web spam detection: principles and algorithms 13(2):50–64
9. Shi K, Ali K (2012) Getjar mobile application recommendations with very sparse datasets. In: 18th ACM SIGKDD, pp 204–212
10. Yan B, Chen G (2011) AppJoy: personalized mobile application discovery, pp 113–126

L. Chandra Sekhar Reddy received his B.Tech (CSE) from VITS College which is affiliated to JNTU in 2007. M.Tech (CSE) from GNIT, JNTUH in 2014, pursuing his Ph.D. from JJTU, Rajasthan from 2017. He is working at CMR College of Engineering & Technology, Medchal, Hyderabad. His research interests include data mining, big data, and compiler design. He is Oracle certified Java professional.

Dr. D. Murali is presently working as Professor and HOD (CSE), VEMU Institute of Technology, Tirupati, Andhra Pradesh. He completed his B.Tech (CSE) in the year 2002 from Jawaharlal Nehru Technological University, Hyderabad, M.Tech (CS) from JNTU, Hyderabad in the year 2006, and Ph.D. in CSE from JNTU, Hyderabad in the year 2016. Before joining VEMU Institute of Technology, he worked as Professor in CSE in Malla Reddy Engineering College for Women, Hyderabad. He authored his credit. He had published papers in Springer and other reputed journals, 14 journals and 08 conferences. He is a member of CSI and ISTE, IAENG. His areas of interest are formal language and automata theory, digital logic design, C programming and data structures, operating system, software engineering, compiler designing, data mining, and data warehousing.

Dr. J. Rajeshwar is working as Professor and HOD of the Department of Computer Science and Engineering, School of Engineering and Technology, Guru Nanak Institutions Technical Campus (Autonomous). He worked in various positions as Principal, Training and Placement Officer, and TASK/IEG-JKC Coordinator. He has more than 17 years of teaching experience. He obtained his B.Tech (CSE) from JNTU College of Engineering, JNTU, Hyderabad with distinction, M.Tech (CSE) from Osmania University College of Engineering OU, Hyderabad with distinction, and Ph.D. from JNTUH College of Engineering, JNTUH, Hyderabad; his research contribution is in the area of security in mobile ad hoc networks. He is Oracle certified Java professional. He published 1 book, 14 international journals, 4 papers in national conferences, and 2 papers in international conferences. He conducted a couple of short-term courses, Faculty Development Programs (FDPs), seminars, workshops, and delivered few expert lectures. He is expert at guiding projects for undergraduate and postgraduate students, guiding students for paper presentation, project exhibition, and poster presentation.

Context-Aware Middleware Architecture for IoT-Based Smart Healthcare Applications

R. Venkateswara Reddy, D. Murali and J. Rajeshwar

Abstract In recent years, Internet of things (IoT) has become an intelligent computer model in which various things and resources are connected to a range of intelligent solutions such as Bluetooth and Wi-Fi, ZigBee, and GSM. These communication technologies offer connectivity between different IoT devices that can help control and operate devices with the user interface. The development and implementation of these applications are ideas for the next era: IoT has enabled the user to define and design a large number of middlewares to connect the IoT application levels, and one of them is a contextual middleware. Contextual applications are more adaptable to their dynamic changes in the environment, with behavior that attracts more attention from users. Contextual applications are in fact three principles of context-awareness, modeling, and reasoning. The existing approaches are technically focused on the style of architecture, abstraction, the expandability of reasoning, fault tolerance, the identification of services, privacy, security, archiving, the level of awareness of the context, and Big Data analysis. In this article, we focus on improving the security and privacy of middleware and data visualization with cloud-based Big Data analysis. At the end of the document, we discussed the challenges of open research at work.

Keywords Internet of things (IoT) · Middleware · Cloud computing · Big data analytics

R. Venkateswara Reddy (✉)
Department of CSE, Shri JJT University, Jhunjhunu, Rajasthan, India
e-mail: venkatreddyvari@gmail.com

D. Murali
Department of CSE, Vemu Institute of Technology, Tirupathi, India
e-mail: dabbumurali@gmail.com

J. Rajeshwar
Department of CSE, Guru Nanak Institutions Technical Campus,
Khanapur, Ibrahimpatnam, India
e-mail: prof.rajeshwar@gmail.com

© Springer Nature Singapore Pte Ltd. 2019
H. S. Saini et al. (eds.), *Innovations in Computer Science
and Engineering*, Lecture Notes in Networks and Systems 74,
https://doi.org/10.1007/978-981-13-7082-3_64

1   Introduction

IoT's Internet-based smart house (IoT) is a wireless network that collects, inter-
changes, and displays data that is connected to many things (HVAC devices), devices,
buildings, sensors, and many electronic devices [1]. Kelly has been researching how
smart devices can be managed and controlled without human interaction and success
[2]. Heterogeneity, row/argument vulnerability, and characteristic inheritance are
considered as issues that Zhang has opened with security-oriented IoT [3]. Nisha Single describes the Linux version of Raspberry Pi, a trilevel model for context modeling
and system architecture for smart devices [4]. IoT's intelligent home-based equipment is able to dynamically track things and the environment by changing context-
sensitive middleware applications. In general, all context-dependent applications
are implemented using three approaches, namely, Owner Mode, Library/Toolkit,
and context management [5]. Context management performs better than the other
two approaches. The design and development of contextual middleware applications emerged and much architecture have been proposed by previous researchers
[1, 6–8]. Zhang and Baosheng [3] and his team only produced a publication of a
context-sensitive middleware architecture and presented evaluations and compar-
isons without overlapping the existing approach.

The rest of this article is organized in different sections. Section 2 offers work
related to the conscious principle of the context and its architectures. In Sect. 2, we
analyze the system model that is the architecture models based on the cloud and
the context and the new architecture with an improved approach to security, the
visualization of data. In Sect. 3, the techniques for displaying data and their results
were presented. Section 4 opens the research challenges. Section 5 summarizes the
conclusion of the work.

2   System Study

2.1   Framework for the Cloud-Based Context-Aware Internet of Things Services

From the perspective of our home research to smart home applications, we have
noticed the lack of precise definitions of the technologies and models involved in
smart home applications after reviewing our existing tasks and reviewing existing
tasks. Therefore, we have considered the successful adoption of a "smart" view, with
a complete perspective reflecting our view of the integration of many technologies
and techniques in this segment. Smart-x offers a framework that contains items
from different enabled elements to strengthen their ability to build and improve
applications and services (e.g., Smart Health, Smart Agriculture, Smart Grid, Smart
Mobility, etc.). Our policy is based on contextual and context-awareness to solve
the consensus referred to above. From linguistics to computing, the context is a key
element.

The Cambridge dictionary defines "context" as "the situation within which something exists or happens, and that can help explain it"; adding to that Abowd and Dey's definition of context stated in our introduction. We believe that in any case context and contextual understanding will play an important role in the understanding and interpretation of the situation. Context-aware computing is defined by Gartner 3. "Communication and environmental information about people, places and things in the computing genre can be used to predict immediate needs and provide useful, useful and intuitive content, functions and experiences." The life cycle of contextual operating systems is usually four stages: (1) collection and integration phase, (2) modeling and storage phase, (3) reasoning and processing phase, and (4) expansion and integration phase. The first phase refers to ways to integrate contextual data around an entity and to provide accurate data. The second stage deals with the collected data (key–value, anatomy, etc.) and how to store it (relation databases, nosclick databases (This is repository from noslick.com), XML files, etc.). The third step is responsible for processing data to improve better knowledge and more meaningful information (Fig. 1), in the context of collecting the fourth task, interested parties (e.g., services, programs, reactions, etc.) .

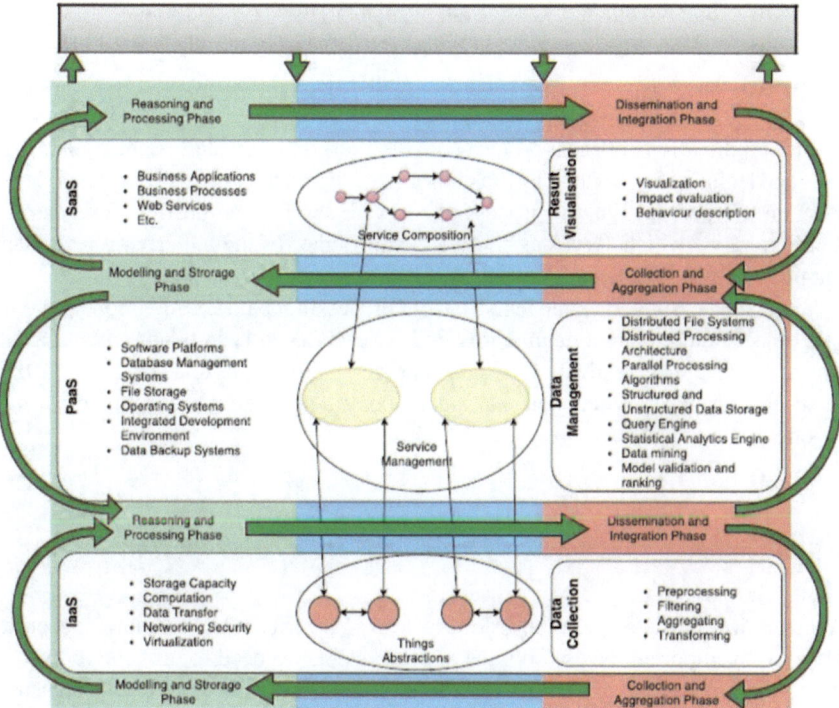

Fig. 1 Cloud-based context-aware internet of things services architecture

In the context of our work on "Smart Home Applications," we believe that we can be firmly credible to IoT data in the IoT, cloud service and Big Data decision-making. From this perspective, we will assign the context.

Management lifecycle to essential subjects provides other cloud computing facilities, IoT and Big Data computing.

Earlier in the movie, we provide service models based on cloud computing, such as service (IaaS), platform (PaaS), and software (SaaS). Smart home applications challenges (e.g., Scalability, Heterogeneity, etc.) provide all the flexibility and effectiveness needed to provide a service that provides challenges.

Infrastructure level

Through IaaS, cloud provides the necessary infrastructure and tools (e.g., processing, storing, networking, etc.) necessary to collect data from physical objects. Cloud is generally referred to as centralized technology, fog, or edge computing [6].

Platform level

Tools that help developers and data professionals in data and services are collected during stage, development, processing, testing, and development level (IaaS). Services are related to services, supervision of services, supervision of services, mapping services of contents, service configuration, service composition, etc. In the Big Data perspective, data management is the way the data is stored (structured or structured), restored, and processed (e.g., data mining, statistical analysis, etc.) (see Fig. 2).

Software level

At this stage, cloud software provides the product regularly to the user (for example, business application, business processes, APIs, etc.). This is usually done by using the tools and techniques provided by the platform at the underlying (platform level) level. From an IoT perspective, the process of software service composition (orchestra or choreography), which serves as a service, call range for the underlying layers and mapped to physical objects.

For example, showing each lego four-out processing, path coordination, and multiple objects and payment details provided for each legend can be interpreted as the time of departure. Visualization of data is processed in graphs to a considerable data perspective, to assist in decision-making, to identify new patterns in data, or to assess the effect of new data.

2.2 Proposed Structure to Improve Security and Privacy

Transmission and data operation is done by a lot of security in the middleware security which is an important issue. To have a safe system, we need to take into consideration, integrity, and availability. Therefore, ubiquitous different security measures must be provided for ubiquitous applications and pervasive environments, such as evidence and evidence, authority modifications, and proof of access control policy and accountability obligations. Confidentiality means that all parts of the IoT healthcare system that access the personal information of the patient must ensure the protection of information specified by unauthorized access.

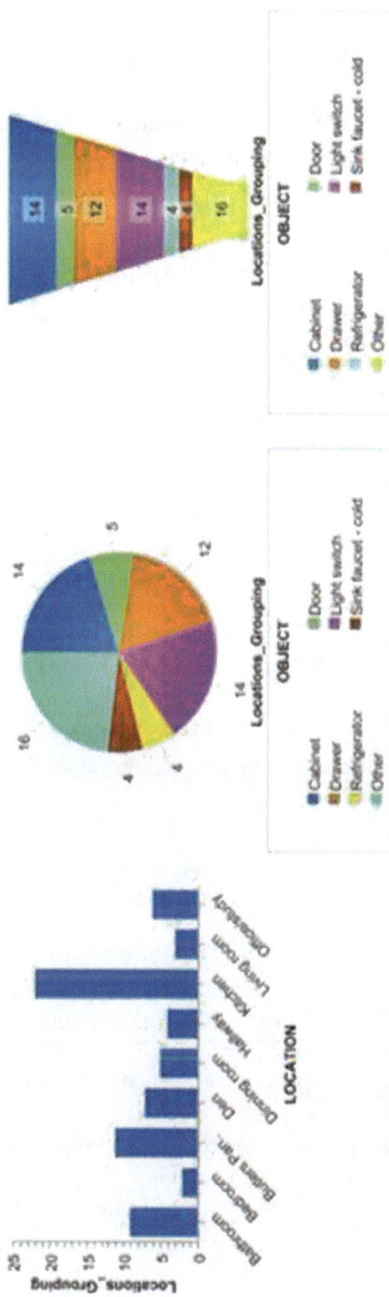

Fig. 2 Data visualization/forecasting of sensor information based on location

2.3   Context Data Security Algorithm (CDSA)

The proposed CDSA guarantees security and privacy of context-aware application data over the cloud repository. We used box to store context data in ECM. CDSA allows the cloud users based on the user id to access, visualize, and search context data. The core context-aware SAAS application algorithm has implemented on Force.com platform using an APEX programming.

3   Data Visualization Techniques for Context-Aware Applications

Context-aware application data forecasting was done with modern analytics such as standard salesforce report and dashboards, wave analytics, and Tableau. To visualize the context, we used two types of reports, namely, summary and matrix-related reports. Figure 2 represents the sensor data gathering based on the location presence. Figure 3 illustrates the data forecasting by grouping sensor ID, heading, category, and subcategory.

4   Open Research Challenges

This section discusses the remaining challenges to address the corresponding IoT device. The purpose of the section is to provide research guidelines to the new investigator in the domain.

4.1   Interoperability

The IoT has three main types of interoperability challenges, in particular, technical, semantic, and pragmatic. The technical challenges are related to the functionality of relevant devices, protocols, and standards to coexist and interoperate in the same computational paradigm, while semantics deals with the capabilities of various IoT components that are responsible for the processing and interpretation of the data exchanged. However, the pragmatic concern is about the capabilities of the components of the system to observe the intentions of the parties. Achieving technical interoperability can be achieved by providing agent-based mediation between devices and IoT standards. Semantic interoperability is a requirement for the computable logic of the machine, the discovery of knowledge, and the federation of data between information systems. Pragmatic interoperability can be achieved through the creative design of predefined specifications of component behavior. In the future, interlayer interoperability solutions are required.

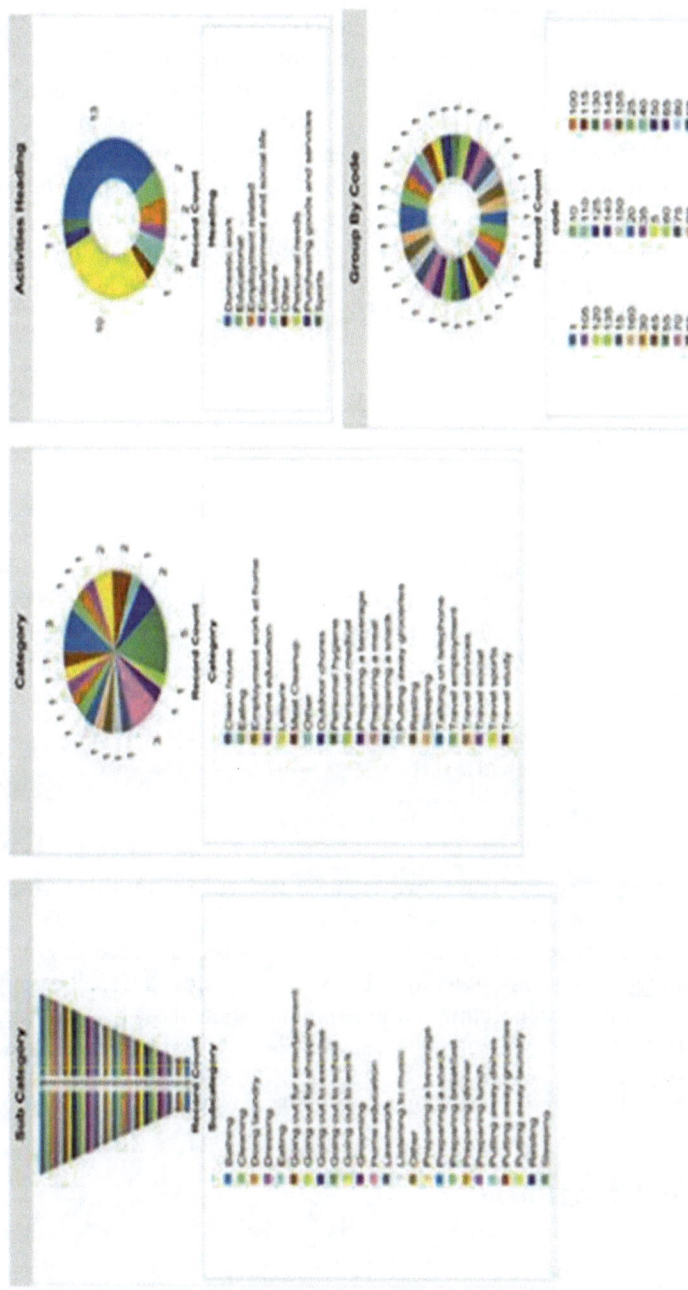

Fig. 3 Data visualization/forecasting of sensor information based on location

4.2 Scalability

The IoT treats many challenges related to significant differences in interacting entities and interaction differences and behaviors. The existing IoT healthcare device formats should be changed to fit billions of smart devices. The scope ownership of IoT healthcare systems will be compiled in two subjects. Initially, IoT devices were developed rapidly. However, the current management protocols do not suit the needs of the IoT device due to its limited capabilities. Second, social connections between equipment owners need to take into consideration, in which some of the IoT systems are individual portable devices. In the future, scalability management protocols are expected to follow social relationships between devices enabling computer services based on specific functions by providing some motivations.

4.3 Flexibility

Since there are many IoT applications, it is a great challenge to provide different IoT applications according to their demands. IoT users usually require dynamic configuration, personalized, value-based services on the fly. In addition, personal, autonomous, and dynamic services can be supported by the construction and use of compatible, sensitive, and reconstructive multiservice network structures. In the future, declaratory service specification models are required to build a future network service structure.

4.4 Energy Efficiency

Tiny devices are the backbone of IoT. However, these devices have limited processing, memory, and battery capacity. As a result, computational-intensive applications and routing processes cannot be implemented on IoT devices because these devices are lightweight. Energy awareness observation in routing protocols does not exist yet. Although some protocols support low-power communication, these protocols are in development. In the future, IoT energy systems will have good solutions to meet the energy requirements.

4.5 Mobility Management

The mobility of the nodes can create several challenges for IoT networks and protocol capabilities. Current mobility protocols of VANET, MANET, and sensor networks do not work well with simple IoT devices due to intense power and processing restrictions. Mobility management is one of the key programs and has two stages: First, the identity of the movement (the movement of the device that connects to the network's new area). Second, signaling and control messages should be merged to help them understand the nodes in a network. Motion recognition can be achieved through persistent scanning, through passive protocols or passive messages from the movements of the movement protocol. Mobility management is one of the most complex issues in the IO example. Consequently, it should take into account the future IoT structure.

4.6 Security

The diversity of IoT applications and the varied IoT communication infrastructure have many kinds of security challenges. In IoT, security can be ordered. In an ascending manner, the system must follow the safe boot process, access control rules, device authentication procedures, firewalling, and security software. If security is a critical concern in IoT, the appropriate security mechanisms cannot be used for the device and network level (physically and physically). IoT devices have some kind of intelligence to identify and counter possible threats. Fortunately, this does not require a revolutionary approach. Instead, the successful evolution of standards in other networks should be taken into consideration the processing capabilities of intelligent devices in the IoT model.

5 Conclusion

This article focused on the design and implementation of important applications based on the context through the cloud. This document highlights some of the open problems of existing approaches/platforms in the field of security and privacy, and aspects of data visualization. We design and implement the panorama for the security and privacy of the data that are sensitive to the context, taking into account the authentication, authorization, accessibility, and ownership of the data. We have designed and modified reports and panels to show the data of the application plus the graphic representation. The analysis of the presented data is more useful to analyze the IoT healthcare applications related to supervision.

References

1. Atzori L et al (2010) The internet of things: a survey. Comput Netw 2787–2805
2. Kelly SDT et al (2013) Towards the implementation of IOT for environmental condition monitoring in homes. 13(10):3846–3853. ISSN 1558-1748

3. Zhang W, Baosheng Q (2013) Security architecture of the internet of things oriented to perceptual. Int J Comput 2
4. Nisha S, Shilpa S (2016) Smart home system based on IOT. 6(9)
5. Hu PZ et al (2008) Context management system for pervasive computing, pp 17–22
6. Sain M, Lee H, Chung WY (2010) Designing context awareness middleware architecture for personal healthcare information system, pp 7–10, Feb 2010, pp 1650–1654
7. Hyunjung P, Jeehyong L (2005) In a framework of context-awareness for ubiquitous computing middlewares, pp 369–374
8. Kelly SDT et al (2013) Towards the implementation of IOT for environmental condition monitoring in homes. 13(10), 3846–3853. ISSN 1558–1748
9. Wang CD, Mo XL, Wang HS (2009) An intelligent home middleware system based on context-awareness. In: Proceedings of ICNC'09, Aug 2009, pp 165–169
10. Jianping Y, Yu H, Jiannong C, Xianping T. In middleware support for context- awareness in asynchronous pervasive computing environments, pp 136–143
11. Li X et al (2015) Context-aware middleware architectures: survey and challenges. ISSN 1124-8220
12. Schilit WN (1995) A system architecture for context-aware mobile computing USA
13. Kim JD, Son I, Balk DK (2012) Onto: an ontological context-aware model based on 5WLH, p 11
14. Van Bunningen, Feng L, Apers PMG (2005) Context for ubiquitous data management. In: Proceedings of UDM 2005, Washington, DC, USA, 4 Ap 2005, pp 17–24
15. Schilit B, Adams N, Want R (1991) Context-aware computing applications, pp 85–90
16. Gonsalves B, Pereira Filho JG, Andreão RV (2008) EDGWARE: an ECG markup language for ambulatory telemonitoring and decision making support, pp 37–43
17. Henricksen K, Indulska J (2004) A software engineering framework for context-aware pervasive computing, Mar 2004, pp 77–86

R. Venkateswara Reddy received his B.Tech (CSE) from AITS college which is affiliated to JNTUH in 2007, M.Tech (CSE) from Hindustan University in 2011, and pursuing his Ph.D. from JJTU, Rajasthan from 2017. Presently, he is working as Assistant Professor in CMR College of Engineering & Technology, Medchal, Telangana, Hyderabad. His research interests include cloud computing, data mining, and big data.

Dr. D. Murali is presently working as Professor and HOD (CSE) in VEMU Institute of Technology, Tirupati, Andhra Pradesh. He completed his B.Tech (CSE) in the year 2002 from Jawaharlal Nehru Technological University, Hyderabad, M.Tech (CS) from JNTU, Hyderabad in the year 2006, and Ph.D. in CSE from JNTU, Hyderabad in the year 2016. Before joining in VEMU Institute of Technology, he worked as Professor in CSE in Malla Reddy Engineering College for Women, Hyderabad. He authored his credit. He had published papers in Springer and other reputed journals, 14 Journals and 08 conferences. He is a member of CSI and ISTE, IAENG. His areas of interest are formal language and automata theory, digital logic design, C programming and data structures, operating system, software engineering, compiler designing, data mining, and data warehousing.

Dr. J. Rajeshwar is working as Professor and HOD of the Department of Computer Science and Engineering, School of Engineering and Technology Guru Nanak Institutions Technical Campus (Autonomous). He worked in the various positions as Principal, Training and Placement Officer, and TASK/IEG-JKC Coordinator. He has more than 17 years of teaching experience. He obtained his B.Tech (CSE) from JNTU College of Engineering, JNTU, Hyderabad with distinction, M.Tech (CSE) from Osmania University College of Engineering OU, Hyderabad with distinction, and Ph.D. from JNTUH College of Engineering, JNTUH, Hyderabad. His research contribution is in the area of security in mobile ad hoc networks. He is Oracle certified Java professional. He published 1 book, 14 international journals, 4 papers in national conferences, and 2 papers in international conferences. He conducted a couple of short-term courses, Faculty Development Programs (FDPs), seminars, workshops, and delivered few expert lectures. He is expert at guiding projects for undergraduate and postgraduate students, guiding students for paper presentation, project exhibition, and poster presentation.

Anatomization of Document-Based NoSQL Databases

Sai Jyothi Bolla, Sudhir Tirumalasetty and S. Jyothi

Abstract In the era of data science, data is evolving unconditionally beyond breadth and depth. This evolution of data resulted and invoked databases which are schema-less; besides, most of the data are not structured. Many vendors have captured this as design objective for designing NoSQL databases. These databases have been proven to be the best in providing solutions to databases which are schema-less. To model such Big Data containing various data types in various formats, document-based databases are prevailing the most, among NoSQL models. To bring the crust, this paper ignites and anatomizes the dominance of document-based databases among NoSQL databases. Prominent features which reflect the significance of NoSQL document databases are taken into account for evaluating the strength of databases and underline the best database which has the flexible expansion for handling schema-less databases.

Keywords Big Data · Data modeling · Data science · Document databases · NoSQL databases

1 Introduction

In the era of Data Science, various sources of Big Data generate huge volumes, complex, structured, semi-structured, and unstructured data. Due to lack of storage, processing power, and querying capacity, traditional data models do not fit to meet the

S. J. Bolla (✉) · S. Jyothi
Department of Computer Science, Sri Padmavathi Mahila Viswa Vidyalayam, Tirupati, India
e-mail: saidilipyerram@gmail.com

S. Jyothi
e-mail: jyothi.spmvv@gmail.com

S. Tirumalasetty
Department of Computer Science & Engineering, Vasireddy Venkatadri Institute of Engineering, Guntur, India
e-mail: sudhir2918@gmail.com

© Springer Nature Singapore Pte Ltd. 2019   569
H. S. Saini et al. (eds.), *Innovations in Computer Science
and Engineering*, Lecture Notes in Networks and Systems 74,
https://doi.org/10.1007/978-981-13-7082-3_65

necessity of market. Traditional approaches use flat relations which are not scalable for handling Big Data. These flat relations degrade the performance of Big Data, besides expensive. This lead to the notion of schema-less databases, which aided in the evolution of NoSQL databases for handling Big Data. These databases were in existence before but prominently got signified when NoSQL had started scaling Big Data. As Big Data is evolving, scaling of such voluminous data became easier to handle by NoSQL databases when compared to relational databases.

NoSQL databases ensure BASE (Basically Available, Soft state, Eventually consistency) rather than ACID properties of transactions in relational databases [1, 2]. It is in compliance with the CAP (Consistency, Availability, Partition tolerance) theorem [1]. The prominent feature which made NoSQL more powerful is; it is capable of handling large amounts of unstructured data stored at multiple resources and multiple servers. NoSQL databases are categorized into four categories [3]:

a. Wide-column databases [4, 5],
b. Key value store [4, 6],
c. Graph database [7, 8], and
d. Document store [4, 9].

The skill of selecting apt database for relevant application plays a vital role in performing analysis effectively.

The following sections include literature about why document-based databases are prevalent over other databases and the available DBDBs in market, followed by comparison of six promising DBDBs with eight benchmarks [10]. Further section includes operations and their histograms. Final section depicts about the conclusion.

2 Why Document-Based Databases (DBDB)?

Big Data contexts are not specific about the type and structure of data, then DBDBs are accountable. This is due to that the collections in DBDBs contain multiple type documents. DBDBs have the power of storing mixed-type objects (documents) containing a wide variety of multi-type values. In other NoSQL, databases support only single-type objects with a wide variety of multi-type values. DBDBs have all benefits of other databases [11]. Querying is not limited by a key; DBDB supports complex querying with multiple multi-type key values.

The benefits of DBDBs are [12, 13]: Flexible Schema, Fast Write, and Fast Query Execution. Mainly the backbone of DBDBs is wide acceptance of multi-type data and documents, and soft computing of queries.

There are more than 40 DBDBs in this era of Data Science [9]. Among those, six prominent DBDBs are taken into account in this paper for comparison. These six databases are pinned based on their usage in the market: Couchbase, CouchDB, DynamoDB, OrientDB, RethinkDB, and MongoDB.

3 Anatomization of DBDBs

To anatomize the DBDBs mentioned in the above section, the following properties are taken into account. They are data model, indexing, replication, consistency, and MapReduce support [14]. These features are qualitative and quantitative; comparison with these features exploits and differentiates DBDBs.

(a) Data Model

The mentioned DBDBs except OrientDB store data in JSON format [15], whereas MongoDB stores data in BSON format [16]. BSON is an extension to JSON and has the capability of storing data with additional data types like object, timestamp, date, time, etc. BSON format is lightweight and allows queries to operate at a lesser time with multilevel indexing support. Even data present at multiple levels can be easily accessed. OrientDB support BLOB [17]; the performance of this database depends on the data type, size of data, usage, etc. Performance decreases with increase of data.

(b) Indexing

Various indexing schemes deployed in various DBDBs are listed in Table 1:
Table 1 exploits that MongoDB is rich in indexing schemes, compared to indexing schemes in other DBDBs and also promotes the execution of complex queries at ease.

(c) Replication

Replication methods in various DBDBs are listed in Table 2:

Table 1 DBDBs with indexing schemes

DBDB name	Type of indexing
CouchDB [20]	Secondary
DynamoDB [21]	Global secondary and local secondary
RethinkDB [22]	Compound (single and multiple fields), secondary, spatial and tree
OrientDB [23]	Tree, hash, text, and spatial
Couchbase [24]	Global secondary, spatial, and text
MongoDB [25]	Field: single and multiple (compound), multi-key, spatial, text, and hash

Table 2 DBDBs with replication methods

DBDB name	Replication method
CouchDB [20]	Master–slave, multi-master
DynamoDB [21]	Cross-region
RethinkDB [22]	Master–slave
OrientDB [23]	Multi-master
Couchbase [24]	Master–slave, multi-master
MongoDB [25]	Master–slave

Master–slave has the probable chance of reducing updates conflicts. Multi-master avoids loading of all writes onto to a single server creating a single point of failure. Cross-region replication automatically replicates tables across two or more AWS regions with full support for multi-master writes. It gives the ability to build fast massively scaled applications. This was introduced by AWS recently for DynamoDB [18].

(d) Consistency

Consistency is crucial in Data Science, i.e., replicated data changed at one location must get reflected with other residing locations. DBDBs use various schemes for maintaining consistency. RethinkDB uses immediate consistency; CouchDB uses eventual consistency; Couchbase, DynamoDB, and MongoDB use eventual and immediate consistency; OrientDB has no prevalent consistency [3].

(e) MapReduce Support

MapReduce is a general framework included in most of the DBDBs. Map is applied over the distributive environment, on each document to filter irrelevant documents and to emit data for all documents of interest. The emitted data is sorted and passed in groups to reduce aggregation. Writing queries for information retrieval in NoSQL databases is intricate and getting information is plodding. To overcome this, MapReduce programs are skilled to get the relevant information from the distributed environment [11]. OrientDB and DynamoDB do not support MapReduce. Couchbase, RethinkDB, CouchDB, and MongoDB support MapReduce [6].

Apart from these parameters, a distinguished feature "In-Memory Capabilities" differentiates DBDB to be quite distinctive [19]. CouchDB and RethinkDB do not support in-memory capabilities. Couchbase has limited in-memory capabilities. OrientDB, DynamoDB, and MongoDB have in-memory capabilities [6].

In view of this study, the further section explores the proposed work framed for evaluating various prominent DBDBs.

4 Proposed Work

In this survey, we looked into data popped out from streaming applications for exploring sentiments. To extract these sentiments, a program is fabricated in Python. The reason for selecting Python is that it is open source, compatible with all DBDBs, and cross-platform runtime environment for developing server-side web-based applications. This .py program uses get_tweets() for extracting sentiments from streaming applications, and the resultant JSON files are exported into every DBDB mentioned. The framework is shown in Fig. 1.

Fig. 1 Streaming framework

Algorithm for acquisition of tweets in Python

```
Step 1:  Acquire the twitter API keys with https://apps.twitter.
         com/.
Step 2:  State the user credentials in the Python code.
Step 3:  Acquire the sentiments.
Step 4:  Eliminate unwanted data associated with
         sentiments.
Step 5:  Load the tweets into .CSV or .TXT file.
Step 6:  Export .CSV or .TXT to the selected DBDBs.
```

5 Results and Comparison

In the above sections, this paper exploited various qualitative features of considered DBDBs. Prominent DBDBs, MongoDB, DynamoDB, Orient DB, Couchbase, RethinkDB, and CouchDB are considered for evaluating these databases over quantitative measures and pins about the distinctive DBDB among these DBDBs. These measures are insertion, updation, deletion, and retrieval times of a unit number of records per millisecond. Three instances of records likely 1000, 5000, and 10000 are taken into account for evaluating the performance of these DBDBs over the mentioned quantitative measures.

The following graphs Figs. 2, 3, 4, and 5 depict the behavior of considered DBDBs.

Figure 2 exploits that MongoDB and DynamoDB prove that they take less amount of time nearly while inserting data when compared to other databases. OrientDB also takes time similar to DynamoDB. With an increase in number of data records, MongoDB proves to the best for insertion.

Figure 3 exploits that MongoDB and DynamoDB prove that they take less amount of time nearly while updating data when compared with other databases. OrientDB also takes time similar to DynamoDB. With an increase in number of data records, MongoDB proves to the best for updation.

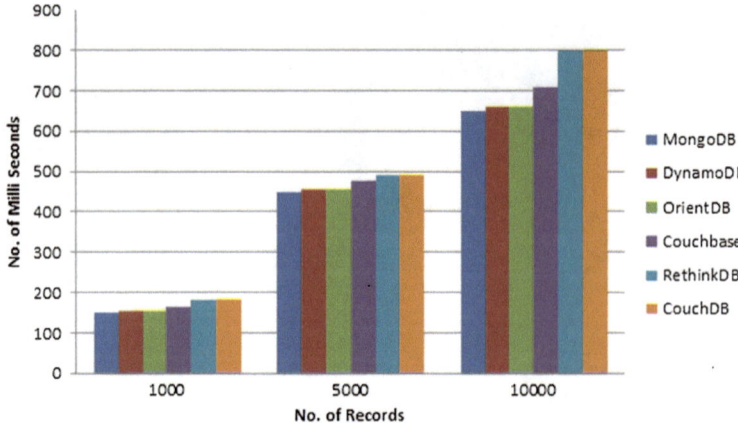

Fig. 2 Insertion time

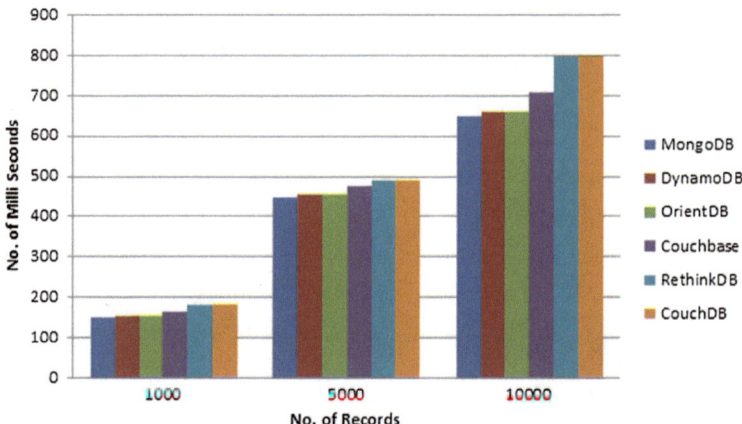

Fig. 3 Updation time

Figure 4 exploits that MongoDB, DynamoDB, and OrientDB prove that they take less amount of time nearly while deleting data when compared to other databases. With an increase in number of data records, MongoDB proves to the best for deletion.

Figure 5 exploits that MongoDB, DynamoDB, and OrientDB prove that they take less amount of time nearly while retrieving data when compared with other databases. With an increase in number of data records, MongoDB proves to the best for retrieving.

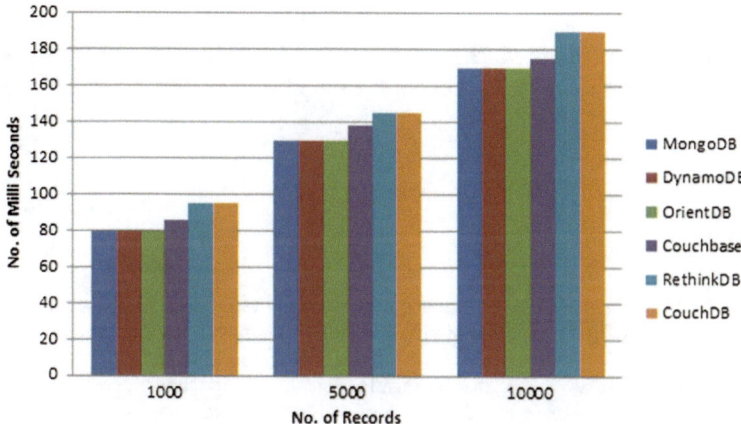

Fig. 4 Deletion time

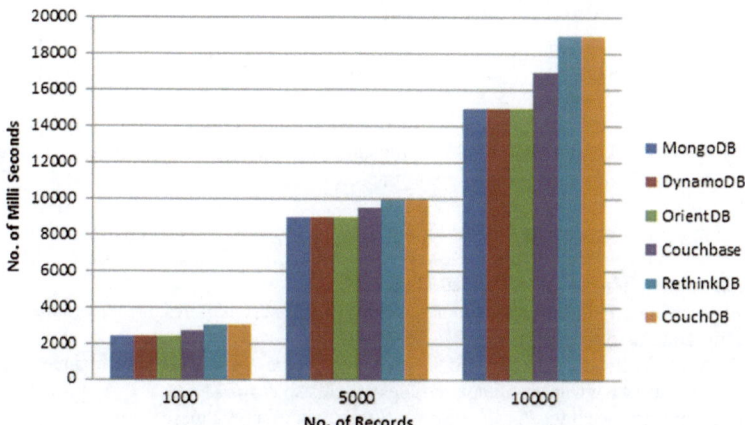

Fig. 5 Retrieval time

6 Conclusion

In this era of Data Science, DBDBs become popular because these databases are fruitful for most of the Big Data applications. This is because DBDBs have the capability of handling data from multilingual resources and multi-type data.

This paper drives that MongoDB, DynamoDB, and OrientDB have "in-memory capabilities." This paved the way of using these databases by various analysts in their applications. Selecting a DBDB to an application depends on the diversity of application. Among these three DBDBs mentioned here, MongoDB has more flexibility, because MapReduce feature is adaptable and complex queries can be expressed at ease. As MapReduce feature is not present in DynamoDB and OrientDB, this makes analyst to pin MongoDB for their applications. Moreover, MongoDB is closely related to SQL.

References

1. Mason RT (2015) NoSQL databases and data modeling techniques for a document-oriented NoSQL database. In: Proceedings of informing science and IT education conference (InSITE) 2015, pp 259–268
2. Roe C (2012) ACID vs. BASE: the shifting pH of database transaction processing. http://www.dataversity.net/acid-vs-base-the-shifting-ph-of-database-transactionprocessing/
3. Tauro CJM, Patil BR, Prashanth KR (2013) A comparative analysis of different NoSQL databases on data model, query model and replication model. In: Proceedings of international conference on "emerging research in computing, information, communication and applications", ERCICA 2013. ISBN 9789351071020
4. Cattell R (2011) Scalable SQL and NoSQL data stores. ACM SIGMOD Rec 39(4):12–27
5. DB-Engines Ranking of Wide Column Stores. https://db-engines.com/en/ranking/wide+column+store
6. DB-Engines Ranking of Key-Value Stores. https://db-engines.com/en/ranking/key-value+store
7. DB-Engines Ranking of Graph DBMS. https://db-engines.com/en/ranking/graph+dbms
8. Angles R, Gutierrez C (2008) Survey of graph database models. ACM Comput Surv (CSUR) 40(1):1–39
9. DB-Engines Ranking of Document Stores. https://db-engines.com/en/ranking/document+store
10. Benchmarking Top NoSQL Databases, Apache Cassandra, Couchbase, HBase, and MongoDB. http://www.endpoint.com/. Accessed 16 Jan 2016
11. Issa A, Schiltz F (2015) Document oriented databases. http://cs.ulb.ac.be/public/_media/teaching/infoh415/student_projects/couchdb.pdf
12. Sullivan D, Sullivan J (2015) NoSQL key-value database simplicity vs. document database flexibility. http://www.informit.com/articles/article.aspx?p=2429466
13. Document Databses: http://basho.com/resources/document-databases/
14. Hashem H, Ranc D (2016) Evaluating NoSQL document oriented data model. In: 2016 4th international conference on future internet of things and cloud workshops. 978-1-5090-3946-3/16 $31.00 © 2016 IEEE. https://doi.org/10.1109/w-ficloud.2016.26
15. Document-Oriented Database (2012). https://en.wikipedia.org/wiki/Document-oriented_database
16. JSON and BSON. https://www.mongodb.com/json-and-bson
17. OrientDB. https://orientdb.com/docs/2.1.x/Binary-Data.html
18. Cross-Region Replication (2012). https://docs.aws.amazon.com/amazondynamodb/latest/developerguide/Streams.CrossRegionRepl.html
19. Girbal A (2013) How to use MongoDB as a pure in-memory DB. https://dzone.com/articles/how-use-mongodb-pure-memory-db
20. CouchDB Data Model Features (2012). https://quabase.sei.cmu.edu /mediawiki/index.php/CouchDB_Data_Model_Features
21. Improving Data Access with Secondary Indexes (2012). https://docs.aws.amazon.com/amazondynamodb/latest/developerguide/SecondaryIndexes.html

22. Using Secondary Indexes in RethinkDB. https://www.rethinkdb.com/docs/secondary-indexes/python/
23. OrientDB Indexes. https://orientdb.com/docs/2.1.x/Indexes.html
24. Indexes. https://developer.couchbase.com/documentation/server/current/indexes/indexing-overview.html
25. Indexes. https://docs.mongodb.com/manual/indexes/

Modelling and Simulation of VOR ILS Receiver Diagnostic Model for Avionics Systems Health Management

Ishrath Tabassum, C. M. Ananda and C. R. Byra Reddy

Abstract Earlier avionics generation utilized federated architecture with individual resources for each application in Line Replaceable Units (LRU). The rapid development of mechanisms, techniques and technology enabled use of the integrated architecture. Avionics system covers the autopilot, communication, indicating and recording systems, warning systems, radar systems, navigation and surveillance systems. Over a period of time, the technology has uplifted the availability and maintainability of avionics systems with rigorous Built In Test (BIT) capability to provide the health information of self to other subsystems. However, BIT capability does not provide sufficient information for complete diagnostic and prognostic management of avionics system. Hence, it is essential to design, develop and implement the avionics system health management methodology in addition to the BIT capability. This paper presents simulation and modelling covering study, understanding, simulation of faults and system behaviour functioning under these simulated faults to establish the complete system behaviour as part of health management.

Keywords ARINC 429 · Aircraft · Line replaceable units · VIR · Diagnostic model · SIMULINK

1 Introduction

Aviation Electronics or Avionics has grown over a period with advancement in technology and ease of adopting the technology in the industry using digital communication. The grouping of the avionics system based on its functionality as given as the

I. Tabassum (✉) · C. R. Byra Reddy
Bangalore Institute of Technology, Bengaluru, India
e-mail: ishrathtabassum3@gmail.com

C. R. Byra Reddy
e-mail: byrareddycr@yahoo.co.in

C. M. Ananda
Council of Scientific and Industrial Research-National Aerospace Laboratories, Bengaluru, India
e-mail: ananda_cm@nal.res.in

© Springer Nature Singapore Pte Ltd. 2019   579
H. S. Saini et al. (eds.), *Innovations in Computer Science
and Engineering*, Lecture Notes in Networks and Systems 74,
https://doi.org/10.1007/978-981-13-7082-3_66

navigation system, communication system, display systems including various cockpit displays, recording systems, audio systems and engine instruments. The major requirements in an aircraft or an avionics system are high reliability, low operating cost, low maintenance cost, high system availability and efficient utilization of system resources. The various problems that might occur due to the use of the complex systems that might have many inherent and hidden faults that might affect the entire system later during operation must be taken care in the design to ensure continued operation. Fault management consists of a set of active functions performing: Detection of off-nominal conditions, Identification of the causes of off-nominal conditions and Response taking actions to remediate off-nominal conditions. Avionics suite in Aircraft [1–3] consists of Very High Frequency (VHF) communications for voice communications from ground stations to flight line crew and vice-versa. VOR/ILS Receiver (VIR) for computations on signals received from ground-based transmitters for determining the aircraft position and for automatic approach and landing. Distance Measuring Equipment (DME) for computations based on received replies following transmitted interrogations determining the aircrafts slant distance from a ground station. Automatic Direction Finder (ADF) for the relative bearing to a selected ground station for direction indication. Air Traffic Control (ATC) Transponder for air traffic tracking and establishing the altitude of the aircraft along with other parameters of aircraft including aircraft tail number and pilot identification (ID). Radio Altimeter (RADALT) for a precise measurement of the aircraft height above terrain. Radio Tuning Units (RTU) for tuning and control the various radios in aircraft. Weather Radar (Wx RDR) for detecting, analysing and displaying the weather ahead of aircraft to a distance of nearly 200 nm on a separate weather radar display system.

An important factor to be considered in avionics is the optimization of reliability and safety of an aircraft. The occurrence of a failure may lead to a critical state in any functional system. To optimize the functioning of an aircraft system and to minimize failure, the reason for the occurrence of failures should be known and the maintenance and diagnostic approaches should be considered [4]. The health awareness is provided when the information is made available of the model-based analysis. The aircraft systems need health management by fault diagnosis. The diagnostic techniques require a detailed modelling for its performance verification [5]. The diagnostic model for VOR ILS receiver (VIR) LRU is modelled, simulated and presented in this paper.

2   Design and Development of Diagnostic Model

Typically avionics in aircraft is connected at hierarchal levels starting from sensors to the processing system to displays in the cockpit [6]. Various sensor systems are connected to the processing system using Remote Data Concentrators to contrite the multiple input data on to a single stream for long distance transmission with minimized wires on a digital communication channel. The generalized block diagram of the interaction between the various LRUs (Line Replaceable Units) and the RDC

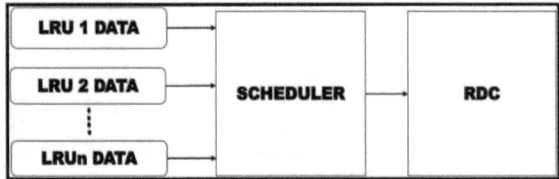

Fig. 1 Generalized block diagram with LRUs in avionics system

(Remote Data Concentrator) through ARINC 429 protocol is shown in Fig. 1. Signals from all the LRUs are combined in the RDC scheduler which periodically sends these signals to the RDC by combining it into one signal. However, BIT capability does not provide sufficient information for the whole system for the complete diagnostic and prognostic management of avionics system [7].

2.1 VIR (VOR/ILS Receiver)

VIR is a short and medium range navigational aid compromising ground-based transmitting stations/beacons and receiving elements carried onboard an aircraft. This will provide enroute information on the bearing of the aircraft from the points at which the stations are geographically located. The VIR navigational aid includes VOR, localizer, marker beacons for recognizing the inner, middle, and outer predefined runway stations, and glideslope. During the Instrument Landing System (ILS) operation, the system will be used for instrument landing and approaches.

2.2 ARINC 429 Protocol

ARINC stands for Aeronautical Radio INC, ARINC 429 is also known as the Mark 33 Digital Information Transfer System (DITS) [8] which is commonly used in all civil aircraft. The source LRU transmits 32-bit words which contain five primary fields and all the unused bits are padded with zeros. There is a minimum separation between the sequential words and it is of 4-bit times with zero voltage. The ARINC 429 32 bit word format is shown in Fig. 2 and when the data words are transmitted on the ARINC bus the bit transmission order is different and is shown in Fig. 3.

Label is also called Information Identifier which contains 8 to 1 bits of the 32-bit word. The data words are transmitted on to the ARINC bus in an order where the first transmitted field is the label (MSB first) and then this follows the remaining bit field (LSB first). The SDI is employed to find which source is transmitting or to find out which receiver the data is sent among the multiple receivers and occupies 9 and 10 bits of the ARINC 429 32 bit word. ARINC 429 defines bits 11 to 29 as those which

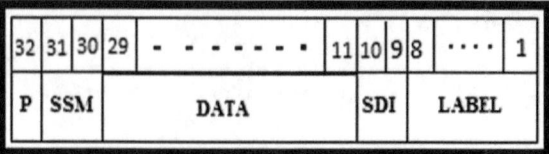

Fig. 2 ARINC 429 32-bit word format

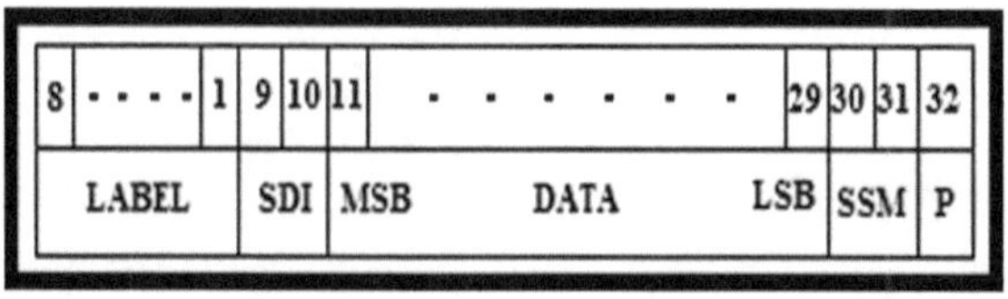

Fig. 3 ARINC 429-bit transmission order

contain the data information of the words. The data types available in ARINC 429 are Binary (BNR), Binary-Coded Decimal (BCD), Maintenance data, acknowledgement and discrete data. The Sign/Status Matrix (SSM) field provides different information and is utilized for the indication of sign or direction of the ARINC words data along with the status of the parameter indicating NORMAL, FAIL, TEST or NCD. The ARINC 429 defines one bit parity which is the most significant bit is configured as odd parity.

2.3 Modelling and Simulation of VOR ILS Receiver (VIR) LRU in Avionics

To start with, the VOR ILS receiver (VIR) LRU is modelled and simulated for its functional interface behaviour and fault/failure behaviour. VIR is a navigational aid takes VOR/ILS input frequency word, VOR/ILS LOC DEV Output word, VOR/ILS G/S DEV Output word, VOR Bearing Output word, ARINC Diagnostic Output word and Equipment ID. The output signal is obtained by encoding the input data onto the ARINC 429 required electrical signal formats with Bi-Phase having HIGH as well LOW signal with NULL in between the transitions. These high and low signals are used as the input to the diagnostic model. The diagnostic model for avionics VIR LRU is shown in Fig. 4.

The diagnostic model consists of the serial decode, hold circuit, data decode label filter and error blocks. The serial decode block takes the high and low signals from the VIR and uses the pulse generation triggered and the negative edge detects blocks to generate the data, data enable and the final out enable signals. The data signal contains the data that is available on the positive pulses of the given signal and the data enable signal contains the data that is available on the positive and negative

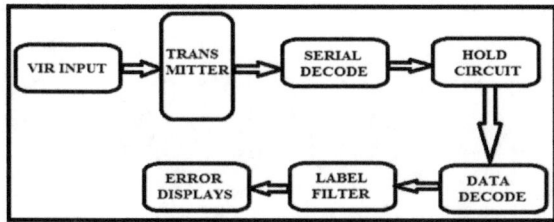

Fig. 4 The general diagnostic model for avionics LRU VIR

pulses separately at the output. The final out enable signal contains the information available both on the positive and negative pulses in a form which can be transmitted on the ARINC 429 bus. The hold circuit block makes use of switches to obtain only one output signal with the use of the three signals obtained at the output of the serial decode block. The data decode block is employed to change the positions of the bits obtained in the output signal from the hold circuit block. The filter block extracts the data, SSM, SDI fields along with the channel and LRU fields and displays the output in the form of binary numbers with which any fault in the system can be described by analysing the various changes in the bits. A further diagnostic model for Equipment ID with its corresponding label is developed and simulated.

3 Results

3.1 Simulation of VIR Diagnostic Model

The VIR diagnostic model blocks described are used to simulate the interface behaviour of VIR using the MATLAB SIMULINK environment. The blocks are designed and developed as shown in Fig. 5 and the simulation of the model give the desired results and information required for the diagnosis of the system. Table 1 shows the input and output data for the VOR system.

The Label, SDI, Data and SSM fields in the form of signals are obtained at the decoded block output for multiple words is as shown in Fig. 6 shows the enlarged view of the same. The yellow colour represents the label, the blue colour represents SDI, the green colour represents the data and the magenta colour represents the SSM.

Each parameter appears in the output for every time after its time delay. A start and stop time for simulation is taken as 1000 ms. The colours purple (dark) and lavender (light) in Fig. 7 represent the extracted high and low signals.

Figure 8 shows the different blocks in a 32-bit ARINC 429 word showing Label, SDI, Data, SSM and parity words being simulated.

The different error display blocks after the simulation of the diagnostic model are shown in Fig. 9 and the obtained output values are displayed in binary format.

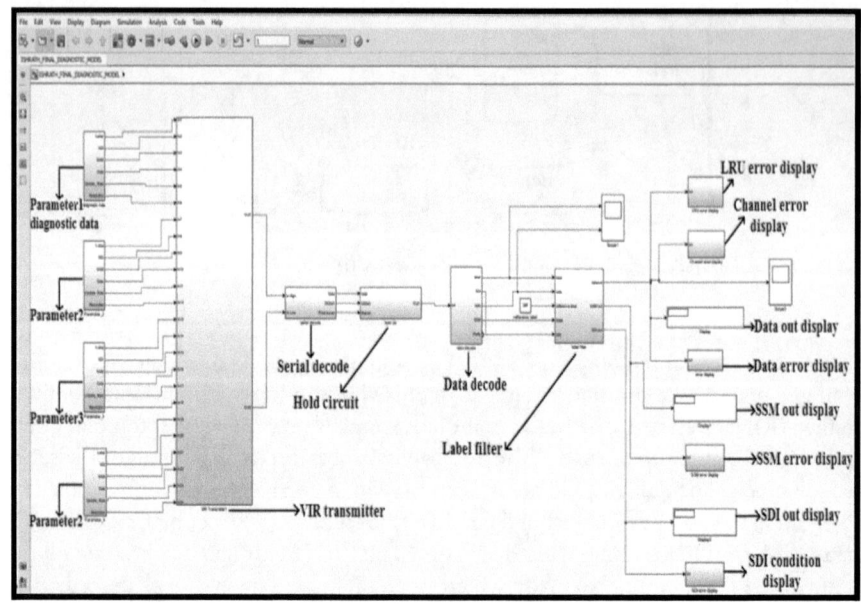

Fig. 5 The diagnostic model for avionics LRU VIR in SIMULINK

Table 1 Diagnostic model input and output with different parameters

Input		Label	Data	SDI	SSM	Update rate	Resolution
	Parameter1	350	1600	1	3	0.05	1
	Parameter2	165	16,000	1	3	0.05	0.1
	Parameter3	270	3584	1	3	0.2	1
	Parameter4	351	2000	1	3	0.25	1
Output	ARINC 429 32 bit word format obtained						
	ARINC 429 bit transmission order obtained						
	Extracted label: 350						
	LRU error	1: LRU failure, 0: no LRU failure					
	Channel error	1: channel failure, 0: no channel failure					
	Data error	1: indicates an error has occurred, 0: indicates no error					
	SSM error	00: failure warning 01: no computed data 10: self-test 11: normal operation					
	SDI condition	00: not used 01: Unit 1 10: Unit 2 11: not used					

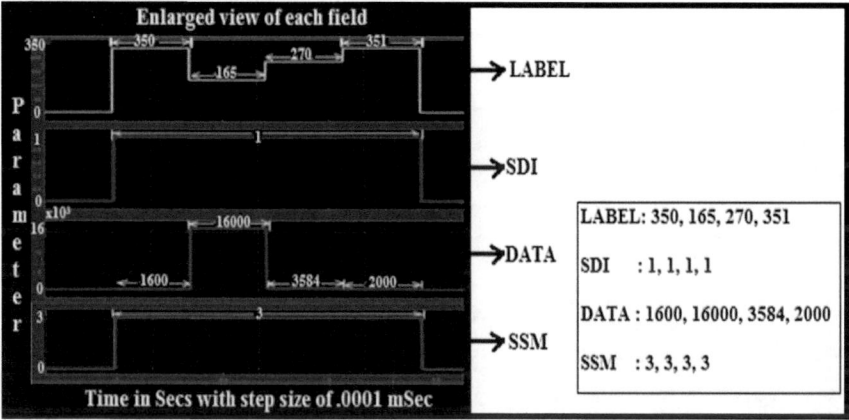

Fig. 6 Enlarged view of signals of the label, SDI, data and SSM from data decode block output and the values of different fields

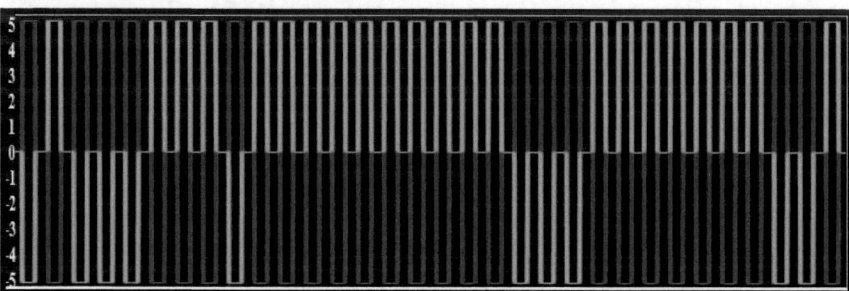

Fig. 7 The high and low output signals to the diagnostic model

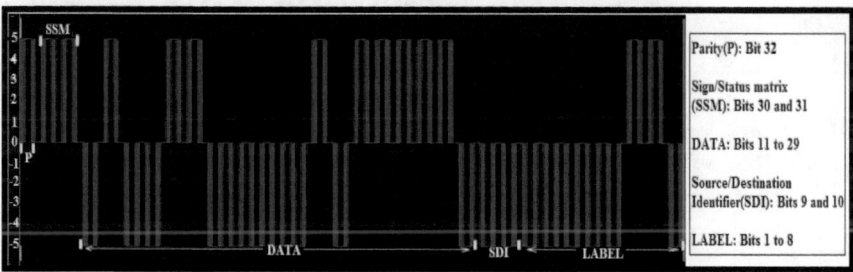

Fig. 8 The output signal with different parameters in a 32-bit ARINC 429 word

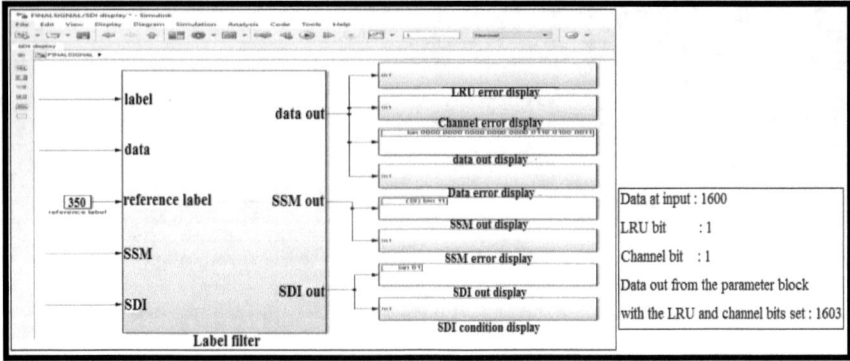

Fig. 9 Error display blocks after the simulation of the diagnostic model

The data given is of value 1600 and the simulated model gives the error by taking the data out into consideration. The diagnostic error value 25 is taken and the error obtained is the (localizer) LOC not locked. The output is shown in Fig. 10.

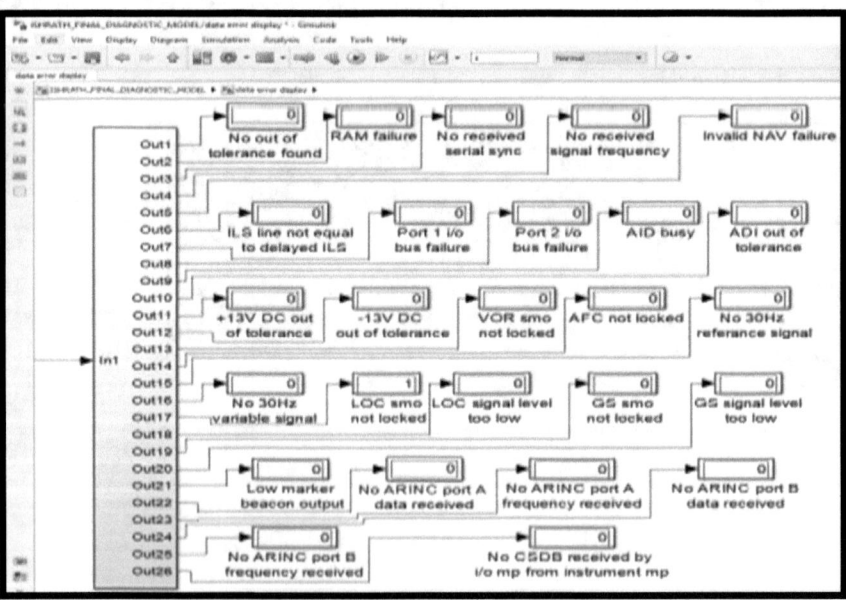

Fig. 10 Data error displayed after simulation

The values are

Label (Bits 8–1)	350 (11101000)
Sign/Status Matrix (SSM)	3 (11)
Source/Destination Identifier (SDI)	1 (01)
Data	1600
Data value displayed	0000 0000 0000 0000 0000 0110 0100 0000
Diag number	25
Diag error value	LOC SMO not locked
Diag value displayed at the output	00011001

The model is simulated and the output for the SDI block is the code 01 and all the conditions are displayed as 0 at the output except one condition which is displayed as 1 as shown in Fig. 11. The different conditions for the codes 00, 01, 10, and 11 are: ALL CALL, unit 1, unit 2, and unit 3 case. The model is simulated and the output for the SSM block is the code 11 and all the conditions or the errors are displayed as 0 at the output except one condition which is displayed as shown in Fig. 11. The different conditions for the codes 00, 01, 10 and 11 are: Failure warning, No Computed Data (NCD), Self-Test, and Normal operation.

Sign/Status Matrix (SSM)	3 (11)
Source/Destination Identifier (SDI)	1 (01)

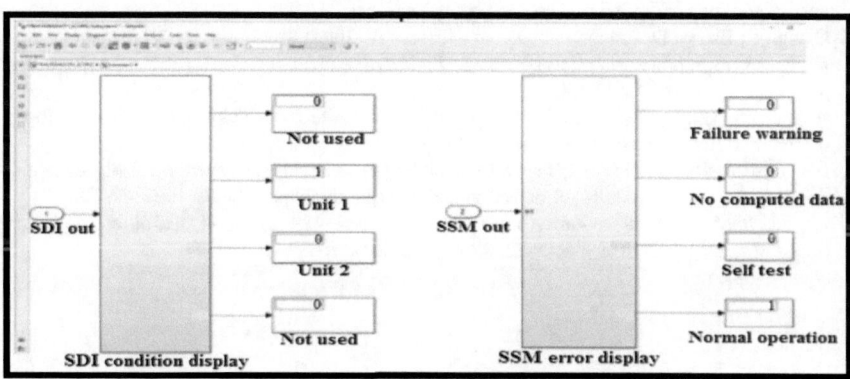

Fig. 11 SSM and SDI displayed after simulation

4 Conclusion

Health of an Avionics System is very important in the aircraft as the failure of aircraft causes endangerment which includes both posing danger to vehicle occupants and those affecting the general public's environment where the vehicle operates. It is very necessary to improve the reliability of an airplane by monitoring its health condition in turn health of all subsystems and hence the Avionics subsystem. The diagnostic model for the avionic LRU has been designed and developed and the simulation of this model is done using Simulink. The effort has been made to model, simulate one LRU in avionics. However to realize the complete Integrated Vehicle Health Management (IVHM) requirement, the complete Avionics system needs to be modelled and similarly, all other systems in aircraft need to be modelled at aircraft level to realize the IVHM.

Acknowledgements Authors would like to thank Mr. Jitendra J. Jadhav, Director NAL for continuous support and motivation. Authors also would like to thank team ALD for their support in the execution of simulation and modelling.

References

1. Ananda CM, Venkatanarayana KG, Preme M, Raghu M (2011) Avionics systems, integration, and technologies of the light transport aircraft. Def Sci J
2. Ananda CM (2007) General aviation light transport aircraft avionics: integration and system tests. In: 2007 IEEE/AIAA 26th digital avionics systems conference
3. Ananda CM et al (2009) General aviation aircraft avionics: integration and system tests. IEEE Aerosp Electron Syst Mag 24(5)
4. Fornlöf V, Galar D, Syberfeldt A, Almgren T, Catelani M, Ciani L (2016) Maintenance, prognostics and diagnostics approaches for aircraft engines
5. Orsagh R, Brown D, Roemer M, Dabney T, Hess A (2004) Prognostic health management for avionics system power supplies. 0-7803-8870-4/05/$20.00© 2005 IEEE IEEEAC paper #1342, Version 1. Updated Oct 27 2004
6. Ananda CM (2007) Civil aircraft advanced avionics architectures an insight into Saras avionics, present and future perspective. 13
7. Macaluso A, Jacazio G (2017) Prognostic and health management system for fly-by-wire electrohydraulic servo actuators for detection and tracking of actuator faults. Procedia CIRP
8. Ananda CM. SARAS avionics and AFCS ARINC 429 bit level definition document. SARAS/ALD/AVIONICS_ARINC/DR_003/SEP 2003 REV 4.0

Development and Integration of Graphical User Interface (GUI) with JUNGO Device Drivers for PCI Express Interface

A. Bepari Nawazish, C. M. Ananda, K. S. Venkatesh and C. Y. Gopinath

Abstract High-bandwidth Video Bus is a video interface protocol defined for high bandwidth, low latency, and uncompressed digital video transmission in avionics systems. This protocol is used for interfacing aircraft system(s) into the cockpit display. Current day technology realizes the systems or Line Replaceable Units (LRU) using Field-Programmable Gate Array (FPGA) based Intellectual Property (IP) Cores for communication between two systems. However, these communication protocols are realized using FPGA and as part of development, the Peripheral Component Interconnect (PCI) express based FPGA embedded modules are used for realization. These embedded modules will be interfaced to the host computers to transmit the data as similar to the communication on aircraft for real LRU. Hence host computers need to be integrated with required GUI application which shall communicate with PCI express bus connecting the FPGA based embedded module. This paper describes the development of the GUI application program for PCI express data simulator using device drivers generated by JUNGO (WinDriver) tool for memory read/write transaction(s). The paper also presents the integration, testing and data transfer for discrete and continuous data over PCI express from the host machine to the embedded FPGA-based module.

Keywords PCI express · JUNGO · Device drivers · Line replaceable units

A. Bepari Nawazish (✉) · C. Y. Gopinath
Bangalore Institute of Technology, Bengaluru, India
e-mail: nawazish2104@gmail.com

C. Y. Gopinath
e-mail: gopinath_cyg@yahoo.com

C. M. Ananda · K. S. Venkatesh
Council of Scientific and Industrial Research-National Aerospace Laboratories, Bengaluru, India
e-mail: ananda_cm@nal.res.in

K. S. Venkatesh
e-mail: venkateshks@nal.res.in

© Springer Nature Singapore Pte Ltd. 2019
H. S. Saini et al. (eds.), *Innovations in Computer Science and Engineering*, Lecture Notes in Networks and Systems 74,
https://doi.org/10.1007/978-981-13-7082-3_67

1 Introduction

Developers face new challenges as there is a requirement for higher bandwidth. As higher bandwidth is achieved by higher data rates that degrade the quality of the signal. The Peripheral Component Interconnect Special Interest Group (PCI-SIG) addresses this challenge by introducing PCI express allowing the system designers to evaluate the performance variation tolerance of the system [1]. PCI express is more likely to operate as a network than a bus. Instead of a single bus handling the data from different sources, PCI express has a switch that operates multiple point-to-point connections. Most of the devices have their own dedicated connection, thus they no longer share bandwidth as they used to do in PCI. PCI express provides excellent speed and reliability with the devices connected to it. Thus, to communicate with these devices, drivers are provided by JUNGO (WinDriver) tool. This tool provides an example code to access these devices from the user mode which performs byte-level Read/Write transaction(s) and reading of all the configuration registers in the windows form application on C# platform. The sample application provided by the JUNGO tool was studied, executed and understood. Based on this learning, an application tailored for the Council of Scientific and Industrial Research-National Aerospace Laboratories (CSIR-NAL), Aerospace Electronics and Systems Division (ALD's) requirement for data communication was attempted. The input required to perform read/write transaction should be in hexadecimal format. Therefore, a text file, image file or a word file needs to be converted to hexadecimal format before a write transaction can be performed and is realized using the data simulator.

2 Related Work

The paper [2], describes the communication between the host computers and Xilinx ZC706 evaluation board containing Zynq-7000 XC7Z045 All Programmable System-on-Chip (APSoc) device using PCI express. It also explains the Linux based application to perform Read/write functions for general purpose programs enabling data transfers through PCI express bus in both the directions. The performance comparison of the data transfer time at different data rates have been detailed. The paper [3] describes direct FPGA to FPGA communication through PCI express without involving memory subsystem of the host computer using Jetstream. The performance results for direct FPGA to FPGA transfers is shown through implementing Finite Impulse Response (FIR) filters. The paper [4] talks about the verification of Host Interface Logic using Xilinx Virtex-4 PCI express core Endpoint, which uses a PCI express core that acts as a root complex connected to the Host Interface Logic. The paper shows how it helps to minimize overall design development time by reducing verification time.

3 PCI Express GUI Interface to Host Using JUNGO Device Drivers

Peripheral Component Interconnect Express (PCI express) is a serial expansion bus standard for connecting a computer to one or more peripheral devices [5]. PCI express provides benefits across these major areas:

- High performance
- Next-generation Input/output (I/O)
- I/O simplification
- Ease of use

Typically, the PCI express data simulator is used to simulate the PCI express data to be communicated over the PCI express bus to other peripherals. One critical requirement is the communication of data from the host computer to any external device like an embedded module with FPGA hosted PCI express. In this case, the data from the host is sent to FPGA and vice versa. In order to realize the same, the PCI express data simulator GUI application integrates the JUNGO drivers and communicates the data to the PCI express bus. Customized PCI express device driver for the embedded module with PCI express interface is generated by JUNGO by collecting complete information of the embedded module like Device Identification (ID), Vendor Identification (ID) and generates the Information file (INF) file with all the libraries for the selected embedded module. Figure 1 shows the device driver generation flow diagram. When the embedded module with FPGA is inserted into the PCI express slot, the Operating System detects the device with the manufacturer details, Device Identification (ID) and Vendor Identification (ID). The information will be displayed in the DriverWizard tool. Select the device and generate and install the INF (Information file). This will install the base drivers for the card. If the 32-bit application program is generated for the 64-bit platform, then copy the "wdapi 1230_32" Dynamic Link Library (dll) file to the debug folder by removing "_32" part of it. If the 64-bit application program is generated on the 64-bit platform then directly run the application program without any changes.

4 JUNGO Tool (WinDriver)

For accessing the hardware, the application/Dynamic Link Library (dll) calls either WinDriver.NET (Network Enabled Technologies) wrapper or the high-level Win-Driver Application Program Interface (API). The hardware is accessed through the native calls of the Operating System given by WinDriver kernel. The WinDriver design limits the performance even if it is running in user mode. Thus, to achieve high performance, the WinDriver Kernel Plug-in needs to be attached with the performance-based modules such as hardware interrupt handler as shown in Fig. 2.

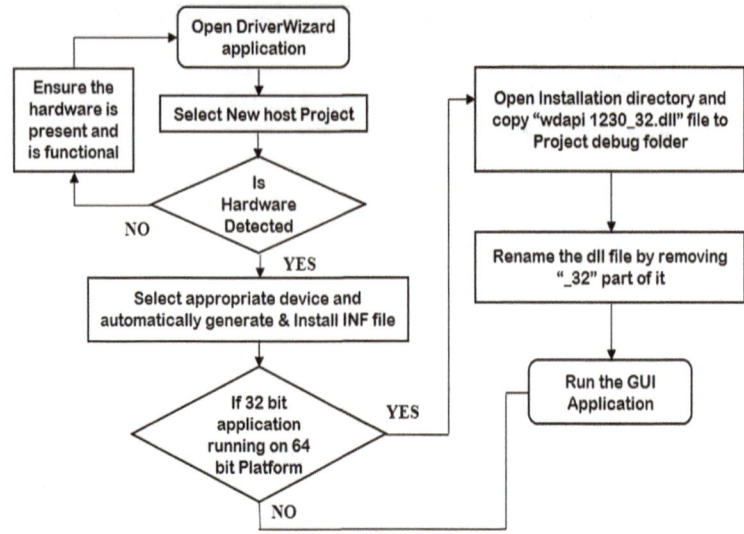

Fig. 1 Device driver generation flowchart

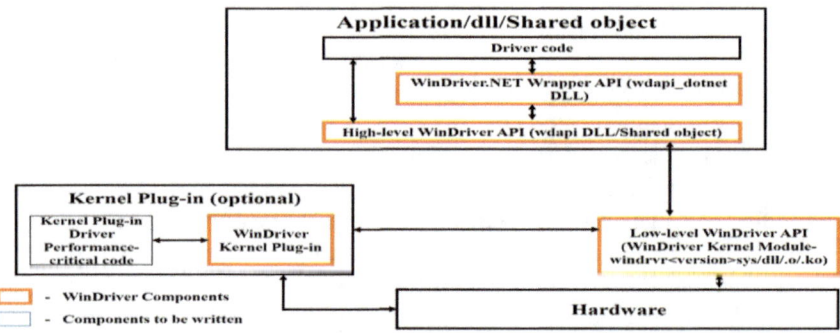

Fig. 2 WinDriver architecture [6]

To obtain maximum performance, the WinDriver kernel calls this module from the kernel code. This helps to achieve kernel performance for particular modules even in the user mode.

4.1 Connecting and Communicating with the Embedded Module Hardware Using Drivers Generated by JUNGO

Figure 3 Shows the communication link between the GUI and the hardware through the PCI express bus. The main entry point for windows device drivers is called as

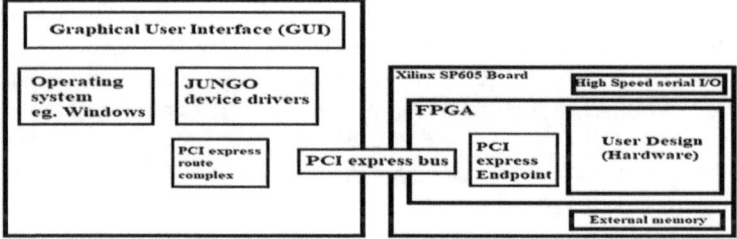

Fig. 3 Communication link between GUI and hardware through PCI express bus

DriverEntry(). A set of initialization protocols needs to be performed only once when the device drivers are loaded for the first time. This initialization needs to be performed by DriverEntry() routine. The driver callback (called as dispatch routines) are called by the Operating System and along with that, the entry functions are also registered. These driver callbacks are the Operating System request to access the services provided by the drivers.

Before the DriverEntry() routine is called, the Information (INF) file connects the hardware and the driver that helps to register the device to work with the driver. The device is identified by the Operating System through the Information (INF) file which is associated with the device connected and depending upon the INF file, the driver's entry point is called. The custom Operating System APIs helps to communicate application GUI and the driver. The application GUI uses CreateFile() function to open a handle to the device. The ReadFile() and WriteFile() functions are used to read and write from/to the device by passing the handle to the file access function [6]. The custom Operating System API DeviceIoControl() function helps the application GUI to send a request to the driver via Input/output control (IOCTL) calls. This custom Operating System mechanism encapsulates the data that is shared between the driver and the GUI via IOCTL calls.

4.2 Modified Graphical User Interface (GUI) for Performing Memory Read/Write Transaction

For the conversion of text data into the hexadecimal string in C#, one of the functions that can be used is StringBuilder. StringBuilder is a dynamic object that allows you to expand the number of characters in the string [7]. First, initialize StringBuilder by passing an integer value in the constructor. This will allocate memory sequentially on the memory stack. This memory allocation will be automatically expanded once it reaches capacity. Now, the text data is converted into a character array and is kept in the memory allotted space. Each character is converted to the hexadecimal string and the converted data is kept in the same memory location with the help of StringBuilder. The StringBuilder doesn't create a new object in the memory instead it expands the memory to accommodate the data [7]. This process is done inside a loop until all the characters are converted into a hexadecimal string.

Drivers generated by JUNGO tool is capable of transmitting one single Double Word (DWORD), whereas the real requirement is to transmit continuous data. The application GUI uses the single DWORD transmitting driver and designed an application to transmit continuous data without any limit in size. This is the critical requirement which has been realized.

In the read/write address form, there are two checkbox options. One will allow the user to select any text file and perform read/write operations and another checkbox will only allow the user to provide manual hexadecimal data for read/write operations. The first checkbox (send a text file) allows the user to browse and select any text file. Further, it populates the contents of the file into the text box (top-left) in the GUI as shown in the Fig. 4. For any write transaction(s) that needs to be performed, the given data should be in the hexadecimal format. So, the conversion of text data needs to be done before providing input for writing into the memory. The text data is converted into a hexadecimal format with the help of the button (TO HEX). The read/write transaction(s) are performed and a log is created to verify the operations that are being performed. If the other checkbox (send manual data) is selected, then the browse button will be disabled and manual hexadecimal data needs to be entered to perform read/write operation(s) as shown in the Fig. 5. The read time and the write time is measured to calculate the bandwidth of the system (highlighted in the red). A verification log is created to verify the text that has been sent has received in the correct format.

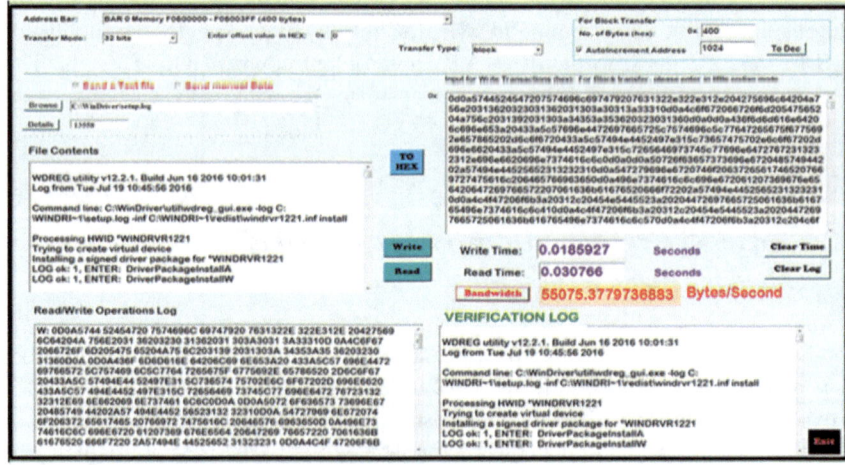

Fig. 4 Read/Write address main form (Text file mode)

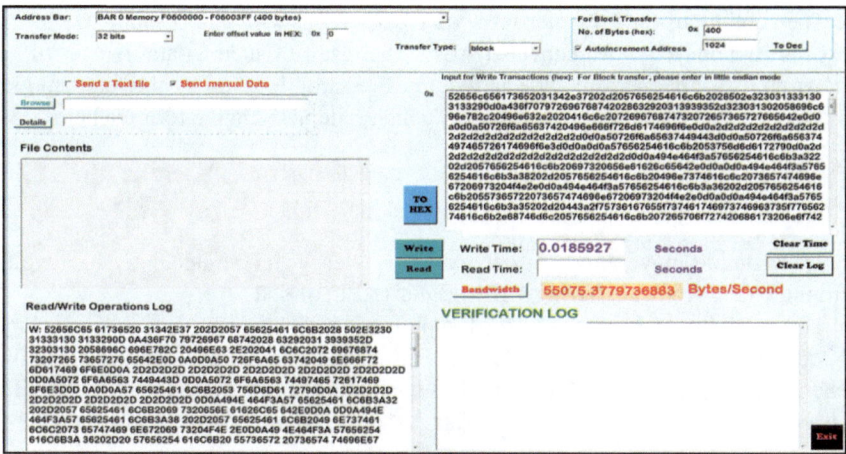

Fig. 5 Read/Write address main form (manual data mode)

5 Results

PCI express data simulator GUI has been developed, integrated with the embedded module and successfully tested for read/write operations. Figure 6 shows the active status of the PCI express embedded module detected the GUI application.

To perform any communication with the device, configuration parameters need to be defined. The address bar provides the selection of Base Address Register (BAR) with the memory allocated for it. The number of bits to be transferred in one cycle will be decided by selecting the required value from the transfer mode. The transfer type will select whether the data should be sent in block or in a non-block mode. The selection of the block mode will activate the block transfer options menu where the number of bytes that needs to be transferred should be assigned. To access PCI express memory, one needs to provide offset value.

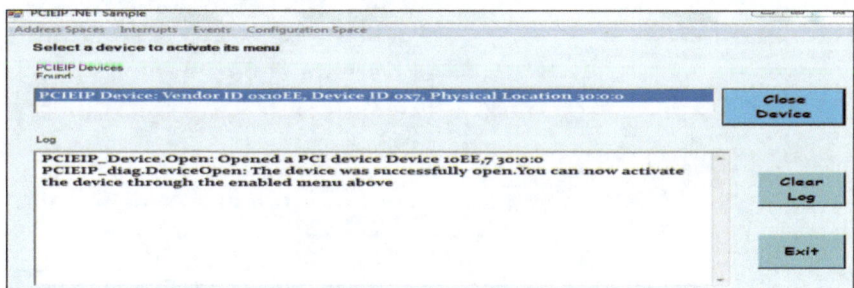

Fig. 6 Main form for opening the handle to the device

The continuous data transfer tests were carried out with a maximum size of the text file of a couple of Megabytes (MBs). The result of such a data transfer file is shown in Fig. 7. Generally, the bandwidth calculation is given by the total amount of data transferred in the stipulated time, which is depicted in the formula below [8].

$$\text{Bandwidth} = \frac{\text{Total data transferred}}{\text{Total time taken}}$$

The example above shows the calculation of latency and the calculated bandwidth (from the GUI) for transferring 1 Kilo Bytes (KBs) of data.

Theoretically, PCI express bus 1.1 supports 2.5 Gig transfers per second which is equal to 312.5 MB per second [9]. As PCI express bus 1.1 uses 8b/10b encoding scheme, due to which 20% of theoretical bandwidth is lost to overhead. Therefore, the theoretical time taken for transferring 1 KB including the overhead would be

$$1024/\left(250 * 10^6\right) = 4.096 \ \mu s$$

Total no. of the clocks as observed on Chipscope Pro Analyzer for transmitting 1 KB of data is 2295. The clock frequency of the FPGA based hardware module (Xilinx Spartans SP605 Evaluation board) is 62.5 MHz.

Therefore, the time taken for transmitting 1 KB of data including the overhead is $2295/(62.5 * 10^6) = 36.72 \ \mu s$.

For every Transaction Layer Packet (TLP) transmission, 4 DWORDS of length 16 bytes are transmitted. In which, 1 DWORD contains 4 bytes data and other three DWORDS of 12 bytes contains Format Type, Length, Request ID, and Address.

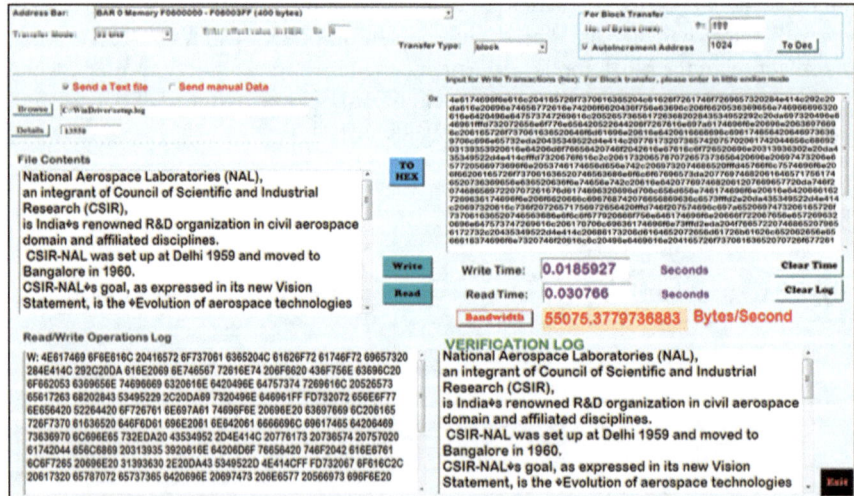

Fig. 7 Read/Write memory transaction windows form application

This format of TLP is followed for complete 256 TLP's that are transmitted for sending 1 KB of data. Thus, total data that is transmitted along with the data will be 4096 bytes. For each TLP, 9 clocks are required for transmission, which contains 4 DWORDS with 4 clock cycles and 5 clocks of turnaround time.

Thus, the achieved throughput including data and header information can be calculated as

$$4096/36.72 = 111.5 \text{ MB/s}$$

Therefore, 44.6% of bandwidth is been utilized for transmitting 1 KB of data. Further study and efforts are being carried out to improve the bandwidth of the PCI express data simulator application.

6 Summary

The PCI express data simulator GUI has been tested successfully for the discrete and continuous data transfer modes. Further work for improving performance is being carried out.

7 Conclusion and Future Work

In this paper, we have studied and generated device drivers from the JUNGO (Win-Driver) tool that helped to communicate with the PCI express card from the user mode through GUI application. Memory read/write transaction(s) were performed by selecting the required BAR, transfer modes and offset values and later verified the data transmission on Chipscope analyzer. The PCI express data simulator GUI is tested for discrete and continuous data successfully and the data has been communicated to the embedded module and verified for correctness. In the future, we intend to enhance the GUI for improved user experience. Efforts for sending video information, receiving video information, and displaying would be of priority.

Acknowledgements The authors would like to thank Mr. Jitendra J. Jadhav, Director CSIR-NAL for continuous support and motivation. Authors also would like to thank team ALD for their support in generation and implementation of the application program with the PCI express.

References

1. Cheng H, Hu J (2010) Research into PCI express device's configuration space on PC platform. In: International conference on computer, mechatronics, control and electronic engineering, vol 1 (2010)
2. Rjabov A, Sudnitson A, Sklyarov V, Skliarova I (2016) Interaction of Zynq-7000 devices with general purpose computers through PCI express: a case study. In: Proceedings of MELECON 2016
3. Vesper M, Koch D, Vipin K, Fahmy SA (2016) Jetstream: an open source high performance PCI express 3 streaming library for FPGA-to-host and FPGA-to-FPGA communication. In: International conference on field programmable logic and applications (FPL)
4. Badhe S, Kulkarni K, Gadre G (2014) Accelerating functional verification of PCI express endpoint by emulating host system using PCI express core. In: International conference on computational systems and communications (ICCSC)
5. Kincaid RK (2010) Signal lens: focus+ context applied to electronic time series. IEEE Trans Vis Comput Graph
6. PCI-e user manual for windows
7. http://www.tutorialsteacher.com/csharp/csharp-stringbuilder
8. Mantripragada SR, Mopuri P (2016) Verifying performance of PCI express in a system for multi giga byte per second data transmission. In: International conference on communication and electronics and systems (ICCES)
9. https://www.tested.com/tech/457440-theoretical-vs-actual-bandwidth-pci-express-and-thunderbolt/

IoT-Based Pipe Burst Detection in Water Distribution Systems

**Suri Shanmukh, Meka Poorna Sai, S. Sai Sri Charan
and Nithya Chidambaram**

Abstract Agriculture is the heart of our vibrant culture which is the backbone of our country's economy. India's population is 1.2 billion now which means that the nonrenewable resources must be used judiciously. Today, most of the places in India suffer from drought due to lack of abundant rainfall. Manual supervision is always needed for traditional farming. Hence, it is the duty of every engineer of the country to contribute to its technological development. This thought was the foundation for our project and the objective of this paper is to automate the irrigation system. The proposed methodology includes a node microcontroller unit (Node MCU) to monitor the water flow and controlling the water supply and communication is done using IOT. The idea of the Internet of Things (IoT) has been integrated with automation to design the proposed "PIPE BURST DETECTION IN WATER DISTRIBUTION SYSTEMS".

Keywords Internet of things (IoT) · Cloud · Node MCU · Sensor-based irrigation · Pipe burst · Pipelines · Water flow sensor

1 Introduction

Agriculture is one of the most important factors which contributes to the growth of the economy of the country where it uses about 70% of the fresh water, mainly from the Underground reserves [1]. But water is one of the most increasing scarce resources,

S. Shanmukh (✉) · M. P. Sai · S. Sai Sri Charan · Nithya Chidambaram
Department of Electronics and Communication Engineering, SASTRA Deemed University,
Thanjavur, Tamil Nadu, India
e-mail: surishanmukh@gmail.com

M. P. Sai
e-mail: mpoornasai@gmail.com

S. Sai Sri Charan
e-mail: charanmsd72@gmail.com

Nithya Chidambaram
e-mail: cnithya@ece.sastra.edu

© Springer Nature Singapore Pte Ltd. 2019
H. S. Saini et al. (eds.), *Innovations in Computer Science
and Engineering*, Lecture Notes in Networks and Systems 74,
https://doi.org/10.1007/978-981-13-7082-3_68

so effective irrigation water management with well timing and the regulating system must be implemented to satisfy the need of the crop without wasting water, soil, and crop nutrients should be done [2, 3]. One of the methods is by measuring the flow of fluid which is used to estimate the productivity of the field. It is also an effective way to keep an eye on the pumping system such as pipe leakage etc. [4]. Flow rate can be measured with a contact type or noncontact type of sensor. Such as ultrasonic sensor, auxiliary fluid flow meter, and Hall Effect sensor [5]. Nevertheless, the life expectancy and maintenance specifications of the flow meters are affected by some constraints. The major task is to choose the specific instrument to that precise application. Although there are many flow meters that can measure the flow rates, they do not have a provision to control the flow rate. Hence, this needs to be supervised by a worker to control the water pumping through the motor which leads to the wastage of manpower [6, 7]. So, applying the necessary amount of water at the right time is the main motto of the Automated Farming, regardless of the presence of worker to power on and off the motor [8]. This is achieved by creating an appliance which will continuously monitor the water flow rate and accordingly furnishing the water to the field [9]. With the advantage of using the Node MCU boards, being open source will reduce the complexity for the programmer. Using GUI. YF-S201 flow sensor sensing of the liquid flow rate of the pipe is done, which in turn controls the total system using Node MCU microcontroller and IOT. This will help the farmers to continuously monitor the water flow, thereby making their work easy and the productivity of the crop will be more.

2 Proposed Method

The proposed methodology is depicted in Fig. 1.

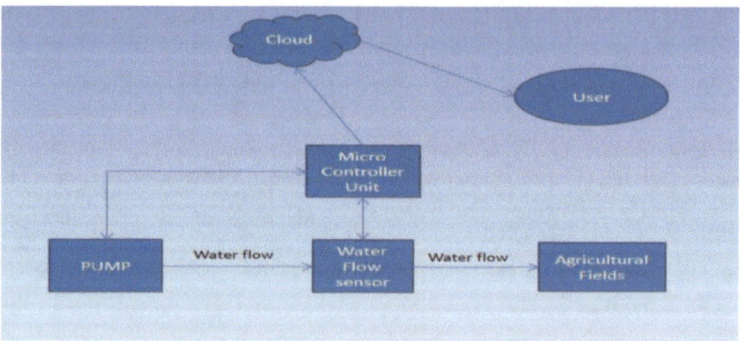

Fig. 1 General block diagram

2.1 Terms and Symbols

Node MCU	Microcontroller.
f	Frequency.
fr	Flow rate.

A flow sensor is placed in line with the water line which is connected to a water pump. The status of the pump can be varied using a microcontroller. The water will be flowing through the sensor, sends a signal to the microcontroller, and the water flow readings will be read for every second. The readings taken will be continuously uploaded into a cloud through the microcontroller which has an inbuilt WiFi module embedded in it and helps in connecting the cloud. The readings will be continuously compared with the threshold and send a signal to the microcontroller to turn off the pump. On turning off the pump through a microcontroller, the cloud sends a message to the user.

2.2 Algorithm

Step1: Start the process.
Step2: Initialize the power supply to the microcontroller and pump.
Step3: Calculate the flowrate(fr) = frequency(f) * 4.5 L/min.
Step4: Check the flow rate (greater than or less than).
Step5: If the flow rate is less than fixed criteria, turn off the pump.
Step6: Configure the message to be sent by the cloud.
Step7: After giving the reset command, the microcontroller comes to the original state.
Step8: Restart the process (if required) or stop the process.

The flow sensor placement will be a key issue in this method. The flow sensor is expected to be placed at the end of the pipeline, where the water will be flowing to the fields. The burst can be detected only for the pipeline being input to the sensor. This method deals with a single pipelined system as a prototype for another pipeline system.

3 Results and Discussion

The obtained values from the sensor enable the system to turn the LED on and off which represents the status of the pump. The experimental setup is shown in Fig. 2 where the water flows through the water flow sensor and gives the value to the Node MCU. Here, Arduino is used as a power source to the Flow sensor. The water flows through the pipe and gets collected in a tumbler (representing an agricultural field). The water flow is manipulated manually by a tap.

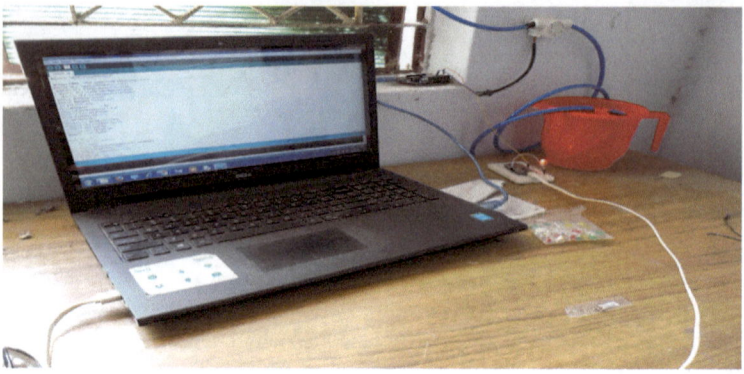

Fig. 2 Experimental setup

The values are graphically represented in Fig. 3 and it depicts that the values of flow rate and the status of the LED are being uploaded into the cloud where value 1 of the pump status represents the LED is on and value 0 represents that the LED is off. When there is a drastic decrease in the flow rate, the LED is turned off by the microcontroller and this can be seen in Fig. 4 where the led represents the status of the motor. In the real-time scenario when a pipe bursts or breaks, there will be a change in flow rate which can be detected and make the pump to turn off thereby conserving water.

On detection of a drastic decrease in flow rate then a message will be sent to the cultivator. Figure 5 represents the configuration of a message that has to be sent to the user. This will contain the value of pump status and the timestamp when the pump was turned off. Additional information can also be sent depending on the type of message that has to be sent. Figure 6 represents the message that has been sent from the cloud to mobile so that we can alert the user about the situation. This will decrease the intervention of farmers and helps in conservation of water.

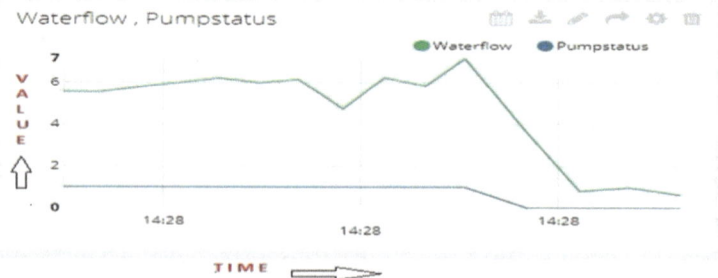

Fig. 3 Graphical representation of uploaded values of flow rate and status of the pump

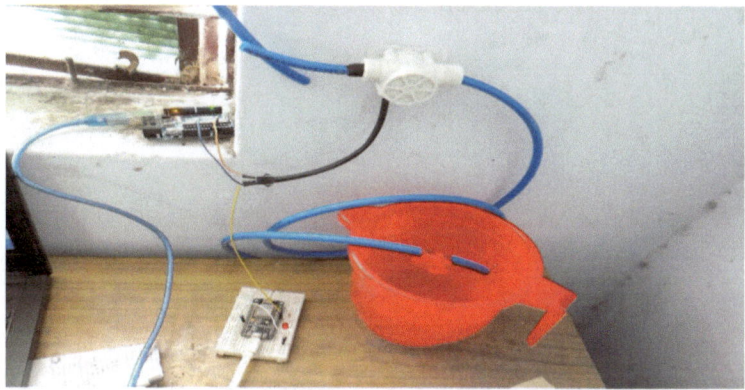

Fig. 4 System condition when burst occurs

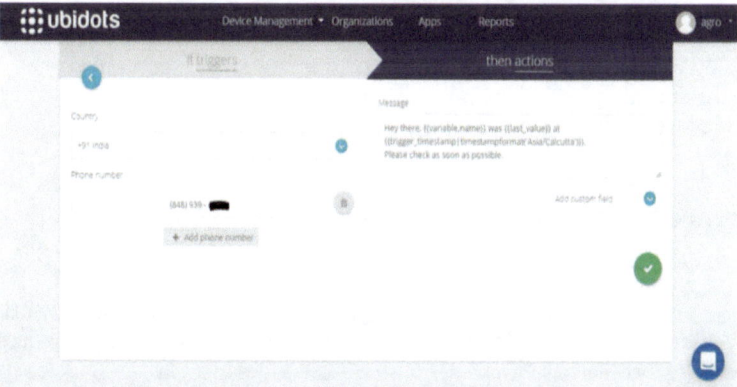

Fig. 5 Configuration of the message to be sent by the cloud

4 Conclusion

The proposed methodology is highly practical and useful in remote farming areas where automation plays a vital role in saving water and proves to be useful in places with water scarcity. Most importantly this ensures soil and fertilizers do not get washed away due to overwatering. Furthermore, this can also be used for the shipment of crude oil underwater where the oil spill can cause catastrophic damage to marine life. The intervention of the farmer has extremely reduced and keeps him notified in case there is a fault in the pumping system. The proposed idea provides a reliable and error detection system for automated farming which could help to conserve water and at the same time reduce the power consumption. This method can further be improvised by using an array of sensors to detect the burst for a multiple pipeline system.

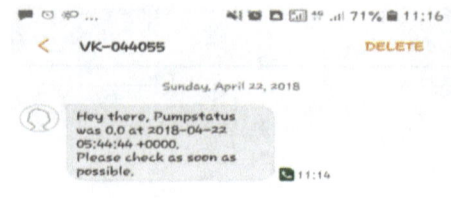

Fig. 6 Message sent to the user

References

1. Bourzac MK (2013) Water: the flow of technology. Nature 501(7468):S4–S6
2. Iyengar RR (2016) The water flow monitoring module. Int J Eng Res Gen Sci 4(3):106–113
3. Sood R, Kaur M, Lenka H (2013) Design and development of automatic water flow meter. Int J Comput Sci Eng Appl 3(3):49–59
4. Ali ML, Ridoy R, Barua U, Alamgir MB (2016) Design and fabrication of a turbine flow meter. In: 11th global engineering, science and technology conference, vol 4, no 1
5. Gatabi JR, Forouzbakhsh F, Darkhaneh HE, Gatabi ZR, Janipour M, Gatabi IR (2010) Auxiliary fluid flowmeter. Eur J Sci Res 42(1)
6. Suresh N et al (2014) Raspberry Pi based liquid flow monitoring and control. Int J Res Eng Technol 3(7):122–125
7. Uma JSS (2015) Human interaction pattern mining using enhanced artificial bee colony algorithm s. Int J Innov Res Comput Commun Eng 3(10):10131–10138
8. Kansara K, Zaveri V, Shah S, Delwadkar S, Jani K (2015) Sensor based automated irrigation system with IOT: a technical review sensor based automated irrigation system with IOT: a technical review. Int J Comput Sci Inf Technol 6(6):5331–5333
9. Harshita Chugh AS, Singh D, Shaik S (2017) IOT based smart irrigation system. Int Res J Eng Technol 4(9):7–11

Author Index

© Springer Nature Singapore Pte Ltd. 2019
H. S. Saini et al. (eds.), *Innovations in Computer Science
and Engineering*, Lecture Notes in Networks and Systems74,
https://doi.org/10.1007/978-981-13-7082-3